MULTICOMPONENT MASS TRANSFER

Wiley Series in Chemical Engineering

Bird, Stewart and Lightfoot: TRANSPORT PHENOMENA

Brownell and Young: PROCESS EQUIPMENT DESIGN: VESSEL DESIGN

Felder and Rousseau: ELEMENTARY PRINCIPLES OF CHEMICAL PROCESSES, 2nd Edition

Franks: MODELING AND SIMULATION IN CHEMICAL ENGINEERING

Froment and Bischoff: CHEMICAL REACTOR ANALYSIS AND DESIGN, 2nd Edition

Gates: CATALYTIC CHEMISTRY

Henley and Seader: EQUILIBRIUM-STAGE SEPARATION OPERATIONS IN CHEMICAL ENGINEERING

Hill: AN INTRODUCTION TO CHEMICAL ENGINEERING KINETICS AND REACTOR DESIGN

Jawad and Farr: STRUCTURAL ANALYSIS AND DESIGN OF PROCESS EQUIPMENT, 2nd edition

Levenspiel: CHEMICAL REACTION ENGINEERING, 2nd Edition

Malanowski and Anderko: MODELLING PHASE EQUILIBRIA: THERMODYNAMIC BACKGROUND AND PRACTICAL TOOLS

Reklaitis: INTRODUCTION TO MATERIAL AND ENERGY BALANCES

Sandler: CHEMICAL AND ENGINEERING THERMODYNAMICS, 2nd Edition

Seborg, Edgar, and Mellichamp: PROCESS DYNAMICS AND CONTROL

Smith and Corripio: PRINCIPLES AND PRACTICE OF AUTOMATIC PROCESS CONTROL

Taylor and Krishna: MULTICOMPONENT MASS TRANSFER

Ulrich: A GUIDE TO CHEMICAL ENGINEERING PROCESS DESIGN AND ECONOMICS

Welty, Wicks and Wilson: FUNDAMENTALS OF MOMENTUM, HEAT AND MASS TRANSFER, 3rd Edition

MULTICOMPONENT MASS TRANSFER

Ross Taylor
Professor of Chemical Engineering
Clarkson University
Potsdam, New York

R. Krishna
Professor of Chemical Engineering
University of Amsterdam
Amsterdam, The Netherlands

JOHN WILEY & SONS, INC.

New York · Chichester · Brisbane · Toronto · Singapore

IBM, IBM PC, IBM PC/XT, IBM PC/AT, IBM PS/2, and
PC-DOS are trademarks of International Business Machines, Inc.

Mathcad is a registered trademark of MathSoft, Inc.

MS-DOS is a registered trademark of Microsoft, Inc.

Windows is a trademark of Microsoft, Inc.

This text is printed on acid-free paper.

Library of Congress Cataloging in Publication Data:
Taylor, Ross, 1954–
    Multicomponent mass transfer/Ross Taylor and R. Krishna.

        p. cm. —(Wiley series in chemical engineering)
    Includes bibliographical references and indexes.
    ISBN 0-471-57417-1 (acid-free)
    1. Mass transfer. I. Krishna, R. II. Title. III. Series.
TP156.M3T39 1993
660'.28423—dc20   92-40667

Printed in the United States of America

10 9 8 7 6 5

PREFACE

Chemical engineers frequently have to deal with multicomponent mixtures; that is, systems containing three or more species. Conventional approaches to mass transfer in multicomponent mixtures are based on an assumption that the transfer flux of each component is proportional to its own driving force. Such approaches are valid for certain special cases.

- Diffusion in a two component (i.e., binary) mixture.
- Diffusion of dilute species in a large excess of one of the components.
- The case in which all of the components in a mixture are of a similar size and nature.

The following questions arise.

- Does the presence of three or more components in the system introduce additional complications unpredicted by binary mass transfer theory alone?
- If the answer to the above question is in the affirmative, how can the problem of multicomponent mass transport be tackled systematically?
- Do the transport processes of mass and heat interact with each other in normal chemical engineering operations?

Though the first question has been in the minds of chemical engineers for a long time (Walter and Sherwood in 1941 raised doubts about the equalities of the component efficiencies in multicomponent distillation), it has been established beyond doubt in the last two decades that multicomponent systems exhibit transport characteristics completely different from those of a simple binary system. Furthermore, procedures have been developed to extend the theory of binary mass transfer to multicomponent systems in a consistent and elegant way using matrix formulations; such formulations have also been incorporated into powerful computational algorithms for equipment design taking into account simultaneous heat transfer effects. These advanced models have been incorporated into design software for distillation, absorption, extraction, and condensation equipment. This is one example where commercial application has apparently preceded a formal academic training in this subject even at the graduate level.

This textbook is our attempt to address two needs:

1. The needs of the academic community for a reference text on which to base advanced lectures at the graduate level in transport phenomena or separation processes.
2. The requirements of a process design or research engineer who wishes to use rigorous multicomponent mass transfer models for the simulation and design of process equipment.

This textbook has grown out of our research and teaching efforts carried out separately and collaboratively at The University of Manchester in England, Clarkson University in the United States, Delft University of Technology, and The Universities of Groningen and Amsterdam in the Netherlands, The Royal Dutch Shell Laboratory in Amsterdam, and The Indian Institute of Petroleum.

This textbook is not designed as a first primer in mass transfer theory; rather, it is meant to follow an undergraduate program of lectures wherein the theory of mass transfer and fundamentals of transport phenomena have already been covered.

The 15 chapters fall into three parts. Part I (Chapters 1–6) deals with the basic equations of diffusion in multicomponent systems. Chapters 7–11 (Part II) describe various models of mass and energy transfer. Part III (Chapters 12–15) covers applications of multicomponent mass transfer models to process design.

Chapter 1 serves to remind readers of the basic continuity relations for mass, momentum, and energy. Mass transfer fluxes and reference velocity frames are discussed here. Chapter 2 introduces the Maxwell–Stefan relations and, in many ways, is the cornerstone of the theoretical developments in this book. Chapter 2 includes (in Section 2.4) an introductory treatment of diffusion in electrolyte systems. The reader is referred to a dedicated text (e.g., Newman, 1991) for further reading. Chapter 3 introduces the familiar Fick's law for binary mixtures and generalizes it for multicomponent systems. The short section on transformations between fluxes in Section 1.2.1 is needed only to accompany the material in Section 3.2.2. Chapter 2 (The Maxwell–Stefan relations) and Chapter 3 (Fick's laws) can be presented in reverse order if this suits the tastes of the instructor. The material on irreversible thermodynamics in Section 2.3 could be omitted from a short introductory course or postponed until it is required for the treatment of diffusion in electrolyte systems (Section 2.4) and for the development of constitutive relations for simultaneous heat and mass transfer (Section 11.2). The section on irreversible thermodynamics in Chapter 3 should be studied in conjunction with the application of multicomponent diffusion theory in Section 5.6.

Chapter 4 suggests usable procedures for estimating diffusion coefficients in multicomponent mixtures. Chapters 5 and 6 discuss general methods for solution of multicomponent diffusion problems. Chapter 5 develops the linearized theory taking account of multicomponent interaction effects, whereas Chapter 6 uses the conventional effective diffusivity formulations. We considered it appropriate to describe both of these approaches and to give the readers a flavor of the important differences in their predictions. We stress the inadequacy of the effective diffusivity approach in several cases of practical importance. It is a matter of continuing surprise to us that the effective diffusivity approach is still being used in the published literature in situations where it is clearly inapplicable. By delineating the region of applicability of the effective diffusivity model for multicomponent mixtures and pointing to the likely pitfalls in misapplying it, we hope that we will be able to warn potential users.

In the five chapters that make up Part II (Chapters 7–11) we consider the estimation of rates of mass and energy transport in multicomponent systems. Multicomponent mass transfer coefficients are defined in Chapter 7. Chapter 8 develops the multicomponent film model, Chapter 9 describes unsteady-state diffusion models, and Chapter 10 considers models based on turbulent eddy diffusion. Chapter 11 shows how the additional complication of simultaneous mass and energy transfer may be handled.

Chapter 12 presents models of mass transfer on distillation trays. This material is used to develop procedures for the estimation of point and tray efficiencies in multicomponent distillation in Chapter 13. Chapter 14 uses the material of Chapter 12 in quite a different way; in an alternative approach to the simulation and design of distillation and absorption columns that has been termed the nonequilibrium stage model. This model is applicable to liquid–liquid extraction with very little modification. Chapter 15 considers the design of mixed vapor condensers.

A substantial portion of the material in this text has been used in advanced level graduate courses at The University of Manchester, Clarkson University, The Universities of Amsterdam, Delft, Groningen and Twente in the Netherlands, and The University of Bombay in India. For a one semester course at the graduate level it should be possible to cover all of the material in this book. In our experience the sequence of presentation of the chapters is also well suited to lecture courses.

We have included three appendices to provide the necessary mathematical background. Appendix A reviews matrix algebra. Appendix B deals with solution of coupled linear differential equations; this material is essential for the solution of multicomponent diffusion problems. Appendix C presents two numerical methods for solving systems of nonlinear algebraic equations; these algorithms are used to compute rates of mass transfer in multicomponent systems and in the solution of the design equations for separation equipment. We have usually found it necessary to include almost all of this material in our advanced level courses; either by setting aside time at the start of the course or by introducing the necessary mathematics as it is needed.

We also feel that portions of the material in this book ought to be taught at the undergraduate level. We are thinking, in particular, of the materials in Section 2.1 (the Maxwell–Stefan relations for ideal gases), Section 2.2 (the Maxwell–Stefan equations for nonideal systems), Section 3.2 (the generalized Fick's law), Section 4.2 (estimation of multicomponent diffusion coefficients), Section 5.2 (multicomponent interaction effects), and Section 7.1 (definition of mass transfer coefficients) in addition to the theory of mass transfer in binary mixtures that is normally included in undergraduate courses.

A special feature of this book is the large number of numerical examples that have been worked out in detail. With very few exceptions these examples have been based on actual physicochemical data and many have direct relevance in equipment design. The worked examples can be used by the students for self-study and also to help digest the theoretical material.

To gain a more complete understanding of the models and procedures discussed it is very important for students to undertake homework assignments. We strongly encourage students to solve at least some of the exercises by hand, although we recognize that a computer is essential for any serious work in multicomponent mass transfer. We have found equation solving packages to be useful for solving most of the simpler mass transfer problems. For some problems these packages are not yet sufficiently powerful and it is necessary to write special purpose software (e.g., for distillation column simulation or for condenser design).

Our research and teaching efforts in multicomponent mass transfer have been strongly influenced by two people. The late Professor George Standart of the University of Manchester who impressed upon us the importance of rigor and elegance. Professor Hans Wesselingh of the University of Groningen motivated us to present the material in a form more easily understandable to the beginner in this area. It is left to our readers to judge how well we have succeeded in achieving both rigor and simplicity.

R. TAYLOR
R. KRISHNA

Potsdam, New York
Amsterdam, The Netherlands
June 1993

A NOTE ON SOFTWARE

Multicomponent mass transfer calculations are sufficiently demanding that one really requires computer software if one is to make more than one such calculation. The examples in this book were solved with a variety of software packages. Almost all of the computational examples were solved first using software that we created specifically for this purpose. A library of Fortran 77 routines for performing multicomponent mass transfer calculations is available from R. Taylor. These routines can be made to work with any number of components and are easily incorporated into other programs. We have checked all of our original calculations by repeating the examples using software that has been designed for mathematical work. We have used several such packages in the course of our work. With the exception of the design examples in Chapters 14 and 15, all of the examples have been solved using Mathcad for DOS (Version 2.5) from MathSoft. A disk containing our Mathcad files is provided with this book.

The distillation design examples in Chapter 14 were solved using a software package called *ChemSep* (Kooijman and Taylor, 1992). *ChemSep* (or an equivalent software package) will be needed for solving some the exercises. Information on the availability of *ChemSep* can be obtained from R. Taylor.

ACKNOWLEDGMENTS

The authors would like to express their appreciation to: Gulf Publishing Company, Houston, TX for permission to base portions of this textbook on the authors contribution entitled *Multicomponent Mass Transfer: Theory and Applications*, which we published in *Handbook of Heat and Mass Transfer*, edited by N. P. Cheremisinoff, 1986; H. L. Toor, E. U. Schlünder, and A. Gorak kindly provided copies of experimental data (some of it unpublished) that we have used in creating a number of examples, figures, and exercises; H. A. Kooijman for creating the software that allowed us to prepare several of the illustrations shown in this book (including the three dimensional plots of diffusion coefficients in Chapter 4); Norton Chemical Process Products Corporation of Stow, Ohio for supplying the photographs of packing elements in Chapter 12; and BP Engineering for permission to include several industrial applications of the nonequilibrium model in Chapter 14.

R. T.
R. K.

CONTENTS

NOMENCLATURE

a_i	Weighting factor (Chapter 1) [various]
a_i	Activity of component i in solution $[-]$
a'	Interfacial area per unit volume of vapor $[m^2/m^3]$
$\bar{a}$	Interfacial area per unit volume of liquid $[m^2/m^3]$
a	Interfacial area per unit volume of dispersion $[m^2/m^3]$
a_p	Specific surface area of packing $[m^2/m^3]$
a'_I	Interfacial area per unit volume of vapor in bubble formation zone (Sections 12.1 and 12.2) $[m^2/m^3]$
$a'_{II,k}$	Interfacial area per unit volume of vapor in kth bubble population (Sections 12.1 and 12.2) $[m^2/m^3]$
A_n	Eigenvalue in Kronig–Brink model
A^+	Damping constant in van Driest mixing length model $[-]$
A_b	Active bubbling area on tray $[m^2]$
A_h	Hole area of sieve tray $[m^2]$
A_c	Cross-sectional area $[m^2]$
A	Interfacial area in batch extraction cell $[m^2]$
$A(y^+)$	Quantity defined by Eqs. 10.3.4
$[A]$	Matrix defined by Eqs. 8.5.21 and 8.5.22 $[s/m^2]$
$[A(\xi)]$	Matrix defined by Eq. 9.3.8
b	Weir length per unit bubbling area (Section 12.1) $[m^{-1}]$
B	Channel base (Section 12.3) [m]
B	Inverse of binary diffusion coefficient $[s/m^2]$
$[B]$	Matrix function of inverted binary diffusion coefficients defined by Eqs. 2.1.21 and 2.1.22 $[s/m^2]$
$[B^n]$	Matrix function of inverted binary diffusion coefficients defined by Eqs. 2.4.10 and 2.4.11 $[s/m^2]$
$[B^{uV}]$	Transformation matrix defined by Eqs. 1.2.21 $[-]$
$[B^{Vu}]$	Transformation matrix defined by Eqs. 1.2.23 $[-]$
$[B^{uo}]$	Transformation matrix defined by Eqs. 1.2.25 $[-]$
$[B^{ou}]$	Transformation matrix defined by Eqs. 1.2.27 $[-]$
$[B(y^+)]$	Matrix defined by Eq. 10.4.7
c_i	Molar density of component i $[mol/m^3]$
c_t	Mixture molar density $[mol/m^3]$
C_{pi}	Specific heat of component i [J/kg]. Also, molar heat capacity of component i [J/mol]
C_p	Specific heat of mixture [J/kg]. Also, molar heat capacity of mixture [J/mol]
Ca	Capillary number (Section 12.3) $[-]$

d	Characteristic length of contacting device [m]
d	Diameter [m]
d_i	Driving force for mass diffusion $[m^{-1}]$
d_I	Diameter of jet in bubble formation zone (Sections 12.1 and 12.2) [m]
$d_{II,k}$	Diameter of bubble in kth bubble population (Sections 12.1 and 12.2) [m]
d_{eq}	Equivalent diameter [m]
d_p	Nominal packing size [m]
$Đ$	Maxwell–Stefan diffusivity for pair i–j $[m^2/s]$
$Đ_{ij, x_k \to 1}$	Limiting value of Maxwell–Stefan diffusivity for pair i–j when x_k tends to unity $[m^2/s]$
$Đ_{ij}^\circ$	Infinite dilution diffusivity for component i present in trace amounts in component j $[m^2/s]$
$D_{i, eff}$	Effective diffusivity of component i in multicomponent mixture $[m^2/s]$
D_{turb}	Turbulent eddy diffusivity $[m^2/s]$
D_i^T	Thermal diffusion coefficient [kg/m s]
D_{ref}	Reference value for diffusion coefficient $[m^2/s]$
$[D]$	Matrix of Fick diffusion coefficients $[m^2/s]$
$[D^\circ]$	Matrix of Fick diffusion coefficients in mass average velocity reference frame $[m^2/s]$
$[D^V]$	Matrix of Fick diffusion coefficients in volume average velocity reference frame $[m^2/s]$
$[D']$	Matrix of Fick diffusion coefficients relative to a reference diffusivity [−]
$[D_{turb}]$	Matrix of turbulent diffusion coefficients $[m^2/s]$
$\hat{D}_i$	ith eigenvalue of $[D]$ $[m^2/s]$
(e_i)	ith eigenvector of $[D]$ [−]
E	Energy flux in stationary coordinate frame of reference $[W/m^2]$
$\mathscr{E}$	Energy transfer rate (Chapter 14) [W]
E	Energy balance equation (Chapters 11, 13, 14, and 15) $[W/m^2, W]$
E_o	Overall efficiency (Eq. 13.1.1) [−]
E^{MV}	Murphree tray efficiency (Eq. 13.1.2) [−]
E_{OV}	Murphree point efficiency (Eq. 13.1.3) [−]
f	Fanning friction factor [−]
f_i	Fugacity of component i [Pa]
F	Discrepancy functions [various]
F_j	Flow rate of feed stream [mol/s]
$f_{II,k}$	Fraction of vapor in kth bubble population (Sections 12.1 and 12.2) [−]
$\mathscr{F}$	Faraday's constant $[9.65 \times 10^4$ C/mol]
$[f]$	Matrix function [various] (Chapter 5)
F_s	F-factor based on superficial velocity $[kg^{1/2}/m^{1/2}$ s]
$F(N_t)$	Function of total molar flux (Section 8.4) [−]
Fo	Fourier number [−]
Fr	Froude number [−]
g	Acceleration due to gravity $[9.81$ m/s$^2]$
G_{ij}	Chemical potential—composition derivative (Eq. 3.3.9) [J/mol]
G_{ij}	Parameter in NRTL model (Appendix D)

h	Heat transfer coefficient (Eq. 11.3.2) [W/m^2 K]
$\hbar$	Heat transfer coefficient (Eq. 11.3.1) [m/s]
h_f	Froth or dispersion height [m]
h_I	Height of bubble formation zone (Sections 12.1 and 12.2) [m]
$h_{II,k}$	Height of bubbling zone (Sections 12.1 and 12.2) [m]
h_w	Weir height [m]
H	Height of packing (Section 12.3) [m]
$\mathbb{H}_V$	Height of a transfer unit for the vapor (Section 12.3) [m]
$\mathbb{H}_L$	Height of a transfer unit for the liquid (Section 12.3) [m]
$\mathbb{H}_{OV}$	Overall height of a transfer unit (Section 12.3) [m]
$[\mathbb{H}_V]$	Matrix of heights of transfer units for the vapor (Section 12.3) [m]
$[\mathbb{H}_L]$	Matrix of heights of transfer units for the liquid (Section 12.3) [m]
$[\mathbb{H}_{OV}]$	Matrix of overall heights of transfer units (Section 12.3) [m]
HETP	Height equivalent to a theoretical plate [m]
$\bar{H}_i$	Partial specific enthalpy [J/kg] Also, partial molar enthalpy of component i [J/mol]
$\Delta H_{\mathrm{vap},i}$	Latent heat of vaporization of component i [J/mol]
i	Current [amps]
I	Referring to interphase or interface
$[H]$	Matrix of transport coefficients (Chapter 3) [J/mol m^2 s]
$[I]$	Identity matrix [$-$]
$\boldsymbol{I}$	Unit tensor [$-$]
j_D	Chilton–Colburn j factor for mass transfer [$-$]
j_H	Chilton–Colburn j factor for heat transfer [$-$]
j_m	Roots of zero-order Bessel function (Chapter 9) [$-$]
j	Mass diffusion flux relative to the mass average velocity [kg/m^2 s]
j^u	Mass diffusion flux relative to the molar average velocity [kg/m^2 s]
j^V	Mass diffusion flux relative to the volume average velocity [kg/m^2 s]
j^r	Mass diffusion flux relative to velocity of component r [kg/m^2 s]
$j_{i,\mathrm{turb}}$	Turbulent diffusion flux of component i [kg/m^2 s]
J	Molar diffusion flux relative to the molar average velocity [mol/m^2 s]
J^V	Molar diffusion flux relative to the volume average reference velocity [mol/m^2 s]
J^r	Molar diffusion flux relative to the velocity of component r [mol/m^2 s]
J^v	Molar diffusion flux relative to the mass average reference velocity [mol/m^2 s]
$J_{i,\mathrm{turb}}$	Molar turbulent diffusion flux of component i [mol/m^2 s]
$\hat{J}_i$	Pseudodiffusion flux (Chapter 5) [mol/m^2 s]
J_0	Zero-order Bessel function [$-$]
$[J]$	Jacobian matrix [various]
k_B	Boltzmann constant [1.38048 J/K]
k	Mass transfer coefficient in a binary mixture [m/s]
K_i	Equilibrium ratio (K value) for component i [$-$]
$[K]$	Diagonal matrix of the first $n-1$ K values [$-$]
$K_{i,\mathrm{eff}}$	"Effective" volumetric mass transfer coefficient [s^{-1}] or [h^{-1}]

$[K]$ Matrix of volumetric mass transfer coefficients (Section 5.6) $[s^{-1}]$

K_{OV} Overall mass transfer coefficient in a binary mixture $[m/s]$

$[k]$ Matrix of multicomponent mass transfer coefficients $[m/s]$

$k_{i,\text{eff}}$ Pseudobinary (effective) mass transfer coefficient of component i in a mixture $[m/s]$

$[K_{OV}]$ Matrix of multicomponent overall mass transfer coefficients $[m/s]$

$\mathcal{K}_i$ Equivalent conductivity of component i (Section 2.4)

$\mathcal{K}$ Equivalent conductivity of mixture (Section 2.4)

ℓ Generalized characteristic length $[m]$

ℓ Mixing length describing turbulent transport (Chapter 10) $[m]$

ℓ^+ Reduced mixing length $[-]$

ℓ_i Component flow of a liquid $[mol/s]$

Le   Lewis number $[-]$

L Liquid flow rate $[mol/s]$

m Mass of molecule $[kg]$

$\overline{M}$ Molar mass of mixture $[kg/mol]$

M_i Molar mass of component i $[kg/mol]$

M_i Moles of i in batch extraction cell $[mol]$

M_t Total moles of mixture in batch extraction cell $[mol]$

$[M]$ Matrix of equilibrium constants (Eq. 7.3.5) $[-]$

M_{ij}^T Component material balance equation (Chapters 13 and 14) $[mol/s]$

M_{ij} Component material balance equation (Chapters 13 and 14) $[mol/s]$

M_V Mass flow of vapor $[kg/s]$

M_L Mass flow of liquid $[kg/s]$

n Number of components in the mixture $[-]$

n_i Mass flux component i referred to a stationary coordinate reference frame $[kg/m^2\,s]$

n_t Mixture total mass flux referred to a stationary coordinate reference frame $[kg/m^2\,s]$

N_i Molar flux of component i referred to a stationary coordinate reference frame $[mol/m^2\,s]$

N_t Mixture molar flux referred to a stationary coordinate reference frame $[mol/m^2\,s]$

Nu   Nusselt number $[-]$

$\mathbb{N}_V$ Number of transfer units for the vapor phase in binary system $[-]$

$\mathbb{N}_L'$ Number of transfer units for the liquid phase in a binary system $[-]$

$[\mathbb{N}_V]$ Matrix of numbers of transfer units for the vapor phase $[-]$

$[\mathbb{N}_L]$ Matrix of numbers of transfer units for the liquid phase $[-]$

$[\mathbb{N}_{OV}]$ Matrix of overall number of transfer units $[-]$

$\mathcal{N}_V$ Number of transfer units for a vapor defined by Eq. 12.1.42 $[-]$

$\mathcal{N}_L'$ Number of transfer units for a liquid defined by Eq. 12.1.42 $[-]$

$\mathcal{N}$ Mass transfer rate (Chapter 14) $[mol/s]$

P Pressure $[Pa]$

P Perimeter in structured packing (Section 12.3) $[m]$

P_j Pressure drop equation (Chapter 14)

p	Sieve tray hole pitch (Section 12.1) [m]
p	Pressure tensor (Chapter 1) [Pa]
$[P]$	Modal matrix $[-]$
P_i^s	Vapor pressure of component i [Pa]
p_i	Partial pressure of component i [Pa]
P_i	Parachor (Section 4.2) [$g^{1/4}$ cm^3/mol $s^{1/2}$]
Pr	Prandtl number $[-]$
Pr_{turb}	Turbulent Prandtl number $[-]$
q	Conductive heat flux [W/m^2]. Also, integer parameter $[-]$
q_{turb}	Turbulent contribution to the conductive heat flux [W/m^2]
Q_I	Unaccomplished equilibrium in bubble formation zone (Section 12.1) $[-]$
$Q_{II,k}$	Unaccomplished equilibrium in kth bubble population (Section 12.1) $[-]$
$[Q]$	Matrix describing unaccomplished equilibrium (Section 12) $[-]$
$[Q_I]$	Matrix of unaccomplished equilibrium in bubble formation zone (Section 12.2) $[-]$
$[Q_{II,k}]$	Matrix describing unaccomplished equilibrium in kth bubble population (Section 12.2) $[-]$
Q_i	Equilibrium equation (Chapters 11–15) $[-]$
Q_j	Heat duty (Chapters 13–14) [W]
Q_L	Volumetric liquid flow rate [m^3/s]
Q_V	Volumetric vapor flow rate [m^3/s]
r	Coordinate direction or position [m]
r_0	Inner edge of film [m]
r_0	Radius of spherical particle [m]
r_δ	Outer edge of film [m]
r_i	Radius of molecule in Eq. 4.1.7 [m]
R	Gas constant [8.314 J/mol K]
R_i	Radius of gyration (Section 4.2 only) [nm]
(R)	Vector of rate equations [mol/m^2s or mol/s]
$[R]$	Matrix function of inverted binary mass transfer coefficients defined by Eqs. 8.3.25 [s/m]
$[R^{OV}]$	Inverse of $[K_{OV}]$ [s/m]
Re	Reynolds number $[-]$
R_w	Inner radius of tube wall [m]
R_w^+	Reduced tube radius $[-]$
R	Mass transfer rate equation
s	Surface renewal frequency [s^{-1}]
S	Quantity defined by Eq. 4.2.4
S	Structured packing channel side (Section 12.3) [m]
S	Summation equation $[-]$
Sc	Schmidt number $[-]$
$[Sc]$	Matrix of Schmidt numbers $[-]$
Sc_{turb}	Turbulent Schmidt number $[-]$
Sh	Sherwood number $[-]$
$[Sh]$	Matrix of Sherwood numbers $[-]$

St	Stanton number $[-]$
St_H	Stanton number for heat transfer $[-]$
$[St]$	Matrix of Stanton numbers $[-]$
t	Time [s]
t_i	Transference number (Section 2.4) $[-]$
t_e	Exposure time [s]
t_V	Vapor residence time (Section 12.1) [s]
t_L	Liquid residence time (Section 12.1) [s]
t_I	Residence time in bubble formation zone (Section 12.1) [s]
$t_{II,k}$	Residence time in kth bubble population (Section 12.1) [s]
T	Temperature [K]
u_i	Velocity of diffusion of species i [m/s]
u	Molar average reference velocity [m/s]
u^V	Volume average reference velocity [m/s]
u^I	Velocity of the interface [m/s]
u^*	Friction velocity [m/s]
$\bar{u}$	Average flow velocity [m/s]
u^+	Reduced velocity $[-]$
u_I	Velocity of vapor in bubble formation zone (Sections 12.1 and 12.2) [m/s]
$u_{II,k}$	Rise velocity of kth bubble population (Section 12.1 and 12.2) [m/s]
u_s	Superficial velocity [m/s]
u_{sf}	Superficial velocity at flooding [m/s]
U	Internal energy (Section 1.3)
U	Velocity [m/s]
v	Mass average mixture velocity [m/s]
V_i	Molar volume at normal boiling point (Section 4.1) [m³/mol]
$\bar{V}_i$	Partial molar volume [m³/mol]
$\bar{V}_t$	Mixture molar volume [m³/mol]
v_i	Molar flow rate of component i [mol/s]
V	Molar flow rate of mixture [mol/s]
V_0	Volume of bulb in two-bulb diffusion cell (Chapter 5) [m²]
V_ℓ	Volume of bulb in two-bulb diffusion cell (Chapter 5) [m²]
W	Weir length (Section 12.1) [m]
$[W]$	Matrix of mass transfer coefficients: $[W] = [\beta][k]$ [m/s]

Greek Letters

α	Relative froth density (Section 12.1) $[-]$
α_e	Parameter in Bennett method for pressure drop (Eq. 12.1.27)
α_{ij}	Multicomponent thermal diffusion factors $[-]$
β	Cell constant in two-bulb diffusion cell (Eq. 5.4.6) [m⁻²]
$[\beta]$	Bootstrap matrix $[-]$
γ_i	Activity coefficient of component i in solution $[-]$
Γ	Liquid flow per unit length of perimeter (Section 12.3) [kg/m³ s]
Γ	Thermodynamic factor for binary system (Eq. 2.2.12) $[-]$

$[\Gamma]$	Matrix of thermodynamic factors with elements defined by Eqs. 2.2.5 $[-]$
δ	Distance from interface [m]
δ_{ij}	Kronecker delta, 1 if $i = k$, 0 if $i \neq k$ $[-]$
ε	Void fraction $[-]$
ζ	Rate of production of field quantity in bulk fluid mixture (Section 1.3)
ζ_I	Rate of production of field quantity at the interface (Section 1.3)
ζ	Combined variable, $\zeta = z/\sqrt{4t}$ (Chapter 9)
η	Dimensionless distance $[-]$
κ	Maxwell–Stefan mass transfer coefficient in a binary mixture (Eqs. 8.3.26 and 8.8.16) [m/s]
Λ	Dimensionless parameters $[-]$
Λ	Stripping factor (Chapters 12 and 13) $[-]$
λ_i	Difference between component molar enthalpies (Eq. 11.5.13) [J/mol]
λ	Parameter in mixing length models (Section 10.2)
λ	Molecular thermal conductivity [W/m K]
λ_{turb}	Turbulent thermal conductivity [W/m K]
λ_n	Eigenvalue in the Kronig–Brink model $[-]$
μ_i	Molar chemical potential of component i [J/mol]
μ_i	Viscosity of component i (Section 4.1) [Pa s]
μ	Molecular (dynamic) viscosity of mixture [Pa s]
μ_{turb}	Turbulent eddy viscosity [Pa s]
ν	Molecular kinematic viscosity of mixture [m^2/s]
ν_{turb}	Turbulent eddy kinematic viscosity [m^2/s]
ν_i	Determinacy coefficients for species i [various]
$\bar{\nu}$	Mole fraction weighted sum of component determinacy coefficients (Section 8.5) [various]
ξ	Unit normal directed from phase "x" to phase "y" $[-]$. Also, dimensionless distance along dispersion or column height $[-]$
ξ	Ratio of component mass flux to total mass flux (Sections 10.3 and 10.4) $[-]$
Ξ	Correction factor for high fluxes in binary mass transfer $[-]$. Also, correction, factor for high fluxes in explicit methods $[-]$
$\Xi_{i,eff}$	Correction factor for high fluxes in pseudobinary (effective diffusivity) methods $[-]$
Ξ_H	Correction factor for the effect of high fluxes on the heat transfer coefficient $[-]$
$[\Xi]$	Matrix of high flux correction factors $[-]$
ρ_i	Mass density of component i [kg/m^3]
ρ_t	Mixture mass density [kg/m^3]
σ	Rate of entropy production (Chapter 2) [J/m^3 s K]
σ_{diff}	Rate of entropy production due to diffusion (Chapter 2) [J/m^3 s K]
σ	Characteristic diameter of molecule (Section 4.1) [Å]
σ	Surface tension [N/m]
σ_c	Critical surface tension (Section 12.3) [N/m]
τ	Shear stress [Pa]
τ_0	Shear stress at the wall [Pa]
τ_{turb}	Turbulent shear stress [Pa]

τ	Stress tensor [Pa]
τ_{ij}	Parameter in NRTL activity coefficient model [$-$]
ϕ	Mass transfer parameter defined by Eq. 8.2.6 (Section 8.2) [$-$]
ϕ_i	Volume fraction (Section 4.1) [$-$]
ϕ_i	Fugacity coefficient (Chapter 2) [$-$]
ϕ	Electrical potential (Section 2.4) [V]
ϕ	Association parameter in Eq. 4.1.8
ϕ	fractional free area (Section 12.1) [$-$]
(ϕ)	Column matrix of dimensionless mass transfer parameters (Section 8.2) [$-$]
$\boldsymbol{\Phi}$	Nonconvective flux of field quantity (Chapter 1) [various]
Φ	Mass transfer rate factor for binary mass transfer (Eqs. 8.2.5, 9.2.3, and 10.3.10) [$-$]
Φ	Mass transfer rate factor for explicit methods (Eq. 8.5.13) [$-$]
$\Phi_{i,\,\mathrm{eff}}$	Mass transfer rate factor in pseudobinary (effective diffusivity) methods [$-$]
Φ_H	Heat transfer rate factors [$-$]
$[\Phi]$	Matrix of mass transfer rate factors [$-$]
$\psi(t)$	Surface age distribution [s^{-1}]
ψ	Referring to any field variable (Section 1.3) [various]
$[\Psi]$	Matrix of mass transfer rate factors in linearized film model (Eq. 8.4.4) [$-$]
$[\Psi]$	Matrix of mass transfer rate factors in turbulent diffusion model (Eq. 10.3.9) [$-$]
ω_i	Mass fraction of component i [$-$]
$[\omega]$	Diagonal matrix of mass fractions [$-$]
Ω	Angular velocity [rad/s]
Ω	Function defined by Eq. 12.1.15 [$-$]
$[\Omega]$	Matrix function [$-$]
$[\Omega]$	Matrizant [various definitions]
θ	Angle
$[\Theta]$	Matrix of rate factors for nonideal systems [$-$]
χ	Arbitrary independent variable
(χ)	Vector defined by Eq. 9.3.4 (Section 9.3)

Subscripts

av	Denotes that suitably averaged properties are used in the determination of the indicated parameter
b	Bulk phase property
d	Dominant eigenvalue
E	Quantity entering zone under consideration
eff	Pseudobinary or "effective" parameter
H	Parameter relevant to heat transfer
I	Referring to the interface
I	Referring to bubble formation zone (Sections 12.1 and 12.2)
II, k	Referring to kth bubble population in bubble rise zone (Sections 12.1 and 12.2)

i	Component i property or parameter
i, j, k	Component indices, stage or section numbers (j only)
L	Quantity leaving zone under consideration
m	Mean value. Also, refers to the mass average velocity
n	nth component
O	Overall parameter. Also, denotes reduced energy and heat conduction fluxes
OV	Overall parameter referred to the vapor phase
ref	Denotes reference quantity
t	Referring to total mixture
T, P	Constant temperature and pressure
x	Referring to the "x" phase
y	Referring to the "y" phase
δ	Quantity evaluated at position $\eta = \delta$
0	Quantity evaluated at position $\eta = 0$
∞	Quantity evaluated at long time or long distance

Superscripts

C	Referring to the coolant
F	Referring to the feed
I	Referring to the interface
(k)	Denotes iteration number
L	Referring to the liquid phase
m	Referring to the mass average velocity
V	Referring to the vapor–gas phase
v	Referring to the volume average velocity
W	Referring to the wall
x	Referring to the "x" phase
y	Referring to the "y" phase
$'$	Referring to the $'$ phase
$''$	Referring to the $''$ phase
$\circ$	Referring to mass average reference velocity frame
$\bullet$	Referring to finite transfer rates

Miscellaneous

$\overline{}$	Overall denotes partial molar property. Also, averaged parameter
$\hat{}$	Eigenvalue of corresponding matrix

Mathematical Symbols

∇	Gradient
Δ	Difference operator
lim	Limit

Matrix Operations and Notation

( )	Column matrix
[ ]	Square matrix
[ ]$^{-1}$	Inverse of a square matrix
( )^T	Row matrix
| |	Determinant of a square matrix
tr[ ]	Trace of matrix

MULTICOMPONENT MASS TRANSFER

PART I
Molecular Diffusion

1 Preliminary Concepts

The reader should not be intimidated by the great generality expressed by the vectorial character of these equations, because a simple one-dimensional approximation is almost always used in applications. (But it is hard to resist the lure of cheap generality when writing down equations.)
—E. A. Mason and H. K. Lonsdale (1990)

1.1 CONCENTRATION MEASURES

In the description of the interphase mass transfer process, a variety of measures for constituent concentrations, mixture reference velocities, and diffusion fluxes (with respect to the arbitrarily defined mixture velocity) are used. Table 1.1 summarizes the most commonly used concentration measures together with a number of other quantities that will be needed from time to time.

1.2 FLUXES

If u_i denotes the velocity of component i (with respect to a stationary coordinate reference frame) then the *mass flux* of that species is defined by

$$n_i = \rho_i u_i \tag{1.2.1}$$

and has units of kilograms per meter squared per second (kg/m^2 s). If we sum the component fluxes we obtain

$$n_t = \sum_{i=1}^{n} n_i = \rho_t v \tag{1.2.2}$$

where we have defined the *mass average velocity* v by

$$v = \sum_{i=1}^{n} \omega_i u_i \tag{1.2.3}$$

The *molar flux* of species i is defined by

$$N_i = c_i u_i \tag{1.2.4}$$

which has units of moles per meter squared per second (mol/m^2 s). The *total molar flux* is the sum of these quantities

$$N_t = \sum_{i=1}^{n} N_i = c_t u \tag{1.2.5}$$

3

TABLE 1.1   Concentration Measures and Other Thermodynamic Mixture Parameters

$c_i \, [\text{mol/m}^3]$	Molar density of i; $c_i = \rho_i / M_i$
$c_t \, [\text{mol/m}^3]$	Mixture molar density; $c_t = \sum\limits_{i=1}^{n} c_i$
$x_i \, [-]$	Mole fraction of i; $x_i = c_i / c_t$; $\sum\limits_{i=1}^{n} x_i = 1$
$\rho_i \, [\text{kg/m}^3]$	Mass density of i; $\rho_i = c_i M_i$
$\rho_t \, [\text{kg/m}^3]$	Mixture mass density; $\rho_t = \sum\limits_{i=1}^{n} \rho_i$
$\omega_i \, [-]$	Mass fraction of i; $\omega_i = \rho_i / \rho_t$; $\sum\limits_{i=1}^{n} \omega_i = 1$
$M_i \, [\text{kg/mol}]$	Molar mass of i
$\bar{V}_i \, [\text{m}^3/\text{mol}]$	Partial molar volume of species i; $\sum\limits_{i=1}^{n} x_i \bar{V}_i = \bar{V}_t$
$\bar{V}_t \, [\text{m}^3/\text{mol}]$	Mixture molar volume; $\bar{V}_t = 1/c_t$
$\phi_i \, [-]$	Volume fraction of species i; $\phi_i = c_i \bar{V}_i$
$f_i \, [\text{N/m}^2]$	Fugacity of i
$\mu_i \, [\text{J/mol}]$	Molar chemical potential of species i

where we have defined the *molar average velocity* u

$$u = \sum_{i=1}^{n} x_i u_i \tag{1.2.6}$$

It is the calculation of these fluxes (particularly the molar ones), which is our main concern. However, before getting down to business we need to define a few more fluxes; in particular, the *diffusion flux*, which is the flux of species i relative to the flux of the mixture as a whole. The definition of this flux raises the first of our problems, which mixture velocity are we going to use? We have already introduced two, v and u, and there are others that we have not discussed yet. The literature on diffusion would be a good deal simpler if there were only one way to define diffusion fluxes. For each choice of reference velocity there are at least two different diffusion fluxes that we could define, mass fluxes and molar fluxes.

Perhaps an example will help to clarify the situation. If we choose v as the reference velocity, then the mass diffusion flux with respect to the mass average velocity is

$$j_i = \rho_i(u_i - v) \tag{1.2.7}$$

and

$$\sum_{i=1}^{n} j_i = 0 \tag{1.2.8}$$

The mass flux n_i is related to the mass diffusion flux as

$$n_i = j_i + \rho_i v = j_i + \omega_i n_t \tag{1.2.9}$$

On the other hand, we could choose u as the reference velocity and define molar diffusion fluxes relative to it as

$$J_i = c_i(u_i - u) \tag{1.2.10}$$

with

$$\sum_{i=1}^{n} J_i = 0 \tag{1.2.11}$$

The molar flux N_i is related to the molar diffusion flux by

$$N_i = J_i + c_i u = J_i + x_i N_t \tag{1.2.12}$$

These are the most commonly encountered sets of fluxes; other sets could be defined. We could, for example, define a mass diffusion flux relative to the molar average velocity or a molar diffusion flux relative to the mass average velocity. Still other choices of reference velocity are sometimes used; for example, the volume average velocity u^V

$$u^V = \sum_{i=1}^{n} c_i \overline{V}_i u_i = \sum_{i=1}^{n} \phi_i u_i \tag{1.2.13}$$

where ϕ_i is the volume fraction of species i defined in Table 1.1.

Table 1.2 summarizes the most commonly used reference velocities.

Let us define an arbitrary reference velocity u^a

$$u^a = \sum_{i=1}^{n} a_i u_i \tag{1.2.14}$$

where a_i is a weighting factor that satisfies the requirement

$$\sum_{i=1}^{n} a_i = 1 \tag{1.2.15}$$

We now define a mass diffusion flux relative to this arbitrary reference velocity j_i^a by

$$j_i^a = \rho_i(u_i - u^a) \tag{1.2.16}$$

Not all of these diffusion fluxes are independent; on summing these fluxes over the n species we find

$$\sum_{i=1}^{n} \frac{a_i}{\omega_i} j_i^a = 0 \tag{1.2.17}$$

In a similar way we define the molar diffusion flux relative the velocity u^a by

$$J_i^a = c_i(u_i - u^a) \tag{1.2.18}$$

TABLE 1.2   Reference Velocities^a

u^a	Arbitrary mixture velocity, weighting factor a_i

$$u^a = \sum_{i=1}^{n} a_i u_i \qquad \sum_{i=1}^{n} a_i = 1$$

v	Mass average mixture velocity, weighting factor ω_i

$$v = \sum_{i=1}^{n} \omega_i u_i \qquad \sum_{i=1}^{n} \omega_i = 1$$

u	Molar averaged reference velocity, weighting factor x_i

$$u = \sum_{i=1}^{n} x_i u_i \qquad \sum_{i=1}^{n} x_i = 1$$

u^V	Volume averaged reference velocity, weighting factor ϕ_i

$$u^V = \sum_{i=1}^{n} \phi_i u_i \qquad \sum_{i=1}^{n} \phi_i = 1$$

u_r	Velocity of species r, weighting factor δ_{ir}, where δ_{ir} is the Kronecker delta

$$\delta_{ir} = 1 \quad \text{if } i = r$$

$$\delta_{ir} = 0 \quad \text{if } i \neq r$$

^{a}Units are in meters per second [m/s].

The sum of these fluxes gives

$$\sum_{i=1}^{n} \frac{a_i}{x_i} J_i^a = 0 \tag{1.2.19}$$

The diffusion fluxes defined earlier are seen to be special cases of the more general definitions presented above. Table 1.3 summarizes the most commonly encountered diffusion fluxes and Table 1.4 summarizes those fluxes that are measured with respect to a laboratory fixed coordinate reference frame.

1.2.1   Transformations Between Fluxes

It will sometimes prove necessary to transform fluxes from one reference velocity to another. We give some examples of the required relations here.

To relate the molar diffusion flux relative to the volume average velocity to the molar diffusion flux relative to the molar average reference velocity we use the transformation

$$J_i = \sum_{k=1}^{n-1} B_{ik}^{uV} J_k^V \tag{1.2.20}$$

TABLE 1.3   Diffusion Fluxes[a]

j_i^a Mass diffusion flux relative to arbitrary reference velocity

$$j_i^a = \rho_i(u_i - u^a) \qquad \sum_{i=1}^{n} \frac{a_i}{\omega_i} j_i^a = 0$$

j_i Mass diffusion flux relative to mass average velocity

$$j_i = \rho_i(u_i - v) \qquad \sum_{i=1}^{n} j_i = 0$$

j_i^u Mass diffusion flux relative to molar average velocity

$$j_i^u = \rho_i(u_i - u) \qquad \sum_{i=1}^{n} \frac{x_i}{\omega_i} j_i^u = 0$$

j_i^V Mass diffusion flux relative to volume average velocity

$$j_i^V = \rho_i(u_i - u^V) \qquad \sum_{i=1}^{n} \frac{\phi_i}{\omega_i} j_i^V = 0$$

j_i^r Mass diffusion flux relative to component r velocity

$$j_i^r = \rho_i(u_i - u_r) \qquad j_r^r = 0$$

J_i^a Molar diffusion flux relative to arbitrary reference velocity

$$J_i^a = c_i(u_i - u^a) \qquad \sum_{i=1}^{n} \frac{a_i}{x_i} J_i^a = 0$$

J_i Molar diffusion flux relative to molar average velocity

$$J_i = c_i(u_i - u) \qquad \sum_{i=1}^{n} J_i = 0$$

J_i^V Molar diffusion flux relative to volume average velocity

$$J_i^V = c_i(u_i - u^V) \qquad \sum_{i=1}^{n} \bar{V}_i J_i^V = 0$$

J_i^v Molar diffusion flux relative to mass average velocity

$$J_i^v = c_i(u_i - v) \qquad \sum_{i=1}^{n} \frac{\omega_i}{x_i} J_i = 0$$

J_i^r Molar diffusion flux relative to component r velocity

$$J_i^r = c_i(u_i - u_r) \qquad J_r^r = 0$$

Units are kilograms per meter squared per second (kg/m^2 s) for mass diffusion fluxes and moles per meter squared per second (mol/m^2 s) for molar diffusion fluxes.

TABLE 1.4   Fluxes With Respect to a Laboratory Fixed Frame of Reference[a]

n_i	Mass flux relative to stationary coordinates

$$n_i = \rho_i u_i = j_i + \rho_i v$$

n_t	Total mass flux relative to stationary coordinates

$$n_t = \sum_{i=1}^{n} n_i = \rho_t v$$

N_i	Molar flux relative to stationary coordinates

$$N_i = c_i u_i = J_i + c_i u$$

N_t	Total molar flux relative to stationary coordinates

$$N_t = \sum_{i=1}^{n} N_i = c_t u$$

[a]Mass fluxes units are kilograms per meter squares per second $(kg/m^2 \ s)$; molar fluxes units are moles per meter squared per second $(mol/m^2 \ s)$.

where the coefficients B_{ik}^{uV} are defined by

$$B_{ik}^{uV} = \delta_{ik} - x_i\left(1 - \bar{V}_k/\bar{V}_n\right) \tag{1.2.21}$$

The inverse transformation is

$$J_i^V = \sum_{k=1}^{n-1} B_{ik}^{Vu} J_k \tag{1.2.22}$$

where the coefficients B_{ik}^{Vu} are defined by

$$B_{ik}^{Vu} = \delta_{ik} - x_i\left(\bar{V}_k - \bar{V}_n\right)/\bar{V}_t \tag{1.2.23}$$

To transform the mass diffusion flux relative to the mass average velocity j_i to the mass diffusion flux relative to the molar average velocity j_i^u we use

$$j_i^u = \sum_{k=1}^{n-1} B_{ik}^{uo} j_k \tag{1.2.24}$$

where the coefficients B_{ik}^{uo} are given by

$$B_{ik}^{uo} = \delta_{ik} - \omega_i\left(\frac{x_k}{\omega_k} - \frac{x_n}{\omega_n}\right) \tag{1.2.25}$$

The inverse transformation is

$$j_i = \sum_{k=1}^{n-1} B_{ik}^{ou} j_k^u \qquad (1.2.26)$$

where the B_{ik}^{ou} are

$$B_{ik}^{ou} = \delta_{ik} - \omega_i \left(1 - \frac{\omega_n x_k}{x_n \omega_k} \right) \qquad (1.2.27)$$

To change the units in which the flux is expressed requires further manipulation.

$$c_t x_i j_i^u = \rho_t \omega_i J_i \qquad (1.2.28)$$

For more on the use of these transformations see Section 3.2.4 and Exercises 1.1, and 1.2.

1.3 BALANCE RELATIONS FOR A TWO-PHASE SYSTEM INCLUDING A SURFACE OF DISCONTINUITY

Let us consider a two-phase system including a surface of discontinuity (phase interface). Let x and y represent the two phases. For example, y may refer to the gas phase and x to the liquid phase in a two-phase system. Let the number of components in each phase be n. Let I represent the phase interface and the unit normal directed from phase x to y. The system considered is shown pictorially in Figure 1.1. Our immediate task is to develop the balance relations describing the interphase transport processes taking place is this system.

During interphase mass transfer, concentration gradients will be set up across the interface. The concentration variations in the bulk phases x and y will be described by differential equations; whereas at the interface I, we will have jump conditions or boundary conditions. Standart (1964) and Slattery (1981) give detailed discussions of these relations for the transport of mass, momentum, energy, and entropy. It will not be possible to give here the complete derivations and the reader is, therefore, referred to these sources. A masterly treatment of this subject is also available in the article by Truesdell and Toupin (1960), which must be compulsory reading for a serious researcher in transport phenomena.

The equations of change for each fluid phase and the "jump" balance conditions that must be met at the interface are summarized in Table 1.5. There is an important restriction on the equations in Table 1.5; the effect of chemical reactions in the bulk fluid phase has been neglected. For all of the applications considered in this book this neglect is justified.

Our first major task is the description of the interfacial mass transfer process and, therefore, we shall examine further the equations for continuity of species i and the equation for conservation of total mass of mixture.

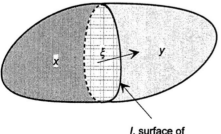

I, surface of
discontinuity

Figure 1.1. Pictorial representation of a two-phase system showing a surface of discontinuity or interface.

TABLE 1.5 Balance Relations for a Two-Phase System

Equation of Change for Any Conserved Property

$$\frac{\partial(\rho_t \Psi)}{\partial t} + \nabla \cdot \{\rho_t \Psi v\} + \nabla \cdot \Phi = \zeta \tag{1.3.1}$$

where

Ψ = an arbitrary field quantity per unit mass of mixture

ζ = the rate of production of field per unit volume of bulk phase

Φ = a nonconvective flux of the field quantity through external bounding surface S

ρ_t = the mass density of bulk fluid mixture

v = the mass average velocity of fluid mixture

The tensorial order of the flux Φ is one higher than that of the field quantity Ψ.

Jump Balance Relation for the Interface I

$$\xi \cdot \left\{ \Phi^y + \rho_t^y \Psi^y (v^y - u^I) - \Phi^x - \rho_t^x \Psi^x (v^x - u^I) \right\} = \zeta^I \tag{1.3.2}$$

where

ξ = the unit normal to I directed from the x to y phase

ζ^I = the rate of production of field quantity per unit area at the interface I

u^I = the velocity of the interface I

All of the foregoing quantities are functions of position and time.

A. Balance of Species i (no chemical reactions in bulk phase)

$$\Psi = \omega_i \qquad \Phi = j_i \qquad \zeta = 0$$

B. Conservation of Total Mass of Mixture

$$\Psi = 1 \qquad \Phi = 0 \qquad \zeta = 0 \qquad \zeta^I = 0$$

C. Conservation of Linear Momentum

$$\Psi = v \qquad \Phi = p \qquad \zeta = \sum_{i=1}^{n} \rho_i \bar{F}_i \qquad \zeta^I = 0$$

where

$p = pI + \tau$ is the pressure tensor

p = the thermodynamic pressure

τ = the stress tensor

I = the unit tensor

$\bar{F}_i$ = the body force acting per unit mass of species i

ζ = the sum of body forces $\bar{F}_i$

D. Energy Balance

$$\Psi = U + \tfrac{1}{2} v \cdot v \qquad \Phi = q + \sum_{i=1}^{n} \bar{H}_i j_i + p \cdot v$$

$$\zeta = \sum_{i=1}^{n} \rho_i \bar{F}_i \cdot u_i \qquad \zeta^I = 0$$

where

U = the internal energy per unit mass of mixture

q = the conductive heat flux

$\bar{H}_i$ = the partial specific enthalpy of component i

The differential balance relation for continuity of mass of species i is

$$\frac{\partial \rho_i}{\partial t} + \nabla \cdot \{\rho_i v\} = -\nabla \cdot j_i \tag{1.3.3}$$

For the total mixture we have

$$\frac{\partial \rho_t}{\partial t} + \nabla \cdot \{\rho_t v\} = \frac{\partial \rho_t}{\partial t} + \rho_t \nabla \cdot v + v \cdot \nabla \rho_t = 0 \tag{1.3.4}$$

If we denote the mass average velocity following derivative as

$$\frac{d}{dt} \equiv \frac{\partial}{\partial t} + v \cdot \nabla \tag{1.3.5}$$

then it is easy to show that Eq. 1.3.3 can be simplified to the form

$$\rho_t \frac{d\omega_i}{dt} \equiv \rho_t \left\{ \frac{\partial \omega_i}{\partial t} + v \cdot \nabla \omega_i \right\} = -\nabla \cdot j_i \tag{1.3.6}$$

Equation 1.3.6 can be expressed in terms of the mass fluxes, n_i, as

$$\frac{\partial \rho_i}{\partial t} + \nabla \cdot n_i = 0 \tag{1.3.7}$$

The differential mass balance for continuity of mass of the mixture is

$$\frac{\partial \rho_t}{\partial t} + \nabla \cdot n_t = 0 \tag{1.3.8}$$

Equation 1.3.6 can also be written in molar units as

$$c_t \frac{dx_i}{dt} \equiv c_t \left\{ \frac{\partial x_i}{\partial t} + u \cdot \nabla x_i \right\} = -\nabla \cdot J_i \tag{1.3.9}$$

where we use the mole average velocity following derivative. Only $n - 1$ of the Eqs. 1.3.9 are independent because the mole fractions x_i sum to unity and the molar diffusion fluxes J_i sum to zero (see Table 1.3). Exactly analogous relations will hold for the mole fractions y_i in phase y.

Equation 1.3.9 can be expressed in terms of the component molar fluxes N_i as

$$\frac{\partial c_i}{\partial t} + \nabla \cdot N_i = 0 \tag{1.3.10}$$

The differential balance expressing conservation of total moles of mixture is obtained by summing Eqs. 1.3.10 for all components to give (recall that we do not consider chemical reactions occurring in the bulk phases)

$$\frac{\partial c_t}{\partial t} + \nabla \cdot N_t = 0 \tag{1.3.11}$$

If we choose to represent the diffusion fluxes with respect to the volume average velocity u^V then the differential balance relations take the form

$$\frac{\partial c_i}{\partial t} + \nabla \cdot c_i u^V = -\nabla \cdot J_i^V \tag{1.3.12}$$

It must be emphasized here that Eq. 1.3.12 cannot be simplified to a form analogous to the Eqs. 1.3.6 or 1.3.9 because there is no law of conservation of volume.

We shall normally use mole fractions as composition measures, the molar average reference velocity u and the molar diffusion fluxes (with respect to u) J_i to describe the diffusion process within a given phase (Chapters 5–9). Molar quantities are not particularly convenient when we have to solve the equations of continuity of mass in conjunction with the equations of motion. The latter are best expressed in the mass average frame. We shall, in fact, switch to the use of mass fractions, the mass average reference velocity v and mass diffusion fluxes (with respect to the mass average velocity) j_i in our discussion of turbulent mass transfer (Chapter 10). The volume average reference velocity u^V is a favorite among physical chemists who use this reference in the interpretation of diffusion data in, for example, stirred cells. However, there is no conservation of volume, in general, and this choice usually is not convenient for chemical engineering purposes. Consider the relative simplicity of Eq. 1.3.9, using u, in comparison to Eq. 1.3.12 for the corresponding choice of u^V. We shall return to discuss this topic again when we consider various choices of the driving force for diffusion (Section 3.1.2).

In addition to the differential Eqs. 1.3.9, which apply to the bulk phases, the following boundary condition must be satisfied at the interface I, provided there are no interface (surface) chemical reactions.

$$\xi \cdot c_i^x \left(u_i^x - u^I \right) = \xi \cdot c_i^y \left(u_i^y - u^I \right) \qquad i = 1, 2, \ldots, n \qquad (1.3.13)$$

that is, the normal component of the flux of component i with respect to the interface must be continuous across the phase boundary. If the interface itself is stationary (i.e., $u^I = 0$), then Eq. 1.3.13 can be written as

$$\xi \cdot N_i^x = \xi \cdot N_i^y \qquad i = 1, 2, \ldots, n \qquad (1.3.14)$$

where N_i^x is the molar flux of component i in phase x in a stationary, laboratory fixed, coordinate reference frame (Table 1.4). Equation 1.3.14 merely states that the normal component of the flux N_i is a phase invariant.

For mass transfer with surface chemical reaction (as, e.g., in a tube wall catalytic reactor), Eqs. 1.3.2 and 1.3.14 yield

$$\xi \cdot N_i^x = \zeta_i^I \qquad (1.3.15)$$

The molar fluxes N_i appear in engineering design models. One of the main objectives of this text is to consider ways in which these fluxes may be calculated from a knowledge of the hydrodynamics and transport properties of the system. The boundary conditions (Eq. 1.3.13), or the simplified versions (Eqs. 1.3.14 and 1.3.15), are well known to chemical engineers (see, e.g., the book by Bird et al., 1960), but it is instructive to follow the general derivations of these relations [see Standart (1964) and Slattery (1981)].

1.4 SUMMARY

This first chapter has been somewhat in the nature of a housekeeping exercise with regard to the definitions of mass and molar fluxes, reference velocities, and transformations from one reference frame to another. We shall not have occasion to use all of these definitions, but they have been included for reasons of completeness. Any reader who is interested in furthering their knowledge on this topic must refer to De Groot and Mazur (1962).

2 The Maxwell–Stefan Relations

Das Studium der Maxwell'schen Abhandlung ist nicht leicht.
>—J. Stefan (1871) commenting on Maxwell's (1866) work

Diffusion is the intermingling of the atoms or molecules of more than one species; it is the inevitable result of the random motions of the individual molecules that are distributed throughout space. The development of a rigorous kinetic theory to describe this intermingling in gas mixtures is one of the major scientific achievements of the nineteenth century. A simplified kinetic theory of diffusion, adapted from Present (1958), is the main theme of Section 2.1. More rigorous (and complicated) developments are to be found in the books by Hirschfelder et al. (1964), Chapman and Cowling (1970), and Cunningham and Williams (1980). An extension to cover diffusion in nonideal fluids is developed thereafter.

2.1 DIFFUSION IN IDEAL GAS MIXTURES

2.1.1 The Mechanics of Molecular Collisions

Let us first consider the mechanics of collisions between an average molecule of species 1 and an average molecule of species 2. The molecule of species 1 has velocity u_1 and the molecule of species 2 has velocity u_2.

The momentum of these two average molecules is $m_1 u_1$ and $m_2 u_2$ where m_1 and m_2 are the masses of the respective molecules. The total momentum of the pair of molecules is $m_1 u_1 + m_2 u_2$. This total momentum of the pair of molecules is conserved on collision. That is, if u_1' and u_2' represent the velocities after collision, then the law of conservation of momentum requires that

$$m_1(u_1 - u_1') + m_2(u_2 - u_2') = 0 \qquad (2.1.1)$$

The momemtum transferred from species 1 to species 2 is the left-hand member of Eq. 2.1.1

$$\begin{array}{l}\text{Momentum transferred} \\ \text{from a molecule of} \\ \text{1 to a molecule of 2} \\ \text{through collision}\end{array} = \begin{array}{l}\text{momentum of 1} \\ \text{before collision}\end{array} - \begin{array}{l}\text{momentum of 1} \\ \text{after collision}\end{array}$$

$$= m_1(u_1 - u_1') \qquad (2.1.2)$$

Our next task is to calculate the average velocity after collision u_1'. This calculation requires us to make some statement about the type of collision undergone by the two molecules. In an *inelastic collision* two bodies collide and stick together. Momentum is conserved in this collision but the kinetic energy of the bodies is not usually conserved. In an *elastic collision* the two bodies collide and then move apart again. Momentum must be conserved in this type of collision also but the important difference between an inelastic

13

collision and an elastic collision is that, in the latter type, the kinetic energy of the center of mass of the two bodies is conserved. Most collisions between real bodies are somewhere between these two extremes. Billiard balls undergo elastic collisions and, in many ways, molecules behave like billiard balls.

When we consider all of the ways in which two hard spheres can approach each other, collide, and separate, we find that the average velocity after collision u_1' is the velocity of the center of mass of the pair of molecules. This velocity is defined by

$$u_c \equiv \frac{(m_1 u_1 + m_2 u_2)}{m_1 + m_2} = u_1' \qquad (2.1.3)$$

If this result is not immediately obvious, you should remember that if two molecules approach each other along the line joining their centers they will collide, exchange momentum, and retrace their paths. However, this is only one out of an infinite number of paths that the two molecules can have before and after collision. If we sit on the center of mass of the pair of molecules and move with velocity u_c, then any one direction of approach and rebound is just as likely to occur as any other direction. We can justify this conclusion with the necessary mathematics if we have to [Present (1958) does so]; for now, let us accept the result and continue.

It is now possible to complete the calculation of the momentum exchanged in a single collision between a molecule of species 1 and a molecule of species 2.

Momentum transferred
from a molecule of 1
to a molecule of 2
through collision
$$= m_1(u_1 - u_1')$$

$$= m_1(u_1 - u_c)$$

$$= \frac{m_1 m_2(u_1 - u_2)}{m_1 + m_2} \qquad (2.1.4)$$

This result demonstrates that if two molecules of the same type collide, then there is no net loss of momentum of the molecules of that type (any other result would be rather unexpected). This result is important because it indicates that the total momentum lost by the molecules of 1 depends only on collisions between molecules of 1 and other types of molecules. The rate at which this momentum transfer occurs depends on how frequently these different molecules collide.

2.1.2 Derivation of the Maxwell–Stefan Equation for Binary Diffusion

The molecules of these other species "get in the way" of the molecules of species 1 (say) and, in effect, exert a drag on them in much the same way that a pipe exerts a frictional drag on the fluid flowing through it. The analogy with pipe-flow does not end here; an analysis of diffusion may be carried out in essentially the same way that we may derive, for example, Poiseuille's equation for the rate of fluid flow in a pipe—through the application of Newton's second law.

The sum of the   The rate of change
forces acting on α of the momentum
a system   of the system

We shall apply this law to the control volume shown in Figure 2.1.

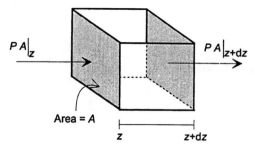

Figure 2.1. Control volume for derivation of Maxwell–Stefan relations.

Momentum can enter and leave this volume due to the motion of the molecules across the boundary walls. However, if the control volume moves with the molar average velocity of the mixture u, then the flow of molecules into the volume across any of the surfaces in Figure 2.1 is exactly balanced by an equal flow of molecules out of the volume across the same surface. There is no net momentum change due to this movement of molecules.

Within the control volume the molecules of species 1 may lose (or gain) momentum each time they collide with the atoms or molecules of the other species. Accounting for the momentum exchange on collision is one part of the elementary kinetic theory of diffusion that follows.

The forces acting on the control volume include surface forces, such as pressure forces, and shear stresses caused by velocity gradients and body forces, such as gravity. We consider only the pressure forces in the analysis that follows; the effects of other forces, such as gravity and electric fields, will be discussed in more detail in a later chapter (and from a rather different viewpoint too). Actually, the system as a whole is assumed to be at constant pressure and there is, therefore, no net force acting on the mixture as a whole.

The rate at which collisions occur between molecules of species 1 and molecules of species 2 depends on the number of species 1 molecules per unit volume $c_1 = c_t x_1$ and on the number of species 2 molecules per unit volume $c_2 = c_t x_2$. Clearly, the more molecules of both types that are present in the unit volume, the higher the number of collisions will be. Thus,

$$\text{The number of 1–2 collisions} \atop \text{per unit volume per unit time} \propto x_1 x_2$$

A concise statement of the ideas put forward to this point might read

$$\begin{array}{l} \text{The rate of change} \\ \text{of momentum of the} \\ \text{molecules of type 1} \\ \text{per unit volume} \end{array} = \begin{array}{l} \text{average amount of} \\ \text{momentum exchanged} \\ \text{in a single collision} \end{array} \times \begin{array}{l} \text{number of 1–2} \\ \text{collisions per} \\ \text{unit volume} \\ \text{per unit time} \end{array}$$

and we have an expression for the first term on the right and we know that the second term is proportional to $x_1 x_2$. We now turn to the development of the force term.

The net force acting on the left-hand wall is the pressure force PA exerted by the molecules outside this box striking this imaginary surface. The force acting on the species 1 molecules alone is the partial pressure of species 1, $p_1 = Px_1$, multiplied by the area A. Thus, the force acting on the species 1 molecules in the plane z is $Ap_1|_z$; the force acting on the species 1 molecules in the plane at $z + \Delta z$ is $-Ap_1|_{z+\Delta z}$ (note that the system as a whole is isobaric). The net force acting on the species 1 molecules is $A(p_1|_z - p_1|_{z+\Delta z})$.

Dividing by the volume $A \, \Delta z$ and taking the limit as $\Delta z \to 0$ we find

$$\begin{array}{l}\text{Net force acting on type 1}\\\text{molecules per unit volume} = \lim_{\Delta z \to 0} \dfrac{(p_1|_z - p_1|_{z+\Delta z})}{\Delta z}\\\text{in the } z \text{ direction}\end{array}$$

$$= -\frac{dp_1}{dz} \tag{2.1.5}$$

When we add on the contributions from the other two spatial dimensions we have

$$\begin{array}{l}\text{Net force acting on species 1}\\\text{molecules per unit volume}\end{array} = -\nabla p_1 \tag{2.1.6}$$

Combining these ideas gives

$$-\nabla p_1 \propto (x_1 x_2, (u_1 - u_2))$$

To convert a proportionality to an equality we multiply one side by a proportionality coefficient, we shall call it f_{12}, to get

$$\nabla p_1 = -f_{12} x_1 x_2 (u_1 - u_2) \tag{2.1.7}$$

where ∇p_1 is the actual force exerted per unit volume of the mixture trying to move species 1 past (through) the molecules of species 2 at a relative velocity $(u_1 - u_2)$; $x_1 x_2$ is a concentration weight factor and, therefore, f_{12} is analogous to a friction factor or drag coefficient. A pictorial representation of the interaction between the molecules of species 1 and 2 is provided in Figure 2.2.

We may define the proportionality coefficient in Eq. 2.1.7 in any way we like to suit our convenience. Let us, therefore, define an inverse drag coefficient $Đ_{12} = P/f_{12}$ and rewrite Eq. 2.1.7 as

$$d_1 \equiv \left(\frac{1}{P}\right) \nabla p_1 = -\frac{x_1 x_2 (u_1 - u_2)}{Đ_{12}} \tag{2.1.8}$$

where $d_i = (1/P) \nabla p_i$ may be considered to be the driving force for diffusion of species i in an ideal gas mixture at constant temperature and pressure.

Equation 2.1.8 is the Maxwell–Stefan equation for the diffusion of species 1 in a two-component ideal gas mixture. The symbol $Đ_{12}$ is the Maxwell–Stefan (MS) diffusivity.

If we carry out a similar analysis for species 2 we obtain

$$d_2 \equiv \left(\frac{1}{P}\right) \nabla p_2 = -\frac{x_1 x_2 (u_2 - u_1)}{Đ_{21}} \tag{2.1.9}$$

For all the applications considered in this book the system pressure is constant across the diffusion path. This simplification allows us to write $d_i = (1/P) \nabla p_i = \nabla x_i$ and Eqs. 2.1.8 and 2.1.9 simplify to

$$\nabla x_1 = -\frac{x_1 x_2 (u_1 - u_2)}{Đ_{12}}$$

$$\nabla x_2 = -\frac{x_1 x_2 (u_2 - u_1)}{Đ_{21}} \tag{2.1.10}$$

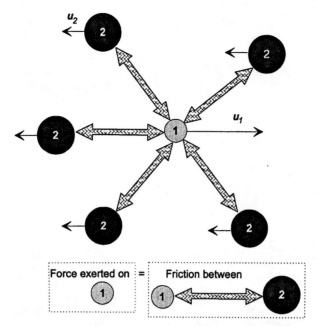

Figure 2.2. Pictorial representation of the interactions between differing kinds of molecules in a two-component system.

Since $\nabla x_1 + \nabla x_2 = 0$ it follows that the Maxwell–Stefan binary diffusion coefficients are symmetric: $Ð_{12} = Ð_{21}$.

2.1.3   The Maxwell–Stefan Equations for Ternary Systems

The setting up of the constitutive relation for a binary system is a relatively easy task because, as pointed out earlier, there is only one independent diffusion flux, only one independent composition gradient (driving force) and, therefore, only one independent constant of proportionality (diffusion coefficient). The situation gets quite a bit more complicated when we turn our attention to systems containing more than two components. The simplest multicomponent mixture is one containing three components, a ternary mixture. In a three component mixture the molecules of species 1 collide, not only with the molecules of species 2, but also with the molecules of species 3. The result is that species 1 transfers momentum to species 2 in 1–2 collisions and to species 3 in 1–3 collisions as well. We already know how much momentum is transferred in the 1–2 collisions and all we have to do to complete the force–momentum balance is to add on a term for the transfer of momentum in the 1–3 collisions. Thus,

$$d_1 = -\frac{x_1 x_2 (u_1 - u_2)}{Ð_{12}} - \frac{x_1 x_3 (u_1 - u_3)}{Ð_{13}} \qquad (2.1.11)$$

The corresponding equations for species 2 and 3 can be obtained from Eq. 2.1.11 by rotating the subscripts 1, 2, and 3.

$$d_2 = -\frac{x_2 x_1 (u_2 - u_1)}{Ð_{21}} - \frac{x_2 x_3 (u_2 - u_3)}{Ð_{23}} \qquad (2.1.12)$$

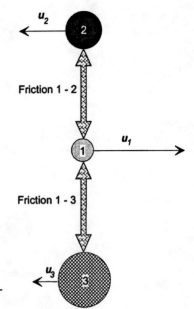

Figure 2.3. Pictorial representation of the interactions between differing kinds of molecules in a ternary system.

and

$$d_3 = -\frac{x_3 x_1 (u_3 - u_1)}{\mathcal{D}_{31}} - \frac{x_3 x_2 (u_3 - u_2)}{\mathcal{D}_{32}} \qquad (2.1.13)$$

At constant pressure the driving forces d_i are equal to the composition gradients ∇x_i. Of the three equations 2.1.11–2.1.13, only two are independent due to the restriction $\nabla x_1 + \nabla x_2 + \nabla x_3 = 0$. It is interesting to note that for a binary system, this restriction is sufficient to prove that $\mathcal{D}_{12} = \mathcal{D}_{21}$. For a multicomponent ideal gas mixture we need a more detailed analysis (Hirschfelder et al., 1964; Muckenfuss, 1973) to show that

$$\mathcal{D}_{ij} = \mathcal{D}_{ji} \qquad (2.1.14)$$

Let us return to Eq. 2.1.11 and consider its physical significance. This equation states that the driving force d_1 of component 1 arises from the frictional drag of molecules of the first constituent moving past (through) those of the constituent 2 with a relative velocity $(u_1 - u_2)$, concentration weight factor $x_1 x_2$, and drag coefficient $1/\mathcal{D}_{12}$ and of the molecules of the first constituent moving past (through) those of constituent 3 with a relative velocity $(u_1 - u_3)$, concentration weight factor $x_1 x_3$, and drag coefficient $1/\mathcal{D}_{13}$. A pictorial representation of the interactions between these three different kinds of molecule is provided by Figure 2.3.

As the molecules of all three constituents are, in general, in relative motion with average velocities u_i, it is hard to see how any simpler formulation will suffice. Equation 2.1.11 reduces to the proper binary equation in the limits $x_3 \to 0$ and $x_2 \to 0$ for the 1–2 and 1–3 binaries, respectively, so that both terms are necessary. It is to be noted that Eq. 2.1.11 does not include a term $(u_2 - u_3)$ for the first constituent as it is not reasonable to assume that the relative velocity of these constituents alone will produce a potential gradient of the first constituent as there would be no direct drag on the molecules of constituent 1. If an additional term of the form $x_1 x_2 x_3 (u_2 - u_3)/E_{123}$ were to be introduced into Eq. 2.1.11, it could be split up into $x_1 x_3 (u_1 - u_3) x_2 / E_{123} - x_1 x_2 (u_1 - u_2) x_3 / E_{123}$ and these terms absorbed into the existing ones with concentration dependent drag coefficients $1/Ð_{13}$ and $1/Ð_{12}$.

Thus, Eqs. 2.1.11–2.1.13 are the only consistent generalization of Eqs. 2.1.9 and 2.1.10 to a ternary mixture, assuming a linear relation between the potential gradients and the constituents' relative velocities.

2.1.4 The Maxwell–Stefan Equations for Multicomponent Systems

For mixtures containing even more species, n say, we just continue to add similar terms for each additional species. The generalization of Eq. 2.1.11 is

$$d_i = - \sum_{j=1}^{n} \frac{x_i x_j (u_i - u_j)}{Ð_{ij}} \tag{2.1.15}$$

Equations 2.1.15 are not yet in the form that is most useful to us; we eliminate the velocities using the definition of the molar fluxes $N_i = c_i u_i$, to get

$$d_i = \sum_{j=1}^{n} \frac{(x_i N_j - x_j N_i)}{c_t Ð_{ij}} \tag{2.1.16}$$

or, in terms of the diffusion fluxes J_i

$$d_i = \sum_{j=1}^{n} \frac{(x_i J_j - x_j J_i)}{c_t Ð_{ij}} \tag{2.1.17}$$

Only $n - 1$ of Eqs. 2.1.16 and 2.1.17 are independent because the ∇x_i sum to zero; the nth component gradient is given by

$$\nabla x_n = -\nabla x_1 - \nabla x_2 - \nabla x_3 \cdots -\nabla x_{n-1}$$
$$= - \sum_{k=1}^{n-1} \nabla x_k \tag{2.1.18}$$

These are the *Maxwell–Stefan* diffusion equations for multicomponent systems. These equations are named after the Scottish physicist James Clerk Maxwell and the Austrian scientist Josef Stefan who were primarily responsible for their development (Maxwell, 1866, 1952; Stefan, 1871). These equations appeared, in more or less the complete form of Eq. 2.1.15, in an early edition of the *Encyclopedia Britannica* (incomplete forms had been published earlier) in a general article on diffusion by Maxwell (see Maxwell, 1952). In addition to his major contributions to electrodynamics and kinetic theory, Maxwell wrote several articles for the encyclopedia. Stefan's 1871 paper is a particularly perceptive one and anticipated several of the multicomponent interaction effects to be discussed later in this book.

2.1.5 Matrix Formulation of the Maxwell–Stefan Equations

It will prove convenient to cast Eqs. 2.1.17 in $n - 1$ dimensional matrix form. First, we write Eq. 2.1.17 as a sum in terms of the J_i. However, since only $n - 1$ of the J_i are independent, we may eliminate J_n using

$$J_n = -J_1 - J_2 - J_3 - \cdots - J_{n-1}$$

$$= -\sum_{i=1}^{n-1} J_i \tag{2.1.19}$$

to get

$$c_t d_i = -B_{ii} J_i - \sum_{\substack{j=1 \\ j \neq i}}^{n-1} B_{ij} J_j \tag{2.1.20}$$

where the coefficients B_{ii} and B_{ij} are defined by

$$B_{ii} = \frac{x_i}{Ð_{in}} + \sum_{\substack{k=1 \\ i \neq k}}^{n} \frac{x_k}{Ð_{ik}} \tag{2.1.21}$$

$$B_{ij} = -x_i \left(\frac{1}{Ð_{ij}} - \frac{1}{Ð_{in}} \right) \tag{2.1.22}$$

Now we may write the $n - 1$ Eqs. 2.1.20 in $n - 1$ dimensional matrix form as

$$c_t(d) = -[B](J) \tag{2.1.23}$$

where $[B]$ is a square matrix of order $n - 1$.

$$[B] \equiv \begin{bmatrix} B_{11} & B_{12} & B_{13} & \cdots & B_{1,n-1} \\ B_{21} & B_{22} & B_{23} & \cdots & B_{2,n-1} \\ \vdots & & & & \vdots \\ B_{n-1,1} & B_{n-1,2} & B_{n-1,3} & \cdots & B_{n-1,n-1} \end{bmatrix}$$

with elements given by Eqs. 2.1.21 and 2.1.22. The column matrix (J) is

$$(J) \equiv \begin{pmatrix} J_1 \\ J_2 \\ \vdots \\ J_{n-1} \end{pmatrix}$$

and (d) is a column matrix of order $n - 1$ defined by

$$(d) \equiv \begin{pmatrix} d_1 \\ d_2 \\ \vdots \\ d_{n-1} \end{pmatrix}$$

Now, if we premultiply Eq. 2.1.23 by the inverse of $[B]$ as follows:

$$c_t[B]^{-1}(d) = -[B]^{-1}[B](J) \tag{2.1.24}$$

which, because $[B]^{-1}[B] = [I]$, simplifies to

$$(J) = -c_t[B]^{-1}(d) \tag{2.1.25}$$

For a two component system all matrices become scalars and Eq. 2.1.25 becomes

$$J_1 = -c_t B^{-1} d_1 \tag{2.1.26}$$

where B is obtained from Eq. 2.1.21 as

$$B = \frac{(x_1 + x_2)}{Ð_{12}} = \frac{1}{Ð_{12}} \tag{2.1.27}$$

Eq. 2.1.27 allows us to rewrite Eq. 2.1.26 as

$$J_1 = -c_t Ð_{12} d_1 \tag{2.1.28}$$

which is just another way of writing Eq. 2.1.8.

Example 2.1.1   Multicomponent Diffusion in a Stefan Tube: An Experimental Test of the Maxwell–Stefan Equations

The Stefan tube, depicted schematically in Figure 2.4, is a simple device sometimes used for measuring diffusion coefficients in binary vapor mixtures. In the bottom of the tube is a pool of quiescent liquid. The vapor that evaporates from this pool diffuses to the top of the tube. A stream of gas across the top of the tube keeps the mole fraction of diffusing vapor there to essentially nothing. The mole fraction of the vapor at the vapor–liquid interface is its equilibrium value.

In an attempt to check the validity of the Maxwell–Stefan equations Carty and Schrodt (1975) evaporated a binary liquid mixture of acetone(1) and methanol(2) in a Stefan tube. Air(3) was used as the carrier gas. In one of their experiments the composition of the vapor at the liquid surface was $x_1 = 0.319$, $x_2 = 0.528$. The pressure and temperature in the vapor phase were 99.4 kPa and 328.5 K, respectively. The length of the diffusion path was

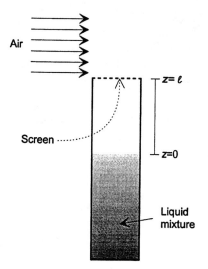

Figure 2.4. Schematic diagram of a Stefan diffusion tube.

0.238 m. The Maxwell–Stefan diffusion coefficients of the three binary pairs are

$$Đ_{12} = 8.48 \text{ mm}^2/\text{s}$$

$$Đ_{13} = 13.72 \text{ mm}^2/\text{s}$$

$$Đ_{23} = 19.91 \text{ mm}^2/\text{s}$$

Calculate the composition profiles predicted by the Maxwell–Stefan equations and compare the results with the experimental data.

SOLUTION   At constant temperature and pressure the molar density c_t and the binary diffusion coefficients are constant and the driving forces are the mole fraction gradients, $d_i = \nabla x_i$. Furthermore, diffusion in the Stefan tube takes place in only one direction, up the tube; there are no radial or circumferential gradients in composition. Thus, the continuity Eqs. 1.3.7 simplify to N_i = constant. The carrier gas (3) diffuses down the tube as the evaporating vapor diffuses up it, but because the gas does not dissolve in the liquid its flux N_3 is zero (i.e., the diffusion flux J_3 of the gas down the tube is exactly balanced by a diffusion induced convective flux $x_3 N_t$ up the tube). The mole fraction of vapor at each end of the tube is kept constant—at the top by the stream of carrier gas sweeping the diffusing vapor away and, at the bottom, by the evaporating pool of liquid. The liquid level falls with time, of course, but since diffusion up the tube is a relatively slow process, the level of liquid at the bottom of the tube falls very slowly. Thus, it is safe to make use of the quasisteady-state assumption; that, at any instant, the flux is given by its steady-state value.

With all of the above assumptions the Maxwell–Stefan relations (Eq. 2.1.16) reduce to a system of first-order linear differential equations

$$\frac{dx_i}{dz} = \sum_{j=1}^{n} \frac{(x_i N_j - x_j N_i)}{c_t Đ_{ij}}$$

An analytical solution of these equations, subject to the boundary conditions

$$z = 0, \quad x_i = x_{i0} \qquad z = \ell, \quad x_i = x_{i\ell}$$

will be derived in Chapter 8. However, for the purposes of this illustration we integrated these equations numerically using a fourth-order Runge–Kutta method. The calculations were started at the interface ($z = 0$) and ended when we had marched a distance equal to the length of the Stefan tube ($z = \ell$). A two-dimensional Newton–Raphson procedure was used to search for the values of N_1 and N_2 ($N_3 = 0$) that allowed us to match the specified composition at the top of the tube. The converged values of the fluxes N_1 and N_2 are

$$N_1 = 1.783 \times 10^{-3} \qquad N_2 = 3.127 \times 10^{-3} \text{ mol/m}^2 \text{ s}$$

The results of the final integration are plotted in Figure 2.5 along with the data from Carty and Schrodt (1975). The agreement between theory and experiment is quite good and support the Maxwell–Stefan formulation of diffusion in multicomponent ideal gas mixtures. This conclusion was also reached by Bres and Hatzfeld (1977) and by Hesse and Hugo (1972). For further analysis of the Stefan diffusion tube see Whitaker (1991). ■

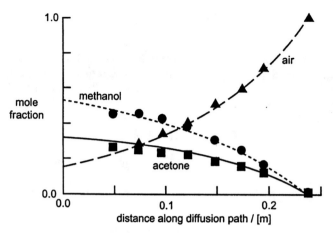

Figure 2.4. Composition profiles in a Stefan diffusion tube. Lines represent calculated profiles; points represent the experimental data of Carty and Schrodt (1975).

2.2 DIFFUSION IN NONIDEAL FLUIDS

The much higher density of liquids and dense gases means that we can no longer safely assume that only binary (two molecule) collisions take place; three (or more) molecule collisions occur sufficiently frequently in liquids and dense gases and contribute to the momentum transfer process. It is, therefore, difficult to develop an analysis of liquid-phase diffusion in complete parallel to that above for gases. However, the physical interpretation of Eq. 2.1.15 applies equally to gases and liquids: If (and only if) the constituents (i and j) are in motion relative to one another and, therefore, moving at different velocities (u_i and u_j), may we expect composition gradients to be set up in the system as a result of the frictional drag of one set of molecules moving through the other. It does not matter whether this frictional drag arises purely from intermolecular collisions as in the kinetic theory of gases or from intermolecular forces acting between the two sets of molecules. Intermolecular forces become dominant in diffusion in liquids and solids.

The force acting on species i per unit volume of mixture tending to move the molecules of species i is $c_i RT d_i$, where d_i is related to the relative velocities ($u_i - u_j$), by

$$d_i = -\sum_{j=1}^{n} \frac{x_i x_j (u_i - u_j)}{Ð_{ij}}$$

$$= \sum_{j=1}^{n} \frac{(x_i N_j - x_j N_i)}{c_t Ð_{ij}} \tag{2.2.1}$$

where $Ð_{ij}$ is the Maxwell–Stefan diffusivity whose physical significance as an inverse drag coefficient is the same as in the ideal gas case. For nonideal fluids d_i, which can be considered to be a driving force, is defined by

$$d_i \equiv \frac{x_i}{RT} \nabla_{T, P} \mu_i \tag{2.2.2}$$

The appearance of chemical potential gradients in these equations should not come as a surprise. Equilibrium is defined by equality of chemical potentials and departures from equilibrium are characterized by the presence of chemical potential gradients. As we shall see in Section 2.3, chemical potential gradients arise in the thermodynamics of irreversible processes as the fundamentally correct driving forces for diffusion. The subscripts T, P are to emphasize that the gradient in Eq. 2.2.2 is to be calculated under constant temperature, constant pressure conditions (pressure gradients and external forces also contribute to d_i, but we shall ignore their influence until we get to Section 2.3). The driving force d_i reduces to $(1/P)\nabla p_i$ for ideal gases, as it should. Also, the sum of the n driving forces vanishes

$$\sum_{i=1}^{n} d_i = 0 \qquad (2.2.3)$$

due to the Gibbs–Duhem restriction (see, e.g., Modell and Reid, 1983); this means that only $n - 1$ driving forces are independent.

Chemical potential gradients are not the easiest of quantities to deal with. For nonideal liquids we may express the driving force d_i in terms of the mole fraction gradients as follows:

$$
\begin{aligned}
d_i &= \frac{x_i}{RT}\nabla_{T,P}\mu_i \\
&= \frac{x_i}{RT}\sum_{j=1}^{n-1}\frac{\partial\mu_i}{\partial x_j}\bigg|_{T,P,\Sigma}\nabla x_j \\
&= \frac{x_i}{RT}\sum_{j=1}^{n-1}RT\frac{\partial\ln\gamma_i x_i}{\partial x_j}\bigg|_{T,P,\Sigma}\nabla x_j \\
&= x_i\sum_{j=1}^{n-1}\left(\frac{\partial\ln x_i}{\partial x_j}+\frac{\partial\ln\gamma_i}{\partial x_j}\bigg|_{T,P,\Sigma}\right)\nabla x_j \\
&= \sum_{j=1}^{n-1}\left(\delta_{ij}+x_i\frac{\partial\ln\gamma_i}{\partial x_j}\bigg|_{T,P,\Sigma}\right)\nabla x_j \\
&= \sum_{j=1}^{n-1}\Gamma_{ij}\nabla x_j \qquad (2.2.4)
\end{aligned}
$$

where γ_i is the activity coefficient of species i in the mixture and where

$$\Gamma_{ij} = \delta_{ij} + x_i\frac{\partial\ln\gamma_i}{\partial x_j}\bigg|_{T,P,\Sigma} \qquad (2.2.5)$$

The symbol Σ is used to indicate that the differentiation of $\ln\gamma_i$ with respect to mole fraction x_j is to be carried out while keeping constant the mole fractions of all other species except the nth. The mole fraction of species n must be eliminated using the fact that the x_i sum to unity. More specifically,

$$\frac{\partial\ln\gamma_i}{\partial x_j}\bigg|_{T,P,\Sigma} = \frac{\partial\ln\gamma_i}{\partial x_j}\bigg|_{T,P,x_k,\,k\neq j=1\cdots n-1}$$

The evaluation of the Γ_{ij} for liquid mixtures from activity coefficient models is discussed at length in Appendix D.

For dense gas mixtures exhibiting deviations from ideal gas behavior the above formulation can be used with the activity coefficient γ_i replaced by the fugacity coefficient ϕ_i.

$$\Gamma_{ij} \equiv \delta_{ij} + x_i\frac{\partial\ln\phi_i}{\partial x_j}\bigg|_{T,P,\Sigma} \qquad (2.2.6)$$

An equation of state needs to be used for the calculation of the molar density c_t and the derivatives of the fugacity coefficients (see, e.g., Walas, 1985).

2.2.1 Matrix Formulation of the Maxwell–Stefan Equations for Nonideal Fluids

It is convenient to cast Eqs. 2.2.1 into $n - 1$ dimensional matrix form as

$$-c_t(d) = [B](J) \tag{2.2.7}$$

where the column matrices (d) and (J) have elements

$$(d) = \begin{pmatrix} d_1 \\ d_2 \\ \vdots \\ d_{n-1} \end{pmatrix} \qquad (J) = \begin{pmatrix} J_1 \\ J_2 \\ \vdots \\ J_{n-1} \end{pmatrix}$$

and where the matrix $[B]$ has elements given by Eqs. 2.1.21 and 2.1.22.

$$B_{ii} = \frac{x_i}{Ð_{in}} + \sum_{\substack{k=1 \\ i \neq k}}^{n} \frac{x_k}{Ð_{ik}} \tag{2.1.21}$$

$$B_{ij} = -x_i \left(\frac{1}{Ð_{ij}} - \frac{1}{Ð_{in}} \right) \tag{2.1.22}$$

Equations 2.2.4 may also be written in $n - 1$ dimensional matrix form as

$$(d) = [\Gamma](\nabla x) \tag{2.2.8}$$

which can be combined with Eq. 2.2.7 to give

$$-c_t[\Gamma](\nabla x) = [B](J) \tag{2.2.9}$$

Equation 2.2.9 is more useful in its inverted form

$$(J) = -c_t[B]^{-1}[\Gamma](\nabla x) \tag{2.2.10}$$

2.2.2 Limiting Cases of the Maxwell–Stefan Equations

Let us now consider some limiting cases of Eqs. 2.2.10. The first important special case is that of diffusion in a two-component mixture. In this case the $n - 1$ dimensional matrices reduce to scalar quantities and we have

$$J_1 = -c_t B^{-1} \Gamma \, \nabla x_i = -c_t Ð \Gamma \, \nabla x_1 \tag{2.2.11}$$

where the thermodynamic factor Γ is obtained from Eq. 2.2.5 as

$$\Gamma = 1 + x_1 \frac{\partial \ln \gamma_1}{\partial x_1} \tag{2.2.12}$$

where it is understood that the mole fractions x_1 and x_2 sum to unity when the partial derivative of $\ln \gamma_1$ is evaluated.

For ideal mixtures the activity and fugacity coefficients are unity, $\gamma_i = 1$, $\phi_i = 1$, and, therefore, $[\Gamma] = [I]$, $(d) = (\nabla x)$, and we recover Eqs. 2.1.25. A subset of this case arises if the mixture is made up of species of (almost) identical size, shape, polarity, ... In this case the Maxwell-Stefan diffusion coefficients are almost equal to one another

$$Ð_{ij} = Ð$$

and so

$$[B]^{-1} \rightarrow Ð[I] \tag{2.2.13}$$

and, therefore

$$(J) = -c_t Ð[I](\nabla x) \tag{2.2.14}$$

or

$$J_i = -c_t Ð \nabla x_i \tag{2.2.15}$$

In many chemical engineering problems we are interested in calculating the transfer rates of a component that is present in a liquid mixture in very low concentrations. Let us identify the trace component by the subscript 1; $x_1 \approx 0$. This means that the coefficients $B_{12}, B_{13}, \cdots, B_{1,n-1}$ of $[B]$ all reduce to zero (see Eqs. 2.1.21 and 2.1.22). Also, the matrix of thermodynamic factors $[\Gamma]$ has the following elements on its first row

$$\Gamma_{11} = 1 \qquad \Gamma_{12} = \Gamma_{13} = \Gamma_{14} = \cdots = \Gamma_{1,n-1} = 0 \tag{2.2.16}$$

With the above simplifications, Eq. 2.2.10 for component 1 reduces to

$$J_1 = -c_t B_{11}^{-1} \nabla x_1 \tag{2.2.17}$$

where B_{11} takes the simplified form

$$B_{11} = \frac{x_2}{Ð_{12}^\circ} + \frac{x_3}{Ð_{13}^\circ} + \cdots + \frac{x_{n-1}}{Ð_{1,n-1}^\circ} \tag{2.2.18}$$

The $Ð_{ij}^\circ$ are the Maxwell-Stefan diffusivities of the i–j pair where species i is present in infinitely dilute concentrations.

Example 2.2.1   Diffusion of Toluene in a Binary Mixture

Consider the diffusion of toluene(1) present in trace amounts in a liquid mixture containing n-tetradecane(2) and n-hexane(3) at a temperature of 25°C. An "effective" diffusivity of the trace toluene defined by

$$Ð_{1,\text{eff}} \equiv -J_1/c_t \nabla x_1$$

was measured by Holmes, et al. (1962); their data is summarized in Table 2.1. Calculate B_{11} and compare its inverse to the measured values of $Ð_{1,\text{eff}}$.

DATA   The infinite dilution coefficients are

$$Ð_{12}^\circ = 1.08 \times 10^{-9} \text{ m}^2/\text{s}$$

$$Ð_{13}^\circ = 4.62 \times 10^{-9} \text{ m}^2/\text{s}$$

SOLUTION   Toluene is present in trace amounts, $x_1 \approx 0$, and so $x_3 \approx 1 - x_2$. We illustrate the calculation of B_{11} at the composition $x_2 = 0.501$ and $x_3 = 0.499$. The

TABLE 2.1   "Effective" Diffusivity of Toluene in a Liquid Mixture
of n-Tetradecane–n-Hexane as a Function of the Mole Fraction
of n-Tetradecane, $x_2{}^a$

x_2	$Ð_{1,\,\text{eff}}$ (measured)	B_{11}^{-1} (from Eq. 2.2.18)
1.000	1.08	1.08
0.803	1.37	1.272
0.672	1.58	1.44
0.501	1.92	1.75
0.336	2.38	2.20
0.215	2.90	2.71
0.113	3.57	3.37
0.000	4.62	4.62

^{a}Units are 10^{-9} m^2/s.

parameter B_{11} follows from Eq. 2.2.18 as

$$B_{11} = \frac{0.501}{1.08 \times 10^{-9}} + \frac{0.499}{4.62 \times 10^{-9}} = 0.572 \times 10^9 \text{ s/m}^2$$

which gives

$$B_{11}^{-1} = 1.75 \times 10^{-9} \text{ m}^2/\text{s}$$

Figure 2.6 shows the predictions of the transport coefficient B_{11}^{-1} from Eq. 2.2.16 as a function of the mole fraction of n-tetradecane. The agreement between measured values of $Ð_{1,\,\text{eff}}$ and predicted values of B_{11}^{-1} is quite good; the lack of better agreement may be attributed to the variation in the liquid viscosity over the composition range covered. The introduction of a factor to correct for the viscosity variation is considered by Perkins and Geankoplis (1969). ∎

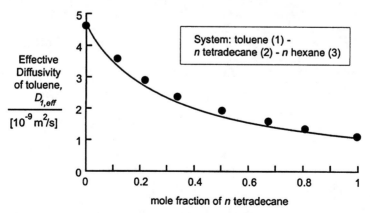

Figure 2.6. "Effective" diffusivity of toluene in a liquid mixture of n-tetradecane–n-hexane as a function of the mole fraction of n-tetradecane, x_2.

2.3 THE GENERALIZED MAXWELL–STEFAN FORMULATION OF IRREVERSIBLE THERMODYNAMICS

Until now, we have considered that the diffusion process took place under essentially isobaric conditions, in nonelectrolyte systems and in the absence of external force fields, such as centrifugal or electric fields. In this section we shall generalize our analysis to include the influence of external force fields. The best starting point for a generalized treatment is the theory of *irreversible thermodynamics*. The treatment below is similar to that given by Lightfoot (1974) but readers will also find the books by de Groot and Mazur (1962) and Haase (1969) very useful.

The purpose of the study of irreversible thermodynamics is to extend classical thermodynamics to include systems in which irreversible processes (e.g., diffusion and heat transfer) are taking place. Such an extension is made possible by assuming that for systems "not too far" from equilibrium the postulate of "local equilibrium" applies "Departures from local equilibrium are sufficiently small that all thermodynamic state quantities may be defined locally by the same relations as for systems at equilibrium."

With the help of this postulate, it is possible to obtain an explicit expression for σ, the rate of entropy production per unit volume due to various irreversible processes taking place within the system (see, e.g., Slattery, 1981). The rate of entropy production due to diffusion is

$$T\sigma_{\text{diff}} = -\sum_{i=1}^{n} \left(\nabla_T \bar{\mu}_i - \bar{F}_i\right) \cdot j_i \geq 0 \tag{2.3.1}$$

where j_i is the mass diffusion flux with respect to the mass average velocity; T is the absolute temperature, $\bar{\mu}_i$ is the specific chemical potential of species i, $\bar{\mu}_i = \mu_i/M_i$; M_i is the molar mass of species i; $\nabla_T \bar{\mu}_i$ represents the isothermal gradient of the specific chemical potential; $\bar{F}_i = F_i/M_i$, where F_i is the external body force per mole of species i.

The requirement that $\sigma_{\text{diff}} \geq 0$ follows from the second law of thermodynamics.

2.3.1 The Generalized Driving Force

Each term contributing to the rate of entropy production may be regarded as the product of two terms, one a "flux" and the other a "driving force." Which is which usually is obvious from the context. However, the assignation of the names flux and driving force to terms in the entropy production rate equation is not always clear cut. As far as we are concerned here the distinction is largely irrelevant; that is, it does not matter which term is regarded as flux and which is the driving force—although it must be noted that there are circumstances where the distinction is very important. It might be more correct to refer to these quantities as dependent and independent variables.

Let us now consider the driving force $\nabla_T \bar{\mu}_i - \bar{F}_i$. This can be rewritten as

$$\nabla_T \bar{\mu}_i - \bar{F}_i = \frac{1}{M_i} \nabla_{T,P} \mu_i + \frac{\bar{V}_i}{M_i} \nabla P - \bar{F}_i \tag{2.3.2}$$

where $\bar{V}_i$ is the partial molar volume of i and $\nabla_{T,P}\mu_i$ represents the isothermal, isobaric gradient of the molar chemical potential. Now, since the diffusion fluxes j_i sum to zero, that is,

$$\sum_{i=1}^{n} j_i = \sum_{i=1}^{n} \rho_i(u_i - v) = 0 \tag{2.3.3}$$

we can add any arbitrary vector to $\nabla_T \bar{\mu}_i - \bar{F}_i$ without altering the value of σ_{diff}. Let us replace $\nabla_T \bar{\mu}_i - \bar{F}_i$ by

$$\nabla_T \bar{\mu}_i - \bar{F}_i - \frac{1}{\rho_t} \nabla P + \sum_{i=1}^{n} \omega_i \bar{F}_i \qquad (2.3.4)$$

where ρ_t is the mixture mass density and ω_i is the mass fraction. The reason for the choice of the arbitrary vector

$$-\frac{1}{\rho_t} \nabla P + \sum_{i=1}^{n} \omega_i \bar{F}_i$$

is that the conservation of linear momentum gives

$$-\frac{1}{\rho_t} \nabla P + \sum_{i=1}^{n} \omega_i \bar{F}_i = \frac{dv}{dt} + \nabla \cdot \tau \qquad (2.3.5)$$

and for a system at mechanical equilibrium, that is, no velocity gradients, the right-hand side of Eq. 2.3.5 vanishes giving

$$\frac{1}{\rho_t} \nabla P = \sum_{i=1}^{n} \omega_i \bar{F}_i \qquad \text{(mechanical equilibrium)} \qquad (2.3.6)$$

that is, the pressure gradients are balanced by the external body forces. In systems of chemical engineering interest, mechanical equilibrium is established faster than diffusion equilibrium and Eq. 2.3.5 is reasonably well obeyed. Thus, in replacing $\nabla_T \bar{\mu}_i - \bar{F}_i$ by Eq. 2.3.4, we are essentially subtracting a vanishing vector from $\nabla_T \bar{\mu}_i - \bar{F}_i$. With this modification to the driving force and utilizing Eq. 2.3.6 we get

$$T\sigma_{\text{diff}} = -\sum_{i=1}^{n} \left(\frac{1}{M_i} \nabla_{T,P}\mu_i + \frac{\bar{V}_i}{M_i} \nabla P - \frac{1}{\rho_t} \nabla P + \sum_{j=1}^{n} \omega_j \bar{F}_j - \bar{F}_i \right) \cdot j_i \qquad (2.3.7)$$

or

$$\sigma_{\text{diff}} = -c_t R \sum_{i=1}^{n} d_i \cdot (u_i - v) \geqq 0 \qquad (2.3.8)$$

where we have used the defining relations $j_i \equiv \rho_i(u_i - v)$ and

$$c_t RT d_i \equiv c_i \nabla_{T,P}\mu_i + \left(c_i \bar{V}_i - \omega_i \right) \nabla P - \rho_i \left(\bar{F}_i - \sum_{j=1}^{n} \omega_j \bar{F}_j \right) \qquad (2.3.9)$$

The physical interpretation of $c_t RT d_i$ is that it represents the force acting on species i per unit volume of mixture tending to move species i relative to the solution. The quantity $c_i \bar{V}_i$ represents the volume fraction of species i, ϕ_i, and so we may rewrite Eq. 2.3.9 as

$$c_t RT d_i \equiv c_i \nabla_{T,P}\mu_i + (\phi_i - \omega_i) \nabla P - \rho_i \left(\bar{F}_i - \sum_{j=1}^{n} \omega_j \bar{F}_j \right) \qquad (2.3.10)$$

which shows that a pressure gradient can effect a separation in a mixture provided there is a difference between its volume and mass fractions. For ideal gas mixtures Eq. 2.3.10 simplifies to give

$$d_i \equiv \nabla x_i + (x_i - \omega_i)\frac{\nabla P}{P} - \frac{\rho_i}{P}\left(\bar{F}_i - \sum_{j=1}^{n} \omega_j \bar{F}_j\right) \qquad (2.3.11)$$

It is sometimes convenient to express Eqs. 2.3.10 and 2.3.11 in terms of the external body force exerted per mole of i; the corresponding equations are

$$c_t R T d_i \equiv c_i \nabla_{T,P}\mu_i + (\phi_i - \omega_i)\nabla P - \left(c_i F_i - \omega_i \sum_{j=1}^{n} c_j F_j\right) \qquad (2.3.12)$$

and, for ideal gases,

$$d_i \equiv \nabla x_i + (x_i - \omega_i)\frac{\nabla P}{P} - \frac{1}{P}\left(c_i F_i - \omega_i \sum_{j=1}^{n} c_j F_j\right) \qquad (2.3.13)$$

What we have achieved so far is to express the rate of entropy production σ_{diff} due to mass diffusion in terms of a convenient driving force $c_t R T d_i$ per unit volume of mixture. Equation 2.3.8 shows that the rate of entropy production is a sum of the products of two quantities; the force acting on i, per unit volume, tending to move i relative to the mixture and the relative velocity of the movement of i with respect to the mixture; σ_{diff} is, therefore, the dissipation due to diffusion.

2.3.2 The Generalized Maxwell–Stefan Equations

Insertion of any other reference mixture velocity in place of v in Eq. 2.3.8 will not alter the value of σ_{diff}, an expected and pleasing result because only relative motions of the various species are important in the description of the diffusion process. In chemical engineering applications it is often convenient to choose the molar average mixture velocity u and so we may write

$$\sigma_{\text{diff}} = -c_t R \sum_{i=1}^{n} d_i \cdot (u_i - u) \geq 0 \qquad (2.3.14)$$

Eliminating the molar average velocity u using Eq. 1.2.6 allows us to write Eq. 2.3.14 in terms of the relative velocities $(u_i - u_j)$ as

$$\sigma_{\text{diff}} = -\frac{c_t R}{n} \sum_{i=1}^{n} \sum_{j=1}^{n} d_i \cdot (u_i - u_j) \geq 0 \qquad (2.3.15)$$

We may use Eq. 2.3.15 as the starting point for developing our constitutive relations rather than the conventional Eq. 2.3.8.

The first postulate of irreversible thermodynamics is that the fluxes (or dependent variables) are directly proportional to the driving forces (or independent variables). [Actually, it may be shown that the assumption of local equilibrium follows from the assumption of a linear relation between the fluxes and driving forces (Truesdell, 1969).] If we take the d_i as dependent variables and the $(u_i - u_j)$ as independent variables we may, therefore, write a linear constitutive relation for diffusion as follows:

$$d_i = - \sum_{j=1}^{n} \beta_{ij}(u_i - u_j) \qquad (2.3.16)$$

with β_{ij} a coefficient of proportionality. We may, of course, define the coefficient of proportionality in any way we like to suit our purposes; we, therefore, introduce a "new" coefficient $\beta_{ij} = x_i x_j / Đ_{ij}$, which allows us to rewrite Eq. 2.3.16 as

$$d_i = - \sum_{j=1}^{n} \frac{x_i x_j (u_i - u_j)}{Đ_{ij}} \qquad (2.3.17)$$

Equations 2.3.17 are the generalized Maxwell–Stefan (GMS) relations and the $Đ_{ij}$ are the Maxwell–Stefan diffusion coefficients we encountered earlier. These equations are more useful when expressed in terms of the molar fluxes $N_i = c_i u_i$,

$$d_i = \sum_{j=1}^{n} \frac{(x_i N_j - x_j N_i)}{c_t Đ_{ij}} \qquad (2.3.18)$$

The positive definiteness requirement of σ_{diff} allows us to derive certain restrictions on the values of the $Đ_{ij}$. If we substitute the GMS relation for d_i into Eq. 2.3.15 we obtain a very neat and compact expression for the rate of entropy production due to diffusion

$$\sigma_{\text{diff}} = \frac{1}{2} c_t R \sum_{i=1}^{n} \sum_{j=1}^{n} \frac{x_i x_j}{Đ_{ij}} |(u_i - u_j)|^2 \geqq 0 \qquad (2.3.19)$$

which is quite remarkable for the absence of any thermodynamic factors. Equation 2.3.19 was derived by Hirschfelder, Curtiss, and Bird (HCB) (1964) for ideal gas mixtures; the generalization to nonideal fluids was carried out by Standart et al. (1979) using the HCB treatment as a consistent basis.

For mixtures of ideal gases the $Đ_{ij}$ are composition independent and taking into consideration that Eq. 2.3.19 is valid for all compositions and values of $(u_i - u_j)$, the positive definite condition can only be satisfied if

$$Đ_{ij} \geqq 0 \qquad |\text{ideal gases}| \qquad (2.3.20)$$

a result derived by HCB (cf. their Eq. 11.2-46).

For nonideal liquid mixtures the $Đ_{ij}$ are composition dependent (as shall be discussed in detail in Chapter 4) and without complete information as to the nature of the composition dependence a result analogous to Eq. 2.3.20 cannot be derived. A more restrictive result for the set of infinitely dilute diffusivities follows from the application of the second law restriction Eq. 2.3.19.

$$Đ_{ij}^{\circ} \geqq 0 \qquad |\text{nonideal fluids}| \qquad (2.3.21)$$

which is the diluted analog of the HCB result for ideal gases.

Since the $Đ_{ij}$ are defined in terms of d_i and component velocity differences $(u_i - u_j)$, both of which are independent of the reference velocity frame, the $Đ_{ij}$ are, therefore, reference frame independent.

A second postulate of irreversible thermodynamics is that the coefficients $Đ_{ij}$ are symmetric

$$Đ_{ij} = Đ_{ji} \qquad (2.3.22)$$

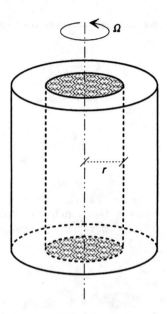

Figure 2.7. Schematic diagram of an ultracentrifuge.

Equation 2.3.22 expresses the Onsager reciprocal relations (ORR) discussed briefly in Section 3.3. For ideal gases, this symmetry relation can be obtained from the kinetic theory of gases (HCB, 1964; Muckenfuss, 1973).

Equations 2.3.18 together with Eqs. 2.3.10 defining the generalized driving force are the starting point for the analysis of diffusion in systems where external force fields influence the process: the ultracentrifuge, for example, in electrolyte systems and in porous media where pressure gradients become important. We examine the first two of these topics in the Sections 2.3.3 and 2.4.

2.3.3 An Application of the Generalized Maxwell–Stefan Equations—The Ultracentrifuge

To illustrate the formulation discussed in Section 2.3.2 let us consider diffusion in the presence of an imposed force field. The centrifuge is a device that subjects a fluid mixture to a centrifugal force; the ultracentrifuge subjects the fluid to extremely high forces simply by spinning at a very high rate. Figure 2.7 is a schematic of such a device.

The centrifugal force exerted on a unit mass of component i in a multicomponent mixture is

$$\tilde{F}_i = \Omega^2 r \tag{2.3.23}$$

where r is the distance from the axis of rotation and Ω is the angular velocity; $\Omega = 2\pi f$, where f is the rotational speed (revolutions per second, rps). If we use Eq. 2.3.23 in Eq. 2.3.10 we find

$$c_t RT d_i \equiv c_i \nabla_{T,P} \mu_i + (\phi_i - \omega_i) \nabla P - \rho_i(\Omega^2 r - \Omega^2 r) \tag{2.3.24}$$

The last term on the right-hand side of Eq. 2.3.24 cancels to leave

$$c_t RT d_i = c_i \nabla_{T,P} \mu_i + (\phi_i - \omega_i) \nabla P \tag{2.3.25}$$

Mechanical equilibrium is established quickly in relation to thermodynamic equilibrium in an ultracentrifuge. At mechanical equilibrium we have (from Eq. 2.3.6)

$$\nabla P = \rho_t \sum_{i=1}^{n} \omega_i \tilde{F}_i = \rho_t \Omega^2 r \qquad (2.3.26)$$

On eliminating ∇P from Eq. 2.3.25 we have

$$c_t RT d_i \equiv c_i \nabla_{T,P} \mu_i + (\phi_i - \omega_i)\rho_t \Omega^2 r \qquad (2.3.27)$$

At equilibrium the driving forces and fluxes vanish; $d_i = 0$, $(u_i - u_j) = 0$. Thus,

$$c_i \nabla_{T,P} \mu_i = (\omega_i - \phi_i)\rho_t \Omega^2 r \qquad (2.3.28)$$

The chemical potential gradients are more conveniently expressed in terms of the composition gradients as

$$\frac{x_i}{RT} \nabla_{T,P} \mu_i = \sum_{j=1}^{n-1} \Gamma_{ij} \nabla x_j \qquad (2.2.4a)$$

where the thermodynamic factors Γ_{ij} are given by

$$\Gamma_{ij} = \delta_{ij} + x_i \frac{\partial \ln \gamma_i}{\partial x_j} \qquad (2.2.5)$$

On combining Eqs. 2.3.27 and 2.2.4a we have

$$\sum_{j=1}^{n-1} \Gamma_{ij} \nabla x_j = (\omega_i - \phi_i)\frac{\overline{M}}{RT}\Omega^2 r \qquad (2.3.29)$$

where $\overline{M}$ is the mean molar mass of the mixture.

For a two component mixture we may simplify Eq. 2.3.29 as follows:

$$\Gamma \frac{dx_1}{dr} = (\omega_1 - c_1 \overline{V}_1)\frac{\overline{M}}{RT}\Omega^2 r \qquad (2.3.30)$$

where $\Gamma = (1 + x_1 \partial \ln \gamma_1 / \partial x_1)$. For dilute solutions the thermodynamic factor Γ is approximately unity (cf. discussion in Section 2.2.2) and Eq. 2.3.30 simplifies to

$$\frac{dx_1}{dr} = (\omega_1 - c_1 \overline{V}_1)\frac{\overline{M}}{RT}\Omega^2 r \qquad (2.3.31)$$

Integration of Eqs. 2.3.29 or the simplified forms (Eqs. 2.3.30 and 2.3.31) yields the equilibrium composition distribution in the centrifuge (see Examples 2.3.1 and 2.3.2). From Eq. 2.3.30 we see that the ultracentrifuge induces a separation only if the volume fraction ($\phi_i \equiv c_i \overline{V}_i$) is different from the mass fraction (ω_i). For dilute aqueous solutions $\overline{M}/RT$ is of the order 10^{-5} s^2/m^2. Thus, for r of about 0.1 m we need an angular velocity Ω of about 1000 inverse seconds or approximately 175 rps in order to obtain a measurable separation. The ultracentrifuge is used for the determination of molecular weights of proteins and for the separation of isotopes. Cullinan and Lenczyck (1969) proposed that the ultracentrifuge be used to determine the thermodynamic factor Γ for nonideal systems; this procedure is, however, very expensive and time consuming.

Example 2.3.1   Ultracentrifugation of a Binary Liquid Mixture

An equimolar mixture of benzene (1) and carbon tetrachloride (2) is placed in a sedimentation cell in an ultracentrifuge and rotated at 30,000 rpm. The outer radius of the cell (r_1) is 100 mm and the depth of the liquid in the cell (of diameter 12.5 mm) is 40 mm. The cell is maintained at a temperature of 20°C.

Estimate the separation achieved at equilibrium and the time required to attain 99% of the equilibrium value.

DATA

Molar mass of benzene: $M_1 = 0.0781$ kg/mol.

Molar mass of carbon tetrachloride: $M_2 = 0.1538$ kg/mol.

Density of equimolar mixture: $\rho_t = 1252$ kg/m^3.

Partial molar volume of benzene in equimolar mixture

$$\bar{V}_1 = 89 \times 10^{-6} \text{ m}^3/\text{mol}$$

The liquid-phase Maxwell–Stefan diffusion coefficient

$$Đ = 1.45 \times 10^{-9} \text{ m}^2/\text{s}$$

The activity coefficient of benzene in solution is given by

$$\ln \gamma_1 = 0.14 x_2^2$$

SOLUTION   The composition distribution at equilibrium is given by Eq. 2.3.30.

$$\Gamma \frac{dx_1}{dr} = \left(\omega_1 - c_1 \bar{V}_1 \right) \frac{\bar{M}}{RT} \Omega^2 r \qquad (2.3.30)$$

Assuming all terms in this expression to be constant (other than x_1 and r, of course) allows us to integrate Eq. 2.3.30 from the surface of the liquid $r = r_0$ to the end of the centrifuge $r = r_1$ (Fig. 2.7). The result is

$$\Gamma \Delta x_1 = \left(\omega_1 - c_1 \bar{V}_1 \right) \frac{\bar{M}}{RT} \Omega^2 \bar{r} \Delta r \qquad (2.3.32)$$

where $\bar{r}$ is the average radius, $\frac{1}{2}(r_0 + r_1)$; Δr is the difference $(r_1 - r_0)$; Δx_1 is the difference in mole fraction of component 1 at position $r = r_1$ and the mole fraction of component 1 at $r = r_0$. Equation 2.3.32 is the result we needed; now for the calculations.

To evaluate the thermodynamic factor Γ we write $\ln \gamma_1$ in terms of x_1 only and differentiate with respect to x_1 to give

$$\Gamma = 1 + x_1 \frac{\partial \ln \gamma_1}{\partial x_1}$$

$$= 1 - 0.28 x_1 (1 - x_1)$$

At the average (initial) composition, $x_1 = 0.5$

$$\Gamma = 1 - 0.28 \times 0.5 \times (1 - 0.5) = 0.93$$

The mass fraction ω_1 is calculated next

$$\omega_1 = \frac{x_1 M_1}{x_1 M_1 + x_2 M_2} = 0.3368$$

The molar concentration of species 1 may now be calculated

$$c_1 = \rho_1/M_1$$
$$= \omega_1 \rho_t/M_1$$
$$= 0.3368 \times 1252/0.0781$$
$$= 5399 \text{ mol benzene}/\text{m}^3 \text{ mixture}$$

and the volume fraction follows as

$$\phi_1 = c_1 \overline{V}_1$$
$$= 5399 \times 89 \times 10^{-6}$$
$$= 0.4805 \text{ m}^3 \text{ benzene}/\text{m}^3 \text{ mixture}$$

The angular velocity Ω is calculated as follows

$$\Omega = 2\pi f$$
$$= 2\pi \times 500$$
$$= 3141.6 \text{ s}^{-1}$$

The remaining terms on the right-hand side of Eq. 2.3.32 are evaluated as follows:

$$\bar{r} = \tfrac{1}{2}(r_0 + r_1) = 80 \text{ mm} = 80 \times 10^{-3} \text{ m}$$
$$\Delta r = r_1 - r_0 = 100 - 60 = 40 \text{ mm} = 40 \times 10^{-3} \text{ m}$$
$$\frac{\overline{M}}{RT} = \frac{0.5 \times 0.0781 + 0.5 \times 0.1538}{8.3144 \times 293.15}$$
$$= 4.757 \times 10^{-5} \text{ kg/J}$$

Finally, we may calculate the separation at equilibrium from Eq. 2.3.32 to be

$$\Delta x_1 = (0.3368 - 0.4805) \times 4.757 \times 10^{-5} \times 3141.6^2 \times 0.08 \times 0.04$$
$$= -0.2322$$

This means that benzene, which is the lighter component, tends to concentrate preferentially at the center of the centrifuge, whereas carbon tetrachloride, which is the heavier component, tends to concentrate at the periphery.

The time required to reach α close to equilibrium is given by (Cullinan and Lenczyck, 1969)

$$t_\alpha = \frac{\Delta r^2 \beta_0}{8 Ð \Gamma} \ln\left(\frac{\beta_0}{1 - \alpha}\right)$$

where β_0 is obtained from the formula

$$\beta_n = \frac{2}{\left(\pi\left(n + \frac{1}{2}\right)\right)^2}$$

Thus, for $n = 0$,

$$\beta_0 = \frac{2 \times 4}{\pi^2} = \frac{8}{\pi^2}$$

The product $Ð\Gamma$ is found to be

$$Ð\Gamma = 1.45 \times 10^{-9} \times 0.93 = 1.349 \times 10^{-9} \text{ m}^2/\text{s}$$

So, for 99% approach to equilibrium $\alpha = 0.99$ and we find

$$t_\alpha = \frac{0.04^2 \times 8}{\pi^2 \times 1.349 \times 10^{-9}} \ln\left(\frac{8}{0.01\pi^2}\right)$$

$$= 528,376 \text{ s}$$

Thus, the time required to reach 99% of equilibrium is about 150 h (or a little over 6 days!). ∎

Example 2.3.2   Separation of Uranium Isotopes with a Gaseous Ultracentrifuge

Consider the separation of $U^{235}F_6$ (1) and $U^{238}F_6$ (2) in a gas centrifuge that has an internal diameter of 60 mm and rotates at 40,000 rpm, at a temperature of 20°C. If y and x refer to the mole fractions of the lighter isotope at radii $r = 0$ and $r = r_1$, respectively, determine the value of the separation factor:

$$\alpha = \frac{y(1 - x)}{x(1 - y)}$$

DATA   Molar masses $M_1 = 0.34915$ kg/mol $M_2 = 0.35215$ kg/mol.

SOLUTION   At equilibrium the composition gradient is given by Eq. 2.3.31

$$\frac{dx_1}{dr} = \left(\omega_1 - c_1\bar{V}_1\right)\frac{\bar{M}}{RT}\Omega^2 r \tag{2.3.31}$$

We may rewrite the term $(\omega_1 - c_1\bar{V}_1)\bar{M}$ as follows:

$$\left(\omega_1 - c_1\bar{V}_1\right)\bar{M} = (\omega_1 - x_1)(x_1M_1 + x_2M_2)$$

$$= \left(\frac{x_1M_1}{x_1M_1 + x_2M_2} - x_1\right)(x_1M_1 + x_2M_2)$$

$$= x_1(M_1 - x_1M_1 - x_2M_2)$$

$$= x_1(1 - x_1)(M_1 - M_2)$$

Equation 2.3.31 may now be simplified to

$$\frac{dx_1}{x_1(1 - x_1)} = (M_1 - M_2)\frac{\Omega^2}{RT}r\,dr$$

and is to be integrated subject to the boundary conditions

$$r = 0 \qquad x_1 = y$$
$$r = r_1 \qquad x_1 = x$$

to give

$$\ln\left(\frac{x}{y}\right) - \ln\left(\frac{1 - x}{1 - y}\right) = (M_1 - M_2)\frac{\Omega^2 r_1^2}{2RT}$$

or

$$\frac{x(1 - y)}{y(1 - x)} = \exp\left[(M_1 - M_2)\frac{\Omega^2 r_1^2}{2RT}\right]$$

or

$$\alpha = \exp\left[-(M_1 - M_2)\frac{\Omega^2 r_1^2}{2RT}\right]$$

The angular velocity Ω is calculated as

$$\Omega = \frac{2\pi\,40{,}000}{60}$$

$$= 4188.8 \text{ s}^{-1}$$

and with $R = 8.3144$ J/mol K and $r_1 = 0.06$ m we calculate the separation factor α as

$$\alpha = \exp\left(-\frac{(0.34915 - 0.35215) \times 4188.8^2 \times 0.06^2}{(2 \times 8.3144 \times 293.15)}\right)$$

$$= 1.0396$$

Despite the low value of the separation factor, ultracentrifugation is a viable commercially used technique for separation of the isotopes of uranium. In view of the small separation factor and low capacity per unit, a commercial plant will have a few million centrifuges (Von Halle, 1980; Voight, 1982)! ■

2.4 DIFFUSION IN ELECTROLYTE SYSTEMS

There are many applications in chemical engineering where diffusion of charged species is involved. Examples include ion exchange, metals extraction, electrochemical reactors, and membrane separations. There is an excellent textbook in this area (Newman, 1991). Here we will be content to show that the treatment of electrolyte diffusion follows naturally from the generalized treatment of diffusion given in Section 2.3.

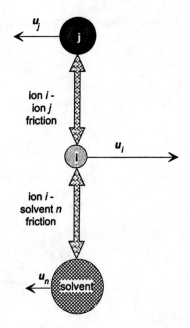

Figure 2.8. Pictorial representation of ion–ion and ion–solvent interactions in an electrolyte solution.

In mixtures of electrolytes the generalized Maxwell–Stefan equations

$$d_i = -\sum_{j=1}^{n} \frac{x_i x_j (u_i - u_j)}{Ð_{ij}} \qquad (2.3.17)$$

$$= \sum_{j=1}^{n} \frac{(x_i N_j - x_j N_i)}{c_t Ð_{ij}} \qquad (2.3.18)$$

are written for each of the ionic and nonionic (i.e., solvent) species in the system. The inverse Maxwell–Stefan diffusion coefficients have the same physical significance as the ones introduced in Section 2.2; they represent the friction experienced by the i–j pair whether or not they be ion–ion or ion–solvent interactions (Fig. 2.8).

The x_i in these equations denote ionic mole fractions. In general, the ionic mole fractions will differ from the undissociated electrolyte mole fractions. To illustrate this fact consider an aqueous solution of sulfuric acid. Let us take 1 m³ of solution with c_s kmol of H_2SO_4 and c_w kmol of H_2O. The mole fraction of the undissociated species are

$$x_{H_2SO_4} = \frac{c_s}{c_w + c_s} \qquad x_{H_2O} = \frac{c_w}{c_w + c_s}$$

On complete dissociation of acid

$$H_2SO_4 \rightarrow 2H^+ + SO_4^{2-}$$

Thus

$$c_H = 2c_s \qquad c_{H_2O} = c_w \qquad c_{SO_4} = c_s$$

where the subscripts H and SO₄ refer to the H^+ and SO_4^{2-} ions, respectively. The mole fractions of the various species are

$$x_H = \frac{2c_s}{3c_s + c_w} \qquad x_{SO_4} = \frac{c_s}{3c_s + c_w} \qquad x_{H_2O} = \frac{c_w}{3c_s + c_w}$$

Note that the mole fraction of H_2O has "decreased" when considered in terms of ionic species. This is an important point to bear in mind and the reader is advised to study Newman (1991) for further discussion. For mixed ion systems there will be contributions to c_i from various ionic species. For example, in the system with mixed salts HCl and $BaCl_2$ the concentration of chloride ion, $c_{Cl^-} = c_{HCl} + 2c_{BaCl_2}$ (see Example 2.4.2).

The generalized driving force d_i is defined by Eq. 2.3.12

$$c_t RT d_i \equiv c_i \nabla_{T,P} \mu_i + (\phi_i - \omega_i) \nabla P - \left(c_i F_i - \omega_i \sum_{j=1}^{n} c_j F_j \right) \qquad (2.3.12)$$

where ϕ_i is the volume fraction $c_i \overline{V}_i$.

The external body force per mole acting on species i, F_i, is given by (see Newman, 1991)

$$F_i = -z_i \mathscr{F} \nabla \phi \qquad (2.4.1)$$

where z_i is the ionic charge of the species (e.g., $z_H = +1$; $z_{SO_4} = -2$), $\mathscr{F}$ is Faraday's constant $= 9.65 \times 10^4$ C/mol $= 9.65 \times 10^7$ C/kmol, and ϕ is the electrical potential measured in volts (1 C $= 1$ A $\times 1$ s and 1 V $\times 1$ C $= 1$ J). With the force term given by Eq. 2.4.1 and neglecting the pressure diffusion term ($\nabla P = 0$), Eq. 2.3.12 becomes

$$c_t RT d_i \equiv c_i \nabla_{T,P} \mu_i + \left(c_i z_i - \omega_i \sum_{j=1}^{n} c_j z_j \right) \mathscr{F} \nabla \phi \qquad (2.4.2)$$

Except in regions close to electrode surfaces, where there will be charge separation (the double layer phenomena), the condition of electroneutrality is met (see Newman, 1991, for a detailed discussion of this topic):

$$\sum_{i=1}^{n} c_i z_i = 0 \qquad (2.4.3)$$

that is, there is no net electrical body force acting on the mixture as a whole. The generalized driving force, therefore, simplifies still further to

$$c_t RT d_i \equiv c_i \nabla_{T,P} \mu_i + c_i z_i \mathscr{F} \nabla \phi \qquad (2.4.4)$$

The chemical potential gradient may be expressed in terms of mole fraction and activity coefficient gradients as shown in Section 2.2. Activity coefficient models for electrolyte systems are discussed by, for example, Newman (1991) and Zemaitis et al. (1986).

As in the treatment of diffusion in nonionic systems it is usual to define diffusion fluxes J_i with respect to a specified reference velocity. For diffusion in electrolyte systems the most commonly used reference velocity is the solvent velocity u_n.

$$J_i^n = N_i - c_i u_n \qquad (2.4.5)$$

With this choice of reference velocity the flux of species n, J_n^n, is zero (Table 1.3)

$$J_n^n = 0 \tag{2.4.6}$$

with

$$\sum_{i=1}^{n} J_i^n \neq 0 \tag{2.4.7}$$

The generalized Maxwell–Stefan equations may be written in terms of these diffusion fluxes as

$$d_i = \sum_{j=1}^{n} \frac{\left(x_i J_j^n - x_j J_i^n \right)}{c_t \mathcal{D}_{ij}} \tag{2.4.8}$$

Equations 2.4.8 may be written in the following equivalent form (cf. Eq. 2.1.23)

$$c_t d_i = - \sum_{j=1}^{n-1} B_{ij}^n J_j^n \tag{2.4.9}$$

where the B_{ij}^n are defined by (cf. Eqs. 2.1.21 and 2.1.22)

$$B_{ii}^n = \sum_{\substack{k=1 \\ i \neq k}}^{n} \frac{x_k}{\mathcal{D}_{ik}} \qquad i = 1, 2, \ldots, n-1 \tag{2.4.10}$$

$$B_{ij}^n = -x_i / \mathcal{D}_{ij} \qquad i \neq j = 1, 2, \ldots, n-1 \tag{2.4.11}$$

Equations 2.4.9 may be written in compact $n-1$ dimensional matrix form as

$$c_t(d) = -[B^n](J^n) \tag{2.4.12}$$

or

$$(J^n) = -c_t[B^n]^{-1}(d) \tag{2.4.13}$$

Equations 2.4.12 describing diffusion in concentrated electrolyte solutions are the counterparts of Eqs. 2.1.23 and 2.2.7 for diffusion in ideal gases and nonideal, nonelectrolyte systems, respectively.

2.4.1 The Nernst–Planck Equation

In dilute electrolyte systems the driving force d_i reduces to

$$d_i = \nabla x_i + x_i z_i \frac{\mathcal{F}}{RT} \nabla \phi \tag{2.4.14}$$

and the matrix $[B^n]$ degenerates to a diagonal matrix with elements given by

$$B_{ii}^n = 1/\mathcal{D}_{in}^{\circ} \qquad B_{ij}^n = 0 \qquad (i \neq j) \tag{2.4.15}$$

The superscript $^\circ$ signifies the infinite dilution limit. Equations 2.4.13 for the diffusion fluxes simplifies to

$$J_i^n = -c_t Ð_{in}^\circ \, \nabla x_i - c_i z_i Ð_{in}^\circ \frac{\mathscr{F}}{RT} \nabla \phi \qquad (2.4.16)$$

The molar fluxes N_i are found by combining Eqs. 2.4.5 with Eq. 2.4.16

$$N_i = -c_t Ð_{in}^\circ \, \nabla x_i - c_i z_i Ð_{in}^\circ \frac{\mathscr{F}}{RT} \nabla \phi + c_i u_n \qquad (2.4.17)$$

The three "contributions" to N_i are termed

Diffusion $- c_t Ð_{in}^\circ \, \nabla x_i$

Migration $- c_i z_i Ð_{in}^\circ \dfrac{\mathscr{F}}{RT} \nabla \phi$

Convection $+ c_i u_n$

Equation 2.4.17 is known as the Nernst–Planck equation.

In the electrochemical literature it is traditional to use molar concentration gradient driving forces and the most commonly used form of the Nernst–Planck equation is

$$N_i = -Ð_{in}^\circ \, \nabla c_i - c_i z_i Ð_{in}^\circ \frac{\mathscr{F}}{RT} \nabla \phi + c_i u_n \qquad (2.4.18)$$

We have shown that the Nernst–Planck equation is only a limiting case of the generalized Maxwell–Stefan equations. Nevertheless, many ionic systems of interest are dilute and the Nernst–Planck equation is widely used.

Even when the system is dilute, the diffusing ionic species are "coupled" to one another in a very interesting manner. This coupling arises out of the constraint imposed by the electroneutrality condition. Equation 2.4.3 can be differentiated to give

$$\sum_{i=1}^{n} z_i \nabla c_i = 0 \qquad (2.4.19)$$

which means that there are only $n - 2$ independent composition gradients describing the system.

Example 2.4.1   Diffusion in the System KCl – H₂O at 25°C

Consider diffusion in the system $KCl - H_2O$ at 25°C. Potassium chloride is a strong electrolyte and complete dissociation into K^+ and Cl^- ions will take place:

$$KCl \rightarrow K^+ + Cl^-$$

The species involved in the diffusion process are

$$1 = K^+$$
$$2 = Cl^-$$
$$3 = H_2O \text{ (undissociated)}$$

1. Determine the elements of $[B^n]$ at a salt concentration of 1 kmol/m^3.
2. To what concentration level of the salt can one "safely" use the Nernst–Planck dilute solution approximation Eq. 2.4.18, instead of the generalized Maxwell–Stefan diffusion equations?

DATA   The Maxwell–Stefan diffusion coefficient $\mathcal{D}_{12}$ at various salt concentrations, taken from a figure in Newman (1991), is tabulated below.

c_{KCl} [mol/m^3]	c_{H_2O} [kmol/m^3]	x_1, x_2	$10^{-10}/\mathcal{D}_{12}$ [s/m^2]
0.1	55.5	1.8×10^{-6}	67.0
1	55.5	1.8×10^{-5}	28.6
10	55.5	1.8×10^{-4}	10.0
100	55.5	1.8×10^{-3}	2.86
1000	55.5	1.74×10^{-2}	0.55

The Maxwell–Stefan diffusion coefficients $\mathcal{D}_{13}$ and $\mathcal{D}_{23}$ are virtually identical and, over the concentration range of interest, independent of concentration at 2×10^{-9} m^2/s.

SOLUTION   As a basis for our calculations we consider 1 m^3 of H$_2$O. This solution will contain 1000 kg of H$_2$O or 55.5 (= 1000/18) kmol H$_2$O. Thus, $c_{H_2O} = 55.5$ kmol/m^3. The concentration of K$^+$ and Cl$^-$ ions will equal the concentration of KCl.

$$c_1 = c_2 = c_{KCl}$$

Thus, the total concentration of ions and solvent will be

$$c_t = c_1 + c_2 + c_{H_2O}$$

$$= 2c_{KCl} + c_{H_2O}$$

The mole fractions of ions and solvent are evaluated as

$$x_1 = c_{KCl}/(2c_{KCl} + c_{H_2O})$$

$$x_2 = c_{KCl}/(2c_{KCl} + c_{H_2O})$$

$$x_3 = 1 - x_1 - x_2$$

Thus, at $c_{KCl} = 1$ kmol/m^3 the mole fractions are

$$x_1 = 1/(2 + 55.5)$$

$$= 0.0174$$

$$= x_2$$

$$x_3 = 0.9652$$

and the Maxwell–Stefan diffusivity $\mathcal{D}_{12}$ is

$$\mathcal{D}_{12} = (1/0.55) \times 10^{-10} = 0.1818 \times 10^{-9} \text{ m}^2/\text{s}$$

For a ternary mixture the elements of $[B^n]$ are evaluated from Eqs. 2.4.10 and 2.4.11 as follows:

$$B_{11}^n = \frac{x_2}{\mathcal{D}_{12}} + \frac{x_3}{\mathcal{D}_{13}}$$

$$= \frac{0.0174}{0.1818 \times 10^{-9}} + \frac{0.9652}{2 \times 10^{-9}}$$

$$= 0.5782 \times 10^9 \text{ s/m}^2$$

$$B_{12}^n = -x_1/\mathcal{D}_{12}$$

$$= -0.0174/0.1818 \times 10^{-9}$$

$$= -0.09556 \times 10^9 \text{ s/m}^2$$

$$B_{21}^n = -x_2/\mathcal{D}_{12}$$

$$= -0.0174/0.1818 \times 10^{-9}$$

$$= -0.09556 \times 10^9 \text{ s/m}^2$$

$$B_{22}^n = \frac{x_1}{\mathcal{D}_{12}} + \frac{x_3}{\mathcal{D}_{23}}$$

$$= 0.5782 \times 10^9 \text{ s/m}^2$$

and we see that $B_{11}^n = B_{22}^n$ and $B_{12}^n = B_{21}^n$; a result that follows from the equality of the Maxwell–Stefan diffusivities, $\mathcal{D}_{13}$ and $\mathcal{D}_{23}$, and from the equality of the mole fractions x_1 and x_2.

To answer the second question we repeat the above calculations for other values of the salt concentration c_{KCl}. The results are summarized below.

c_{KCl} [mol/m^3]	$-B_{12}^n$ [10^6 s/m^2]	B_{11}^n [10^9 s/m^2]	$-B_{12}^n/B_{11}^n$ [–]
0.1	1.206	0.5012	0.0024
1	5.148	0.5051	0.0102
10	17.993	0.5178	0.0348
100	51.295	0.5495	0.0934
1000	95.560	0.5782	0.1653

The ratio of cross-coefficient to the main coefficient, B_{12}^n/B_{11}^n, is less than 5% if the salt concentration is less than 10 mol/m^3. So the Nernst–Planck equation can be used "safely" below $c_{KCl} = 10$ mol/m^3.

Many industrially important electrochemical systems have salt concentrations less than 10 mol/m^3. This explains the widespread use of the Nernst–Planck equations. ∎

2.4.2 Conductivity, Transference Numbers, and The Diffusion Potential

Each species j carries with it a current

$$\mathcal{F} z_j N_j$$

and the current carried by the mixture is

$$i = \mathscr{F} \sum_{j=1}^{n-1} z_j N_j \tag{2.4.20}$$

with the units amp per meter squared (A/m^2). The solvent, species n, carries no charge, $z_n = 0$, so the summation in Eq. 2.4.20 is taken over $n - 1$ terms. With the N_i given by the Nernst–Planck equations (Eq. 2.4.18) we have

$$i = -\mathscr{F} \sum_{j=1}^{n-1} z_j Đ_{jn}^\circ \, \nabla c_j - \frac{\mathscr{F}^2}{RT} \nabla \phi \sum_{j=1}^{n-1} c_j z_j^2 Đ_{jn}^\circ \tag{2.4.21}$$

In proceeding with the development, it is convenient to define the following quantities:

1. Equivalent conductivity of species j

$$\mathscr{K}_j = \frac{\mathscr{F}^2}{RT} c_j z_j^2 Đ_{jn}^\circ \tag{2.4.22}$$

2. Equivalent conductivity of the mixture

$$\mathscr{K} = \sum_{j=1}^{n-1} \mathscr{K}_j$$

$$= \frac{\mathscr{F}^2}{RT} \sum_{j=1}^{n-1} c_j z_j^2 Đ_{jn}^\circ \tag{2.4.23}$$

3. The transference number of species j

$$t_j = \frac{\mathscr{K}_j}{\mathscr{K}} \tag{2.4.24}$$

With the above definitions an expression for the current i carried by the mixture can be obtained

$$i = -\mathscr{F} \sum_{j=1}^{n-1} z_j Đ_{jn}^\circ \, \nabla c_j - \mathscr{K} \, \nabla \phi \tag{2.4.25}$$

which may be rearranged to give the electrical potential as

$$\nabla \phi = -\frac{i}{\mathscr{K}} - \frac{\mathscr{F}}{\mathscr{K}} \sum_{j=1}^{n-1} z_j Đ_{jn}^\circ \, \nabla c_j \tag{2.4.26}$$

Equation 2.4.26 shows that even when no current is carried by the mixture (i.e., $i = 0$), there exists a finite electrical potential

$$\nabla \phi = -\frac{\mathscr{F}}{\mathscr{K}} \sum_{j=1}^{n-1} z_j Đ_{jn}^\circ \, \nabla c_j \tag{2.4.27}$$

termed the diffusion potential. Substituting this expression for $\nabla \phi$ in the Nernst–Planck relationship (Eq. 2.4.18) we find

$$N_i = -Ð_{in}^\circ \nabla c_i + c_i u_n + c_i z_i Ð_{in}^\circ \frac{\mathscr{F}}{RT} \frac{i}{\mathscr{K}}$$
$$+ c_i z_i Ð_{in}^\circ \frac{\mathscr{F}^2}{RT} \frac{1}{\mathscr{K}} \sum_{j=1}^{n-1} z_j Ð_{jn}^\circ \nabla c_j \tag{2.4.28}$$

Equation 2.4.28 can be written a little more compactly in terms of the transference numbers t_i

$$N_i = -Ð_{in}^\circ \nabla c_i + c_i u_n + \frac{t_i}{z_i \mathscr{F}} i$$
$$+ \frac{t_i}{z_i} \sum_{j=1}^{n-1} z_j Ð_{jn}^\circ \nabla c_j \tag{2.4.29}$$

Now, for "pure diffusion" as is encountered in processes such as ion exchange or extraction there is no current flowing through the mixture

$$i = 0 \tag{2.4.30}$$

and the molar flux of each ionic species is given by

$$N_i = -Ð_{in}^\circ \nabla c_i + c_i u_n + \frac{t_i}{z_i} \sum_{j=1}^{n-1} z_j Ð_{jn}^\circ \nabla c_j \tag{2.4.31}$$

2.4.3 Effective Ionic Diffusivities

It is common to define an "effective ionic diffusivity" by

$$N_i = -\mathscr{D}_{i,\text{eff}} \nabla c_i + c_i u_n \tag{2.4.32}$$

An expression for the effective ionic diffusivity $\mathscr{D}_{i,\text{eff}}$ may be obtained by setting equal the right-hand sides of Eqs. 2.4.31 and 2.4.32.

$$\mathscr{D}_{i,\text{eff}} = Ð_{in}^\circ - \frac{t_i}{z_i} \sum_{j=1}^{n-1} z_j Ð_{jn}^\circ \frac{\nabla c_j}{\nabla c_i} \tag{2.4.33}$$

The effective ionic diffusivity $\mathscr{D}_{i,\text{eff}}$ is seen to depend on

1. The infinite dilution MS diffusivities

$$Ð_{in}^\circ \,(i = 1, 2, \ldots, n - 1).$$

2. The charge numbers of all species

$$z_i \,(i = 1, 2, \ldots, n - 1).$$

3. The concentration gradients of all ionic species

$$\nabla c_i \,(i = 1, 2, \ldots, n - 1).$$

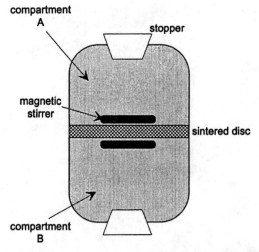

Figure 2.9. Schematic diagram of a two compartment diffusion cell. The experiments by Vinograd and McBain (1941) on diffusion in electrolyte systems were carried out in apparatus of this kind.

Example 2.4.2   *Diffusion in an Aqueous Solution of HCl and $BaCl_2$*

Vinograd and McBain (1941) investigated the diffusion of electrolytes and their ionic species using a two compartment diffusion cell similar to the one depicted in Figure 2.9. The solution in one compartment was pure water while the other contained an aqueous electrolyte solution. Diffusion took place through the pores of a sintered glass disk that separated the two compartments. Vinograd and McBain experimented with a variety of salts and mixtures of salts; the experiments that we are concerned with in this example involved the system $HCl - BaCl_2 - H_2O$. The HCl and $BaCl_2$ dissociate as follows:

$$HCl \rightarrow H^+ + Cl^-$$
$$BaCl_2 \rightarrow Ba^{2+} + 2Cl^-$$

We are required to compute the effective diffusivities of the ions in solution and compare with the experimental values shown in Figure 2.10.

DATA   The components in the mixture will be numbered as follows:

$$1 = H^+$$
$$2 = Cl^-$$
$$3 = Ba^{2+}$$
$$4 = H_2O$$

The charge numbers of these species are

$$z_1 = 1 \quad z_2 = -1 \quad z_3 = 2 \quad z_4 = 0$$

The infinite dilution Maxwell–Stefan diffusion coefficients for the ions in water are (from Newman, 1991).

$$Ð_{14}^\circ = 9.3 \times 10^{-9} \text{ m}^2/\text{s}$$
$$Ð_{24}^\circ = 2.0 \times 10^{-9} \text{ m}^2/\text{s}$$
$$Ð_{34}^\circ = 0.85 \times 10^{-9} \text{ m}^2/\text{s}$$

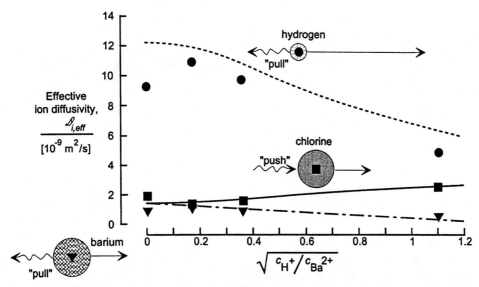

Figure 2.10. Comparison of calculated experimental values of the effective ionic diffusivity. Data from Vinograd and McBain (1941).

SOLUTION   The effective diffusivities may be calculated from Eq. 2.4.33; we approximate the ratio of concentration gradients $\nabla c_j / \nabla c_i$ by the ratio of concentration differences $\Delta c_j / \Delta c_i$; thus:

$$\mathscr{D}_{i,\text{eff}} = Ð_{in}^\circ - \frac{t_i}{z_i} \sum_{j=1}^{n-1} z_j Ð_{jn}^\circ \frac{\Delta c_j}{\Delta c_i}$$

The concentrations of the ions in compartment 1 are

$$c_1 = c_{\text{HCl}}$$
$$c_2 = c_{\text{HCl}} + 2c_{\text{BaCl}_2}$$
$$c_3 = c_{\text{BaCl}_2}$$

As a basis for calculating the effective diffusivities we make $c_{\text{BaCl}_2} = 1 \text{ kmol/m}^3$. We further define the concentration ratio r as

$$r = c_1/c_3 = c_{\text{HCl}}/c_{\text{BaCl}_2}$$

and the concentrations of ions in solution are given in terms of r as

$$c_1 = r \qquad c_2 = (2 + r) \qquad c_3 = 1 \text{ kmol/m}^3$$

Compartment 2 contains pure H_2O so the concentration of ions in compartment 2 is zero. Thus, the concentration differences are equal to the concentrations in compartment 1

$$\Delta c_1 = r \text{ kmol/m}^3$$
$$\Delta c_2 = (2 + r) \text{ kmol/m}^3$$
$$\Delta c_3 = 1 \text{ kmol/m}^3$$

Let us illustrate the calculation of the effective diffusivities at $r = 1$. The concentrations of ions in compartment 1 become

$$c_1 = 1 \qquad c_2 = 3 \qquad c_3 = 1 \text{ kmol/m}^3$$

and the concentration differences are

$$\Delta c_1 = 1 \qquad \Delta c_2 = 3 \qquad \Delta c_3 = 1 \text{ kmol/m}^3$$

The equivalent conductivity of the three ionic species are calculated from Eq. 2.4.22 as

$$\begin{aligned}
\mathcal{K}_1 &= (\mathcal{F}^2/RT) c_1 z_1^2 \mathcal{D}_{14}^\circ \\
&= (9.65 \times 10^7)^2 \times 1 \times 1^2 \times 9.3 \times 10^{-9}/(8314.3 \times 298.15) \\
&= 34.94 \\
\mathcal{K}_2 &= (\mathcal{F}^2/RT) c_2 z_2^2 \mathcal{D}_{24}^\circ \\
&= (9.65 \times 10^7)^2 \times 3 \times 2^2 \times 2.0 \times 10^{-9}/(8314.3 \times 298.15) \\
&= 22.54 \\
\mathcal{K}_3 &= (\mathcal{F}^2/RT) c_3 z_3^2 \mathcal{D}_{34}^\circ \\
&= (9.65 \times 10^7)^2 \times 1 \times (-1)^2 \times 0.85 \times 10^{-9}/(8314.3 \times 298.15) \\
&= 12.77
\end{aligned}$$

The equivalent conductivity of the mixture is

$$\begin{aligned}
\mathcal{K} &= \mathcal{K}_1 + \mathcal{K}_2 + \mathcal{K}_3 \\
&= 34.94 + 22.54 + 12.77 \\
&= 70.25
\end{aligned}$$

The transference numbers are given by Eqs. 2.4.24

$$\begin{aligned}
t_1 &= \mathcal{K}_1/\mathcal{K} \\
&= 34.94/70.25 \\
&= 0.4973 \\
t_2 &= \mathcal{K}_2/\mathcal{K} \\
&= 22.54/70.25 \\
&= 0.3208 \\
t_3 &= \mathcal{K}_1/\mathcal{K} \\
&= 12.77/70.25 \\
&= 0.1818
\end{aligned}$$

It will simplify subsequent calculations if we introduce a quantity ξ

$$\begin{aligned}
\xi &= z_1 \mathcal{D}_{14}^\circ \Delta c_1 + z_2 \mathcal{D}_{24}^\circ \Delta c_2 + z_3 \mathcal{D}_{34}^\circ \Delta c_3 \\
&= 1 \times 9.3 \times 10^{-9} \times 1 + (-1) \times 2.0 \times 10^{-9} \times 3 + 2 \times 0.85 \times 10^{-9} \times 1 \\
&= 5.0 \times 10^{-9} \text{ kmol/m s}
\end{aligned}$$

We may finally calculate the effective ionic diffusivities as

$$\mathscr{D}_{1,\text{eff}} = Đ^{\circ}_{14} - t_1 \xi / z_1 \Delta c_1$$

$$= 9.3 \times 10^{-9} - 0.4973 \times 5.0 \times 10^{-9}/(1 \times 1)$$

$$= 6.813 \times 10^{-9} \text{ m}^2/\text{s}$$

$$\mathscr{D}_{2,\text{eff}} = Đ^{\circ}_{24} - t_2 \xi / z_2 \Delta c_2$$

$$= 2.0 \times 10^{-9} - 0.3208 \times 5.0 \times 10^{-9}/(-1 \times 3)$$

$$= 2.535 \times 10^{-9} \text{ m}^2/\text{s}$$

$$\mathscr{D}_{3,\text{eff}} = Đ^{\circ}_{34} - t_3 \xi / z_3 \Delta c_3$$

$$= 0.85 \times 10^{-9} - 0.1818 \times 5.0 \times 10^{-9}/(2 \times 1)$$

$$= 0.395 \times 10^{-9} \text{ m}^2/\text{s}$$

Additional calculations of the effective ionic diffusivities are shown in Figure 2.10 as a function of the square root of the concentration ratio r. The experimentally determined effective diffusivities are shown in the same figure for comparison. The agreement between theory and experiment is very good, especially for the Cl^- and Ba^{2+} ions. The theory overestimates the effective diffusivity of the H^+ ions but the decrease in the effective diffusivity of the H^+ ions as the concentration ratio increases is predicted correctly.

The important features of mixed ion diffusion are brought out very clearly in the calculations. The rapidly diffusing H^+ ions are slowed down by the electrostatic "pull" being exerted on them by the more slowly diffusing Cl^- ions (Fig. 2.10). At the same time the Cl^- ions are accelerated by the H^+ ions. The Ba^{2+} ions, which have a low diffusion coefficient already, diffuse even slower because of the constraint of electroneutrality. ■

The consequences of mixed ion diffusion effects in chemical engineering are felt in the following areas:

1. Metals extraction; see Tunison and Chapman (1976) and Van Brocklin and David (1972).
2. Absorption of HCl by NaOH (Sherwood and Wei, 1955).
3. Ion exchange (Helfferich, 1962).
4. Electrodialysis (Wesselingh and Krishna, 1990).

3 Fick's Law

It is a striking symptom of the common ignorance in this field that not one of the phenomenological schemes which are fit to describe the general case of diffusion is widely known.
—L. Onsager (1945)

At about the same time that Maxwell was developing his kinetic theory of gases, Thomas Graham, Adolf Fick, and others were attempting to uncover the basic diffusion equations through experimental studies involving binary mixtures (Cussler (1976) provides a brief history of the early work on diffusion). The result of Fick's work was the "law" that now bears his name (Fick, 1855a, b) and discussed in Section 3.1. A generalization of Fick's law to cover diffusion in multicomponent systems is the subject of Section 3.2. The irreversible thermodynamics (IT) formulation for multicomponent diffusion, pioneered by Onsager (1931), will also be developed and compared with the Fick formulation in Section 3.3.

3.1 DIFFUSION IN BINARY MIXTURES: FICK'S FIRST LAW

Let us consider in more detail diffusion in a simple system made up of components 1 and 2. Let u_1 and u_2 represent the velocities of transfer of components 1 and 2 and $u = x_1 u_1 + x_2 u_2$ represents the molar average velocity of the mixture. If c_1 and c_2 are the molar concentrations of 1 and 2 and c_t is the total mixture molar concentration, then the diffusion flux J_1 is usually related to the mole fraction gradient by the constitutive relation

$$J_1 = c_1(u_1 - u) = -c_t D_{12} \nabla x_1 \tag{3.1.1}$$

which is Fick's first law of diffusion. The Fick diffusion coefficient is D_{12}. An analogous relation may also be written for component 2.

$$J_2 = c_2(u_2 - u) = -c_t D_{21} \nabla x_2 \tag{3.1.2}$$

It is easy to confirm that since $J_1 + J_2 = 0$ and $x_1 + x_2 = 1$ we must have

$$D_{12} = D_{21} = D \quad \text{(say)} \tag{3.1.3}$$

that is, there is only one diffusion coefficient describing the molecular diffusion process in a binary mixture. There is also only one independent driving force ∇x_1 and only one independent flux J_1. Equation 3.1.1 defines the Fick diffusion coefficient.

3.1.1 Fick Diffusion Coefficients

A few typical values of the Fick diffusion coefficients are listed in Table 3.1. Although it may not be discerned from this small sample of values, the diffusion coefficient in an ideal gas mixture is independent of the mixture composition, inversely proportional to pressure, and varies with the absolute temperature to around the 1.5 power. More extensive listings are provided by Reid et al. (1987) and by Cussler (1984). The most comprehensive collection of

TABLE 3.1   Diffusion Coefficients in Binary Gas Mixtures at 101.3 kPa

System	Temperature [K]	D [10^{-5} m^2/s]
Air–CO$_2$	276.2	1.42
Air–H$_2$O	289.1	2.82
Air–benzene	298.2	0.96
Air–2-propanol	299.1	0.99
CO–N$_2$	273.0	1.77
CO–O$_2$	273.0	1.85
H$_2$–N$_2$	297.2	7.79
H$_2$–He	298.2	11.32
N$_2$–O$_2$	273.2	1.81
N$_2$–H$_2$O	307.5	2.56

data on gaseous diffusion coefficients is a review by Marrero and Mason (1972). The range of values in Table 3.1 is slightly more than one decade.

Diffusion coefficients in binary liquid mixtures are of the order 10^{-9} m^2/s. Unlike the diffusion coefficients in ideal gas mixtures, those for liquid mixtures can be strong functions of concentration. We defer illustration of this fact until Chapter 4 where we also consider models for the correlation and prediction of binary diffusion coefficients in gases and liquids.

3.1.2   Alternative Forms of Fick's Law

In place of the molar diffusion flux with respect to the molar average velocity J_1 we may use the diffusion fluxes J_i^V in the volume average reference velocity frame u^V, in which case Fick's law takes the form

$$J_1^V = c_1\left(u_1 - u^V\right) = -D\,\nabla c_1 \tag{3.1.4}$$

where we use the molar concentration gradient driving force, ∇c_1. Equation 3.1.4 is, in fact, the most commonly used form of the binary constitutive relationship. However, this form is not the most convenient to use in practical design problems because under nonisothermal conditions, the molar concentration gradients will vary with composition and temperature, thus

$$\nabla c_1 = \left(c_t + x_1 \frac{\partial c_t}{\partial x_1}\right)\nabla x_1 + x_1 \frac{\partial c_t}{\partial T}\nabla T$$

$$= c_t \frac{\overline{V}_2}{\overline{V}_t}\nabla x_1 + x_1 \frac{\partial c_t}{\partial T}\nabla T \tag{3.1.5}$$

where $\overline{V}_i$ is the partial molar volume of component i and $\overline{V}_t$ is the mixture molar volume. The use of molar concentration gradients as driving forces is not to be recommended because

- Molar concentrations c_i are not suggested by solution theories as convenient concentration variables (even in ideal solutions) to represent the thermodynamically based activity a_i.

- It is not true that $\overline{V}_2/\overline{V}_t \rightarrow 1$ for small concentration gradients, that is, the simple relation $\nabla c_1 = c_t \nabla x_1$ holds if and only if c_t is constant and not approximately even for a dilute solute.

- The presence of the temperature gradient term ∇T is indeed disturbing and the second term of Eq. 3.1.5 can be very large for gases—leading to the "hot radiator paradox" mentioned by Sherwood et al. (1975, p. 15). Thus, as pointed out by these authors, use of the molar concentration gradient driving force will predict the existence of a diffusion flux, J_1^V, in a system of uniform composition subject to a temperature gradient.

The third commonly encountered form of Fick's first law is in the mass average reference velocity frame v.

$$j_1 = \rho_1(u_1 - v) = -\rho_t D \nabla \omega_1 \tag{3.1.6}$$

This form of Fick's law is most convenient when we must solve the mass continuity equations simultaneously with the equations of motion (see, e.g., Chapter 10).

The three diffusion coefficients D defined in Eqs. 3.1.1, 3.1.4, and 3.1.6 are identical (Bird et al., 1960).

3.2 THE GENERALIZED FICK'S LAW

For the binary systems already discussed, we may regard Eq. 3.1.1 as a linear relationship between the independent flux J_1 and driving force ∇x_1. For a ternary mixture there are two independent fluxes (J_1, J_2) and two independent driving forces $(\nabla x_1, \nabla x_2)$. Thus, assuming a linear relationship between the fluxes and composition gradients, we may write

$$J_1 = -c_t D_{11} \nabla x_1 - c_t D_{12} \nabla x_2 \tag{3.2.1}$$

$$J_2 = -c_t D_{21} \nabla x_1 - c_t D_{22} \nabla x_2 \tag{3.2.2}$$

Here we see that J_1 and J_2 depend on both of the independent mole fraction gradients ∇x_1 and ∇x_2. The D_{ij} in Eqs. 3.2.1 and 3.2.2 are the multicomponent diffusion coefficients; note that four of them are needed to characterize a ternary system. These coefficients are not to be confused with the binary diffusion coefficient in Eq. 3.1.1; they may take positive or negative signs and they are not, in general, symmetric $(D_{12} \neq D_{21})$. Also, the multicomponent D_{ij} do not have the physical significance of the binary Fick diffusivity in that the D_{ij} do not reflect the i–j interactions. Furthermore, the numerical values of the D_{ij} depend on the particular choice of system numbering.

For n-component systems there are $n - 1$ independent diffusion fluxes and composition gradients and we simply continue to add terms and equations. Thus,

$$J_1 = -c_t D_{11} \nabla x_1 - c_t D_{12} \nabla x_2 \cdots -c_t D_{1,n-1} \nabla x_{n-1}$$

$$J_2 = -c_t D_{21} \nabla x_1 - c_t D_{22} \nabla x_2 \cdots -c_t D_{2,n-1} \nabla x_{n-1}$$

$$\vdots$$

$$J_i = -c_t D_{i1} \nabla x_1 - c_t D_{i2} \nabla x_2 \cdots -c_t D_{i,n-1} \nabla x_{n-1} \tag{3.2.3}$$

$$\vdots$$

$$J_{n-1} = -c_t D_{n-1,1} \nabla x_1 - c_t D_{n-2,2} \nabla x_2 \cdots -c_t D_{n-1,n-1} \nabla x_{n-1}$$

No additional equation is needed for J_n, which is given in terms of the other diffusion fluxes by Eq 1.2.11. Each of Eqs. 3.2.3 can be written in the following algebraic form:

$$J_i = -c_t \sum_{k=1}^{n-1} D_{ik} \nabla x_k \qquad (3.2.4)$$

3.2.1 Matrix Representation of the Generalized Fick's Law

The set of $n - 1$ equations (Eqs. 3.2.3 or 3.2.4) is more conveniently written in $n - 1$ dimensional matrix notation

$$(J) = -c_t[D](\nabla x) \qquad (3.2.5)$$

where (J) represents a column matrix of molar diffusion fluxes

$$(J) = \begin{pmatrix} J_1 \\ J_2 \\ \vdots \\ J_{n-1} \end{pmatrix}$$

(∇x) represents a column matrix of composition gradients with $n - 1$ elements

$$(\nabla x) = \begin{pmatrix} \nabla x_1 \\ \nabla x_2 \\ \vdots \\ \nabla x_{n-1} \end{pmatrix} = \nabla(x)$$

The nth component diffusion flux J_n is not independent and is obtained from Eq. 2.1.19. The nth component gradient is given by Eq. 2.1.18.

The matrix $[D]$ of Fick diffusion coefficients is a square matrix of dimension $n - 1 \times n - 1$.

$$[D] = \begin{bmatrix} D_{11} & D_{12} & D_{13} & \cdots & D_{1,n-1} \\ D_{21} & D_{22} & D_{23} & \cdots & D_{2,n-1} \\ \vdots & & & & \vdots \\ D_{n-1,1} & D_{n-1,2} & D_{n-1,3} & \cdots & D_{n-1,n-1} \end{bmatrix}$$

It is important to note that for multi-(n-)-component diffusion, the nondiagonal or off-diagonal elements or cross-coefficients D_{ij} $(i \neq j = 1, 2, \ldots, n - 1)$ are, in general, nonzero.

For a ternary system ($n = 3$), the matrix representation of the generalized Fick's law (Eq. 3.2.5) is two dimensional. Using the property of matrix multiplication we recover Eqs. 3.2.1 and 3.2.2 for the molecular diffusion fluxes J_1 and J_2.

The reader should satisfy himself/herself that the three formulations (Eqs. 3.2.3, 3.2.4, and 3.2.5) are entirely equivalent to one another. It is not only in the interests of economy and elegance of presentation that we shall consistently prefer the matrix formulation (Eq. 3.2.4); we shall see later that matrix formulations lend themselves to easy manipulations and in many cases the n-component mass transfer relations can be written down as $n - 1$ dimensional matrix analogs of the corresponding binary mass transfer relationships (Chapter 5).

3.2.2 Alternative Forms of the Generalized Fick's Law

There are three forms of the generalized Fick's law in common use:

1. Molar flux with respect to molar average velocity.

$$(J) = -c_t[D](\nabla x) \tag{3.2.5}$$

2. Mass flux with respect to mass average velocity.

$$(j) = -\rho_t[D^\circ](\nabla \omega) \tag{3.2.6}$$

3. Molar flux with respect to volume average velocity.

$$(J^V) = -[D^V](\nabla c) \tag{3.2.7}$$

For binary systems all matrices contain just a single element and Eqs. 3.2.5–3.2.7 reduce to Eqs. 3.1.1, 3.1.6, and 3.1.4, respectively. As noted earlier, the three binary coefficients, D, D°, and D^V, are equal (Bird et al., 1960). For the general multicomponent case, the three matrices defined above are, in general, different from one another (as indicated in the next section). Cullinan (1965) has shown that the eigenvalues of $[D]$, $[D^\circ]$, and $[D^V]$ are, correspondingly, equal to one another. The eigenvalues of $[D]$ are the roots of the determinantal equation (cf. Eq. A.4.5)

$$\left| [D] - \hat{D}[I] \right| = 0 \tag{3.2.8}$$

For an n-component system, Eq. 3.2.8 reduces to an $(n-1)$th-order polynomial in $\hat{D}$, giving $n-1$ eigenvalues: $\hat{D}_1, \hat{D}_2, \ldots$. For a ternary system, Eq. 3.2.8 is a quadratic polynomial and the two roots $\hat{D}_1$ and $\hat{D}_2$ can be found from

$$\begin{aligned} \hat{D}_1 &= \tfrac{1}{2}\left\{\mathrm{tr}[D] + \sqrt{\mathrm{disc}[D]}\right\} \\ \hat{D}_2 &= \tfrac{1}{2}\left\{\mathrm{tr}[D] - \sqrt{\mathrm{disc}[D]}\right\} \end{aligned} \tag{3.2.9}$$

where

$$\mathrm{tr}[D] = D_{11} + D_{22}$$
$$|D| = D_{11}D_{22} - D_{12}D_{21}$$

are the trace and determinant of $[D]$. The term $\mathrm{disc}[D]$ is the discriminant of the determinantal polynomial (Eq. 3.2.8)

$$\mathrm{disc}[D] = (\mathrm{tr}[D])^2 - 4|D|$$

3.2.3 Multicomponent Fick Diffusion Coefficients

The Fick diffusion coefficients may be termed practical in the sense that the binary coefficient D and the corresponding multicomponent diffusion coefficients can be obtained from composition profiles measured in a diffusion apparatus. The measurement of binary and multicomponent diffusion coefficients, a subject with an extensive literature, is beyond the scope of this book. The interested reader is referred to Dunlop et al. (1972), Cussler (1976) and Tyrrell and Harris (1984) for descriptions of techniques and summaries of experimental results. Most experimental data are reported for $[D^V]$. This matrix must be

TABLE 3.2   Fick Diffusion Coefficients in the System
Acetone(1)–Benzene(2)–Methanol(3) at 25°C[a]

x_1	x_2	D_{11}^V	D_{12}^V	D_{21}^V	D_{22}^V
0.350	0.302	3.819	0.420	−0.561	2.133
0.766	0.114	4.400	0.921	−0.834	2.680
0.553	0.190	4.472	0.962	−0.480	2.569
0.400	0.500	4.434	1.866	−0.816	1.668
0.299	0.150	3.192	0.277	−0.191	2.368
0.206	0.548	3.513	0.665	−0.602	1.948
0.102	0.795	3.502	1.204	−1.130	1.124
0.120	0.132	3.115	0.138	−0.227	2.235
0.150	0.298	3.050	0.150	−0.269	2.250

[a]The units of D_{ij}^V are in 10^{-9} m^2/s.

transformed to $[D]$ or $[D°]$ in order for it to be useful in the applications we consider later in this book. More on this topic below.

To give an indication of the magnitude of the cross-coefficients that may sometimes be encountered in practice we present in Table 3.2 some of the data of Alimadadian and Colver (1976) for $[D^V]$ for the system acetone(1)–benzene(2)–methanol(3) at 25°C and, in Table 3.3, some of the data of Cullinan and Toor (1965) for the system acetone(1)–benzene(2)–carbon tetrachloride(3).

It is clear from this small selection of data that the matrix of multicomponent diffusion coefficients may be a complicated function of the composition of the mixture. The matrix $[D]$ is generally nonsymmetric, except for two special cases identified below. The cross coefficients D_{ik} ($i \neq k$) can be of either sign; indeed it is possible to alter the sign of these cross-coefficients by altering the numbering of the components.

There are circumstances where the matrix $[D]$ is diagonal and the diffusion flux of species i is independent of the composition gradients of the other species. For an ideal mixture made up of chemically similar species the matrix of diffusion coefficients degenerates to a scalar times the identity matrix, that is,

$$[D] = D[I] \quad \text{(special)} \tag{3.2.10}$$

The system toluene–chlorobenzene–bromobenzene is one where this simplification applies (Burchard and Toor, 1962).

TABLE 3.3   Fick Diffusion Coefficients in the System
Acetone(1)–Benzene(2)–Carbon Tetrachloride(3) at 25°C[a]

x_1	x_2	D_{11}^V	D_{12}^V	D_{21}^V	D_{22}^V
0.2989	0.3490	1.887	−0.213	−0.037	2.255
0.1496	0.1499	1.598	−0.058	−0.083	1.812
0.1497	0.6984	1.961	0.013	−0.149	1.929
0.6999	0.1497	2.330	−0.432	0.132	2.971
0.0933	0.8967	3.105	0.550	−0.780	1.860
0.2415	0.7484	3.069	0.603	−0.638	1.799
0.4924	0.4972	2.857	0.045	−0.289	2.471
0.7432	0.2466	3.251	−0.011	−0.301	2.896
0.8954	0.0948	3.475	−0.158	0.108	3.737

[a]The units of D_{ij}^V are in 10^{-9} m^2/s.

As the concentration of species i approaches zero, the off-diagonal elements D_{ik} $(i \neq k)$ also approach zero. Thus, for $n - 1$ components infinitely diluted in the nth $(x_i$ $(i = 1, \ldots, n - 1)$ close to 0), we find that all the cross-coefficients D_{ik} $(i \neq k)$ vanish. In this case, however, the diagonal elements D_{ii} are not necessarily equal to one another. Dilute solutions occur sufficiently often for this special case to be of some practical importance.

The prediction of the Fick matrix $[D]$ from fundamental data is considered in Chapter 4.

3.2.4   Transformation of Multicomponent Diffusion Coefficients from One Reference Velocity Frame to Another

To relate $[D]$ in the molar average velocity reference frame to the mass average reference frame $[D^\circ]$ we must use Eq. 1.2.26 (in matrix form) to transform Eq. 3.2.6 to the molar average velocity reference frame and Eq. 1.2.28 to change the units in which the flux is expressed. We also need the analogous relations to transform mass fraction gradients to mole fraction gradients (Exercise 1.5). The end result is the following similarity transformation

$$[D^\circ] = [B^{uo}]^{-1}[\omega][x]^{-1}[D][x][\omega]^{-1}[B^{uo}]$$

$$= [B^{ou}][\omega][x]^{-1}[D][x][\omega]^{-1}[B^{ou}]^{-1} \tag{3.2.11}$$

where $[x]$ is a diagonal matrix whose nonzero elements are the mole fractions x_i. The matrix $[\omega]$ is also diagonal with nonzero elements that are the mass fractions ω_i. Since $[x]$ and $[\omega]$ are diagonal matrices, their inverses are easy to compute $[x]^{-1}$, for example, is diagonal with elements that are the reciprocals of the mole fractions: $1/x_i$. The matrices $[B^{uo}]$ and $[B^{ou}]$ have elements defined by Eqs. 1.2.25 and 1.2.27, respectively.

$$B_{ik}^{uo} = \delta_{ik} - \omega_i\left(\frac{x_k}{\omega_k} - \frac{x_n}{\omega_n}\right) \tag{1.2.25}$$

$$B_{ik}^{ou} = \delta_{ik} - \omega_i\left(1 - \frac{\omega_n x_k}{x_n \omega_k}\right) \tag{1.2.27}$$

It is interesting to note that the matrix $[B^{ou}]$ is the inverse of $[B^{uo}]$; that is, $[B^{ou}] = [B^{uo}]^{-1}$ as may be proved using the Sherman–Morrison formula (see Ortega and Rheinbolt, 1970, p. 50 and Exercise 1.2).

To relate $[D]$ to the volume average velocity reference frame $[D^V]$ we use another similarity transformation

$$[D^V] = [B^{Vu}][D][B^{Vu}]^{-1}$$

$$= [B^{Vu}][D][B^{uV}] \tag{3.2.12}$$

where the matrices $[B^{Vu}]$ and $[B^{uV}]$ have elements

$$B_{ik}^{Vu} = \delta_{ik} - x_i(\bar{V}_k - \bar{V}_n)/\bar{V}_t \tag{1.2.23}$$

$$B_{ik}^{uV} = \delta_{ik} - x_i(1 - \bar{V}_k/\bar{V}_n) \tag{1.2.21}$$

The fact that $[B^{uV}] = [B^{Vu}]^{-1}$ may be proved using the Sherman–Morrison formula.

It follows from Eqs. 3.2.11 and 3.2.12 and $[D]$, $[D^\circ]$, and $[D^V]$ will not, in general, be equal.

The proof that the eigenvalues of the three $[D]$ matrices are equal follows immediately from Eqs. 3.2.11 and 3.2.12. The eigenvalues of two matrices, $[A]$ and $[B]$ say, that are related by a nonsingular similarity transformation $[A] = [P]^{-1}[B][P]$ are equal (see, e.g., Amundson, 1966; Appendix A.4). The equality of the three binary diffusion coefficients defined by Eqs. 3.1.1, 3.1.4, and 3.1.6 also follows directly from Eqs. 3.2.11 and 3.2.12.

Example 3.2.1 *Fick Diffusion Coefficients for the System Acetone–Benzene–Methanol*

The Fick diffusion coefficients for the system acetone(1)–benzene(2)–methanol(3) in the volume average reference velocity frame are given in Table 3.2. Calculate the elements of $[D]$ in the molar average reference velocity frame.

DATA   The partial molar volumes $\bar{V}_i$ for acetone, benzene, and methanol are

$$\bar{V}_1 = 74.1 \times 10^{-6} \text{ m}^3/\text{mol}$$

$$\bar{V}_2 = 89.4 \times 10^{-6} \text{ m}^3/\text{mol}$$

$$\bar{V}_3 = 40.7 \times 10^{-6} \text{ m}^3/\text{mol}$$

SOLUTION   The matrix $[D]$ is related to $[D^V]$ by the inverse of Eq. 3.2.12. Thus, the first step is the calculation of the transformation matrix $[B^{Vu}]$ and its inverse $[B^{uV}]$ from Eqs. 1.2.23 and 1.2.21, respectively. For the first line of data in Table 3.2 $\bar{V}_t$ is found to be

$$\bar{V}_t = x_1\bar{V}_1 + x_2\bar{V}_2 + x_3\bar{V}_3$$

$$= 0.350 \times 74.1 \times 10^{-6} + 0.302 \times 89.4 \times 10^{-6} + 0.348 \times 40.7 \times 10^{-6}$$

$$= 67.1 \times 10^{-6} \text{ m}^3/\text{mol}$$

The elements of $[B^{Vu}]$ follow as:

$$B_{11}^{Vu} = 1 - x_1(\bar{V}_1 - \bar{V}_3)/\bar{V}_t$$

$$= 1 - 0.350 \times (74.1 - 40.7)/67.1$$

$$= 0.8258$$

$$B_{12}^{Vu} = -x_1(\bar{V}_2 - \bar{V}_3)/\bar{V}_t$$

$$= -0.350 \times (89.4 - 40.7)/67.1$$

$$= -0.2540$$

$$B_{21}^{Vu} = -x_2(\bar{V}_1 - \bar{V}_3)/\bar{V}_t$$

$$= -0.302 \times (74.1 - 40.7)/67.1$$

$$= -0.1503$$

$$B_{22}^{Vu} = 1 - x_2(\bar{V}_2 - \bar{V}_3)/\bar{V}_t$$

$$= 1 - 0.302 \times (89.4 - 40.7)/67.1$$

$$= 0.7808$$

The inverse of this matrix can be calculated directly or from Eqs. 1.2.21 as follows:

$$B_{11}^{uV} = 1 - x_1\left(1 - \bar{V}_1/\bar{V}_3\right)$$
$$= 1 - 0.350 \times (1 - 74.1/40.7)$$
$$= 1.2872$$
$$B_{12}^{uV} = -x_1\left(1 - \bar{V}_2/\bar{V}_3\right)$$
$$= -0.350 \times (1 - 89.4/40.7)$$
$$= 0.4188$$
$$B_{21}^{uV} = -x_2\left(1 - \bar{V}_1/\bar{V}_3\right)$$
$$= -0.302 \times (1 - 74.1/40.7)$$
$$= 0.2478$$
$$B_{22}^{uV} = 1 - x_2\left(1 - \bar{V}_2/\bar{V}_3\right)$$
$$= 1 - 0.302 \times (1 - 89.4/40.7)$$
$$= 1.3614$$

Now, we can compute $[D]$ directly from the inverse of Eq. 3.2.12

$$[D] = [B^{Vu}]^{-1}[D^V][B^{Vu}]$$
$$= [B^{uV}][D^V][B^{Vu}]$$

The result is

$$[D] = \begin{bmatrix} 3.651 & -0.069 \\ -0.300 & 2.303 \end{bmatrix} \times 10^{-9} \text{ m}^2/\text{s}$$

It is interesting to note the change in sign of the cross-coefficients D_{12}^V and D_{12}.
The eigenvalues of $[D]$ may be computed from Eq. 3.2.9 as follows:

$$\text{tr}[D] = D_{11} + D_{22}$$
$$= 3.651 \times 10^{-9} + 2.303 \times 10^{-9}$$
$$= 5.954 \times 10^{-9} \text{ m}^2/\text{s}$$
$$|D| = D_{11}D_{22} - D_{12}D_{21}$$
$$= 3.651 \times 10^{-9} \times 2.303 \times 10^{-9} - (-0.069 \times 10^{-9}) \times (-0.300 \times 10^{-9})$$
$$= 8.388 \times 10^{-18} \text{ m}^4/\text{s}^2$$

We will also need to evaluate the discriminant

$$\text{disc}[D] = (\text{tr}[D])^2 - 4|D|$$
$$= (5.954 \times 10^{-9})^2 - 4 \times 8.388 \times 10^{-18}$$
$$= 1.898 \times 10^{-18} \text{ m}^4/\text{s}^2$$

The eigenvalues of $[D]$ may now be evaluated as

$$\hat{D}_1 = \tfrac{1}{2}\{\text{tr}[D] + \sqrt{\text{disc}[D]}\}$$
$$= \tfrac{1}{2}\{5.954 \times 10^{-9} + \sqrt{1.898 \times 10^{-18}}\}$$
$$= 3.665 \times 10^{-9} \text{ m}^2/\text{s}$$
$$\hat{D}_2 = \tfrac{1}{2}\{\text{tr}[D] - \sqrt{\text{disc}[D]}\}$$
$$= \tfrac{1}{2}\{5.954 \times 10^{-9} - \sqrt{1.898 \times 10^{-18}}\}$$
$$= 2.287 \times 10^{-9} \text{ m}^2/\text{s}$$

We leave it as an exercise for our readers to calculate the eigenvalues of $[D^V]$. These values should be equal to the eigenvalues of $[D]$. ■

3.3 IRREVERSIBLE THERMODYNAMICS AND THE GENERALIZED FICK'S LAW

Support for the form of the generalized Fick's law of the preceding section can be found in irreversible thermodynamics introduced in Section 2.3. The treatment below follows that of Stewart and Prober (1964), but readers will also find the books by de Groot and Mazur (1962) and Haase (1969), and the review of Kirkaldy (1970) very useful.

The starting point for our analysis is an expression for the rate of entropy production per unit volume. For isothermal, isobaric processes in the absence of external force fields, the rate of entropy production due to diffusion is given by (cf. Eq. 2.3.1)

$$T\sigma_{\text{diff}} = -\sum_{i=1}^{n} \nabla_{T,p}\mu_i \cdot J_i \geq 0 \tag{3.3.1}$$

where J_i is the molar diffusion flux with respect to the molar average reference velocity. The rate of entropy production σ is seen to be a sum of scalar, or dot, products of two quantities; one of these is the diffusion flux and the other, the chemical potential gradient, may be interpreted as the "driving force" for diffusion. The second law of thermodynamics requires σ to be positive definite, $0 \geq 0$. In terms of independent fluxes and forces, Eq. 3.3.1 takes the form

$$T\sigma_{\text{diff}} = -\sum_{i=1}^{n-1} \nabla_{T,P}(\mu_i - \mu_n) \cdot J_i \geq 0 \tag{3.3.2}$$

or in $n - 1$ dimensional matrix notation

$$T\sigma_{\text{diff}} = -\nabla_{T,P}(\mu - \mu_n)^T \cdot (J) \tag{3.3.3}$$

At equilibrium both the fluxes and the driving forces vanish simultaneously giving

$$\sigma = 0 \quad |\text{equilibrium}| \tag{3.3.4}$$

We now postulate a linear relationship between independent fluxes and driving forces

$$c_t \nabla_{T,P}(\mu_i - \mu_n) = -\sum_{k=1}^{n-1} H_{ik} J_k \tag{3.3.5}$$

Equation 3.3.5 can be written in matrix form as

$$-c_t \nabla_{T,P}(\mu - \mu_n) = [H](J) \tag{3.3.6}$$

where $[H]$ is the Onsager matrix of coefficients. The matrix $[H]$ is positive definite because of the second law restriction $\sigma \geq 0$.

A second postulate of irreversible thermodynamics is that the matrix $[H]$ is symmetric

$$[H] = [H]^T \quad \text{or} \quad H_{ik} = H_{ki} \tag{3.3.7}$$

Equation 3.3.7 expresses the Onsager reciprocal relations (ORR), named after Lars Onsager who first established the principles of irreversible thermodynamics (Onsager, 1931). The ORR have been the subject of many journal papers receiving support as well as criticism, the latter from, in particular, Coleman and Truesdell (1960) and Truesdell (1969). We shall assume the validity of the ORR in the development that follows.

The gradients $\nabla_{T,P}(\mu_i - \mu_n)$ can be expressed in terms of the mole fraction gradients $\nabla(x)$ by the relation

$$\nabla_{T,P}(\mu - \mu_n) = [G]\nabla(x) \tag{3.3.8}$$

where $[G]$ is the Hessian matrix of the Gibbs free energy function; its elements G_{ij} are given by

$$G_{ij} = \frac{\partial^2 G}{\partial x_i \partial x_j} = \frac{\partial(\mu_i - \mu_n)}{\partial x_j} = \frac{\partial(\mu_j - \mu_n)}{\partial x_i} = G_{ji} \tag{3.3.9}$$

The matrix $[G]$ is symmetric. Its elements may be obtained from activity coefficient models in much the same way that the matrix $[\Gamma]$ is obtained. Expressions for the G_{ij} for some models of the excess Gibbs energy are given in Appendix D.

On combining Eqs. 3.3.6 and 3.3.8 we see that

$$(J) = -c_t[H]^{-1}[G](\nabla x) \tag{3.3.10}$$

Comparison with Eq. 3.2.5 shows that the Fick matrix $[D]$ is related to the Onsager coefficient matrix $[H]$ by

$$[D] = [H]^{-1}[G] \tag{3.3.11}$$

Now, the matrix $[H]$ is symmetric (from the ORR) and positive definite (from the second law requirement that $\sigma_{\text{diff}} \geq 0$). In addition, for a thermodynamically stable fluid the matrix $[G]$ is symmetric and positive definite (see Section 3.3.1). The implication of Eq. 3.3.11 is that the matrix $[D]$ is positive definite, so that all its eigenvalues are real and positive (Kirkaldy, 1970).

The condition for real and positive eigenvalues $\hat{D}_1$ and $\hat{D}_2$, for a ternary system can be expressed as (Kirkaldy, 1970; Yao, 1966, Eqs. 3.2.9)

$$D_{11} + D_{22} > 0$$
$$D_{11}D_{22} - D_{12}D_{21} > 0 \tag{3.3.12}$$
$$(D_{11} - D_{22})^2 + 4D_{12}D_{21} > 0$$

It is interesting to note that thermodynamic stability considerations do not require the diagonal elements D_{11} and D_{22} to be individually positive. If recourse is made to the kinetic theory of gases, it can be shown that the main coefficients are individually positive, that is,

$$D_{11} > 0 \qquad D_{22} > 0 \qquad\qquad (3.3.13)$$

All available experimental measurements of the D_{ik} suggest the general validity of requirement (Eq. 3.3.13) (see Cussler, 1976; Dunlop et al., 1972; Tyrrell and Harris, 1984).

Example 3.3.1 Calculation of the Onsager Coefficients

The matrix of Fick diffusion coefficients in the molar average reference velocity frame for the system acetone(1)–benzene(2)–carbon tetrachloride(3) at a temperature of 25°C and composition $x_1 = 0.70$, $x_2 = 0.15$, $x_3 = 0.15$ has been obtained from the experimental data of Cullinan and Toor (1965) as

$$[D] = \begin{bmatrix} 2.354 & -0.471 \\ 0.089 & 2.947 \end{bmatrix} \times 10^{-9} \text{ m}^2/\text{s}$$

Calculate the matrix of Onsager coefficients $[H]$.

DATA The Hessian matrix of the Gibbs free energy $[G]$ may be calculated with the nonrandom two liquid (NRTL) model. The NRTL parameters are

$$\tau_{12} = -0.46504 \qquad \tau_{21} = 0.76432 \qquad \alpha_{12} = \alpha_{21} = 0.2$$
$$\tau_{13} = -0.42790 \qquad \tau_{31} = 1.5931 \qquad \alpha_{13} = \alpha_{31} = 0.2$$
$$\tau_{23} = -0.51821 \qquad \tau_{32} = 0.7338 \qquad \alpha_{23} = \alpha_{32} = 0.2$$

SOLUTION The Hessian $[G]$ is calculated from Eq. D.3.4 and the NRTL model equations in Table D.8. The result is

$$[G]/RT = \begin{bmatrix} 6.856 & 6.229 \\ 6.229 & 13.323 \end{bmatrix}$$

Note that $[G]/RT$ is dimensionless and symmetric; also, the cross-coefficients G_{12} and G_{21} are a large fraction of the main coefficients G_{11} and G_{22}. Multiplying each element of the above matrix by RT gives

$$[G] = \begin{bmatrix} 16.995 & 15.441 \\ 15.441 & 33.026 \end{bmatrix} \times 10^2 \text{ J/mol}$$

Equation 3.3.11 can be rearranged to

$$[H][D] = [G]$$

which we solved directly to give

$$[H] = \begin{bmatrix} 6.981 & 6.355 \\ 6.099 & 12.18 \end{bmatrix} \times 10^{12} \text{ J s/mol m}^2$$

Note that the ORR are not satisfied precisely pointing to experimental inaccuracies in the measured data. ∎

3.3.1 Diffusion in the Region of a Critical Point

Consider diffusion in a binary liquid mixture exhibiting an upper critical solution temperature (UCST) or lower critical solution temperature (LCST) (see Fig. 3.1). Let us take a mixture at the "critical" composition x_{ic} at point A just above the UCST. Any concentration fluctuation at A will tend to be "smeared" out due to the effects of diffusion in this homogeneous mixture. On the other hand, any fluctuation of a system at point B, infinitesimally below the UCST, will lead to separation in two phases. Similarly, the mixture at point D, just below the LCST is stable whereas the mixture at point C, just above the LCST is unstable and will separate into two phases.

For thermodynamically stable binary systems the second derivative of the Gibbs free energy with respect to the mole fraction x_1 is positive

$$G_{11} \equiv \frac{\partial^2 G}{\partial x_1^2} > 0$$

while

$$G_{11} < 0$$

implies thermodynamic instability, that is, phase splitting. The locus of points where G_{11} goes to 0 is the spinodal curve and is the boundary between the metastable and unstable regions. At the critical point itself both G_{11} and the third derivative of G with respect to x_1 are equal to zero. For a detailed discussion of these points see, for example, Modell and Reid (1983).

It follows that the Fick diffusion coefficient must tend towards zero as the spinodal curve is approached. This has been experimentally confirmed for a few systems, the data of Haase and Siry (1968) for the systems water–triethylamine and n-hexane–nitrobenzene are shown in Figs. 3.2 and 3.3 (see, also, Claesson and Sundelöf, 1957; Myerson and Senol, 1984). Vitagliano et al. (1980) and Clark and Rowley (1986) determined spinodal compositions by extrapolating diffusivity data to zero.

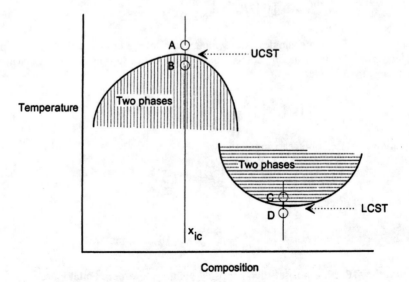

Figure 3.1. Upper and lower critical solution points.

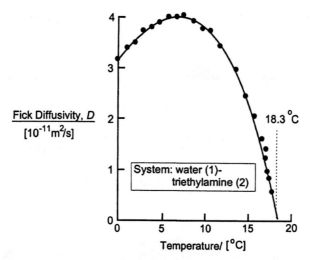

Figure 3.2. Fick diffusion coefficient D as a function of temperature for the system water–triethyl-amine. Measured data for Fick diffusivity D at constant composition = critical composition $x_2 = 0.0874$. Critical temperature = 18.3°C. Data from Haase and Siry (1968).

For thermodynamic stability in a multicomponent system the matrix $[G]$ must be positive definite. Thus

$$|G| > 0 \qquad\qquad (3.3.14)$$

and each of the eigenvalues of $[G]$ is positive definite. Negative eigenvalues of $[G]$ imply thermodynamic instability; that is, phase splitting. At the critical point

$$|G| = 0 \qquad\qquad (3.3.15)$$

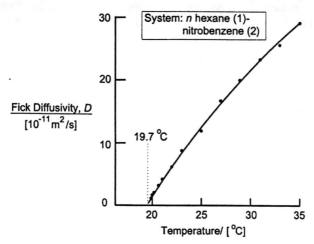

Figure 3.3. Fick diffusion coefficient as a function of temperature for the system n-hexane–nitrobenzene. Measured data for Fick diffusivity D at constant composition $x_2 = 0.42$. Critical temperature = 19.7°C. Data from Haase and Siry (1968).

Now, the determinant of $[D]$ is related to the determinant of $[G]$ by

$$|D| = |G|/|H| \tag{3.3.16}$$

Thus, at the critical point we must have

$$|D| = 0 \quad \text{(critical point)} \tag{3.3.17}$$

and one of the eigenvalues of $[D]$ must vanish.

For a ternary system the requirement $|D| = 0$ at the critical point implies that

$$|D| = D_{11}D_{22} - D_{12}D_{21} = 0 \tag{3.3.18}$$

and the eigenvalues of $[D]$ are (see Eq. 3.2.9)

$$\hat{D}_1 = 0 \qquad \hat{D}_2 = D_{11} + D_{22} \tag{3.3.19}$$

If $|D| = 0$, the cross-coefficients must match the main coefficients in magnitude—implying large coupling effects.

The eigenvectors corresponding to the eigenvalues (Eq. 3.3.19) are, respectively,

$$(e_1) = \begin{pmatrix} 1 \\ -\dfrac{D_{11}}{D_{12}} \end{pmatrix} \quad \text{and} \quad (e_2) = \begin{pmatrix} 1 \\ -\dfrac{D_{22}}{D_{12}} \end{pmatrix} \tag{3.3.20}$$

We shall now show that the first eigenvector (e_1) is in a direction parallel to the limiting tie line in the vicinity of P. To do this, we begin with the condition of equilibrium between two phases $'$ and $''$ for which the following relation must be satisfied

$$\mu_i(x_1', x_2') = \mu_i(x_1'', x_2'') \qquad i = 1, 2, 3 \tag{3.3.21}$$

Consider now two phases, also in mutual equilibrium, the compositions of which differ from the corresponding phases above by infinitesimal amounts. Considered as variations of the above, the chemical potentials of these new phases must satisfy

$$d\mu_i' = d\mu_i'' \qquad i = 1, 2, 3 \tag{3.3.22}$$

The Gibbs–Duhem restriction for constant temperature and pressure can be written for a ternary system as

$$x_1'd\mu_1' + x_2'd\mu_2' + x_3'd\mu_3' = x_1''d\mu_1'' + x_2''d\mu_2'' + x_3''d\mu_3'' \tag{3.3.23}$$

The mole fraction x_3 can be eliminated using $x_3 = 1 - x_1 - x_2$ to give

$$d(\mu_i' - \mu_3') = d(\mu_i'' - \mu_3'') \qquad i = 1, 2 \tag{3.3.24}$$

and

$$\begin{aligned} x_1'd(\mu_1' - \mu_3') &+ x_2'd(\mu_2' - \mu_3') \\ &= x_1''d(\mu_1'' - \mu_3'') + x_2''d(\mu_2'' - \mu_3'') \end{aligned} \tag{3.3.25}$$

which relations can be rearranged in the form

$$\frac{x_2'' - x_2'}{x_1'' - x_1'} = -\frac{d(\mu_1' - \mu_3')}{d(\mu_2' - \mu_3')} = -\frac{d(\mu_1'' - \mu_3'')}{d(\mu_2'' - \mu_3'')} \tag{3.3.26}$$

The differentials $d(\mu_i - \mu_3)$ may be written in terms of composition fluctuations as

$$d(\mu - \mu_3) = [G]d(x) \tag{3.3.27}$$

and so Eqs. 3.3.22, written in terms of dx_i, take the form

$$[G']d(x') = [G'']d(x'') \tag{3.3.28}$$

and

$$\frac{x_2'' - x_2'}{x_1'' - x_1'} = -\frac{G_{11}'dx_1' + G_{12}'dx_2'}{G_{21}'dx_1' + G_{22}'dx_2'} = -\frac{G_{11}''dx_1'' + G_{12}''dx_2''}{G_{21}''dx_1'' + G_{22}''dx_2} \tag{3.3.29}$$

with $G_{12} = G_{21}$.

As the critical point is approached, the various $[G]$ matrices become equal

$$[G'] = [G''] = [G] \tag{3.3.30}$$

and

$$dx_i' = dx_i'' = dx_i \tag{3.3.31}$$

so Eq. 3.3.29 takes the limiting form

$$\frac{dx_2}{dx_1} = -\frac{G_{11} + G_{12}(dx_2/dx_1)}{G_{12} + G_{22}(dx_2/dx_1)} \tag{3.3.32}$$

Solving the resultant quadratic equation for dx_2/dx_1 we obtain

$$\frac{dx_2}{dx_1} = \frac{G_{12}}{G_{22}} + \frac{G_{12}^2 - G_{11}G_{22}}{G_{22}} = -\frac{G_{12}}{G_{22}} = -\frac{G_{11}}{G_{22}} \tag{3.3.33}$$

because $G_{11}G_{22} - G_{12}^2 = 0$ at P. The derivative dx_2/dx_1 represents the slope of the limiting tie-line.

Now using the relation $[D] = [H]^{-1}[G]$ we can show that

$$\frac{D_{21}}{D_{22}} = -\frac{D_{11}}{D_{12}} = -\frac{G_{11}}{G_{22}} \tag{3.3.34}$$

after invoking the requirement of $|H| = 0$ at equilibrium. Since the slope of the eigenvector (e_1) is $-D_{11}/D_{12}$ (cf. Eq. 3.3.35), it follows that (e_1) is parallel to the limiting tie-line.

The implications of $|D| = 0$ at the plait point for composition trajectories in this region will be discussed later (Section 5.6.2).

Vitagliano et al. (1978) attempted to test Eq. 3.3.19 with the system water–chloroform–acetic acid; their data are summarized on the triangular diagram (Fig. 3.4). The first five

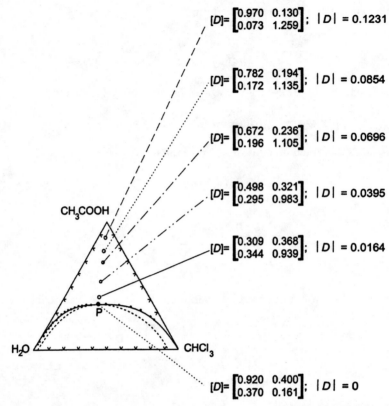

$$[D]= \begin{bmatrix} 0.970 & 0.130 \\ 0.073 & 1.259 \end{bmatrix}; \quad |D| = 0.1231$$

$$[D]= \begin{bmatrix} 0.782 & 0.194 \\ 0.172 & 1.135 \end{bmatrix}; \quad |D| = 0.0854$$

$$[D]= \begin{bmatrix} 0.672 & 0.236 \\ 0.196 & 1.105 \end{bmatrix}; \quad |D| = 0.0696$$

$$[D]= \begin{bmatrix} 0.498 & 0.321 \\ 0.295 & 0.983 \end{bmatrix}; \quad |D| = 0.0395$$

$$[D]= \begin{bmatrix} 0.309 & 0.368 \\ 0.344 & 0.939 \end{bmatrix}; \quad |D| = 0.0164$$

$$[D]= \begin{bmatrix} 0.920 & 0.400 \\ 0.370 & 0.161 \end{bmatrix}; \quad |D| = 0$$

Figure 3.4. Phase diagram of the system water–chloroform–acetic acid at 25°C. Diffusion coefficient data are expressed in the units 10^{-9} m^2/s. Also shown in the figure are values of the determinant of $[D]$. Data from Vitagliano et al. (1978).

data sets fall on a more or less straight line that intersects the two-phase boundary at the plait point. Note that the cross-coefficients do indeed match the main coefficients in magnitude. The last point on Figure 3.4 represents extrapolated data for the plait point. It is very difficult to measure multicomponent diffusion coefficients at the plait point because the accuracy of most measurement techniques decreases as the two-phase boundary is approached.

4 Estimation of Diffusion Coefficients

The coefficient of interdiffusion of two liquids must be considered as depending on all the physical properties of the mixture according to laws which must be ascertained only by experiment.
—J. C. Maxwell writing in the Encyclopedia Brittanica. See his collected papers (1952).

Thus far we have introduced two different constitutive relations along with their respective diffusion coefficients, the Maxwell–Stefan $Đ$ and the Fick D. Here we show how these coefficients are related to each other and present a sample of values of these coefficients determined experimentally.

For process engineering calculations it is almost inevitable that experimental values of D or $Đ$, even if available in the literature, will not cover the entire range of temperature, pressure, and concentration that is of interest in any particular application. It is, therefore, important that we be able to predict these coefficients from fundamental physical and chemical data, such as molecular weights, critical properties, and so on. Estimation of gaseous diffusion coefficients at low pressures is the subject of Section 4.1.1, the correlation and prediction of binary diffusion coefficients in liquid mixtures is covered in Sections 4.1.3–4.1.5. We do not intend to provide a comprehensive review of prediction methods since such are available elsewhere (Reid et al., 1987; Ertl et al., 1974; Danner and Daubert, 1983); rather, it is our purpose to present a selection of methods that may be useful in engineering calculations.

While the thermodynamic treatments of diffusion in Sections 2.3 and 3.3 provide some useful information on the multicomponent diffusion coefficients, it does not solve our most important problem, how do we predict these coefficients? Multicomponent diffusivity data is not an item that we have in abundance and there are no correlations of multicomponent diffusivity data that we might use. It is the Maxwell–Stefan Eqs. 2.1.24 for ideal gases or Eq. 2.2.9 for nonideal fluids that come to our aid (Section 4.2).

4.1 DIFFUSION COEFFICIENTS IN BINARY MIXTURES

4.1.1 Relationship Between Fick and Maxwell–Stefan Diffusion Coefficients

The Maxwell–Stefan equation for diffusion in a two component system is Eq. 2.2.11.

$$J_1 = -c_t B^{-1} \Gamma \nabla x_1 = -c_t Đ \Gamma \nabla x_1 \qquad (2.2.11)$$

where Γ is given by Eq. 2.2.12. Equation 2.2.11 is to be compared to Fick's law (Eq. 3.1.1)

$$J_1 = -c_t D \nabla x_1 \qquad (3.1.1a)$$

We see that, for a binary system, the Fick diffusivity D and the Maxwell–Stefan diffusivity

$Ð$ are related by

$$D = B^{-1}\Gamma = Ð\Gamma \qquad (4.1.1)$$

The correlation and prediction of Fick and Maxwell–Stefan diffusion coefficients is discussed in the sections that follow. The Fick D incorporates two aspects: (1) the significance of an inverse drag ($Ð$) and (2) thermodynamic nonideality (Γ). Consequently, the physical interpretation of the Fick D is less transparent than for the Maxwell–Stefan diffusivity.

For ideal systems Γ is unity and the Fick D and the Maxwell–Stefan $Ð$ are identical.

$$D = B^{-1} = Ð \qquad |\text{ideal}| \qquad (4.1.2)$$

4.1.2 Estimation of Diffusion Coefficients in Gas Mixtures

A more rigorous kinetic theory than that in Chapter 2 not only supplies us with the proper form of the constitutive relations for multicomponent diffusion, it also provides an explicit relation for the binary diffusion coefficient. A slightly simplified version of the kinetic theory result is

$$D = CT^{3/2}\frac{\sqrt{\{(M_1 + M_2)/M_1M_2\}}}{P\sigma_{12}^2\Omega_D} \qquad (4.1.3)$$

where

D = diffusion coefficient $[\text{m}^2/\text{s}]$
C = 1.883×10^{-2}
T = absolute temperature [K]
P = pressure [Pa]
σ = characteristic length [Å]
Ω_D = diffusion collision integral $[-]$
M_i = molar mass of component i [g/mol]

The parameter Ω_D, the diffusion collision integral, is a function of k_BT/ε, where k_B is the Boltzmann constant and ε is a molecular energy parameter. Values of Ω_D, tabulated as a function of k_BT/ε, have been published (Hirschfelder et al., 1964; Bird et al., 1960). Neufeld et al., (1972) correlated Ω_D using a simple eight parameter equation that is suitable for computer calculations (see, also, Danner and Daubert, 1983; Reid et al., 1987). Values of σ and ε/k_B (which has units of kelvin) can be found in the literature—for only a few species—or estimated from critical properties (Reid et al., 1987; Danner and Daubert, 1983). The mixture σ is calculated as the arithmetic average of the pure component values. The mixture ε is taken to be the geometric average of the pure component values.

A number of empirical or semiempirical correlations for estimating gaseous diffusion coefficients have also been developed. These include the method of Wilke and Lee (1955), which is based on Eq. 4.1.3 with $C = 0.02199 - 0.00507\sqrt{\{(M_1 + M_2)/M_1M_2\}}$.

A correlation due to Fuller et al. (1966, 1969) is recommended by Reid et al. (1987) and by Danner and Daubert (1983).

$$D = CT^{1.75}\frac{\sqrt{\{(M_1 + M_2)/M_1M_2\}}}{P\left\{\sqrt[3]{V_1} + \sqrt[3]{V_2}\right\}^2} \qquad (4.1.4)$$

TABLE 4.1   Diffusion Volumes in Fuller–Schettler–Giddings Correlation Parameters

Atomic and Molecular Diffusion Volume Increments

C	15.9	F	14.7
H	2.31	Cl	21.0
O	6.11	Br	21.9
N	4.54	I	29.8
S	22.9	Aromatic ring	-18.3
		Heterocyclic ring	-18.3

Diffusion Volumes for Some Simple Molecules

He	2.67	CO	18.0
Ne	5.98	CO_2	26.7
Ar	16.2	N_2O	35.9
Kr	24.5	NH_3	20.7
Xe	32.7	H_2O	13.1
H_2	6.12	SF_6	71.3
D_2	6.84	Cl_2	38.4
N_2	18.5	Br_2	69.0
O_2	16.3	SO_2	41.8
Air	19.7		

[a] From Fuller et al. (1969).

With T in kelvin (K), P in pascals (Pa), M_1 and M_2 in grams per mole (g/mol) and $C = 1.013 \times 10^{-2}$, D will be in square meters per second (m^2/s). The terms V_1 and V_2 are molecular diffusion volumes and are calculated by summing the atomic contributions in Table 4.1.

Examples illustrating the use of Eq. 4.1.1 and 4.1.2 are given by Reid et al. (1977, 1987) and by Danner and Daubert (1983). The same authors describe methods for estimating Fick diffusion coefficients for gases at high pressure.

4.1.3   Diffusion Coefficients in Binary Liquid Mixtures

Diffusion coefficients in binary liquid mixtures can be strong functions of composition. To illustrate this fact we have plotted experimental data for a few systems in Figure 4.1. The Maxwell–Stefan coefficient $Đ$ also is shown in Figure 4.1. To obtain the Maxwell–Stefan coefficients we have divided the Fick D by the thermodynamic factor Γ

$$Đ = D/\Gamma \qquad (4.1.5)$$

We have calculated Γ using the activity coefficient models in Appendix D with parameters from the literature (see Examples 4.1.2 and 4.1.3).

It is clear from these figures that the Fick D shows a significantly greater variation with concentration than does the Maxwell–Stefan $Đ$. A particularly extreme example of the strong composition dependence of D is afforded by the system methanol–n-hexane in the vicinity of the spinodal curve. The experimental data for this system (obtained by Clark and Rowley, 1986) are plotted in Figure 4.1(d); the values of D vary by a factor of almost 20! The Maxwell–Stefan $Đ$, calculated from Eq. 4.1.5 varies by a factor of only 1.5.

To explain the rather striking difference in the behavior of the Fick D and Maxwell–Stefan $Đ$ we recall from Section 3.3.1 that the Fick D vanishes along the spinodal curve because G_{11}, the Hessian of the Gibbs free energy, vanishes on that curve. Now, the thermodynamic factor Γ may be expressed in terms of G_{11} as

$$\Gamma = x_1 x_2 G_{11}/RT \qquad (4.1.6)$$

It follows that the thermodynamic factor Γ vanishes wherever G_{11} goes to 0. The parameter Γ for the methanol–n-hexane system, calculated from the NRTL model given in Table D.5 using parameters given by Clark and Rowley (1986), is shown in Figure 4.2. It can be seen that the thermodynamic factor for this system comes close to zero at a mole fraction of methanol of 0.52. The composition dependence of Γ closely follows that of the Fick D (compare the shapes of the curves for D in Figure 4.1(d) and Γ in Figure 4.2) with the result that the Maxwell–Stefan $Đ$ ($= D/\Gamma$) is much less concentration dependent.

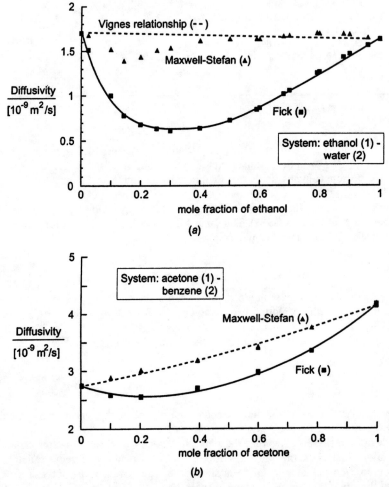

(a)

(b)

Figure 4.1. (a) Concentration dependence of the Fick diffusivity D and the Maxwell–Stefan $Đ$ for the system ethanol(1)–water(2). Data from Tyn and Calus (1975b). (b) Composition dependence of Fick D and Maxwell–Stefan $Đ$ for the system acetone(1)–benzene(2). Data from Anderson et al. (1958) and Cullinan and Toor (1965). (c) Composition dependence of Fick D and Maxwell–Stefan $Đ$ for diffusion in triethylamine(1)–water(2). Data from Dudley and Tyrell (1973). (d) Fick diffusion coefficient for the system methanol–n-hexane at 40°C measured by Clark and Rowley (1986).

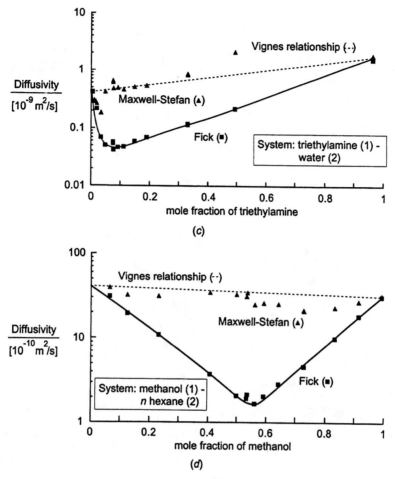

(c)

(d)

Figure 4.1. (*Continued*).

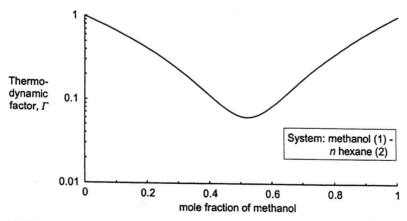

Figure 4.2. Thermodynamic factor Γ for the system methanol–n-hexane at 40°C. The parameter Γ computed from the NRTL model with parameters from Clark and Rowley (1986).

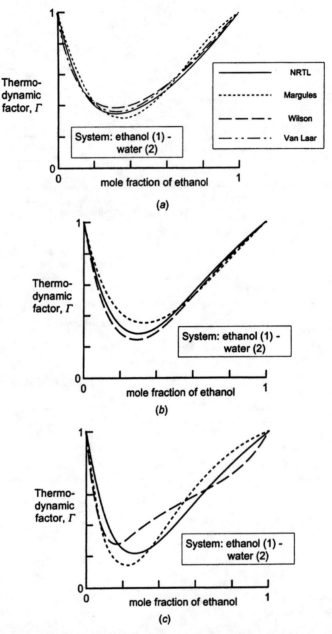

Figure 4.3. (*a*) Thermodynamic factor for the system ethanol–water at 40°C obtained from different activity coefficient models. Parameters from Gmehling and Onken (1977ff Vol. I/la p. 133). (*b*) Thermodynamic factor for the system ethanol–water at 50°C obtained using the NRTL equation using parameters fitted to isothermal vapor–liquid equilibrium data. Parameters from Gmehling and Onken (1977ff): —— Vol. I/la p. 116; ··· Vol. I/1 p. 191; --- Vol. I/1 p. 171. (*c*) Thermodynamic factor for the system ethanol–water at 40°C obtained using the NRTL equation using parameters fitted to isobaric vapor–liquid equilibrium data. Parameters from Gmehling and Onken (1977ff): —— Vol. I/la p. 133; ··· Vol. I/la p. 138; ---; Vol. I/1 p. 162.

It should be noted that the Maxwell–Stefan $Ð$ calculated from Eq. 4.1.5 can be quite sensitive to the model used to compute Γ, an observation first made by Dullien (1971). One of the reasons for this sensitivity is that Γ involves the first derivative of the activity coefficient with respect to composition. Activity coefficient model parameters are fitted to vapor–liquid equilibrium (VLE) data (see, e.g., Prausnitz et al., 1980; Gmehling and Onken, 1977). Several models may provide estimates of ln γ_i that give equally good fits of the vapor–liquid equilibrium data but that does not mean that the first derivatives of ln γ_i (and, hence, Γ) will be all that close. To illustrate this fact we have calculated the thermodynamic factor, Γ, for the system ethanol–water with several different models of ln γ_i. The results are shown in Figure 4.3 (a). The interaction parameters used in these calculations were fitted to one set of VLE data as identified in the figure caption. Similar illustrations for other systems are provided by Taylor and Kooijman (1991).

The Maxwell–Stefan diffusion coefficients calculated from Eq. 4.1.5 may also be sensitive to the parameters used in the calculation of Γ. Different sets of parameters obtained by fitting different sets of equilibrium data may give quite different values of Γ (and, hence, of $Ð$) as shown in Figures 4.3(b) and (c) where we plot Γ for the ethanol–water system with the NRTL model using several different sets of interaction parameters. Figure 4.3(b) was prepared using three different sets of parameters fitted to three different sets of isothermal VLE data. Figure 4.3(c) was prepared using three different sets of parameters fitted to three different sets of isobaric VLE data. Whenever possible $Ð$ should be calculated with Γ obtained using activity coefficient parameters fitted to equilibrium data obtained at the same temperature as the diffusion data. Parameters fitted to constant pressure equilibrium data may not give good estimates of $Ð$ at constant temperature.

More comprehensive collections of liquid diffusivity data can be found in the reviews by Johnson and Babb (1956) and a two-part review by Dullien and co-workers (Ghai et al., 1973; Ertl et al., 1974). The book by Tyrrell and Harris (1984) is a good place to begin a search for experimental measurements of D.

4.1.4 Estimation of Diffusion Coefficients in Dilute Liquid Mixtures

As the mole fraction of either component in a binary mixture approaches unity, the thermodynamic factor Γ approaches unity and the Fick D and the Maxwell–Stefan $Ð$ are equal. This result is shown clearly in Figures 4.1–4.3. The diffusion coefficients obtained under these conditions are the infinite dilution diffusion coefficients and given the symbol $Ð°$.

The Stokes–Einstein equation is a purely theoretical method of estimating $Ð°$.

$$Ð°_{12} = \frac{k_B T}{6\pi\mu_2 r_1} \tag{4.1.7}$$

where $Ð°_{12}$ is the diffusion coefficient of species 1 infinitely diluted in species 2, k_B is Boltzmann's constant, μ_2 is the viscosity of the solvent, and r_1 is the radius of the diffusing molecule. This simple relation is valid only if the molecules of the diffusing species are very large compared to the solvent molecules (Evans et al., 1981), this restriction being one of the assumptions made in its derivation. Despite this limitation on solvent size, Eq. 4.1.7 has provided a useful starting point for a number of semiempirical correlations of infinite dilution diffusivities.

One of the best known methods, due to Wilke and Chang (1955), is

$$Ð°_{12} = 7.4 \times 10^{-8} \frac{(\phi_2 M_2)^{1/2} T}{\mu_2 V_1^{0.6}} \tag{4.1.8}$$

where

D_{12}° = diffusion coefficent of species 1 (the solute) present in infinitely low concentration in species 2 (the solvent) [cm^2/s]

M_2 = molar mass of the solvent [g/mol]

T = temperature [K]

μ = viscosity [mPa s = cP]

TABLE 4.2   Diffusion Coefficients at Infinite Dilution

A: General Correlations for Polar and NonPolar Systems[a]

1. Tyn–Calus (1975a) correlation[b]

$$D_{12}^{\circ} = 8.93 \times 10^{-8} V_1^{1/6} V_2^{-1/3} \mu_2^{-1} (P_2/P_1)^{0.6} T$$

2. Hayduk–Minhas (1982) correlations[b]

$$D_{12}^{\circ} = 1.55 \times 10^{-8} V_2^{-0.23} \mu_2^{-0.92} P_2^{0.5} P_1^{-0.42} T^{1.29}$$

$$D_{12}^{\circ} = 6.915 \times 10^{-10} \mu_2^{-0.19} R_1^{0.2} R_2^{-0.4} T^{1.70}$$

3. Siddiqi–Lucas (1986) correlation

$$D_{12}^{\circ} = 9.89 \times 10^{-8} \mu_2^{-0.907} V_1^{-0.45} V_2^{0.265} T$$

4. Hayduk–Minhas (1982) correlation for *n*-alkanes

$$D_{12}^{\circ} = 13.3 \times 10^{-8} V_1^{-0.71} \mu_2^{(10.2/V_1 - 0.791)} T^{1.47}$$

5. King et al. (1965) correlation

$$D_{12}^{\circ} = 4.4 \times 10^{-8} \mu_2^{-1} (V_2/V_1)^{1/6} (\Delta H_2/\Delta H_1)^{1/2} T$$

B: Correlations for Aqueous Mixtures[c]

6. Hayduk–Laudie (1974) correlation

$$D_{12}^{\circ} = 13.26 \times 10^{-5} \mu_2^{-1.14} V_1^{-0.589}$$

7. Hayduk–Minhas (1982) correlation

$$D_{12}^{\circ} = 1.25 \times 10^{-8} (V_1^{-0.19} - 0.292) \mu_2^{(9.58/V_1 - 1.12)} T^{1.52}$$

8. Siddiqi–Lucas (1986) correlation

$$D_{12}^{\circ} = 2.98 \times 10^{-7} \mu_2^{-1.026} V_1^{-0.5473} T$$

[a]Notation [units]: D_{12}° = infinite dilution diffusion coefficient [cm^2/s]; T = temperature [K]; μ = viscosity [cP = mPa s]; V = molar volume at the normal boiling point [cm^3/mol]; R = radius of gyration [nm]; P = parachor [g$^{1/4}$cm^3/(mol s$^{1/2}$)] and; ΔH = latent heat of vaporization at the normal boiling point [any consistent units].

[b]Water should be treated as a dimer; that is, parachor and molar volumes should be doubled. Organic acid solutes should be treated as dimers except when water, methanol, or butanol is the solvent. For nonassociating solutes in monohydroxy alcohols, the solvent parachor and molar volume should be multiplied by $8\mu_2$.

[c]Water is species 2 in these correlations.

V_1 = molar volume of Solute 1 at its normal boiling point [cm^3/mol]

ϕ_2 = association factor for the solvent (2.26 for water, 1.9 for methanol, 1.5 for ethanol, and 1.0 for unassociated solvents)

The value of 2.26 for the association factor for water was found by Hayduk and Laudie (1974) to give better results than the value of 2.6 that was suggested by Wilke and Chang.

A number of other useful (and sometimes more accurate) correlations have been developed; some of them are listed in Table 4.2. Of the correlations for nonaqueous mixtures, those of King et al. (1965), Hayduk and Minhas (1982), Tyn and Calus (1975a), and Siddiqi and Lucas (1986) gave about the same average error when evaluated against 1275 measured diffusivities (Siddiqi and Lucas, 1986). Siddiqi and Lucas report that their own correlation for diffusion of a solute in water had a noticeably lower average error than any other correlation when evaluated against 658 measured diffusivities. This does not, however, mean that the Siddiqi–Lucas correlation is best for all aqueous systems (as Example 4.1.1 demonstrates). Erkey et al. (1990) developed a correlation for predicting infinite dilution diffusion coefficients in alkanes. Wong and Hayduk (1990) compare correlations of estimating infinite dilution diffusivities in *n*-alkane mixtures, for dissolved gases in organic solvents and for dissolved gases in water. Comparative reviews (including example calculations) of selected correlations published prior to 1986 are by Hayduk (1986) and by Reid et al. (1987).

Example 4.1.1   Diffusion of Alcohols Infinitely Diluted in Water

The diffusivities of methanol and 2-propanol at infinite dilution in water were measured by Matthews and Akgerman (1988). Use their data, given in Table 4.3, to provide a spot check

TABLE 4.3   Infinite Dilution Diffusion Coefficients of Methanol and 2-Propanol in Water [a]

Diffusivity Data of Matthews and Akgerman (1988)			
T(°C)	$\mathcal{D}^\circ_{MeOH-H_2O}$	$\mathcal{D}^\circ_{2\text{-}PrOH-H_2O}$	μ_{H_2O} [b]
30	1.83	1.43	0.814
56	3.42	2.21	0.504
81	4.91	3.32	0.351
120	7.73	5.82	0.230

	$\mathcal{D}^\circ_{MeOH-H_2O}$ *Computed from Various Correlation*		
T(°C)	Wilke-Chang	Hayduk-Laudie	Hayduk-Minhas	Siddiqi-Lucas
30	1.87	1.86	1.77	1.44
56	3.28	3.21	3.07	2.56
81	5.06	4.85	4.75	4.00
120	8.58	7.84	8.11	6.85

	$\mathcal{D}^\circ_{2\text{-}PrOH-H_2O}$ *Computed from Various Correlations*		
T(°C)	Wilke-Chang	Hayduk-Laudie	Hayduk-Minhas	Siddiqi-Lucas
30	1.27	1.27	1.29	1.01
56	2.22	2.19	2.36	1.80
81	3.43	3.31	3.80	2.80
120	5.81	5.35	6.79	4.80

[a] The units of $\mathcal{D}^\circ_{ij}$ are in 10^{-9} m^2/s. The units of viscosity are cP [10^{-3} Pa s]

[b] Viscosity computed from DIPPR correlation (Daubert and Danner, 1985).

on the correlations in Table 4.1(B) and the Wilke–Chang equation (Eq. 4.1.8).

DATA   The molar volumes are

$$V_{MeOH} = 41.94 \text{ cm}^3/\text{mol}$$
$$V_{PrOH} = 80.28 \text{ cm}^3/\text{mol}$$

The viscosity of water is given in Table 4.3.

SOLUTION   The Wilke–Chang equation is selected to illustrate the calculation of the infinite dilution diffusivities. The association factor ϕ for water is 2.26

$$\mathcal{D}_{12}^{\circ} = 7.4 \times 10^{-8} \frac{(\phi M_2)^{1/2} T}{\mu_2 V_1^{0.6}}$$

$$= 7.4 \times 10^{-8} \times \frac{(2.26 \times 18)^{1/2} \times 303.15}{0.814 \times 41.94^{0.6}}$$

$$= 1.87 \times 10^{-5} \text{ cm}^2/\text{s}$$

The infinite dilution diffusivities computed from all four correlations are given in Table 4.3 below the experimental values. From the calculated results we see that for the methanol–water system the Hayduk–Laudie method gives good estimates over the temperature range of these data. The Wilke–Chang method (with $\phi = 2.26$) gives good results at the lower temperatures. For the 2-propanol system these two correlations change places in the order of merit. The Hayduk–Minhas correlation comes in as third best with results that would probably still be adequate for many engineering purposes. The Siddiqi–Lucas correlation consistently underpredicts the diffusivities. ■

4.1.5   Estimation of Diffusion Coefficients in Concentrated Liquid Mixtures

Most methods for predicting $\mathcal{D}$ in concentrated solutions attempt to combine the infinite dilution coefficients $\mathcal{D}_{12}^{\circ}$ and $\mathcal{D}_{21}^{\circ}$ in a simple function of composition. The simplest expression

$$\mathcal{D}_{12} = x_2 \mathcal{D}_{12}^{\circ} + x_1 \mathcal{D}_{21}^{\circ} \tag{4.1.9}$$

proposed by Caldwell and Babb (1956) is recommended by Danner and Daubert (1983).

Vignes (1966) suggested that the composition dependence of $\mathcal{D}$ can be expressed by a relation of the form

$$\mathcal{D}_{12} = (\mathcal{D}_{12}^{\circ})^{x_2} (\mathcal{D}_{21}^{\circ})^{x_1} \tag{4.1.10}$$

This formula is recommended by Reid et al. (1987).

The success of the Vignes relationship can be judged in Figure 4.1, where the dashed lines represent Eq. 4.1.10. It would appear from the results in Figure 4.1 that Eq. 4.1.10 is not always as good as had been shown earlier by Vignes himself. However, as noted earlier, the Maxwell–Stefan diffusion coefficient can be quite sensitive to the correlation used to calculate the activity coefficients (Dullien, 1971). Thus, it may be dangerous to draw definitive conclusions from the limited number of data shown here. The Vignes equation is less succesful for mixtures containing an associating component (e.g., an alcohol). Alternative prediction methods need to be developed for such systems (see, e.g., McKeigue and Gulari, 1989; Rutten, 1992).

Several modifications of the Vignes relation have also been proposed; Leffler and Cullinan (1970), for example, included viscosity in the relation as follows:

$$Ð_{12}\mu = (Ð_{12}^{\circ}\mu_2)^{x_2}(Ð_{21}^{\circ}\mu_1)^{x_1} \qquad (4.1.11)$$

This expression is sometimes recommended for predicting the Fick diffusivity D instead of the Maxwell–Stefan diffusivity (Danner and Daubert, 1983).

A number of studies suggested that the composition dependence of the Fick diffusivity can be reasonably well represented by an equation with the form

$$D = D^{\circ}\Gamma^m \qquad (4.1.12)$$

where D° is a function of the infinite dilution coefficients $Ð_{12}^{\circ}$ and $Ð_{21}^{\circ}$ and the composition of the mixture. In terms of D°, the Maxwell–Stefan diffusivity $Ð$ would be given by

$$Ð = D^{\circ}\Gamma^{m-1} \qquad (4.1.13)$$

Rathbun and Babb (1966) used Eq. 4.1.9 for D° and found that a value of $m = 0.6$ gave a good fit to data for a few systems that exhibited positive deviations from Raoult's law, whereas $m = 0.3$ worked well for systems having negative deviations from Raoult's law. Kosanovich and Cullinan (1976) found that an exponent of 0.5 on the thermodynamic factor reproduced the concentration dependence of several nonideal binary mixtures quite nicely.

Siddiqi and Lucas (1986) evaluated many of the foregoing ways of accounting for the composition dependence of D (along with a number of other methods). For 79 mixtures of two nonpolar components they found that Eq. 4.1.12 with $m = 0.4$ and Eq. 4.1.9 for D° was best. However, the average errors were not much worse with $m = 0.6$ (the Rathbun and Babb value) or 0.5. For 54 mixtures with a single polar component, the Rathbun and Babb method fared noticeably better than any other methods tested and an exponent of 0.5 on Γ came in second best. For 43 mixtures of two polar components they found best results were obtained by Eq. 4.1.12 with $m = 1$ and D° given by

$$D^{\circ} = \phi_2 Ð_{12}^{\circ} + \phi_1 Ð_{21}^{\circ} \qquad (4.1.14)$$

where ϕ is the volume fraction. However, for this class of mixtures, the advantage over the methods of Vignes and Leffler and Cullinan was not significant. It should also be noted that the combination of Eq. 4.1.9 and 4.1.12 with $m = 0.5$ would have given results acceptable for most engineering applications.

Of course, for specific systems, any one of the above (or other) models may give the best results and it pays to check the correlation against experimental data if that is possible. For example, Dullien and Asfour (1985) found that the Fick D for regular solutions may be well represented by

$$D_{12}/\mu = (Ð_{12}^{\circ}/\mu_2)^{x_2}(Ð_{21}^{\circ}/\mu_1)^{x_1} \qquad (4.1.15)$$

It must be remembered that the sensitivity of the predicted Fick D to the thermodynamic model used to calculate Γ, as well as to the model parameters, is more than enough to make all of the above findings subject to some uncertainty. We suggest the Vignes method, Eq. 4.1.10, and shall use it consistently throughout this book.

Example 4.1.2 Diffusion Coefficients for the System Acetone–Benzene

Estimate the Maxwell–Stefan and Fick diffusion coefficients for an acetone(1)–benzene(2) mixture of composition $x_1 = 0.7808$ at a temperature of 25°C. The NRTL equation may be used to estimate the thermodynamic correction factor Γ.

DATA   The infinite dilution diffusivities are (Anderson et al., 1958)

$$\mathcal{D}^{\circ}_{12} = 2.75 \times 10^{-9} \text{ m}^2/\text{s}$$
$$\mathcal{D}^{\circ}_{21} = 4.15 \times 10^{-9} \text{ m}^2/\text{s}$$

The NRTL parameters for the acetone–benzene system at 25°C are available in the data collection of Gmehling and Onken (1977ff Vol. I/3 + 4, p. 208). Their dimensionless values are

$$\tau_{12} = -0.1189 \qquad \tau_{21} = 0.6482 \qquad \alpha_{12} = \alpha_{21} = 0.3029$$

SOLUTION   We will use the Vignes method to estimate the MS diffusivity. Substituting the numerical values for the infinite dilution diffusivities and the mole fraction into Eq. 4.1.10 gives

$$\mathcal{D}_{12} = \left(\mathcal{D}^{\circ}_{12}\right)^{x_2}\left(\mathcal{D}^{\circ}_{21}\right)^{x_1}$$
$$= \left(2.75 \times 10^{-9}\right)^{0.2192} \times \left(4.15 \times 10^{-9}\right)^{0.7808}$$
$$= 3.792 \times 10^{-9} \text{ m}^2/\text{s}$$

The NRTL equation for the activity coefficient γ_i is presented in Table D.5. The thermodynamic factor Γ is given by

$$\Gamma = 1 - 2x_1 x_2 \left(\tau_{21} G_{21}^2 / S_1^3 + \tau_{12} G_{12}^2 / S_2^3\right)$$

where

$$S_1 = x_1 + x_2 G_{21} \qquad S_2 = x_2 + x_1 G_{12}$$

and

$$G_{12} = \exp(-\alpha_{12}\tau_{12}) \qquad G_{21} = \exp(-\alpha_{21}\tau_{21})$$

Substituting the values of the parameters τ_{12}, τ_{21}, and α_{12} into the above expressions from Table D.5 gives

$$\Gamma = 0.871$$

The Fick diffusivity may now be computed from the product of $\mathcal{D}$ and Γ as

$$D = \mathcal{D}\Gamma$$
$$= 3.792 \times 10^{-9} \times 0.871$$
$$= 3.30 \times 10^{-9} \text{ m}^2/\text{s}$$

which compares reasonably well with the experimental value of D at this composition of 3.35×10^{-9} m^2/s (Anderson et al., 1956). Additional data of Anderson et al. and of Cullinan and Toor (1965) are shown in Figure 4.1(b).   ∎

Example 4.1.3   Diffusion Coefficients in the System Ethanol–Water

Estimate the diffusivities of an ethanol(1)–water(2) mixture at 40°C and $x_1 = 0.68$.

DATA   The infinite dilution diffusivities are (Tyn and Calus, 1975b)

$$\mathcal{D}^{\circ}_{12} = 1.70 \times 10^{-9} \text{ m}^2/\text{s}$$
$$\mathcal{D}^{\circ}_{21} = 1.64 \times 10^{-9} \text{ m}^2/\text{s}$$

The NRTL parameters for ethanol–water at 40°C are taken from the collection of Gmehling and Onken (1977ff Vol. I/1 p. 172).

$$\tau_{12} = -0.02188 \qquad \tau_{21} = 1.6139 \qquad \alpha_{12} = \alpha_{21} = 0.2946$$

SOLUTION The Maxwell–Stefan diffusivity is computed using the Vignes equation (Eq. 4.1.10) as

$$\begin{aligned} \DJ_{12} &= (\DJ_{12}^{\circ})^{x_2}(\DJ_{21}^{\circ})^{x_1} \\ &= (1.70 \times 10^{-9})^{0.32} \times (1.64 \times 10^{-9})^{0.68} \\ &= 1.659 \times 10^{-9} \text{ m}^2/\text{s} \end{aligned}$$

The parameter Γ is computed from the NRTL equation as illustrated in Example 4.1.2 with the result

$$\Gamma = 0.610$$

The Fick diffusivity may now be computed as

$$\begin{aligned} D &= \DJ\Gamma \\ &= 1.659 \times 10^{-9} \times 0.610 \\ &= 1.012 \times 10^{-9} \text{ m}^2/\text{s} \end{aligned}$$

The data of Tyn and Calus (1975b) covering the entire range of compositions is shown in Figure 4.1(a). Their experimental value of D at the conditions used in this illustration is 1.02×10^{-9} m^2/s, which compares quite well with the value predicted using the Vignes method. ∎

4.2 ESTIMATION OF MULTICOMPONENT DIFFUSION COEFFICIENTS

Comparison of the Maxwell–Stefan formulation

$$(J) = -c_t[B]^{-1}[\Gamma](\nabla x) \tag{2.2.10}$$

with the generalized Fick's law

$$(J) = -c_t[D](\nabla x) \tag{3.2.5}$$

shows that the matrix $[D]$ and the product $[B]^{-1}[\Gamma]$ are equivalent

$$[D] \equiv [B]^{-1}[\Gamma]$$

Strictly speaking, the rules of matrix algebra do not allow us, on the basis of Eqs. 3.2.5 and 2.2.10, to assert that $[D]$ and $[B]^{-1}[\Gamma]$ are equal. The equality of these two matrices is an assumption, albeit the only reasonable way to relate the Fick diffusion coefficients D_{ij} to the Maxwell–Stefan diffusion coefficients $\DJ_{ij}$. The equality

$$[D] = [B]^{-1}[\Gamma] \tag{4.2.1}$$

will be used throughout this book, just as it is in the literature on multicomponent mass transfer.

Equation 4.2.1 is an important result for it allows us to predict the Fick matrix $[D]$ from information on the binary Maxwell–Stefan diffusivities $\DJ_{ij}$ and activity coefficients.

4.2.1 Estimation of Multicomponent Diffusion Coefficients for Gas Mixtures

For ideal gases the thermodynamic matrix $[\Gamma]$ reduces to the identity matrix and Eq. 4.2.1 becomes

$$[D] = [B]^{-1} \qquad |\text{ideal}| \tag{4.2.2}$$

As discussed briefly in Section 4.1.2, the diffusion coefficients of the binary pairs $Đ_{ij}$ can be estimated from the kinetic theory of gases or from an appropriate correlation to a reasonable degree of accuracy, particularly for nonpolar molecules. The matrix of diffusion coefficients may therefore be calculated using Eq. 4.2.1 or, for a ternary system, directly from the relations derived below.

For a ternary system $[B]$ is of order 2 with elements given by

$$B_{11} = \frac{x_1}{Đ_{13}} + \frac{x_2}{Đ_{12}} + \frac{x_3}{Đ_{13}}$$

$$B_{12} = -x_1\left(\frac{1}{Đ_{12}} - \frac{1}{Đ_{13}}\right)$$

$$B_{21} = -x_2\left(\frac{1}{Đ_{12}} - \frac{1}{Đ_{23}}\right) \tag{4.2.3}$$

$$B_{22} = \frac{x_1}{Đ_{12}} + \frac{x_2}{Đ_{23}} + \frac{x_3}{Đ_{23}}$$

The inversion of $[B]$ can be carried out explicitly using Equations A.3.9–A.3.10 as shown below.

The determinant of $[B]$ is

$$|B| = B_{11}B_{22} - B_{12}B_{21} = S/Đ_{12}Đ_{13}Đ_{23} \tag{4.2.4}$$

where

$$S = x_1Đ_{23} + x_2Đ_{13} + x_3Đ_{12} \tag{4.2.5}$$

The cofactor matrix is

$$[B^c]^T = \begin{bmatrix} B_{22} & -B_{21} \\ -B_{12} & B_{11} \end{bmatrix}^T = \begin{bmatrix} B_{22} & -B_{12} \\ -B_{21} & B_{11} \end{bmatrix} \tag{4.2.6}$$

Thus, the elements of $[D]$ are

$$
\begin{aligned}
D_{11} &= Đ_{13}(x_1Đ_{23} + (1-x_1)Đ_{12})/S \\
D_{12} &= x_1Đ_{23}(Đ_{13} - Đ_{12})/S \\
D_{21} &= x_2Đ_{13}(Đ_{23} - Đ_{12})/S \\
D_{22} &= Đ_{23}(x_2Đ_{13} + (1-x_2)Đ_{12})/S
\end{aligned} \tag{4.2.7}
$$

It is easy to check that if the binary diffusion coefficients $Đ_{ij}$ are equal; $Đ_{ij} = Đ$, for all i, j permutations; then the matrix of diffusion coefficients degenerates to a scalar times the identity matrix (cf. Eq. 3.2.10)

$$[D] = Đ[I] \qquad (\text{special}) \tag{4.2.8}$$

For two components infinitely diluted in a third (x_1 and x_2 close to zero), we find the cross-coefficients D_{12} and D_{21} vanish and the main coefficients D_{11} and D_{22} are given simply by

$$D_{ii} = Ð_{i3} \quad i = 1, 2 \tag{4.2.9}$$

These results, obtained for a dilute ternary vapor mixture may be generalized with the help of Eq. 4.2.2 to a mixture of any number of components where one component is present in a large excess; $x_n \to 1$, $x_i \to 0$, $i = 1, 2, \cdots, n - 1$ (see discussion below Eq. 3.2.10)

$$D_{ii} = Ð_{in} \quad D_{ij} = 0 \tag{4.2.10}$$

Example 4.2.1   The Structure of the Fick Matrix [D] When All of the Binary Diffusion Coefficients Are Nearly Equal

The Maxwell–Stefan diffusion coefficients for three binary gas pairs at a temperature of 273 K and a pressure of 101 kPa are listed below:

oxygen–nitrogen	$Ð = 18.1 \text{ mm}^2/\text{s}$
oxygen–carbon monoxide	$Ð = 18.5 \text{ mm}^2/\text{s}$
nitrogen–carbon monoxide	$Ð = 17.7 \text{ mm}^2/\text{s}$

Investigate the structure of the Fick matrix [D] for the ternary system oxygen(1)–nitrogen(2)–carbon monoxide(3).

SOLUTION   For the purposes of this illustration [D] is calculated from Eqs. 4.2.7 with $x_1 = x_2 = 0.3$, $x_3 = 0.4$.

$$S = x_1 Ð_{23} + x_2 Ð_{13} + x_3 Ð_{12}$$
$$= 0.3 \times 17.7 + 0.3 \times 18.5 + 0.4 \times 18.1$$
$$= 18.1 \text{ mm}^2/\text{s}$$

$$D_{11} = Ð_{13}(x_1 Ð_{23} + (1 - x_1)Ð_{12})/S$$
$$= 18.5 \times (0.3 \times 17.7 + (1 - 0.3) \times 18.1)/18.1$$
$$= 18.38 \text{ mm}^2/\text{s}$$

$$D_{12} = x_1 Ð_{23}(Ð_{13} - Ð_{12})/S$$
$$= 0.3 \times 17.7 \times (18.5 - 18.1)/18.1$$
$$= 0.117 \text{ mm}^2/\text{s}$$

$$D_{21} = x_2 Ð_{13}(Ð_{23} - Ð_{12})/S$$
$$= 0.3 \times 18.5 \times (17.7 - 18.1)/18.1$$
$$= -0.123 \text{ mm}^2/\text{s}$$

$$D_{22} = Ð_{23}(x_2 Ð_{13} + (1 - x_2)Ð_{12})/S$$
$$= 17.7 \times (0.3 \times 18.5 + (1 - 0.3) \times 18.1)/18.1$$
$$= 17.82 \text{ mm}^2/\text{s}$$

We may write these results in matrix form as

$$[D] = \begin{bmatrix} 18.38 & 0.12 \\ -0.12 & 17.82 \end{bmatrix} mm^2/s$$

As might be expected from the discussion above, this matrix is dominated by the diagonal elements that are two orders of magnitude greater than the off-diagonal elements. Moreover, the diagonal elements themselves are approximately the same. It, therefore, seems reasonable to use just one average value of $Đ$ in the calculation of $[D]$. In this example $[D] = 18.1[I]$ mm^2/s is a good approximation to $[D]$. ∎

Example 4.2.2   [D] for Dilute Gas Mixtures

2-Propanol(1) and water vapor(2) are condensing on the cooled surface of a vertical tube. Nitrogen (3) is also present in the vapor mixture fed to the condenser. At the temperature and pressure in the condenser the diffusion coefficients of the three binary gas pairs are

$$Đ_{12} = 15.99 \text{ mm}^2/s$$

$$Đ_{13} = 14.43 \text{ mm}^2/s$$

$$Đ_{23} = 38.73 \text{ mm}^2/s$$

Near the exit from the condenser the composition in the vapor mixture is $x_1 = 0.005$, $x_2 = 0.003$, $x_3 = 0.992$. Calculate the matrix of multicomponent diffusion coefficients at this composition.

SOLUTION The matrix $[D]$ is calculated directly from Eqs. 4.2.7 as demonstrated in Example 4.2.1. The result here is

$$[D] = \begin{bmatrix} 14.43 & -0.019 \\ 0.061 & 38.45 \end{bmatrix} mm^2/s$$

As in our last illustration, this matrix is dominated by the diagonal elements. Since the mixture is almost pure nitrogen with only traces of the condensable vapors remaining we may let the mole fraction of nitrogen approach unity, $x_3 \rightarrow 1$, and the mole fractions of 2-propanol and water go to zero, $x_1 \rightarrow 0$, $x_2 \rightarrow 0$ and approximate $[D]$ as

$$[D] = \begin{bmatrix} 14.43 & 0.0 \\ 0.0 & 38.73 \end{bmatrix} mm^2/s \qquad ∎$$

Example 4.2.3   Composition Dependence of the Fick Matrix [D]

The vapor-phase catalytic dehydrogenation of ethanol to acetaldehyde involves the diffusion of ethanol to the catalyst surface where it reacts to produce acetaldehyde and hydrogen. Under typical reactor conditions (temperature = 548 K, pressure = 101.3 kPa) the binary diffusivities of the three binary pairs encountered are

$$Đ_{12} = 147.7 \text{ mm}^2/s$$

$$Đ_{13} = 25.0 \text{ mm}^2/s$$

$$Đ_{23} = 142.4 \text{ mm}^2/s$$

where component 1 is acetaldehyde, component 2 is hydrogen, and component 3 is ethanol.

The composition of the bulk gas may be taken to be $x_1 = 0.20$, $x_2 = 0.20$, $x_3 = 0.60$. At the catalyst surface ethanol is consumed by the reaction that produces acetaldehyde. The mole fraction of ethanol there will be much lower than in the bulk gas. For this illustration assume the gas composition at the catalyst surface to be $x_1 = 0.584$, $x_2 = 0.328$, $x_3 = 0.088$. Investigate the structure of $[D]$ and calculate some representative values.

SOLUTION The diffusivities $Ð_{23}$ and $Ð_{12}$ are nearly equal and we will make this assumption in the estimation of the multicomponent diffusion coefficients. With $Ð_{23} = Ð_{12} = Ð$ we calculate, with reference to Eqs. 4.2.7, the elements of $[D]$ as

$$S = x_1 Ð_{23} + x_2 Ð_{13} + x_3 Ð_{12}$$
$$= (1 - x_2)Ð + x_2 Ð_{13}$$
$$D_{11} = Ð Ð_{13}/S$$
$$D_{12} = x_1 Ð(Ð_{13} - Ð)/S$$
$$D_{21} = 0$$
$$D_{22} = Ð(x_2 Ð_{13} + (1 - x_2)Ð)/S$$

Note that D_{21} is zero regardless of the composition of the mixture. Hence, the flux of hydrogen (2) depends on its own composition gradient ∇x_2 only

$$J_2 = -c_t D_{22} \nabla x_2 = -c_t Ð \nabla x_2$$

No similar simplification is possible for acetaldehyde, the flux of which is given by Eq. 3.2.1.

At the bulk gas composition we estimate $[D]$ using the above expressions to be

$$[D] = \begin{bmatrix} 29.95 & -28.77 \\ 0.0 & 145.05 \end{bmatrix} \text{mm}^2/\text{s}$$

($Ð$ was taken to be the arithmetic average of $Ð_{12}$ and $Ð_{13}$ in this and the next calculation.)

At the interface composition $[D]$ is

$$[D] = \begin{bmatrix} 34.32 & -96.23 \\ 0.0 & 145.05 \end{bmatrix} \text{mm}^2/\text{s}$$

At this composition the ratio $|D_{12}/D_{11}|$ exceeds unity suggesting that the flux of acetaldehyde will be strongly influenced by the composition gradient of species 2 (hydrogen).

The ratio D_{12}/D_{11} is given by

$$\frac{D_{12}}{D_{11}} = \frac{x_1 Ð_{23}(Ð_{13} - Ð_{12})}{Ð_{13}(x_1 Ð_{23} + (1 - x_1)Ð_{12})}$$
$$= \frac{1 - Ð_{12}/Ð_{13}}{1 + (1 - x_1)Ð_{12}/x_1 Ð_{23}}$$

For an ideal gas mixture in which $Ð_{12}$ and $Ð_{23}$ are similar the ratio D_{12}/D_{11} becomes

$$D_{12}/D_{11} = x_1(1 - Ð_{12}/Ð_{13})$$

When $Ð_{12}/Ð_{13} = 1$ the ratio $D_{12}/D_{11} = 0$ regardless of the composition of the mixture (this corresponding to the case of three equal, or nearly equal, binary diffusivities discussed

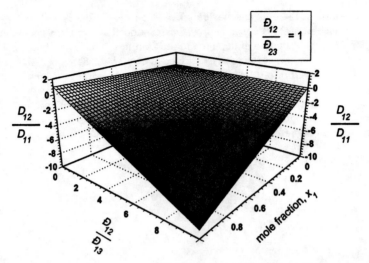

Figure 4.4. The ratio D_{12}/D_{11}, which are the elements of the Fick matrix $[D]$, as a function of the binary pair Maxwell–Stefan diffusion coefficients $Ð_{12}/Ð_{13}$.

in Example 4.2.1). When the ratio $Ð_{12}/Ð_{13}$ is different from unity, then the cross-coefficients may be significant with respect to D_{11}, especially for high values of the mole fraction x_1. The ratio D_{12}/D_{11} may also assume negative values and the absolute values of this ratio may approach or even exceed unity as shown in the example calculation above and in Figure 4.4 where D_{12}/D_{11} is shown as a function of the ratio of the binary pair Maxwell–Stefan diffusion coefficients $Ð_{12}/Ð_{13}$. ∎

Example 4.2.4   Effect of Component Numbering on the Fick Matrix

The preceding examples illustrated some of the ways in which the Fick matrix $[D]$ depends on the composition of the mixture and on the binary (Maxwell–Stefan) diffusion coefficients. One further property of $[D]$ is that both the sign and magnitude of the elements of $[D]$ depends on the order in which the components are numbered. Wesselingh (1985) provided a dramatic and elegant illustration of this fact for the system H_2–N_2–CCl_2F_2. At a temperature of 298 K and a pressure of 101.3 kPa the diffusion coefficients of the three binary pairs that make up the mixture are

$$
\begin{array}{lll}
H_2\text{–}N_2 & Ð = 77.0 \text{ mm}^2/\text{s} \\
N_2\text{–}CCl_2F_2 & Ð = 8.1 \text{ mm}^2/\text{s} \\
CCl_2F_2\text{–}H_2 & Ð = 33.1 \text{ mm}^2/\text{s}
\end{array}
$$

The ratio of the largest to the smallest of these coefficients is nearly an order of magnitude; thus we may expect the Fick matrix $[D]$ to show a strong composition dependency.

Compute the matrix $[D]$ at the composition $x_N = 0.4$, $x_C = 0.25$ and $x_H = 0.35$. The subscripts N, C, and H refer to N_2, CCl_2F_2, and H_2, respectively.

SOLUTION   For each of the three possible choices for "component 3" there are two ways to order the components. We will calculate the elements of $[D]$ for three ways of ordering the components obtained by rotating the component numbers in order. With the compo-

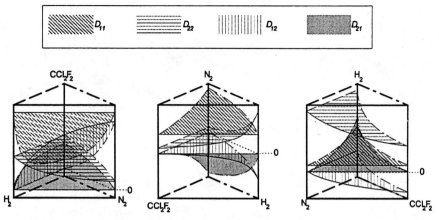

Figure 4.5. Four Fick diffusion coefficients plotted as a function of composition for the system $H_2-N_2-CCl_2F_2$ for the three possible choices of "component 3."

nents ordered as N = 1, C = 2, H = 3 we calculate $[D]$ as

$$[D] = \begin{bmatrix} 39.45 & 25.82 \\ 13.62 & 23.73 \end{bmatrix} mm^2/s$$

With the components ordered as C = 1, H = 2, N = 3 we find

$$[D] = \begin{bmatrix} 10.11 & -13.63 \\ 3.52 & 53.08 \end{bmatrix} mm^2/s$$

Finally, with the components ordered as H = 1, N = 2, C = 3 we have

$$[D] = \begin{bmatrix} 49.55 & -3.52 \\ -25.82 & 13.63 \end{bmatrix} mm^2/s$$

These three matrices are quite different but all three represent the one system at the same conditions of temperature, pressure, and composition. It may be verified from Eq. 3.2.5 that, for a given set of driving forces ∇x_i, all three choices of $[D]$ will yield the same set of fluxes regardless of the component numbering (as is only to be expected, of course). It is interesting to note that, even though these three matrices appear to be quite unrelated, their eigenvalues are the same (for a formal proof that this will always be the case see, e.g., Taylor, 1981a).

We repeated Wesselingh's (1985) calculations of the $[D]$ matrix as a function of composition with the results shown in Figure 4.5. With hydrogen as "component 3," all the D_{ij} are positive regardless of the composition of the mixture. In complete contrast, the cross-coefficient D_{21} is always negative if CCl_2F_2 is component 3 and the coefficient D_{12} is always negative if nitrogen is component 3. Once again, we can see that the cross-coefficients can overshadow the main coefficients in absolute value in certain composition regions but not with all choices of component 3; in this example the main coefficients are always larger than the cross-coefficients if hydrogen is component 3. In fact, it is always possible to order the components so that this is the case; the trick is to choose the species with the smallest molecule as "component n." The cross-coefficients are zero along one side of the ternary composition triangle (this should not come as a surprise; consult Eq. 4.2.7 again). The main coefficients D_{11} and D_{22} are always positive regardless of the ordering of the components or of the composition of the mixture (see discussion around inequalities Eq. 3.3.13). The component numbering is, to a very large extent, arbitrary. Thus, the Fick matrix

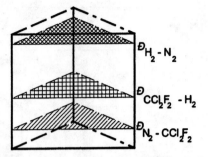

Figure 4.6. The three Maxwell–Stefan diffusion coefficients in the system H_2–N_2–CCl_2F_2. Note that the Maxwell–Stefan coefficients are independent of composition.

$[D]$ is, within limits set by the values of the Maxwell–Stefan diffusivities, also arbitrary. It is worth noting here that the Maxwell–Stefan diffusion coefficients $\mathcal{D}_{ij}$ are completely independent of the composition of the mixture. This result is emphasized in Figure 4.6 (after Wesselingh, 1985).

Another aspect of the choice of component numbering relates to the accuracy of the calculations; it is not a good idea to calculate small fluxes by subtracting two "large" numbers. Put another way, if the flux of component i is expected to be small, do not label this as "solvent" species n; Include i as one of the first $n - 1$ species. Greater accuracy is obtained by choosing n to be the species with the highest concentration. ∎

Example 4.2.5   Prediction of Multicomponent Diffusion Coefficients in the Mass Average Reference Velocity Frame

Estimate the Fick diffusion coefficients in the mass average velocity reference frame for a mixture containing carbon monoxide (1), hydrogen (2), methane (3), and water vapor (4) flowing inside a tubular reactor at a point where the mole fractions are

$$x_1 = 0.05 \qquad x_2 = 0.75382 \qquad x_3 = 0.09809 \qquad x_4 = 0.09809$$

DATA   Maxwell–Stefan diffusivities of the binary pairs

$$\mathcal{D}_{12} = \mathcal{D}_{23} = \mathcal{D}_{24} = 13.5 \times 10^{-6} \text{ m}^2/\text{s}$$

$$\mathcal{D}_{13} = \mathcal{D}_{14} = \mathcal{D}_{34} = 4.0 \times 10^{-6} \text{ m}^2/\text{s}$$

The molar masses of the four components are

$$M_1 = 0.02801 \qquad M_2 = 0.002016 \text{ kg/mol}$$

$$M_3 = 0.01604 \qquad M_4 = 0.018015 \text{ kg/mol}$$

SOLUTION   To predict multicomponent Fick diffusivities in the mass average velocity reference frame we combine Eq. 4.2.2 to predict the Fick diffusivities in the molar average velocity reference frame

$$[D] = [B]^{-1}$$

with Eq. 3.2.11 to transform between velocity reference frames to give

$$[D^o] = [B^{ou}][\omega][x]^{-1}[B]^{-1}[x][\omega]^{-1}[B^{uo}]$$

where we have made use of the fact that $[B^{uo}] = [B^{ou}]^{-1}$.

Our first calculation is to convert the mole fractions to mass fractions. The formulas for converting between mole and mass fractions are given in Table 1.1 and the results of the conversion are

$$\omega_1 = 0.2237 \qquad \omega_2 = 0.2427 \qquad \omega_3 = 0.2513 \qquad \omega_4 = 0.2822$$

The elements of $[B]$ are defined by Eqs. 2.1.21 and 2.1.22. The calculation of the first row of $[B]$ is illustrated below.

$$B_{11} = \frac{x_1}{\mathcal{D}_{14}} + \frac{x_2}{\mathcal{D}_{12}} + \frac{x_3}{\mathcal{D}_{13}} + \frac{x_4}{\mathcal{D}_{14}}$$

$$= \frac{0.05}{4.0 \times 10^{-6}} + \frac{0.75382}{13.5 \times 10^{-6}} + \frac{0.09809}{4.0 \times 10^{-6}} + \frac{0.09809}{4.0 \times 10^{-6}}$$

$$= 1.1174 \times 10^5 \text{ s/m}^2$$

$$B_{12} = -x_1 \left(\frac{1}{\mathcal{D}_{12}} - \frac{1}{\mathcal{D}_{14}} \right)$$

$$. = -0.05 \times \left(\frac{1}{13.5 \times 10^{-6}} - \frac{1}{4.0 \times 10^{-6}} \right)$$

$$= 8.796 \times 10^3 \text{ s/m}^2$$

$$B_{13} = -x_1 \left(\frac{1}{\mathcal{D}_{13}} - \frac{1}{\mathcal{D}_{14}} \right)$$

$$= -0.05 \times \left(\frac{1}{4.0 \times 10^{-6}} - \frac{1}{4.0 \times 10^{-6}} \right)$$

$$= 0 \text{ s/m}^2$$

The remaining elements of $[B]$ are calculated in similar ways with the result

$$[B] = \begin{bmatrix} 0.11738 & 0.00879 & 0.0 \\ 0.0 & 0.07407 & 0.0 \\ 0.0 & 0.01725 & 0.11738 \end{bmatrix} \times 10^6 \text{ s/m}^2$$

The Fick $[D]$ in the molar average velocity reference frame is the inverse of this matrix

$$[D] = \begin{bmatrix} 8.5191 & -1.0116 & 0.0 \\ 0.0 & 13.5000 & 0.0 \\ 0.0 & -1.98846 & 8.5191 \end{bmatrix} \times 10^{-6} \text{ m}^2/\text{s}$$

The elements of the reference frame transformation matrix $[B^{ou}]$ are given by Eq. 1.2.27.

$$B_{ik}^{ou} = \delta_{ik} - \omega_i \left(1 - \frac{\omega_n x_k}{x_n \omega_k} \right)$$

The first row of this matrix has elements given explicitly by

$$B_{11}^{ou} = 1 - \omega_1(1 - x_1\omega_4/x_4\omega_1)$$
$$= 1 - 0.2237 \times (1 - 0.05 \times 0.2822/(0.09809 \times 0.2237))$$
$$= 0.9202$$
$$B_{12}^{ou} = -\omega_1(1 - x_2\omega_4/x_4\omega_2)$$
$$= -0.2237 \times (1 - 0.7538 \times 0.2822/(0.09809 \times 0.2427))$$
$$= 1.7752$$
$$B_{13}^{ou} = -\omega_1(1 - x_3\omega_4/x_4\omega_3)$$
$$= -0.2237 \times (1 - 0.09809 \times 0.2822/(0.09809 \times 0.2513))$$
$$= 0.0275$$

The complete transformation matrix is

$$[B^{ou}] = \begin{bmatrix} 0.9202 & 1.7752 & 0.0275 \\ -0.0866 & 2.9263 & 0.0298 \\ -0.0898 & 1.9947 & 1.0309 \end{bmatrix}$$

The inverse matrix is

$$[B^{uo}] = \begin{bmatrix} 1.0277 & -0.6169 & -0.0096 \\ 0.0301 & 0.3305 & -0.0104 \\ 0.0312 & -0.6932 & 0.9893 \end{bmatrix}$$

The Fick diffusion coefficients in the mass average reference velocity frame $[D^o]$ may now be calculated directly from the expression above as

$$[D^o] = \begin{bmatrix} 8.3828 & -1.4960 & 0.0469 \\ 0.4612 & 13.5835 & -0.1589 \\ -0.1531 & -1.6809 & 8.5718 \end{bmatrix} \times 10^{-6} \text{ m}^2/\text{s}$$

The eigenvalues of this matrix are found to be

$$\hat{D}_1 = 8.519 \times 10^{-6} \text{ m}^2/\text{s}$$
$$\hat{D}_2 = 13.50 \times 10^{-6} \text{ m}^2/\text{s}$$
$$\hat{D}_1 = 8.519 \times 10^{-6} \text{ m}^2/\text{s}$$

which, in fact, are the diagonal elements of $[D]$ in the molar average velocity reference frame. This special result is obtained because two of the columns of $[D]$ contain only one nonzero element, the one on the main diagonal. It is also interesting to note that two of the eigenvalues are equal. This will have interesting consequences when we compute mass transfer rates in this system in Example 10.4.1. ∎

4.2.2 Estimation of Multicomponent Fick Diffusion Coefficients for Liquid Mixtures

Ideally one would like to be able to predict the elements of $[D]$ from a knowledge of the infinite dilution diffusion coefficients $Ð_{ij}^o$. A comparison of the generalized Fick's law (Eq. 3.2.5), with the Maxwell–Stefan equations (Eq. 2.2.10) shows that, for a nonideal system, the matrix of Fick diffusion coefficients is related to the Maxwell–Stefan diffusion coefficients by

$$[D] = [B]^{-1}[\Gamma] \tag{4.2.11}$$

This, or some other similar relation, is the starting point for developing methods of predicting $[D]$.

For ternary systems the inverse of $[B]$ is given by the right-hand sides of Eq. 4.2.7. This matrix must be postmultiplied by $[\Gamma]$ to obtain the elements of $[D]$ as

$$D_{11} = \left\{ Ð_{13}(x_1 Ð_{23} + (1 - x_1) Ð_{12}) \Gamma_{11} + x_1 Ð_{23}(Ð_{13} - Ð_{12}) \Gamma_{21} \right\} / S$$

$$D_{12} = \left\{ Ð_{13}(x_2 Ð_{23} + (1 - x_1) Ð_{12}) \Gamma_{12} + x_1 Ð_{23}(Ð_{13} - Ð_{12}) \Gamma_{22} \right\} / S$$

$$D_{21} = \left\{ Ð_{23}(x_2 Ð_{13} + (1 - x_2) Ð_{12}) \Gamma_{21} + x_2 Ð_{13}(Ð_{23} - Ð_{12}) \Gamma_{11} \right\} / S \tag{4.2.12}$$

$$D_{22} = \left\{ Ð_{23}(x_2 Ð_{13} + (1 - x_2) Ð_{12}) \Gamma_{22} + x_2 Ð_{13}(Ð_{23} - Ð_{12}) \Gamma_{12} \right\} / S$$

with S given by Eq. 4.2.5.

This method of predicting $[D]$ is illustrated in Example 4.2.6.

4.2.3 Estimation of Maxwell–Stefan Diffusion Coefficients for Multicomponent Liquid Mixtures

In order to use the procedure of Section 4.2.2 to predict $[D]$ we need the Maxwell–Stefan diffusivities of each binary pair in the multicomponent mixture.

There are few published values of Maxwell–Stefan diffusivities in multicomponent liquids. Figure 4.7 shows the binary Maxwell–Stefan diffusion coefficients for the ternary mixture 2-propanol(1)–water(2)–glycerol(3). The values of the diffusion coefficients for the binary systems water–glycerol and 2-propanol–glycerol are also shown in this figure, which is adapted from Riede and Schlünder (1991). As would be expected (hoped), the binary coefficients in the ternary mixture are similar to those in the respective binaries. Lightfoot et al. (1962) found the same situation for the glycine–water–potassium chloride system at low concentrations of glycine and potassium chloride.

There are few methods for predicting the Maxwell–Stefan diffusivities in multicomponent liquid mixtures. The methods that have been suggested are based on extensions of the techniques proposed for binary systems discussed in Section 4.1.5 (see, e.g., the works of Cullinan and co-workers, 1966–1975; Bandrowski and Kubaczka, 1982; Kosanovich, 1975). The Vignes equation, for example, may be generalized as follows (Wesselingh and Krishna, 1990; Kooijman and Taylor, 1991).

$$Ð_{ij} = \prod_{k=1}^{n} \left(Ð_{ij, x_k \to 1} \right)^{x_k} \tag{4.2.13}$$

where the $Ð_{ij,\, x_k \to 1}$ are the limiting values of the Maxwell–Stefan diffusivities in a mixture where component k is present in a very large excess.

Equation 4.2.13 is shown in Figure 4.8 for a ternary system where it becomes clear that the limiting diffusivities are, in fact, the Maxwell–Stefan diffusion coefficients at the corners of the diffusivity–composition surface. Equation 4.2.13 should reduce to the binary Vignes

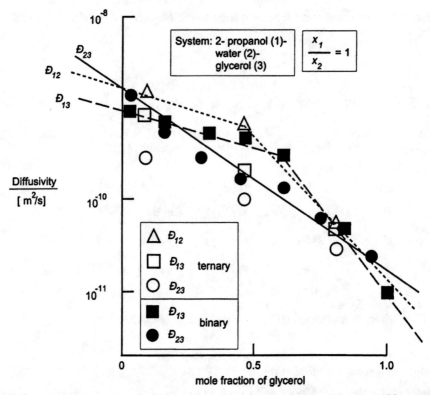

Figure 4.7. Maxwell–Stefan diffusion coefficients for the ternary system 2-propanol(1)–water(2)–glycerol(3) and the binaries 2-propanol–glycerol and water–glycerol. From Riede and Schlünder (1991).

equation when $(x_i + x_j) \rightarrow 1$ and $x_k \rightarrow 0$ $(k \neq i, j)$. The two limiting diffusivities on the i–j face may, therefore, be identified as the binary i–j infinite dilution diffusivities.

$$\DJ_{ij,\, x_j \rightarrow 1} = \DJ_{ij}^{\circ}$$

$$\DJ_{ij,\, x_i \rightarrow 1} = \DJ_{ji}^{\circ}$$

$$(4.2.14)$$

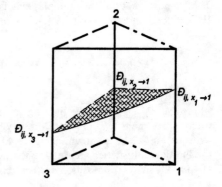

Figure 4.8. Maxwell–Stefan diffusion coefficients as a function of composition for a ternary system showing limiting diffusivities.

Substitution of these limiting values into Eq. 4.2.13 gives

$$\mathcal{D}_{ij} = \left(\mathcal{D}_{ij}^\circ\right)^{x_j}\left(\mathcal{D}_{ji}^\circ\right)^{x_i}\prod_{\substack{k=1\\k\neq i,j}}^{n}\left(\mathcal{D}_{ij,\,x_k\to1}\right)^{x_k} \tag{4.2.15}$$

Any model of the Maxwell–Stefan diffusivities based on Eq. 4.2.15 must obey some rules that places certain restrictions on the kind of model that can be used for the limiting diffusivities $\mathcal{D}_{ij,\,x_k\to1}$. First, the final expression for the Maxwell–Stefan diffusivities must be symmetric $(\mathcal{D}_{ij} = \mathcal{D}_{ji})$. In addition, we must have

$$\lim x_i \to 0\left(\mathcal{D}_{ij,\,x_j=0}\right) = \lim x_j \to 0\left(\mathcal{D}_{ij,\,x_i=0}\right) \tag{4.2.16}$$

to avoid discontinuities at the corners of the diffusivity–composition surface.

Wesselingh and Krishna (1990) tentatively suggested the following model for the limiting diffusivities.

$$\mathcal{D}_{ij,\,x_k\to1} = \left(\mathcal{D}_{ij}^\circ\mathcal{D}_{ji}^\circ\right)^{1/2} \tag{4.2.17}$$

When we combine Eqs. 4.2.15 and 4.2.17 we obtain the following elegant expression for the binary Maxwell–Stefan diffusivity in a multicomponent system.

$$\mathcal{D}_{ij} = \left(\mathcal{D}_{ij}^\circ\right)^{(1+x_j-x_i)/2}\left(\mathcal{D}_{ji}^\circ\right)^{(1+x_i-x_j)/2} \tag{4.2.18}$$

Kooijman and Taylor (1991) proposed the following model for the limiting diffusivities

$$\mathcal{D}_{ij,\,x_k\to1} = \left(\mathcal{D}_{ik}^\circ\mathcal{D}_{ki}^\circ\right)^{1/2} \tag{4.2.19}$$

and found that it gave better predictions of multicomponent Fick diffusion coefficients for a few systems than did Eq. 4.2.17 (or any other simple combination of infinite dilution diffusion coefficients) (see, also, Rutten, 1992). For want of a fundamentally sound method, we shall use Eq. 4.2.18 in the remainder of this book.

The infinite dilution diffusivities, $\mathcal{D}_{ij}^\circ$, are positive definite (cf. Eq. 2.3.21) and, consequently, Eq. 4.2.18 leads us to conclude that the $\mathcal{D}_{ij}$ are positive definite everywhere in the composition space.

Example 4.2.6 Prediction of [D] in the System Acetone–Benzene–Carbon Tetrachloride

Cullinan and Toor (1965) presented experimental data for the matrix of Fick diffusion coefficients in the volume average reference velocity frame for the system acetone(1)–benzene(2)–carbon tetrachloride(3) at a temperature of 25°C. Some of their data is shown in Table 3.3. Estimate the Fick matrix [D] at the composition $x_1 = 0.70$, $x_2 = 0.15$, $x_3 = 0.15$ using the predictive method described above.

DATA The infinite dilution diffusion coefficients $\mathcal{D}_{ij}^\circ$ have been estimated as follows [10^{-9} m^2/s]:

$$\mathcal{D}_{12}^\circ = 2.75 \qquad \mathcal{D}_{21}^\circ = 4.15$$
$$\mathcal{D}_{13}^\circ = 1.70 \qquad \mathcal{D}_{31}^\circ = 3.57$$
$$\mathcal{D}_{23}^\circ = 1.42 \qquad \mathcal{D}_{32}^\circ = 1.91$$

The NRTL parameters τ_{ij} and α_{ij} are given in Example 3.3.1.

SOLUTION   The first step is to estimate the Maxwell–Stefan diffusion coefficients from Eq. 4.2.18.

$$\eth_{12} = \eth_{12}^{\circ(1+x_2-x_1)/2} \eth_{21}^{\circ(1+x_1-x_2)/2}$$

$$= (2.75 \times 10^{-9})^{0.225} \times (4.15 \times 10^{-9})^{0.775}$$

$$= 3.783 \times 10^{-9} \ \text{m}^2/\text{s}$$

$$\eth_{13} = \eth_{13}^{\circ(1+x_3-x_1)/2} \eth_{31}^{\circ(1+x_1-x_3)/2}$$

$$= (1.70 \times 10^{-9})^{0.225} \times (3.57 \times 10^{-9})^{0.775}$$

$$= 3.021 \times 10^{-9} \ \text{m}^2/\text{s}$$

$$\eth_{23} = \eth_{23}^{\circ(1+x_3-x_2)/2} \eth_{32}^{\circ(1+x_2-x_3)/2}$$

$$= (1.42 \times 10^{-9})^{0.5000} \times (1.91 \times 10^{-9})^{0.5000}$$

$$= 1.647 \times 10^{-9} \ \text{m}^2/\text{s}$$

The elements of $[B]$ may now be calculated from Eqs. 2.1.21 and 2.1.22 as

$$[B] = \begin{bmatrix} 0.321 & 0.047 \\ 0.051 & 0.367 \end{bmatrix} \times 10^9 \ \text{s}/\text{m}^2$$

The elements of the thermodynamic factor matrix $[\Gamma]$ are determined using the NRTL model as described in Table D.2.

$$[\Gamma] = \begin{bmatrix} 0.788 & -0.088 \\ 0.075 & 1.043 \end{bmatrix}$$

The Fick matrix $[D]$ can now be calculated.

$$[D] = [B]^{-1}[\Gamma]$$

$$= \begin{bmatrix} 2.476 & -0.701 \\ -0.142 & 2.939 \end{bmatrix} \times 10^{-9} \ \text{m}^2/\text{s}$$

At the composition $x_1 = 0.70$, $x_2 = 0.15$, $x_3 = 0.15$ the matrix $[D]$ in the molar average reference frame, recalculated from the reported $[D^V]$ values with the help of Eq. 3.2.12, is

$$[D] = \begin{bmatrix} 2.354 & -0.471 \\ 0.089 & 2.947 \end{bmatrix} \times 10^{-9} \ \text{m}^2/\text{s}$$

$[D]$ predicted by the above method agrees reasonably well with the experimental values except for the coefficient D_{21}. Since cross-coefficients reflect differences in pair diffusivities, it is not uncommon to find that small cross-coefficients are not predicted accurately, even in sign. It may be noted from Example 3.3.1 that the measured D_{ij} do not satisfy the Onsager relations accurately. This could be one reason for the deviation between the predicted and experimental D_{ij} values; the predicted set of coefficients implicitly satisfy the ORR.

The errors in flux calculations that result from any errors in the prediction of $[D]$ are, however, not large, as can be checked by assigning arbitrary values to ∇x_1 and ∇x_2. We draw comfort from the fact that the relatively large cross-coefficient D_{12} has been predicted reasonably well. ∎

4.3   MAXWELL–STEFAN, FICK, AND ONSAGER IRREVERSIBLE THERMODYNAMICS FORMULATIONS: A SUMMARY COMPARISON

The three formulations for multicomponent diffusion are all equivalent to one another and interrelatable

$$[B]^{-1}[\Gamma] \equiv [D] = [H]^{-1}[G] \qquad (4.3.1)$$

The Maxwell–Stefan coefficients $Ð_{ij} = Ð_{ji}$ are the ones that are most easily amenable to physical interpretation in terms of intermolecular friction or drag, retaining this significance even when other forces such as pressure gradients and electrostatic potential gradients coexist. The Maxwell–Stefan $Ð_{ij}$ are, therefore, convenient starting points for diffusivity prediction methods. Furthermore, the Maxwell–Stefan coefficients are well behaved as regards composition dependence in nonideal liquid mixtures, and retain this advantage even in the region of the critical point. The second law of thermodynamics restricts the infinite dilution coefficients, $Ð_{ij}^{o}$, to positive definite values. The Maxwell–Stefan $Ð_{ij}$ are independent of the choice of the reference velocity frame.

The Fick $[D]$ exhibits a complex composition dependence, reflecting as it does both frictional and thermodynamic interactions and the elements D_{ij} cannot be interpreted simply. The elements D_{ij} are dependent on the choice of the reference velocity frame. Since $[D]$ is a product of two positive definite matrices $[H]^{-1}$ and $[G]$, the Fick $[D]$ is positive definite. The Fick $[D]$ is singular at the critical point, which imparts some interesting characteristics to the mass transfer trajectories in that region (see Chapter 5). The symmetry of $[H]$ and $[G]$ matrices places restrictions on the elements of $[D]$ and only $\frac{1}{2}n(n-1)$ of the D_{ij} are independent. The Fick formulation is more easily introduced into the continuity equations and in this sense they are often referred to as being "practical."

To illustrate some of the above points we have carried out calculations of the Maxwell–Stefan coefficients and the Fick matrix $[D]$ over the entire ternary composition triangle for the system acetone–benzene–carbon tetrachloride using the data and methods employed in Example 4.2.6. The Maxwell–Stefan diffusivities predicted using Eq. 4.2.18 are shown as a function of composition in Figure 4.9. The Fick diffusivities are shown in Figure 4.10. These illustrations are analogous to Figures 4.6 and 4.5, respectively. Figures 4.5, 4.9,

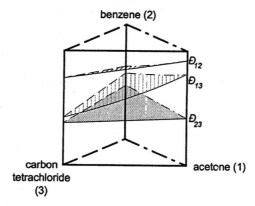

Figure 4.9. The Maxwell–Stefan diffusion coefficients predicted using a generalized Vignes equation for a nonideal liquid mixture.

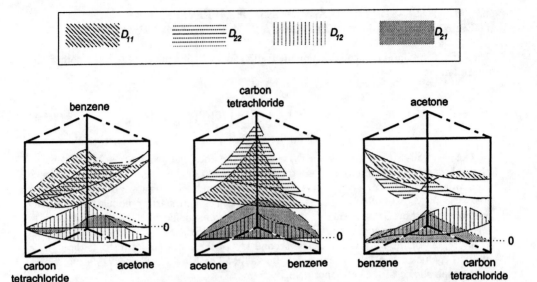

Figure 4.10. The Fick diffusion coefficients as a function of composition and component numbering for a nonideal liquid mixture.

and 4.10 graphically emphasize the well behavedness of the Maxwell–Stefan diffusivities and the complex composition dependence of the Fick D_{ij}, even for ideal gases.

The Onsager IT formulation is particularly useful in the analysis of mass transfer processes in the region of the critical point because it is the Hessian matrix $[G]$ that dictates thermodynamic stability.

We may conclude by saying that in practice we need all three formulations in analyzing multicomponent diffusion in one context or another.

5 Solution of Multicomponent Diffusion Problems: The Linearized Theory

The linearized theory would be of interest even if restricted to small changes in concentration since it gives the essential behavior of multicomponent systems. The fact that it may yield accurate results even for large changes in concentration is encouraging, for higher accuracy will require facing the nonlinear characteristics of the differential equations.

—H. L. Toor (1964a)

In order to analyze multicomponent diffusion processes we must be able to solve the continuity equations (Eq. 1.3.9) together with constitutive equations for the diffusion process and the appropriate boundary conditions. A great many problems involving diffusion in binary mixtures have been solved. These solutions may be found in standard textbooks, as well as in specialized books, such as those by Crank (1975) and Carslaw and Jaeger (1959).

The solution of multicomponent diffusion problems is a little more complicated than the solution of binary diffusion problems because the differential equations governing the process are coupled. In the early 1960s a versatile and powerful method of solving multicomponent diffusion problems was developed independently by Toor (1964a) and by Stewart and Prober (1964). The method they proposed is described and illustrated in this chapter.

5.1 MATHEMATICAL PRELIMINARIES

5.1.1 The Binary Diffusion Equations

For a binary system, the conservation equations (Eq. 1.3.9) may be written in terms of the molar fluxes N_1 and N_t as

$$\frac{\partial c_1}{\partial t} = -\nabla \cdot N_1 = -\nabla \cdot J_1 - \nabla \cdot (N_t x_1) \tag{5.1.1}$$

Inserting Fick's law (Eq. 3.1.1), into Eq. 5.1.1 gives

$$\frac{\partial c_1}{\partial t} + \nabla \cdot (N_t x_1) = \nabla \cdot (c_t D \nabla x_1) \tag{5.1.2}$$

In most cases the general Eq. 5.1.2 is simplified somewhat before being solved. It is common, for example, to assume constant molar density c_t and Fick diffusion coefficient D. Unless these (or other) assumptions are made it may be difficult or impossible to solve

Eq. 5.1.2 analytically. With these assumptions Eq. 5.1.2 simplifies to

$$\frac{\partial x_1}{\partial t} + \nabla \cdot u x_1 = D \nabla^2 x_1 \tag{5.1.3}$$

Furthermore, if the velocity u is zero then Eq. 5.1.3 becomes

$$\frac{\partial x_1}{\partial t} = D \nabla^2 x_1 \tag{5.1.4}$$

an expression sometimes referred to as Fick's second law. This equation is usually used for diffusion in solids or stationary liquids and for equimolar counterdiffusion in gases. Other simplifications of Eq. 5.1.2 can be found in texts by Bird et al. (1960) and by Slattery (1981).

5.1.2 The Multicomponent Diffusion Equations

Both formulations of the constitutive equations for multicomponent diffusion, the Maxwell–Stefan equations and the generalized Fick's law, are most compactly written in matrix form. It might, therefore, be as well to begin by writing the continuity equations (Eq. 1.3.9) in $n - 1$ dimensional matrix form as well

$$\frac{\partial(c)}{\partial t} = -\nabla \cdot (N) = -\nabla \cdot (J) - \nabla \cdot (N_t(x)) \tag{5.1.5}$$

Inserting the generalized Fick's law, (Eq. 3.2.5) into this result gives

$$\frac{\partial(c)}{\partial t} + \nabla \cdot (N_t(x)) = \nabla \cdot (c_t[D](\nabla x)) \tag{5.1.6}$$

Equations 5.1.6 represent a set of $n - 1$ coupled partial differential equations. Since the Fick matrix $[D]$ is a strong function of composition it is not always possible to obtain exact solutions to Eqs. 5.1.6 without recourse to numerical techniques. The basis of the method put forward by Toor and by Stewart and Prober is the assumption that c_t and $[D]$ can be considered constant. (Actually, Toor worked with the generalized Fick's law formulation, whereas Stewart and Prober worked with the Maxwell–Stefan formulation. Toor et al. (1965) subsequently showed the two approaches to be equivalent.) With this assumption Eqs. 5.1.6 reduce to

$$c_t \frac{\partial(x)}{\partial t} + \nabla \cdot (N_t(x)) = c_t[D](\nabla^2 x) \tag{5.1.7}$$

or, equivalently

$$\frac{\partial(x)}{\partial t} + \nabla \cdot (u(x)) = [D](\nabla^2 x) \tag{5.1.8}$$

For binary systems the matrix equations (Eqs. 5.1.7 and 5.1.8) reduce to Eqs. 5.1.2 and 5.1.3.

The theory of Toor and of Stewart and Prober is referred to as the linearized theory of multicomponent mass transfer because the set of nonlinear Eqs. 5.1.6 is linearized to give the set Eq. 5.1.7.

5.1.3 Solving the Multicomponent Equations

The general method of solution that was proposed by Toor and by Stewart and Prober exploits the properties of the modal matrix $[P]$ whose columns are the eigenvectors of $[D]$ (see Appendix A.4). The matrix product

$$[P]^{-1}[D][P] = [\hat{D}] \tag{5.1.9}$$

is a diagonal matrix with elements that are the eigenvalues of $[D]$. Cullinan (1965) showed that this transformation is always possible. Thus, on premultiplying Eqs. 5.1.7 by $[P]^{-1}$ and inserting an identity matrix $[P][P]^{-1}$ between $[D]$ and $\nabla^2(x)$ gives

$$c_t \frac{\partial(\hat{x})}{\partial t} + \nabla \cdot (N_t(\hat{x})) = c_t[\hat{D}](\nabla^2 \hat{x}) \tag{5.1.10}$$

where we have defined "pseudocompositions" $(\hat{x})$ by

$$(\hat{x}) = [P]^{-1}(x) \tag{5.1.11}$$

The generalized Fick's law (Eq. 3.2.5), may also be premultiplied by $[P]^{-1}$ to give

$$(\hat{J}) = -c_t[\hat{D}](\nabla \hat{x}) \tag{5.1.12}$$

where the "pseudodiffusion fluxes" $\hat{J}_i$ are given by

$$(\hat{J}) = [P]^{-1}(J) \tag{5.1.13}$$

Examination of Eqs. 5.1.7–5.1.10 shows that the similarity transformation reduces the original set of $n-1$ coupled partial differential equations to a set of $n-1$ uncoupled partial differential equations in the pseudocompositions. Equation (5.1.10) for the ith pseudocomponent is

$$c_t \frac{\partial \hat{x}_i}{\partial t} + \nabla \cdot N_t \hat{x}_i = c_t \hat{D}_i \nabla^2 \hat{x}_i \tag{5.1.14}$$

with pseudodiffusion fluxes given by a set of uncoupled constitutive relations

$$\hat{J}_i = -c_t \hat{D}_i \nabla \hat{x}_i \tag{5.1.15}$$

If we compare Eqs. 5.1.14 with the conservation equation (Eq. 5.1.2) for a binary system and the pseudo-Fick's law Eq. 5.1.15, with Eq. 3.1.1 then we can see that from the mathematical point of view these pseudomole fractions and pseudofluxes behave as though they were the corresponding variables of a real binary mixture with diffusion coefficient $\hat{D}_i$. The fact that the $\hat{D}_i$ are real, positive, and invariant under changes of reference velocity strengthens the analogy. If the initial and boundary conditions can also be transformed to pseudocompositions and fluxes by the same similarity transformation, the uncoupled equations represent a set of independent binary-type problems, $n-1$ in number. Solutions to binary diffusion problems are common in the literature (see, e.g., Bird et al., 1960; Slattery, 1981; Crank, 1975). Thus, the solution to the corresponding multicomponent problem can be written down immediately in terms of the pseudomole fractions and fluxes. Specifically, if a binary diffusion problem has the solution

$$\frac{(x_1 - x_{1\infty})}{(x_{10} - x_{1\infty})} = f(r, t, D_{12}, N_t) \tag{5.1.16}$$

then the corresponding multicomponent problem has the solution

$$\frac{(\hat{x}_i - \hat{x}_{i\infty})}{(\hat{x}_{i0} - \hat{x}_{i\infty})} = f(r, t, \hat{D}_i, N_t) \tag{5.1.17}$$

or, equivalently,

$$(\hat{x}_i - \hat{x}_{i\infty}) = f(r, t, \hat{D}_i, N_t)(\hat{x}_{i0} - \hat{x}_{i\infty}) \tag{5.1.18}$$

where $\hat{x}_{i0}$ and $\hat{x}_{i\infty}$ are suitably transformed boundary conditions. The precise form of the function f depends on the problem at hand.

In order to recover the solution to the original problem in terms of real mole fractions and fluxes we apply the inverse transformations

$$(x) = [P](\hat{x}) \tag{5.1.19}$$

and

$$(J) = [P](\hat{J}) \tag{5.1.20}$$

When the transformation (Eq. 5.1.19) is applied to Eqs. 5.1.18 we find

$$[P](\hat{x} - \hat{x}_\infty) = [P][\hat{f}][P]^{-1}[P](\hat{x}_0 - \hat{x}_\infty) \tag{5.1.21}$$

where $[\hat{f}]$ is a diagonal matrix whose entries are given by $\hat{f}_{ii} = f(r, t, \hat{D}_i, N_t)$. In view of Eq. 5.1.19, Eq. 5.1.21 simplifies to

$$(x - x_\infty) = [f](x_0 - x_\infty) \tag{5.1.22}$$

where $[f]$ is defined by

$$[f] \equiv [P][\hat{f}][P]^{-1} \tag{5.1.23}$$

The application of the transformation (Eq. 5.1.20) allows us to recover the generalized Fick's law (Eq. 3.2.5); with the composition gradients then obtained by differentiation of Eq. 5.1.22 we have

$$(J) = c_t[\nabla f](x_0 - x_\infty) \tag{5.1.24}$$

with

$$[\nabla f] \equiv [P][\nabla\hat{f}][P]^{-1} \tag{5.1.25}$$

where $[\nabla\hat{f}]$ is a diagonal matrix with elements that are the gradients of the diagonal elements of $[f]$.

The assumption of constant $[D]$, therefore, allows a host of solutions to multicomponent diffusion problems to be obtained quite simply.

5.1.4 Special Relations for Ternary Systems

For ternary systems the computation of the modal matrix $[P]$ is fairly straightforward. The two eigenvalues, $\hat{D}_1$ and $\hat{D}_2$, are given by Eqs. 3.2.9.

$$\hat{D}_1 = \tfrac{1}{2}\{\text{tr}[D] + \sqrt{\text{disc}[D]}\}$$

$$\hat{D}_2 = \tfrac{1}{2}\{\text{tr}[D] - \sqrt{\text{disc}[D]}\}$$

$$(3.2.9)$$

where

$$\text{tr}[D] = D_{11} + D_{22}$$

$$\text{disc}[D] = (\text{tr}[D])^2 - 4|D|$$

$$|D| = D_{11}D_{22} - D_{12}D_{21}$$

are, respectively, the trace, discriminant, and determinant of $[D]$.

Once the eigenvalues are known, the eigenvectors are found by solving the linear system

$$\big[[D] - \hat{D}_i[I]\big](P_i) = (0) \qquad (5.1.26)$$

for (P_i), the ith eigenvector of $[D]$. For the three component case Eq. 5.1.26 simplifies to

$$\begin{bmatrix} D_{11} - \hat{D}_i & D_{12} \\ D_{21} & D_{22} - \hat{D}_i \end{bmatrix} \begin{pmatrix} P_{i_1} \\ P_{i_2} \end{pmatrix} = \begin{pmatrix} 0 \\ 0 \end{pmatrix} \qquad (5.1.27)$$

Carrying out the multiplications required by Eq. 5.1.27 and solving for P_{i_2} in terms of P_{i_1} gives

$$P_{i_2} = -\big(D_{11} - \hat{D}_i\big)P_{i_1}/D_{12} \qquad (5.1.28)$$

which has been derived from the first row or

$$P_{i_2} = -D_{21}P_{i_1}/\big(D_{22} - \hat{D}_i\big) \qquad (5.1.29)$$

which is obtained from the second row.

The parameters P_{11} and P_{22} may take any value; unity is, however, the obvious choice. Thus, if we let $[P]$ be the modal matrix of $[D]$ formed from the eigenvectors of $[D]$ we may have

$$[P([D])] = \big[P_1(\hat{D}_1), P_2(\hat{D}_2)\big]$$

$$= \begin{bmatrix} 1 & \dfrac{\hat{D}_2 - D_{22}}{D_{21}} \\ \dfrac{\hat{D}_1 - D_{11}}{D_{12}} & 1 \end{bmatrix} \qquad (5.1.30)$$

An alternative structure for $[P]$ is

$$P = \begin{bmatrix} 1 & 1 \\ \dfrac{\hat{D}_1 - D_{11}}{D_{12}} & \dfrac{D_{21}}{\hat{D}_2 - D_{22}} \end{bmatrix} \tag{5.1.31}$$

Other structures for $[P]$ are possible, it all depends on what values we choose for P_{11} and P_{22}.

When the cross-coefficients D_{12} and D_{21} are negligibly small compared to the main coefficients D_{11} and D_{22}, the eigenvalues approach the values of the main coefficients. The matrix $[P]$ (and its inverse) will, therefore, tend towards the identity matrix $[I]$. A matrix $[P]$ that is quite different from $[I]$ is an indication of a strongly coupled system.

5.2   INTERACTION EFFECTS

The presence of nonzero cross-coefficients, $D_{ij} \neq 0$ $(i \neq j)$, in the Fick matrix $[D]$ lends to multicomponent systems characteristics quite different from the corresponding binary system. These characteristics are best illustrated by considering a binary system for which the diffusion flux is given by Eq. 3.1.1

$$J_1 = -c_t D \nabla x_1 \tag{3.1.1a}$$

where D is the binary Fick diffusion coefficient. For a ternary mixture the fluxes are given explicitly by Eqs. 3.2.1 and 3.1.2

$$J_1 = -c_t D_{11} \nabla x_1 - c_t D_{12} \nabla x_2 \tag{3.2.1}$$

$$J_2 = -c_t D_{21} \nabla x_1 - c_t D_{22} \nabla x_2 \tag{3.2.2}$$

where the D_{ij} are the multicomponent Fick diffusion coefficients. Representative composition profiles for a binary system and for a ternary system are shown in Figure 5.1. For a binary system the two composition gradients are equal in magnitude but opposite in sign: $\nabla x_2 = -\nabla x_1$ [Fig. 5.1(a)]. For a multicomponent system, however, the composition gradients of the individual species may be very different [Fig. 5.1(b)]; subject only to the

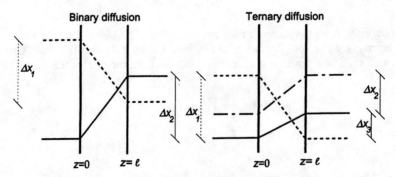

Figure 5.1. Representative composition profiles in (a) binary and (b) ternary systems. Note that for all systems $\sum \nabla x_i = 0$. For binary systems this means that $\nabla x_2 = -\nabla x_1$ but a similar restriction does not exist for the ternary system.

requirement that the composition gradients must sum over all species to zero. Indeed, we see that is possible for one component, number 1 say, to have a vanishingly small composition gradient, $\nabla x_1 = 0$, at the same time that other components have nonzero, and possibly quite large, driving forces for diffusion. Since, in general, D_{12} is not zero, it follows that we may have a diffusion flux of component 1 even in the absence of a composition gradient for component 1

$$J_1 = -c_t D_{12} \nabla x_2 \qquad (\nabla x_1 = 0) \tag{5.2.1}$$

This phenomenon is known as *osmotic diffusion* following Toor (1957) who investigated these interaction effects. It is also possible for species 1 not to diffuse at all if the first and second members of the right-hand side of Eq. 3.2.1 are precisely equal in magnitude but opposite in sign

$$J_1 = 0 \qquad (c_t D_{11} \nabla x_1 = -c_t D_{12} \nabla x_2) \tag{5.2.2}$$

This condition is known as a *diffusion barrier* (Toor, 1957). Finally, it is possible for species 1 to diffuse in a direction opposite to that indicated by its own concentration gradient.

$$J_1/(-\nabla x_1) < 0 \qquad (D_{11} \nabla x_1 < D_{12} \nabla x_2 \quad \text{and} \quad \nabla x_2/\nabla x_1 < 0) \tag{5.2.3}$$

This is known as *reverse diffusion* (Toor, 1957).

These interaction effects are illustrated in Figure 5.2, where we have plotted the diffusion flux J_1 as a function of $-\nabla x_1$ for a binary and for a hypothetical ternary system. Fick diffusion coefficients and the gradient of component 2, ∇x_2, are considered independent of composition for the purposes of drawing these diagrams. Notice that the line that represents Fick's law for binary systems in Figure 5.2(a) passes through the origin. For ternary systems, the line that represents Eq. 3.2.1 does not, in general, pass through the origin but is shifted up or down to an extent that depends on the magnitude and sign of $D_{12} \nabla x_2$. The region of reverse diffusion is delimited by the osmotic diffusion point and the

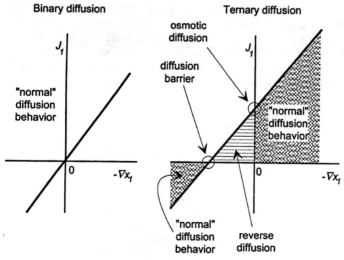

Figure 5.2. The diffusion flux as a function of the composition gradient for (*a*) binary and (*b*) ternary systems. Note the diffusion barrier, the osmotic diffusion point and the region of reverse diffusion that are possible in ternary systems.

diffusion barrier. Outside of this region the direction of diffusion is "normal" but this does not mean that diffusion fluxes are uninfluenced by the other composition gradients.

It should now be clear that diffusional interaction effects cannot occur in binary mixtures or in multicomponent systems where all the cross-coefficients, D_{ij} $(i \neq j)$, vanish. This will be the case in mixtures in which all the binary diffusion coefficients are alike (the system oxygen–nitrogen–carbon monoxide discussed in Example 4.2.1 is a case in point), as well as in mixtures where one component is present in very large excess (see Example 4.2.2). Thus, the first requirement for significant interaction effects is that the cross-coefficients D_{ij} be "large" compared to the main coefficients D_{ii}. That is,

$$|D_{ij}/D_{ii}| \sim O(1) \tag{5.2.4}$$

This condition is satisfied in the systems acetaldehyde–hydrogen–ethanol discussed in Example 4.2.3 and hydrogen–nitrogen–dichlorodifluoromethane used to illustrate Example 4.2.4.

It is not sufficient for the multicomponent diffusion coefficients merely to satisfy the approximation bound in Eq. 5.2.4 in order to have significant diffusional interaction effects; a large cross-diffusion coefficient may be multiplied by a vanishingly small concentration gradient leading to negligible interactions. It is also necessary to have a significant gradient in the mole fraction of species j in order for the flux of species i $(i \neq j)$ to be affected to any extent. A more demanding criterion for significant interaction effects, therefore, is

$$|D_{ij} \nabla x_j / D_{ii} \nabla x_i| \sim O(1) \tag{5.2.5}$$

or, in other words, that the area of reverse diffusion in Figure 5.2(b) be "large."

At first sight it might appear that the second law of thermodynamics is violated for reverse diffusion to occur. This is not so. One process may depart from equilibrium in such a sense as to consume entropy provided it is coupled to another process that produces entropy even faster. This is, of course, the basic principle of any pump, whether it moves water uphill or moves heat towards a higher temperature region. For the second law requirement $\sigma > 0$ to hold it is allowable for σ_1 to be less than zero, corresponding to reverse diffusion for 1, provided σ_2 and σ_3, due to species 2 and 3 diffusion, be such that the overall entropy production rate is positive $(\sigma_1 + \sigma_2 + \sigma_3 > 0)$.

These interaction phenomena have been discussed above in the context of ternary gas diffusion but are typical of the general multicomponent case. The practical implications of these interaction phenomena include such interesting possibilities as negative Murphree point efficiencies in multicomponent distillation as we shall see later (Chapter 13).

5.3 STEADY-STATE DIFFUSION

As our first application of the linearized theory we consider steady-state, one-dimensional diffusion. This is the simplest possible diffusion problem and has applications in the measurement of diffusion coefficients as discussed in Section 5.4. Steady-state diffusion also is the basis of the film model of mass transfer, which we shall discuss at considerable length in Chapter 8. We will assume here that there is no net flux $N_t = 0$. In the absence of any total flux, the diffusion fluxes and the molar fluxes are equal: $N_i = J_i$.

With the above assumptions, the differential mass balance (Eq. 5.1.8), simplifies to

$$\frac{d^2 x_i}{dz^2} = 0 \tag{5.3.1}$$

The mole fractions x_i are known at two planes a distance ℓ apart. The boundary conditions therefore are

$$z = 0, x = x_{i0}; \qquad z = \ell, x = x_{i\ell} \tag{5.3.2}$$

In this case the diffusion equations are already uncoupled so we do not need to use the diagonalization procedure discussed above. The solution to the set of uncoupled linear ordinary differential equations (Eq. 5.3.1) is obtained as

$$\frac{x_i - x_{i0}}{x_{i\ell} - x_{i0}} = z/\ell \tag{5.3.3}$$

Equation 5.3.3 shows that the composition profiles are linear.

The flux in a binary mixture is obtained from Fick's law, (Eq. 3.1.1) with the composition derivative obtained by differentiating Eq. 5.3.3 to give

$$J_1 = \frac{c_t D}{\ell}(x_{10} - x_{1\ell}) \tag{5.3.4}$$

This equation is a special case of a more general result we obtain later (Section 8.2).

For a multicomponent system we write the first $n - 1$ Eqs. 5.3.3 into column matrix form as

$$(x - x_0) = (z/\ell)(x_\ell - x_0) \tag{5.3.5}$$

This matrix equation can be differentiated to give

$$\frac{d(x)}{dz} = (1/\ell)(x_\ell - x_0) \tag{5.3.6}$$

The fluxes in a multicomponent system are calculated from the one-dimensional form of the generalized Fick's law (Eq. 3.2.5) with the composition derivatives given by Eq. 5.3.6 to give

$$(J) = -c_t[D]\frac{d(x)}{dz} = \frac{c_t[D]}{\ell}(x_0 - x_\ell) \tag{5.3.7}$$

Equation 5.3.7 is the proper matrix generalization of Eq. 5.3.4.

Example 5.3.1   Steady-State Diffusion in a Ternary System

For our first multicomponent flux calculation we ask you to calculate the fluxes for the ternary system hydrogen (1)–nitrogen (2)–carbon dioxide (3) under the following conditions:

The boundary compositions are

$$x_{10} = 0.0 \qquad x_{20} = 0.50086 \qquad x_{30} = 0.49914$$
$$x_{1\ell} = 0.50121 \qquad x_{2\ell} = 0.49879 \qquad x_{3\ell} = 0.0$$

The diffusion path length is 85.9 mm.
The temperature is 35.2°C.

The values of the diffusion coefficients of the three binary pairs at 35.2°C and a pressure of 101.3 kPa are as follows:

$$H_2 - N_2 \qquad Đ_{12} = 83.3 \text{ mm}^2/\text{s}$$
$$H_2 - CO_2 \qquad Đ_{13} = 68.0 \text{ mm}^2/\text{s}$$
$$CO_2 - N_2 \qquad Đ_{23} = 16.8 \text{ mm}^2/\text{s}$$

The total molar flux may be assumed to be zero: $N_t = 0$.

SOLUTION The molar density c_t can be estimated from the ideal gas law. At 35.2°C and 101.3 kPa this gives

$$c_t = P/RT$$
$$= 101.3 \times 10^3/(8.3144 \times 308.35)$$
$$= 39.513 \text{ mol/m}^3$$

The arithmetic average composition is

$$x_{1av} = 0.2506 \qquad x_{2av} = 0.4998 \qquad x_{3av} = 0.2496$$

[D] may now be computed from Eqs. 4.2.7 at the average composition

$$[D] = \begin{bmatrix} 7.682 & -0.109 \\ -3.832 & 2.155 \end{bmatrix} \times 10^{-5} \text{ m}^2/\text{s}$$

We now calculate the fluxes from Eq. 5.3.7

$$(J) = \frac{39.513}{0.0859} \begin{bmatrix} 7.682 & -0.109 \\ -3.832 & 2.155 \end{bmatrix} \times 10^{-5} \begin{pmatrix} -0.5012 \\ 0.0021 \end{pmatrix}$$
$$= \begin{pmatrix} -1.771 \\ 0.885 \end{pmatrix} \times 10^{-2} \text{ mol/m}^2 \text{ s}$$

The diffusion fluxes J_i and the molar fluxes N_i are equal in this case because the total molar flux N_t is zero.

Although nitrogen is diffusing "normally" (i.e., in the direction that would be expected on the basis of the composition change for nitrogen alone), the magnitude of the nitrogen flux is far in excess of what we might have anticipated on the basis of the Fick diffusion coefficient D_{22} and the rather small driving force for nitrogen Δx_2. In fact, interaction effects are particularly strong in this system under these conditions. Not only is the criterion in Eq. 5.2.4 satisfied for nitrogen (component 2, but not for hydrogen, component 1)

$$D_{12}/D_{11} = -0.0142$$
$$D_{21}/D_{22} = -1.7783$$

but so, also, is the criterion in Eq. 5.2.5

$$D_{12} \Delta x_2/D_{11} \Delta x_1 = 5.871 \times 10^{-5}$$
$$D_{21} \Delta x_1/D_{22} \Delta x_2 = 430.59$$

This calculation shows that by far the largest contribution to the nitrogen (species 2) flux is the product of the Fick diffusion coefficient D_{21} with the driving force for hydrogen (component 1). ∎

5.4 DIFFUSION IN A TWO BULB DIFFUSION CELL

The two bulb diffusion cell is a simple device that can be used to measure diffusion coefficients in binary gas mixtures. Figure 5.3 is a schematic of the apparatus. Two vessels containing gas mixtures with different compositions are connected by a capillary tube. At the start of the experiment (at $t = 0$), the valve is opened and the gases in the two bulbs allowed to diffuse along the capillary tube. Samples from each bulb are taken after some time and this information is used to calculate the binary diffusion coefficient.

An analysis of binary diffusion in the two bulb diffusion apparatus has been presented by Ney and Armistead (1947) (see, also, Geankoplis, 1972). Their development is extended below for multicomponent systems.

It is assumed that each bulb is at a uniform composition (the composition of each bulb is, of course, different until equilibrium is reached). It is further assumed that the volume of the capillary tube connecting the bulbs is negligible in comparison to the volume of the bulbs themselves. This allows us to express the component material balances around each bulb as follows:

$$c_t V_0 \frac{dx_{i0}}{dt} = -c_t V_\ell \frac{dx_{i\ell}}{dt} = -N_i A \tag{5.4.1}$$

where A is the cross-sectional area of the capillary, x_{i0} is the mole fraction of component i in the left-hand side bulb, and $x_{i\ell}$ is the mole fraction of that component in the right-hand side bulb. The molar flux of species i through the capillary tube N_i and is taken to be positive if from bulb 0 to bulb ℓ. At constant temperature and pressure the molar density of

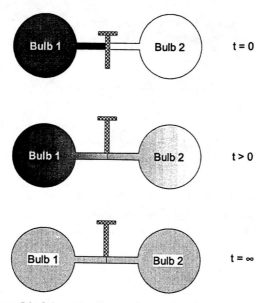

Figure 5.3. Schematic diagram of two bulb diffusion apparatus.

an ideal gas c_t is a constant; thus, there is no volume change on mixing and in the closed system depicted in Figure 5.3 the total flux N_t must be zero.

The composition in each bulb at any time is related to the composition at equilibrium $x_{i\infty}$ by

$$(V_0 + V_\ell)x_{i\infty} = V_0 x_{i0} + V_\ell x_{i\ell} \tag{5.4.2}$$

The compositions at the start of the experiment are, therefore, related by

$$(V_0 + V_\ell)x_{i\infty} = V_0 x_{i0}^\circ + V_\ell x_{i\ell}^\circ \tag{5.4.3}$$

where x° is the mole fraction at time $t = 0$.

5.4.1 Binary Diffusion in a Two Bulb Diffusion Cell

In the analysis of Ney and Armistead it is assumed that, at any instant, the flux N_1 is given by its steady-state value; that is, by Eq. 5.3.4. Thus,

$$c_t V_0 \frac{dx_{10}}{dt} = -c_t \frac{D}{\ell} A(x_{10} - x_{1\ell}) \tag{5.4.4}$$

To eliminate $x_{1\ell}$ from Eq. 5.4.11 we make use of the component material balance around both bulbs, Eqs. 5.4.2 and 5.4.3.

$$\frac{dx_{10}}{dt} = -\beta D(x_{10} - x_{1\infty}) \tag{5.4.5}$$

where β is a cell constant defined by

$$\beta = (V_0 + V_\ell)A/\ell V_0 V_\ell \tag{5.4.6}$$

A similar equation for the mole fraction of component 1 in bulb ℓ may also be derived.

Equation 5.4.5 is easily integrated, starting from the initial condition that at $t = 0$, $x_{10} = x_{10}^\circ$, to give

$$\frac{(x_{10} - x_{1\infty})}{(x_{10}^\circ - x_{1\infty})} = \exp(-\beta Dt) \tag{5.4.7}$$

Hence, if β is known then just one value of x_0 is all that is needed to calculate the diffusivity D. Alternatively, if an accurate value of D is available, Eq. 5.4.7 can be used to calibrate a diffusion cell for later use in measuring diffusion coefficients of other systems.

5.4.2 Multicomponent Diffusion in a Two Bulb Diffusion Cell

To generalize the above analysis for multicomponent systems we rewrite the time-dependent mass balances (Eq. 5.4.1) in $n - 1$ dimensional matrix form as

$$c_t V_0 \frac{d(x_0)}{dt} = -c_t V_\ell \frac{d(x_\ell)}{dt} = -(N)A \tag{5.4.8}$$

Now, with the molar fluxes given by the multicomponent rate relations (Eq. 5.3.7), we may write

$$c_t V_0 \frac{d(x_0)}{dt} = -\frac{c_t [D] A}{\ell}(x_0 - x_\ell) \qquad (5.4.9)$$

The material balance relation (Eq. 5.4.3) is used to eliminate the mole fractions (x_ℓ) to give

$$\frac{d(x_0)}{dt} = -\beta[D](x_0 - x_\infty) \qquad (5.4.10)$$

By assuming $[D]$ to be constant the matrix differential equation may be uncoupled as described above and solved subject to the "initial" conditions:

$$t = t_0, (x_0) = (x_0^\circ) \qquad (5.4.11)$$

to give the pseudocomposition profiles as

$$\frac{(\hat{x}_{i0} - \hat{x}_{i\infty})}{(\hat{x}_{i0}^\circ - \hat{x}_{i\infty})} = \exp\{-\beta \hat{D}_i (t - t_0)\} \qquad (5.4.12)$$

To recover the solution in terms of real mole fractions we group all $n - 1$ of Eqs. 5.4.12 together as

$$(\hat{x}_{i0} - \hat{x}_{i\infty}) = \exp\{-\beta \hat{D}_i (t - t_0)\}(\hat{x}_{i0}^\circ - \hat{x}_{i\infty}) \qquad (5.4.13)$$

and apply the inverse transformation (Eq. 5.1.19) to Eq. 5.4.13 to get

$$(x_0 - x_\infty) = \exp[-\beta[D](t - t_0)](x_0^\circ - x_\infty) \qquad (5.4.14)$$

which is the matrix generalization of the binary result, Eq. 5.4.7. Equations 5.4.14 allow us to calculate the time history of the concentration in each of the two bulbs. The exponential matrix in Eq. 5.4.14 may be calculated using Sylvester's expansion formula (Eq. A.5.17) or using Eq. A.5.28.

Example 5.4.1 *Multicomponent Diffusion in a Two Bulb Diffusion Cell: An Experimental Test of the Linearized Theory*

A set of multicomponent diffusion experiments in a two bulb diffusion cell apparatus was carried out by Duncan and Toor (1962) in an investigation of diffusional interaction effects. The two bulbs in their apparatus had volumes of 77.99 and 78.63 cm³, respectively. The capillary tube joining them was 85.9 mm long and 2.08 mm in diameter. The entire device was placed in a water bath at 35.2°C. The system used by Duncan and Toor was the ternary hydrogen (1)–nitrogen (2)–carbon dioxide (3). The initial concentration in each cell is

$$x_{10} = 0.0 \qquad x_{20} = 0.50086 \qquad x_{30} = 0.49914$$
$$x_{1\ell} = 0.50121 \qquad x_{2\ell} = 0.49879 \qquad x_{3\ell} = 0.0$$

where x_{i0} is the mole fraction of species i in bulb 1 and $x_{i\ell}$ is the mole fraction of species i in bulb 2.

Investigate how the concentrations in each bulb change with time.

DATA The effective ℓ/A ratio for the cell was calculated from binary diffusion experiments to be 25,810 m⁻¹ (Duncan and Toor, 1962). The pressure is 101.3 kPa. The binary diffusivities are given in Example 5.3.1.

SOLUTION   The composition at equilibrium is calculated from Eq. 5.4.3.

$$x_{1\infty} = 0.2516 \qquad x_{2\infty} = 0.4998 \qquad x_{3\infty} = 0.2486$$

The matrix $[D]$ may now be computed from Eqs. 4.2.7 at the equilibrium composition as

$$[D] = \begin{bmatrix} 7.683 & -0.1098 \\ -3.836 & 2.157 \end{bmatrix} \times 10^{-5} \ \text{m}^2/\text{s}$$

The eigenvalues of this matrix are

$$\hat{D}_1 = 7.758 \times 10^{-5} \ \text{m}^2/\text{s} \qquad \hat{D}_2 = 2.082 \times 10^{-5} \ \text{m}^2/\text{s}$$

The modal matrix $[P]$ is calculated from Eq. 5.1.30

$$[P] = \begin{bmatrix} 1 & 0.0196 \\ -0.6849 & 1 \end{bmatrix}$$

The inverse of $[P]$ is

$$[P]^{-1} = \begin{bmatrix} 0.9868 & -0.0193 \\ 0.6759 & 0.9868 \end{bmatrix}$$

Next, we need to compute the transformed boundary conditions $(\hat{x}_{i0}^{\circ})$ and $(\hat{x}_{\infty})$ using Eq. 5.1.11.

$$(\hat{x}_0^{\circ}) = [P]^{-1}(x_0^{\circ})$$
$$= \begin{bmatrix} 0.9868 & -0.0193 \\ 0.6759 & 0.9868 \end{bmatrix} \begin{pmatrix} 0.0 \\ 0.50086 \end{pmatrix}$$
$$= (-0.0097, 0.4942)^T$$

(Note the negative pseudomole fraction for pseudospecies 1!)

$$(\hat{x}_\infty) = [P]^{-1}(x_\infty)$$
$$= \begin{bmatrix} 0.9868 & -0.0193 \\ 0.6759 & 0.9868 \end{bmatrix} \begin{pmatrix} 0.2516 \\ 0.4998 \end{pmatrix}$$
$$= (0.2386, 0.6633)^T$$

The cell constant β is calculated from Eq. 5.4.6 and has the value 0.9895 m^{-2}. Thus, with $(t - t_0) = 4 \ \text{h} = 14{,}400 \ \text{s}$, we compute the pseudomole fractions from Eq. 5.4.13.

$$\hat{x}_{10} = \hat{x}_{1\infty} + \exp\{-\beta \hat{D}_1(t - t_0)\}(\hat{x}_{10}^{\circ} - \hat{x}_{1\infty})$$
$$= 0.2386 + \exp(-0.9895 \times 7.758 \times 10^{-5} \times 14{,}400) \times (-0.0097 - 0.2386)$$
$$= 0.1564$$
$$\hat{x}_{20} = \hat{x}_{2\infty} + \exp\{-\beta \hat{D}_2(t - t_0)\}(\hat{x}_{20}^{\circ} - \hat{x}_{2\infty})$$
$$= 0.6633 + \exp(-0.9895 \times 2.082 \times 10^{-5} \times 14{,}400) \times (0.4942 - 0.6633)$$
$$= 0.5376$$

The real mole fractions follow by applying the inverse transformation (Eq. 5.1.17) to give

$$(x_0) = [P](\hat{x}_0^\circ)$$

$$= \begin{bmatrix} 1 & 0.0196 \\ -0.6849 & 1 \end{bmatrix} \begin{pmatrix} 0.1564 \\ 0.5376 \end{pmatrix}$$

$$= (0.1670, 0.4305)^T$$

The composition in the other bulb can be calculated from the material balance (Eq. 5.4.3). The results of similar calculations at other times are shown in the composition time histories shown in Figure 5.4 along with experimental data from the thesis of Duncan (1960). It can be seen that the agreement between theory and experiment is very good indeed.

The boundary conditions used for the flux calculation in Example 5.3.1 were, in fact, the compositions in the two bulbs at the start of this experiment of Duncan. The initial flux of nitrogen computed in Example 5.3.1 was positive; thus, nitrogen accumulates in the right-hand bulb and is depleted in the left-hand bulb. At the beginning of the experiment, the flux of nitrogen is almost completely due to the interactions with the other two components, since the initial driving force for nitrogen is essentially nothing (Example 5.3.1). However, as nitrogen diffuses from the left-hand bulb to the right-hand bulb, a nonzero driving force of nitrogen is established that will tend to cause nitrogen to diffuse from right to left. In the early part of the experiment this driving force is not large enough to counter the reverse diffusion effect but we see in Figure 5.4 a slowing down in the rate of change of the mole fractions in each bulb. Eventually, the driving force of nitrogen is large enough to overcome the reverse diffusion effect and so the mole fraction of nitrogen in bulb 2 reaches a maximum and then starts to decrease. At the same time, the mole fraction of nitrogen in bulb 1 goes through a minimum and then starts to increase. If diffusional interaction effects did not exist, then the concentration maximum could not have occurred (we shall demonstrate this in Chapter 6).

The matrices of Fick diffusion coefficients differ slightly in Example 5.3.1 from those used here. In this example the equilibrium composition was used to calculate $[D]$, whereas in Example 5.3.1, $[D]$ was calculated at the arithmetic average composition.

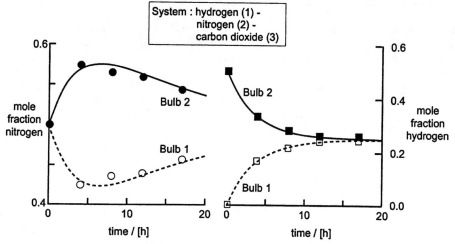

Figure 5.4. Composition–time history in two bulb diffusion cell. Experimental data from Duncan (1960).

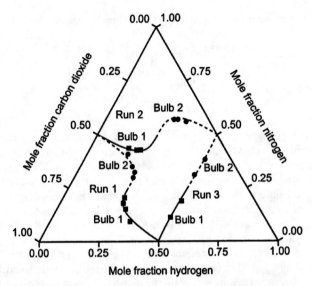

Figure 5.5. Composition profiles in two bulb diffusion cell. Calculation and experimental data for three separate experiments are shown here. Experimental data from Duncan (1960).

We have carried out similar computations covering the entire duration of three similar experiments that were carried out by Duncan and Toor (1962). The results of these calculations are shown in the triangular diagram, Figure 5.5, along with the data of Duncan (1960). We see that for all three experiments theoretical profiles are in good agreement with the data. This experiment (and others like it) provides support for the theoretical considerations of earlier chapters and the successful prediction of the concentration time history in the two bulb diffusion cell is a valuable test of the linearized theory of multicomponent diffusion. ■

5.5 THE LOSCHMIDT TUBE

Another device used to study diffusion and to measure diffusion coefficients is the Loschmidt tube illustrated in Figure 5.6. Two tubes containing fluids with different concentrations are brought together at time $t = 0$ and the fluids allowed to interdiffuse. After some time the tubes are separated and the compositions measured.

An analysis of multicomponent diffusion in a Loschmidt tube was presented by Arnold and Toor (1967). The salient results of their work are summarized below.

The equation governing unsteady-state, one-dimensional, multicomponent diffusion in the Loschmidt tube is

$$\frac{\partial(x)}{\partial t} = [D]\frac{\partial^2(x)}{\partial z^2} \tag{5.5.1}$$

The initial condition states that the concentration in each tube is uniform

$$t \leqq 0 \qquad 0 < z < \ell \qquad (x) = (x_+)$$

$$t \leqq 0 \qquad -\ell < z < 0 \qquad (x) = (x_-) \tag{5.5.2}$$

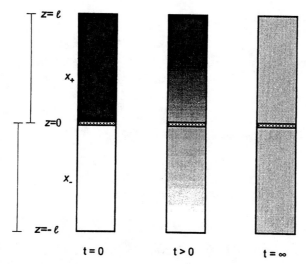

Figure 5.6. Schematic diagram of Loschmidt tube.

After the tubes are brought together diffusion proceeds, but since the tubes are sealed and are of finite length 2ℓ, there can be no mass transfer across the ends of the tube; the boundary conditions therefore are

$$t > 0 \qquad z = \pm \ell \qquad \frac{\partial (x)}{\partial z} = 0 \qquad\qquad (5.5.3)$$

Before presenting the solution to the multicomponent diffusion problem we note that for binary systems the differential Eq. 5.5.1 simplifies to

$$\frac{\partial x_1}{\partial t} = D \frac{\partial^2 x_1}{\partial z^2} \qquad\qquad (5.5.4)$$

This equation can be solved by means of the technique of separation of variables. The composition profiles as a function of time and position are given by

$$(x_1 - x_{1-})/(x_{1+} - x_{1-}) = f(z, t, D) \qquad\qquad (5.5.5)$$

where

$$f(z, t, D) = \left[\frac{1}{2} \pm \frac{1}{\pi} \sum_{k=0}^{\infty} \frac{1}{m} \sin \frac{m\pi z}{\ell} \exp\left\{ -\frac{m^2 \pi^2}{\ell} Dt \right\} \right] \qquad\qquad (5.5.6)$$

where $m = k + \frac{1}{2}$.

In order to solve the multicomponent Eqs. 5.5.1, we use the transformation (Eq. 5.1.9) to uncouple them

$$\frac{\partial \hat{x}_i}{\partial t} = \hat{D}_i \frac{\partial^2 \hat{x}_i}{\partial z^2} \qquad\qquad (5.5.7)$$

which may be solved subject to suitably transformed initial and boundary conditions.

$$
\begin{aligned}
t &\leq 0 & 0 < z < \ell & \quad (\hat{x}) = (\hat{x}_+) \\
t &\leq 0 & -\ell < z < 0 & \quad (\hat{x}) = (\hat{x}_-) \\
t &> 0 & z = \pm \ell & \quad \frac{\partial(\hat{x})}{\partial z} = 0
\end{aligned}
\tag{5.5.8}
$$

The solution for the pseudocomposition profiles as a function of time and position is given by Eq. 5.1.17

$$
\frac{(\hat{x}_i - \hat{x}_{i-})}{(\hat{x}_{i+} - \hat{x}_{i-})} = \hat{f}(z, t, \hat{D}_i)
\tag{5.5.9}
$$

where $\hat{f}(z, t, \hat{D}_i)$ is given by Eq. 5.5.6 with the ith eigenvalue $\hat{D}_i$ replacing the binary diffusivity D.

The concentration profiles in terms of real mole fractions are given by Eqs. 5.1.22 and 5.1.23

$$
(x - x_-) = [P][\hat{f}][P]^{-1}(x_+ - x_-)
\tag{5.5.10}
$$

with $[\hat{f}]$ a diagonal matrix with elements given by $\hat{f}(z, t, \hat{D}_i)$.

The average concentration $(\bar{x})$ in the bottom tube is obtained by integrating Eq. 5.5.10 over the length $-\ell < z < 0$. The result is

$$
(\bar{x} - x_-) = [P][\hat{\bar{f}}][P]^{-1}(x_+ - x_-)
\tag{5.5.11}
$$

where $[\hat{\bar{f}}]$ is another diagonal matrix with elements given by

$$
\hat{\bar{f}}_i = \left[\frac{1}{2} - \frac{1}{\pi^2} \sum_{k=0}^{\infty} \frac{1}{m^2} \exp\left\{-\frac{m^2 \pi^2}{\ell^2} \hat{D}_i t\right\}\right]
\tag{5.5.12}
$$

$(\bar{x})$ is the composition that is measured in this kind of diffusion apparatus.

Example 5.5.1 *Multicomponent Diffusion in the Loschmidt Tube: Another Test of the Linearized Theory*

Arnold and Toor (1967) investigated diffusional interaction effects in a Loschmidt tube of the kind described above. The system they used was methane (1)–argon (2)–hydrogen (3). The diffusion tube had a length of $(\pi/\ell)^2 = 60$ m^{-2}. At the temperature (34°C) and pressure of the experiments (101.3 kPa) the binary diffusion coefficients of the three ternary gas pairs were

$$
CH_4 - Ar = 21.57 \text{ mm}^2/\text{s}
$$

$$
CH_4 - H_2 = 77.16 \text{ mm}^2/\text{s}
$$

$$
Ar - H_2 = 83.35 \text{ mm}^2/\text{s}
$$

The composition in each tube at the start of one of their experiments was as follows:

$$
\begin{aligned}
x_{1-} &= 0.0 & x_{2-} &= 0.509 & x_{3-} &= 0.491 \\
x_{1+} &= 0.515 & x_{2+} &= 0.485 & x_{3+} &= 0.0
\end{aligned}
$$

Calculate the average concentration in the bottom tube after 15 min and plot the change in concentration with time.

SOLUTION   The average composition is

$$x_{1av} = 0.2575 \qquad x_{2av} = 0.4970 \qquad x_{3av} = 0.2455$$

The matrix of multicomponent diffusion coefficients is computed from Eqs. 4.2.7 using the binary diffusivities given above and the average composition. The result is

$$[D] = \begin{bmatrix} 4.442 & 1.832 \\ 3.639 & 6.298 \end{bmatrix} 10^{-5} \text{ m}^2/\text{s}$$

The eigenvalues of this matrix are

$$\hat{D}_1 = 8.114 \times 10^{-5} \text{ m}^2/\text{s} \qquad \hat{D}_2 = 2.626 \times 10^{-5} \text{ m}^2/\text{s}$$

The modal matrix $[P]$ is

$$[P] = \begin{bmatrix} 1 & 1 \\ 2.004 & -0.991 \end{bmatrix}$$

where we have chosen to set P_{11} and P_{12} to unity. The inverse of this matrix is

$$[P]^{-1} = \begin{bmatrix} 0.331 & 0.334 \\ 0.669 & -0.334 \end{bmatrix}$$

With t equal to 15 min, we compute the functions $\bar{f}_i$ from Eq. 5.5.12. We illustrate the calculation using just the first term of the series in Eq. 5.5.12.

$$\hat{\bar{f}}_1 = \frac{1}{2} - \frac{4}{\pi^2} \exp\left\{ -\frac{1}{4} \frac{\pi^2}{\ell^2} \hat{D}_1 t \right\}$$

$$= \frac{1}{2} - \frac{4}{\pi^2} \exp\left\{ -\frac{1}{4} \times 60 \times 8.114 \times 10^{-5} \times 900 \right\}$$

$$= 0.36447$$

$$\hat{\bar{f}}_2 = \frac{1}{2} - \frac{4}{\pi^2} \exp\left\{ -\frac{1}{4} \frac{\pi^2}{\ell^2} \hat{D}_2 t \right\}$$

$$= \frac{1}{2} - \frac{4}{\pi^2} \exp\left\{ -\frac{1}{4} \times 60 \times 2.626 \times 10^{-5} \times 900 \right\}$$

$$= 0.21568$$

(Note that $(\pi/\ell)^2 = 60 \text{ m}^{-2}$). In fact, these results are very close to the converged series.

$$\hat{\bar{f}}_1 = 0.36447 \qquad \hat{\bar{f}}_2 = 0.21383$$

The composition in the bottom tube is then given directly by carrying out the multiplications required by Eq. 5.5.11 from right to left.

$$(\bar{x}) = (0.135, 0.553)^T$$

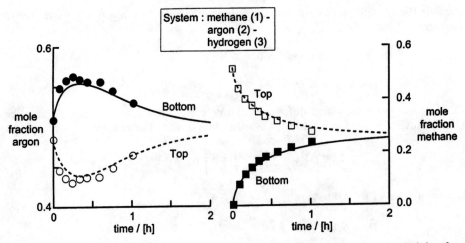

Figure 5.7. Composition–time history in Loschmidt tube diffusion experiment. Experimental data from Arnold (1965).

The composition in the top tube can be calculated simply from the relationship between the compositions at any time and the equilibrium (average) composition.

The mole fractions of argon and methane in both the bottom and top tubes are shown as a function of time in Figure 5.7 together with the experimental measurements from the thesis of Arnold (1965). The composition profiles for three different experiments are shown in the triangular diagram in Figure 5.8. It can be seen that the agreement between the predictions of the linearized equations and the experimental data is excellent. If there were no diffusional interaction effects, the composition profiles would be a straight line between the two end (initial) points in the triangular diagram in Figure 5.7 (see Example 6.5.1)!   ■

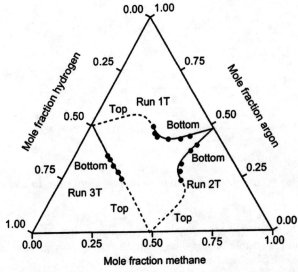

Figure 5.8. Composition profiles in Loschmidt tube. Calculation and experimental data for three separate experiments are shown here. Experimental data from Arnold (1965).

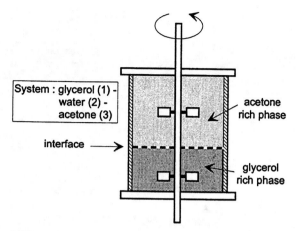

Figure 5.9. Schematic illustration of batch extraction cell used in liquid–liquid mass transfer studies.

5.6 MULTICOMPONENT DIFFUSION IN A BATCH EXTRACTION CELL

Since thermodynamic nonidealities are of the essence for phase separation in liquid–liquid systems, and such nonidealities contribute to multicomponent interaction effects, it may be expected that liquid–liquid extraction would offer an important test of the theories presented in this book. Here, we present some experimental evidence to show the significance of interaction effects in liquid–liquid extraction. The evidence we present is largely based on experiments carried out in a modified Lewis batch extraction cell (Standart et al., 1975; Sethy and Cullinan, 1975; Cullinan and Ram, 1976; Krishna et al., 1985). The analysis we present here is due to Krishna et al. (1985). The experimental system that will be used to demonstrate multicomponent interaction effects is glycerol(1)–water(2)–acetone(1); this system is of Type I. The analysis presented below is the liquid–liquid analog of the two bulb gas diffusion experiment considered in Section 5.4.

The modified Lewis batch extraction cell used by Krishna et al. (1985) to obtain mass transfer data for the system glycerol–water–acetone is illustrated in Figure 5.9. It is basically a single glass cylinder with mixing, contacting, and sampling facilities, with a net capacity of around 6 L. The heavier phase (i.e., glycerol or ″ phase) is introduced at the bottom of the cell, and a horizontal ring and disc is placed on the liquid surface. The lighter phase (acetone or ′ phase) is charged on top. Two turbine stirrers and a set of vertical baffles in each compartment provide the necessary agitation for complete mixing of the phases. Representative samples are withdrawn through sample tubes at suitable intervals (typically 30 min), and their compositions determined.

5.6.1 Equilibration Paths

From the experimental phase compositions at different intervals, the "equilibration paths" can be determined on a ternary diagram. These are parametric curves, one for each phase and with points in the two curves in pairwise correspondence. In the general case where both phases are initially unsaturated, the equilibration paths obtained experimentally (Krishna et al., 1985) are shown typically by Run C of that paper in Figure 5.10. Given enough contact time, the phases will approach and attain equilibrium, becoming mutually saturated. The equilibration paths will, therefore, terminate on the corresponding ends of a tie-line, since metastable states are prevented from arising by the constant agitation.

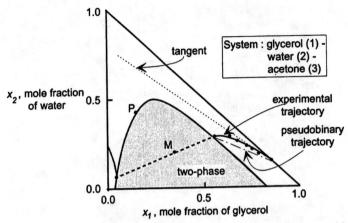

Figure 5.10. Equilibration paths during mass transfer in the system glycerol (1), water (2), and acetone (3) in a batch extraction cell. The point M is the mixture point and P represents the plait point. Experimental data correspond to Run C of Krishna et al. (1985).

Now, the initial amounts and compositions of each phase are fixed and known from experiment. If the comparatively small amounts of sample withdrawn from either phase are neglected, the extraction cell can be considered to be a closed system in which the total and constituent mass are constant and equal to the initial amounts M_{t0}, M_{i0}. At any given time, therefore,

$$M_t' + M_t'' = M_{t0} \tag{5.6.1}$$

$$M_i' + M_i'' = M_{i0} \tag{5.6.2}$$

or, in terms of the "bulk" or average compositions of the phases defined by

$$x_i' M_t' = M_i' \qquad x_i'' M_t'' = M_i'' \tag{5.6.3}$$

the following equations of conservation are obtained

$$M_t' + M_t'' = M_{t0}$$
$$x_i' M_t' + x_i'' M_t'' = M_{i0} = x_{i0} M_{t0} \qquad i = 1, 2 \tag{5.6.4}$$

These are the only independent equations of material balance that can be written down for the ternary system. All the terms in the left-hand side of Eq. 5.6.4 are functions of time. Therefore, we have three equations in two unknowns M_t' and M_t'', and this redundancy provides for a statistical check of the measurements and allows "best values" to be calculated.

Eliminating M_t' and M_t'' between the above relations, a form of the lever rule is obtained

$$\frac{x_2'' - x_2'}{x_1'' - x_1'} = \frac{x_2'' - x_{20}}{x_1'' - x_{10}} \tag{5.6.5}$$

This shows that the straight line that joins any two light- and heavy-phase compositions corresponding to a same instant during the experiments must also pass through the point that represents the overall composition of the system (point M in Fig. 5.10). This must, in particular, be true of the terminal compositions of each phase in mutual equilibrium.

Hence, we see that the equilibrium tie-line is uniquely fixed by the initial masses charged to the cell, since there is only one tie-line that passes by M (otherwise, more than two phases in equilibrium could coexist, which is not possible in a system of Type I as is the case here).

If A represents the interfacial area between the two liquid phases, we can calculate the interfacial fluxes as follows:

$$N_i = \frac{1}{A}\frac{dM_i}{dt} \qquad N_t = \frac{1}{A}\frac{dM_t}{dt} \tag{5.6.6}$$

From the material balance relations (Eqs. 5.6.1–5.6.4) we see that

$$\begin{aligned} N_i' + N_i'' &= 0 \\ N_t' + N_t'' &= 0 \end{aligned} \tag{5.6.7}$$

that is, the rates of transfer of one phase must be the negative of the corresponding rate for the other. The diffusive fluxes $(J_i = N_i - x_i N_t)$ can be calculated for either phase from

$$J_i = \frac{M_t}{A}\frac{dx_i}{dt} \qquad i = 1, 2 \tag{5.6.8}$$

where only two of the J_i are independent.

If the driving forces for mass transfer are taken to be the difference in compositions between the interface (x_{iI}) and bulk fluid (x_i), then the constitutive relations for (J) may be written as

$$(J) = \frac{c_t[D]}{\ell}(x_I - x) \tag{5.6.9}$$

where $[D]$ represents the Fick matrix of diffusion coefficients in the phase under consideration and ℓ is the effective diffusion path length. We may combine Eqs. 5.6.8 and 5.6.9 and write

$$\frac{d(x)}{dt} = [K](x_I - x) \tag{5.6.10}$$

where we have defined a volumetric transfer coefficient matrix

$$[K] = \frac{c_t A}{\ell M_t}[D] \tag{5.6.11}$$

a matrix whose elements have the units of reciprocal seconds (s^{-1}).

If we assume, for the moment, that $[K]$ is time invariant, the differential Eq. 5.6.10 can be solved with the initial condition $t = 0$, $(x) = (x_0)$, to obtain the transient composition trajectories

$$(x_I - x) = [\exp[-[K]t]](x_I - x_0) \tag{5.6.12}$$

Let $[P]$ represent the modal matrix of $[K]$, that is, $[P]$ has the property that

$$[P]^{-1}[K][P] = [\hat{K}] \tag{5.6.13}$$

where $[\hat{K}]$ represents the diagonal matrix

$$[\hat{K}] = \begin{bmatrix} \hat{K}_1 & 0 \\ 0 & \hat{K}_2 \end{bmatrix} \tag{5.6.14}$$

The columns of $[P]$ are the eigenvectors, (e_1) and (e_2), of the matrix $[K]$

$$[P] = [(e_1)(e_2)] \tag{5.6.15}$$

where

$$(e_1) = \begin{pmatrix} 1 \\ -\dfrac{K_{11} - \hat{K}_1}{K_{12}} \end{pmatrix} \qquad (e_2) = \begin{pmatrix} 1 \\ \dfrac{-K_{12}}{K_{22} - \hat{K}_2} \end{pmatrix} \tag{5.6.16}$$

Premultiplying Eq. 5.6.12 by $[P]^{-1}$ we obtain

$$(\hat{x}_I - \hat{x}) = \left[\exp\left[-[\hat{K}]t\right]\right](\hat{x}_I - \hat{x}_0) \tag{5.6.17}$$

which represents a set of two uncoupled equations

$$\hat{x}_{iI} - \hat{x}_i = \exp\{-\hat{K}_i t\}(\hat{x}_{iI} - \hat{x}_{i0}) \qquad i = 1, 2 \tag{5.6.18}$$

in pseudocompositions, defined by $(\hat{x}) = [P]^{-1}(x)$. In the pseudocomposition space $(\hat{x}_1, \hat{x}_2)$, the equilibration paths (Eq. 5.6.18) are straight lines approaching the equilibrium state.

To recover the composition profiles in the "real" (x_1, x_2) space we premultiply Eq. 5.6.17 by $[P]$ giving

$$(x_I - x) = e^{-\hat{K}_1 t}\,\Delta \hat{x}_{10}(e_1) + e^{-\hat{K}_2 t}\,\Delta \hat{x}_{20}(e_2) \tag{5.6.19}$$

where the $\Delta \hat{x}_{i0}$ are (pseudo-) initial driving forces

$$(\Delta \hat{x}_0) = [P]^{-1}(x_I - x_0). \tag{5.6.20}$$

Examination of Eq. 5.6.19 shows that the initial trajectory will be dictated by the dominant eigenvalue, say $\hat{K}_2$, and the initial path will lie along (e_2). As equilibrium is approached, the equilibration path will lie along the "slow" eigenvector (e_1) as illustrated in Figure 5.11.

Example 5.6.1 *Equilibration Paths in a Batch Extraction Cell*

We analyze the composition trajectory in the glycerol-rich phase (G) of the system glycerol(1)–water(2)–acetone(3). The initial composition G_0 is

$$x_{10} = 0.85 \qquad x_{20} = 0.15 \qquad x_{30} = 0.0$$

The composition of the interface (G_I) is

$$x_{1I} = 0.5480 \qquad x_{2I} = 0.2838 \qquad x_{3I} = 0.1682$$

For this set of conditions the matrix of volumetric transfer coefficients $[K]$, defined by Eq. 5.6.11 is (Krishna et al., 1985).

$$[K] = \begin{bmatrix} 0.2327 & -0.3742 \\ -0.1162 & 0.3373 \end{bmatrix} \text{h}^{-1}$$

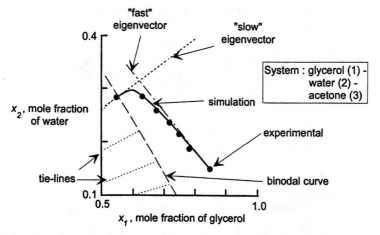

Figure 5.11. Equilibration paths during mass transfer in the system glycerol (1), water (2), and acetone (3) in a batch extraction cell. Experimental data correspond to Run C of Krishna et al. (1985).

Draw the composition trajectories in x_1, x_2 space along with the two eigenvectors (e_i) of $[K]$. Compare with the experimental data of Krishna et al. (1985).

SOLUTION We begin by defining two one-dimensional vectors for the initial and final compositions.

$$(x_0) = \begin{pmatrix} 0.85 \\ 0.15 \end{pmatrix}$$

$$(x_I) = \begin{pmatrix} 0.5480 \\ 0.2838 \end{pmatrix}$$

The first calculation step is the evaluation of the eigenvalues, $\hat{K}_1$ and $\hat{K}_2$, of $[K]$ from the following expressions

$$\hat{K}_1 = \tfrac{1}{2}\left\{ \mathrm{tr}[K] + \sqrt{\mathrm{disc}[K]} \right\}$$

$$\hat{K}_2 = \tfrac{1}{2}\left\{ \mathrm{tr}[K] - \sqrt{\mathrm{disc}[K]} \right\}$$

where

$$\mathrm{tr}[K] = K_{11} + K_{22}$$

$$\mathrm{disc}[K] = (\mathrm{tr}[K])^2 - 4|K|$$

$$|K| = K_{11}K_{22} - K_{12}K_{21}$$

The results are

$$\hat{K}_1 = 0.5 \quad \text{and} \quad \hat{K}_2 = 0.07 \text{ h}^{-1}$$

The corresponding eigenvectors (e_1) and (e_2) are given by Eq. 5.6.16

$$(e_1) = \begin{pmatrix} 1 \\ \dfrac{-K_{11} + \hat{K}_1}{K_{12}} \end{pmatrix} \qquad (e_2) = \begin{pmatrix} 1 \\ \dfrac{-K_{12}}{K_{22} - \hat{K}_2} \end{pmatrix} \qquad (5.6.16)$$

which gives

$$(e_1) = \begin{pmatrix} 1 \\ -0.7143 \end{pmatrix} \qquad (e_2) = \begin{pmatrix} 1 \\ 0.4347 \end{pmatrix}$$

The two eigenvectors are determined only in direction and not in absolute value. For definiteness we may specify that these eigenvectors pass, respectively, through the points G_0 and G_I. In (x_1, x_2) composition space the expressions for the first eigenvector is

$$\frac{x_2 - x_{20}}{x_1 - x_{10}} = \frac{e_{12}}{e_{11}}$$

or, defining x_2 as a function of x_1

$$x_2(x_1) = x_{20} - 0.7143(x_1 - x_{10})$$

The second eigenvector is given by

$$\frac{x_2 - x_{2I}}{x_1 - x_{1I}} = \frac{e_{22}}{e_{21}}$$

or, defining x_2 as a function of x_1

$$x_2(x_1) = x_{2I} + 0.4347(x_1 - x_{1I})$$

These two eigenvectors have been drawn in Figure 5.3.

The actual composition trajectory can be determined either from Eqs. 5.6.19 and 5.6.20 or directly from Eq. 5.6.12, evaluating the exponential matrix as described in Appendix A.4 or A.6. We shall illustrate the computation of the transient composition trajectories by calculating the mole fractions after 1 h from Eq. 5.6.19. In order to calculate the composition trajectory we need to determine the modal matrix $[P]$. The columns of $[P]$ are the eigenvectors (e_1) and (e_2).

$$[P] = \begin{bmatrix} 1 & 1 \\ -0.7143 & 0.4347 \end{bmatrix}$$

The initial composition driving force vector, $(\Delta x) = (x_I - x_0)$, is

$$(\Delta x) = \begin{pmatrix} -0.302 \\ 0.1338 \end{pmatrix}$$

The pseudoinitial driving forces can be determined from Eq. 5.6.20.

$$(\Delta \hat{x}) = \begin{pmatrix} -0.2307 \\ -0.0713 \end{pmatrix}$$

We may now compute the mole fractions after 1 h from Eq. 5.6.19.

$$x_1 = x_{1I} - \exp\{-\hat{K}_2 t\} \Delta \hat{x}_{10} e_{11} - \exp\{-\hat{K}_2 t\} \Delta \hat{x}_{20} e_{12}$$

$$= 0.5480 - \exp(-0.5 \times 1) \times (-0.2307) \times 1$$

$$\quad - \exp(-0.07 \times 1) \times (-0.0713) \times 1$$

$$= 0.7544$$

$$x_2 = x_{2I} - \exp\{-\hat{K}_2 t\} \Delta \hat{x}_{10} e_{21} - \exp\{-\hat{K}_2 t\} \Delta \hat{x}_{20} e_{22}$$

$$= 0.2838 - \exp(-0.5 \times 1) \times (-0.2307) \times (-0.7143)$$

$$\quad - \exp(-0.07 \times 1) \times (-0.0713) \times 0.4347$$

$$= 0.2127$$

The complete composition trajectory is plotted in Figure 5.11 and found to be highly curvilinear. Initially the trajectory is along the first "fast" eigenvector and towards the end of the equilibration process, the trajectory relaxes towards G_I along the second "slow" eigenvector. Notice the sharp turn about in the trajectory as G_I is approached. The theoretical trajectory is in good agreement with the measured data. The highly curvilinear trajectory signifies the strong coupling in the system. ∎

5.6.2 Equilibration Paths in the Vicinity of the Plait Point

Let us now consider the equilibration paths in the vicinity of the plait point P. The Fick matrix $[D]$ is singular at the plait point with eigenvalues given by Eqs. 3.3.19

$$\hat{D}_1 = 0 \qquad \hat{D}_2 = D_{11} + D_{22} \tag{3.3.19}$$

The eigenvectors corresponding to the eigenvalues (Eq. 3.3.19) are, respectively,

$$(e_1) = \begin{pmatrix} 1 \\ -\dfrac{D_{11}}{D_{12}} \end{pmatrix} \quad \text{and} \quad (e_2) = \begin{pmatrix} 1 \\ -\dfrac{D_{22}}{D_{12}} \end{pmatrix} \tag{3.3.20}$$

The matrix $[K]$, which determines the equilibration path is a function of $[D]$, and so the modal matrix of $[K]$ will have the following structure

$$[P] = [(e_1)(e_2)] = \begin{bmatrix} 1 & 1 \\ -D_{11}/D_{12} & D_{22}/D_{12} \end{bmatrix} \tag{5.6.21}$$

With (e_1) and (e_2) as shown above, the equilibration paths (Eq. 5.6.19) simplify to give (recall that $\hat{D}_1 = 0$).

$$x_{1I} - x_1 = \Delta \hat{x}_{10} + \exp\{-\hat{K}_2 t\} \Delta \hat{x}_{20} \tag{5.6.22}$$

$$x_{2I} - x_2 = -\frac{D_{11}}{D_{12}} \Delta \hat{x}_{10} + \frac{D_{22}}{D_{12}} \exp\{-\hat{K}_2 t\} \Delta \hat{x}_{20} \tag{5.6.23}$$

Eliminating $\Delta \hat{x}_{10}$ from the above two equations we obtain the following relation between

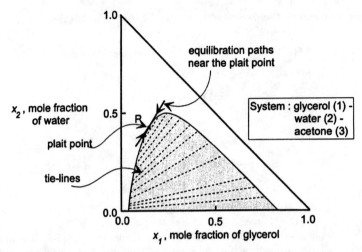

Figure 5.12. Composition trajectory in the region of the critical point in the system glycerol (1), water (2), acetone (3). Source of data: Krishna et al. 1985).

the compositions x_1 and x_2

$$(x_{2I} - x_2) = -\frac{D_{11}}{D_{12}}(x_{1I} - x_1) + \frac{D_{11} + D_{22}}{D_{12}} \exp\{-\hat{K}_2 t\} \Delta \hat{x}_{20} \qquad (5.6.24)$$

which represents a straight line in the (x_1, x_2) space with a slope $dx_2/dx_1 = -D_{11}/D_{12}$ equal to the slope of the limiting tie-line as shown in Section 3.3.1.

Figure 5.12 shows diagramatically the results of actual equilibration runs (Krishna et al., 1985) obtained with the initial phases on the tangent to the binodal curve at the critical point P. The path was indeed linear, and the rates of mass transfer were significantly lower than in other regions of the ternary diagram. The slowness of the mass transfer can be understood because the equilibration process is dictated by the smaller of the two eigenvalues. As the critical point is approached, the smaller eigenvalue tends to vanish.

5.7 THE LINEARIZED THEORY: AN APPRAISAL

The utility of the linearized Eqs. 5.1.7 would seem, on the face of things, to be limited to situations in which the diffusion coefficient matrix does not change significantly as the concentration changes during the diffusion process. Thus, the goodness of the assumption of constant $[D]$, of central importance to the theory, has been fairly thoroughly tested by those responsible for the development of the method as well as by many other investigators.

Obviously, $[D]$, if evaluated at the average concentration as usually recommended, will not change significantly if the concentration gradients are small. Significant variations of the D_{ik} with composition may be found in gas mixtures (as in Examples 4.2.3 and 4.2.4) and in nonideal liquids. For ideal gas systems it is possible to compare the predictions of the linearized equations against exact solutions of the Maxwell–Stefan equations. The calculations of a number of investigators (Stewart and Prober, 1964; Arnold and Toor, 1967; Cotrone and De Giorgi, 1971; Taylor and Webb, 1980b, 1981; Taylor, 1982c; Smith and Taylor, 1983; Krishna et al., 1976; Bandrowski and Kubaczka, 1981; Webb and Sardesai, 1981; Taylor et al., 1985) all show the linearized equations to be of excellent accuracy even in situations involving the highest possible composition gradients. (It turns out that any error in assuming [D] to be constant is usually favorably compensated by other details of the mathematical solution.) It can be concluded from that, at least for the purposes of calculating mass transfer rates in multicomponent mixtures, the linearized theory is almost always adequate for engineering purposes. This conclusion is comforting, since with the exception of the film model of steady-state, one-dimensional diffusion, it has so far proved impossible to obtain exact analytic solutions of the Maxwell–Stefan equations that are easy to use.

While fluxes computed from the linearized equations usually compare favorably with fluxes computed from exact solutions, the same may not always be said for composition profiles (see, e.g., Krishnamurthy and Taylor, 1982). Indeed, the assumption of constant [D] may sometimes lead to physically impossible composition profiles (see, e.g., Gupta and Cooper, 1971).

The linearized theory of Toor (1964a) and of Stewart and Prober (1964) is probably the most important method of solving multicomponent diffusion problems. Very often, the method provides the only practical means of obtaining useful analytical solutions of multicomponent diffusion problems. Additional applications of the method are developed in Chapters 8–10 and still more can be found in the literature [see Cussler (1976), Krishna and Standart (1979) and Taylor (1982c) for sources].

6 Solution of Multicomponent Diffusion Problems: Effective Diffusivity Methods

It takes more time to scotch falsehood and expose fables than it does to set forth something solid and new.

—C. A. Truesdell (1969)

The complexity of the Maxwell–Stefan equations and the generalized Fick's law have lead many investigators to use simpler constitutive relations that avoid the mathematical complexities (specifically, the use of matrix algebra in applications). In this chapter we examine these effective diffusivity or pseudobinary approaches.

6.1 THE EFFECTIVE DIFFUSIVITY

6.1.1 Definitions

The effective diffusivity is defined by assuming that the rate of diffusion of species i depends only on the composition gradient of species; that is,

$$J_i = -c_t D_{i,\text{eff}} \nabla x_i \qquad i = 1, 2, \ldots, n \tag{6.1.1}$$

where $D_{i,\text{eff}}$ is some characteristic diffusion coefficient of species i in the mixture.

There is a second class of effective diffusivity method in which $D_{i,\text{eff}}$ is defined with respect to the molar flux N_i as follows:

$$N_i = -c_t D_{i,\text{eff}} \nabla x_i \qquad i = 1, 2, \ldots, n \tag{6.1.2}$$

We continue the discussion by focusing attention on the more widely used of these two definitions (Eq. 6.1.1).

For two component systems Eq. 6.1.1 reduces to Fick's law of diffusion (Eq. 3.1.1). But does it provide a generally acceptable description of diffusion in multicomponent mixtures? If we sum Eqs. 6.1.1 over the n species, we find in view of the restriction $\Sigma J_i = 0$, that

$$c_t \sum_{i=1}^{n} D_{i,\text{eff}} \nabla x_i = 0 \tag{6.1.3}$$

Eliminating the nth gradient ∇x_n from Eq. 6.1.3 we obtain

$$c_t \sum_{i=1}^{n-1} (D_{i,\text{eff}} - D_{n,\text{eff}}) \nabla x_i = 0 \tag{6.1.4}$$

If we now demand of the $D_{i,\text{eff}}$ that they be independent of the composition gradients (as are the Maxwell–Stefan and Fick diffusion coefficients) then, since each of the $n - 1$ gradients in Eq. 6.1.4 can be varied independently, the only solution possible to Eq. 6.1.4 is that all the effective diffusivities are equal to one another

$$D_{i,\text{eff}} = D_{n,\text{eff}} \qquad i = 1, 2, \ldots, n - 1 \tag{6.1.5}$$

From our knowledge of diffusion in binary and ternary mixtures we can assert that this circumstance, even if likely, is certainly not general (actually, it is not even very likely). Consider, for example, the ternary gas mixture H_2, N_2, and CCl_2F_2. The diffusion coefficients of the binary pairs are given in Example 4.2.4. These values are so different that it really does not seem possible that diffusion in the ternary mixture could be described by a single diffusivity. We may expect a simple result like Eq. 6.1.5 to be true only when all the species making up the mixture are of a similar nature as, for example, in the mixture nitrogen–oxygen–carbon monoxide considered in Example 4.2.1.

6.1.2 Relationship Between Effective, Maxwell–Stefan, and Multicomponent Fick Diffusion Coefficients

Equation 6.1.1 maybe rearranged to give the mole fraction gradient as

$$\nabla x_i = -\frac{J_i}{c_t D_{i,\text{eff}}}$$

$$= -\frac{(N_i - x_i N_t)}{c_t D_{i,\text{eff}}} \tag{6.1.6}$$

which may be equated to the mole fraction gradient given by the Maxwell–Stefan Eqs. 2.1.16 in order to obtain a general expression for $D_{i,\text{eff}}$ as (Bird et al., 1960)

$$D_{i,\text{eff}} = \left[(N_i - N_t x_i) \middle/ \left(N_i \sum_{\substack{j=1 \\ j \neq i}}^{n} \frac{x_j}{\mathcal{D}_{ij}} - \sum_{\substack{j=1 \\ j \neq i}}^{n} \frac{N_j}{\mathcal{D}_{ij}} \right) \right] \tag{6.1.7}$$

Alternatively, by setting the right-hand side of Eq. 6.1.1 equal to the right-hand side of Eq. 3.2.4 we have

$$D_{i,\text{eff}} = \sum_{k=1}^{n-1} D_{ik} \frac{\nabla x_k}{\nabla x_i} \tag{6.1.8}$$

For ternary system Eq. 6.1.8 simplifies to

$$D_{1,\text{eff}} = D_{11} + D_{12} \frac{\nabla x_2}{\nabla x_1} \tag{6.1.9}$$

$$D_{2,\text{eff}} = D_{22} + D_{21} \frac{\nabla x_1}{\nabla x_2} \tag{6.1.10}$$

If the D_{ij} are given by Eqs. 4.2.7, Eqs. 6.1.9 and 6.1.10 are equivalent to formulas presented by Stewart (1954). Stewart suggested that for practical calculations the mole fraction gradients ∇x_i be replaced by mole fraction differences.

Force fitting the Maxwell–Stefan equations into the form of Eq. 6.1.2 gives the following expression for $D_{i,\text{eff}}$ (Kubota et al., 1969)

$$\frac{1}{D_{i,\text{eff}}} = \sum_{\substack{j=1 \\ j \neq i}}^{n} \frac{x_j}{\mathcal{D}_{ij}} \left(1 - \frac{x_i N_j}{x_j N_i}\right) \tag{6.1.11}$$

Equation 6.1.11 is useful when the flux ratios (N_j/N_i) are known and constant along the diffusion path (as is the case when we have diffusion controlled chemical reactions taking place on catalyst surfaces).

It is clear form the above expressions that effective diffusivities do not, in general, have the physical significance of a diffusion coefficient, since they may assume values ranging from minus to plus infinity (Toor and Sebulsky, 1961a). The effective diffusivity changes along the diffusion path as x_i and the ∇x_i change, $D_{i,\text{eff}}$ is zero at a diffusion barrier $(D_{11} \nabla x_1 = D_{12} \nabla x_2)$, is negative in the region of reverse diffusion $(D_{11} \nabla x_1/D_{12} \nabla x_2 < 0)$ and is infinite at the osmotic diffusion point $(\nabla x_1 = 0, J_1 \neq 0)$. Care must therefore be taken when drawing analogies between this quantity and a binary diffusion coefficient. Only when the effective diffusivity is positive, bounded, and not a strong function of composition or fluxes is it possible to draw useful analogies.

6.1.3 Limiting Cases

Despite the fundamental weakness of the effective diffusivity approach to multicomponent diffusion, it has been widely used in the past and so it would seem to be worthwhile to spend some time to delineate the conditions when such an approach is justified. Useful limiting cases of the exact relation (Eq. 6.1.7) include

1. All binary diffusion coefficients equal

$$D_{i,\text{eff}} = \mathcal{D}_{ij} = \mathcal{D} \tag{6.1.12}$$

An example of a mixture that meets this condition is the oxygen–nitrogen–carbon monoxide system considered in Example 4.2.1. However, such mixtures, although they do exist, do not constitute more than a small fraction of the total number of systems of interest.

2. In dilute mixtures where one component is in a large excess we may safely make the approximation $x_n \to 1$; $x_i \to 0$, $i = 1, 2, 3, \ldots, n - 1$. For this case $D_{ii} = \mathcal{D}_{in}$ and $D_{ij} = 0$ $(i \neq j = 1, 2, \ldots, n - 1)$ (cf., Example 4.2.2) and Eqs. 6.1.8 simplify to

$$D_{i,\text{eff}} = \mathcal{D}_{in} \qquad i = 1, 2, \ldots, n - 1 \tag{6.1.13}$$

Dilute solutions are encountered sufficiently often to make this limiting case of some practical importance.

3. When species i diffuses through a $n - 1$ stagnant gases and we have $N_j = 0$, $i \neq j$ (Wilke, 1950)

$$D_{i,\text{eff}} = \left[(1 - x_i) \middle/ \sum_{\substack{j=1 \\ j \neq i}}^{n} \frac{x_j}{\mathcal{D}_{ij}}\right] \tag{6.1.14}$$

Equation 6.1.14 is often used to estimate $D_{i,\text{eff}}$ for all species in a mixture even in cases where none of the components has a vanishing flux (see, e.g., Elnashaie et al., 1988)!

Other simple formulas for $D_{i,\text{eff}}$ include the expression of Burghardt and Krupiczka (1975):

$$\frac{1}{D_{i,\text{eff}}} = \frac{x_i}{Ð_{in}} + \sum_{\substack{k=1 \\ k \neq i}}^{n} \frac{x_k}{Ð_{ik}} \tag{6.1.15}$$

which is equivalent to setting $D_{i,\text{eff}} = 1/B_{ii}$ with B_{ii} defined by Eq. 2.1.21. This really amounts to neglecting the off-diagonal elements in the matrix $[B]$. A similar approach has been taken by Kato et al. (1981) who write

$$D_{i,\text{eff}} = D_{ii} \tag{6.1.16}$$

which amounts to neglecting the off-diagonal elements of the Fick matrix $[D]$ (note that this is not the same as neglecting the off-diagonal elements of $[B]$ although the results would probably be similar).

Example 6.1.1 Computation of the Effective Diffusivity

Calculate effective diffusivities for the H_2–N_2–CCl_2F_2 system discussed in Example 4.2.4. By way of a reminder, the Maxwell–Stefan diffusivities are

$$H_2\text{–}N_2 \qquad Ð = 77.0 \text{ mm}^2/\text{s}$$

$$N_2\text{–}CCl_2F_2 \qquad Ð = 8.1 \text{ mm}^2/\text{s}$$

$$CCl_2F_2\text{–}H_2 \qquad Ð = 33.1 \text{ mm}^2/\text{s}$$

The composition of the mixture is $x_N = 0.4$, $x_C = 0.25$, and $x_H = 0.35$ where the subscripts N, C, and H refer to N_2, CCl_2F_2, and H_2, respectively.

SOLUTION We shall proceed by taking H_2 as component 1, N_2 as component 2, and CCl_2F_2 as component 3.

The method of Wilke yields the following values of the effective diffusion coefficients

$$
\begin{aligned}
D_{1,\text{eff}} &= \frac{(1 - x_1)}{x_2/Ð_{12} + x_3/Ð_{13}} \\[2mm]
&= \frac{(1 - 0.35)}{0.4/77.0 + 0.25/33.1} \\[2mm]
&= 50.99 \text{ mm}^2/\text{s} \\[4mm]
D_{2,\text{eff}} &= \frac{(1 - x_2)}{x_1/Ð_{12} + x_3/Ð_{23}} \\[2mm]
&= \frac{(1 - 0.40)}{0.35/77.0 + 0.25/8.1} \\[2mm]
&= 16.95 \text{ mm}^2/\text{s}
\end{aligned}
$$

The method of Burghardt may be applied as follows:

$$D_{1,\text{eff}} = \cfrac{1}{x_1/Ð_{13} + x_2/Ð_{12} + x_3/Ð_{13}}$$

$$= \cfrac{1}{0.35/33.1 + 0.4/77.0 + 0.25/33.1}$$

$$= 42.88 \text{ mm}^2/\text{s}$$

$$D_{2,\text{eff}} = \cfrac{1}{x_2/Ð_{23} + x_1/Ð_{12} + x_3/Ð_{23}}$$

$$= \cfrac{1}{0.4/8.1 + 0.35/77.0 + 0.25/8.1}$$

$$= 11.79 \text{ mm}^2/\text{s}$$

In the method of Kato et al. we set the effective diffusivities equal to the corresponding diagonal elements of the Fick matrix $[D]$. For a ternary system Eqs. 4.2.7 and 4.2.5 may be combined to give

$$D_{1,\text{eff}} = \frac{Ð_{13}(x_1 Ð_{23} + (1 - x_1)Ð_{12})}{x_1 Ð_{23} + x_2 Ð_{13} + x_3 Ð_{12}}$$

$$= \frac{33.1 \times (0.35 \times 8.1 + (1 - 0.35)) \times 77.0}{0.35 \times 8.1 + 0.4 \times 33.1 + 0.25 \times 77.0}$$

$$= 49.54 \text{ mm}^2/\text{s}$$

$$D_{2,\text{eff}} = \frac{Ð_{23}(x_2 Ð_{13} + (1 - x_2)Ð_{12})}{x_1 Ð_{23} + x_2 Ð_{13} + x_3 Ð_{12}}$$

$$= \frac{8.1 \times (0.4 \times 33.1 + (1 - 0.4)) \times 77.0}{0.35 \times 8.1 + 0.4 \times 33.1 + 0.25 \times 77.0}$$

$$= 13.63 \text{ mm}^2/\text{s}$$

The results of all three methods are summarized below along with the results of the dilute solution limiting case, which is clearly inapplicable here.

Method	$D_{1,\text{eff}}$	$D_{2,\text{eff}}$
Wilke	50.99	16.95
Burghardt	42.88	11.79
Dilute solution limit	33.10	8.10
Kato	49.54	13.63

There is a factor of 2 variation in the values of $D_{2,\text{eff}}$ and a factor of 1.6 in the values of $D_{1,\text{eff}}$. Fluxes calculated using these methods would differ by similar amounts! ∎

6.2 SOLUTION OF MULTICOMPONENT DIFFUSION PROBLEMS USING AN EFFECTIVE DIFFUSIVITY MODEL

As we have already seen, the analysis of a diffusion problem proceeds by solving the conservation equations together with appropriate constitutive relations for the diffusion fluxes. Use of Eq. 6.1.1 for the diffusion flux with Eq. 1.3.9 for the conservation of mass leads to

$$c_t \frac{\partial x_i}{\partial t} + \nabla \cdot (N_t x_i) = \nabla \cdot (c_t D_{i,\text{eff}} \nabla x_i) \tag{6.2.1}$$

Some assumptions regarding the constancy of certain parameters are usually in order to facilitate the solution of the diffusion equations. For the binary diffusion problems discussed in Chapters 5 (as well as later in Chapters 8–10), we assume the binary Fick diffusion coefficient can be taken to be a constant. In the applications of the linearized theory presented in the same chapters, we assume the matrix of multicomponent Fick diffusion coefficients to be constant. If, on the other hand, we use Eq. 6.2.1 to model the diffusion process then we must usually assume constancy of the effective diffusion coefficient $D_{i,\text{eff}}$, if we are to have any hope of obtaining a simple analytical solution. Since the effective diffusivity is a rather strong function of concentration (even stronger functions than are the multicomponent Fick diffusion coefficients D_{ij}), as well as of the molar fluxes, this does not seem to be a particularly good way to go. If we assume constancy of $c_t D_{i,\text{eff}}$, Eq. 6.2.1 simplifies to

$$c_t \frac{\partial x_i}{\partial t} + \nabla \cdot (N_t x_i) = c_t D_{i,\text{eff}} \nabla^2 x_i \tag{6.2.2}$$

or

$$\frac{\partial x_i}{\partial t} + \nabla \cdot (u x_i) = D_{i,\text{eff}} \nabla^2 x_i \tag{6.2.3}$$

Equation 6.2.3 has exactly the same form as Eq. 5.1.3 for binary systems. This means that we may immediately write down the solution to a multicomponent diffusion problem if we know the solution to the corresponding binary diffusion problem simply by replacing the binary diffusivity by the effective diffusivity. We illustrate the use of the effective diffusivity by reexamining the three applications of the linearized theory from Chapter 5: diffusion in the two bulb diffusion cell, in the Loschmidt tube, and in the batch extraction cell.

6.3 STEADY-STATE DIFFUSION

Let us reconsider steady-state, one-dimensional diffusion. A more rigorous analysis of this simple diffusion problem has already been presented in Section 5.3. In that analysis it was found that if the total flux was zero (the assumption made there and here), the mole fraction profiles are straight lines and independent of the diffusion coefficients

$$\frac{x_i - x_{i0}}{x_{i\ell} - x_{i0}} = \frac{z}{\ell} \tag{5.3.3}$$

The molar flux in a binary mixture is given by Eq. 5.3.4; the corresponding result for a multicomponent mixture, assuming an effective diffusivity approach, therefore, is

$$N_i = (c_t D_{i, \text{eff}} / \ell)(x_{i0} - x_{i\ell}) \tag{6.3.1}$$

Example 6.3.1   Computation of the Fluxes with an Effective Diffusivity Model

Let us illustrate the calculation of the effective diffusivity and the molar fluxes for the conditions existing at the start of the two bulb diffusion cell experiment of Duncan and Toor discussed in Examples 5.3.1 and 5.4.1. The components are hydrogen (1), nitrogen (2), and carbon dioxide (3) and the values of the diffusion coefficients of the three binary pairs at 35.2°C and 1-atm pressure were

$$H_2-N_2 \qquad Đ_{12} = 83.3 \text{ mm}^2/\text{s}$$

$$H_2-CO_2 \qquad Đ_{13} = 68.0 \text{ mm}^2/\text{s}$$

$$CO_2-N_2 \qquad Đ_{23} = 16.8 \text{ mm}^2/\text{s}$$

The composition in each bulb at the start of the experiment was

$$x_{10} = 0.0 \qquad x_{20} = 0.50086 \qquad x_{30} = 0.49914$$

$$x_{1\ell} = 0.50121 \qquad x_{2\ell} = 0.49879 \qquad x_{3\ell} = 0.0$$

The capillary tube joining the two bulbs was 85.9 mm long.

SOLUTION   Strictly speaking none of the limiting cases of the two general expressions (Eqs. 6.1.10–6.1.14) applies to the two bulb diffusion cell. Nevertheless, we will proceed to calculate $D_{i, \text{eff}}$ and the molar fluxes through the capillary tube using Wilke's method (Eq. 6.1.14).

The mole fractions at equilibrium are used in the evaluation of the effective diffusivities (from Example 5.4.1)

$$x_{1\infty} = 0.2506 \qquad x_{2\infty} = 0.4998 \qquad x_{3\infty} = 0.2496$$

The effective diffusivities are computed from Wilke's Eq. 6.1.14 as follows:

$$
\begin{aligned}
D_{1, \text{eff}} &= \frac{(1 - x_1)}{(x_2/Đ_{12}) + (x_3/Đ_{13})} \\
&= \frac{(1 - 0.2506)}{(0.4998/83.3) + (0.2496/68.0)} \\
&= 77.49 \text{ mm}^2/\text{s} \\
D_{2, \text{eff}} &= \frac{(1 - x_2)}{(x_1/Đ_{12}) + (x_3/Đ_{23})} \\
&= \frac{(1 - 0.4998)}{(0.2506/83.3) + (0.2496/16.8)} \\
&= 28.00 \text{ mm}^2/\text{s}
\end{aligned}
$$

The molar density c_t is 39.5 mol/m^3. Hence, the fluxes through the tube are

$$N_1 = (c_t D_{1,\text{eff}}/\ell)(x_{10} - x_{1\ell})$$

$$= \frac{39.51 \times 77.5 \times 10^{-6}}{85.9 \times 10^{-3}}(0.0 - 0.50121)$$

$$= -17.86 \times 10^{-3} \text{ mol/m}^2\text{s}$$

$$N_2 = (c_t D_{2,\text{eff}}/\ell)(x_{20} - x_{2\ell})$$

$$= \frac{39.51 \times 28.00 \times 10^{-6}}{85.9 \times 10^{-3}}(0.50086 - 0.49879)$$

$$= 0.0267 \times 10^{-3} \text{ mol/m}^2\text{s}$$

The flux of hydrogen (component 1) is not too different from the flux estimated using the linearized equations in Example 5.3.1. However, the effective diffusivity method predicts a very small flux of nitrogen (component 2), a result quite different from the predictions of the linearized theory. This, of course, is because the effective diffusivity method ignores the contribution due to the driving forces of the other components. We will investigate the consequences of this prediction in Example 6.4.1. ∎

6.4 THE TWO BULB DIFFUSION CELL

We derived expressions for the concentration time history in the two bulb diffusion cell in Section 5.4. Here we present the corresponding problem solved using an effective diffusivity formulation.

With the molar fluxes given by Eqs. 6.3.1 we may write the differential mass balances for the two bulb cell, Eq. 5.4.1 as

$$c_t V_0 \frac{dx_{i0}}{dt} = -\frac{c_t D_{i,\text{eff}}}{\ell}(x_{i0} - x_{i\ell}) \tag{6.4.1}$$

The material balance relation (Eq. 5.4.2) is used to eliminate the mole fractions $x_{i\ell}$

$$\frac{dx_{i0}}{dt} = -\beta D_{i,\text{eff}}(x_{i0} - x_{i\infty}) \tag{6.4.2}$$

where β is the cell constant defined by Eq. 5.4.6. The solution to this differential equation subject to the initial condition (Eq. 5.4.11) is

$$(x_{i0} - x_{i\infty}) = \exp\{-\beta D_{i,\text{eff}}(t - t_0)\}(x_{i0}^\circ - x_{i\infty}) \tag{6.4.3}$$

from which we may calculate the composition at any time t knowing the initial composition, the cell constant and the effective diffusivities.

Example 6.4.1 Diffusion in a Two Bulb Diffusion Cell: A Test of the Effective Diffusivity

Let us now proceed to see if the Wilke effective diffusivity method is able to model the diffusional process in the two bulb diffusion cell experiment of Duncan and Toor (1962). For convenience, we repeat the following information from Example 5.4.1. The two bulbs in the apparatus built by Duncan and Toor had volumes of 77.99 and 78.63 cm^3, respectively. The capillary tube joining them was 85.9 mm long and 2.08 mm in diameter. The entire device was maintained at a temperature of 35.2°C. The system used by Duncan and Toor was hydrogen (1)–nitrogen (2)–carbon dioxide (3). In one of their experiments the initial concentration in each cell was

$$x_{10} = 0.0 \qquad x_{20} = 0.50086 \qquad x_{30} = 0.49914$$
$$x_{1\ell} = 0.50121 \qquad x_{2\ell} = 0.49879 \qquad x_{3\ell} = 0.0$$

where x_{i0} is the mole fraction of species i in bulb 1 and $x_{i\ell}$ is the mole fraction of species i in bulb 2. The values of the diffusion coefficients of the three binary pairs are given in Example 6.3.1.

SOLUTION   The concentration time history can be determined directly from Eq. 6.4.3. We illustrate the procedure by calculating the mole fractions after 4 h (14,400 s).

The effective diffusivities were calculated in Example 6.3.1

$$D_{1,\text{eff}} = 77.49 \text{ mm}^2/\text{s}$$
$$D_{2,\text{eff}} = 28.00 \text{ mm}^2/\text{s}$$

The cell constant β is calculated from Eq. 5.4.6

$$\beta = 0.9895 \text{ m}^{-2}$$

The mole fraction of component 1 may be calculated from Eq. 6.4.3 as

$$
\begin{aligned}
x_{10} &= x_{1\infty} + \exp\{-\beta D_{1,\text{eff}}(t - t_0)\}(x_{10}^\circ - x_{1\infty}) \\
&= 0.2506 + \exp(-0.9895 \times 7.749 \times 10^{-5} \times 14,400) \times (0.0 - 0.2506) \\
&= 0.1682 \\
x_{20} &= x_{2\infty} + \exp\{-\beta D_{2,\text{eff}}(t - t_0)\}(x_{20}^\circ - x_{2\infty}) \\
&= 0.4998 + \exp(-0.9895 \times 2.800 \times 10^{-5} \times 14,400) \times (0.50086 - 0.4998) \\
&= 0.5005
\end{aligned}
$$

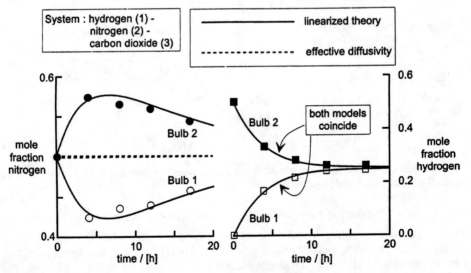

Figure 6.1. Composition–time history in two bulb diffusion cell. Experimental data from Duncan (1960).

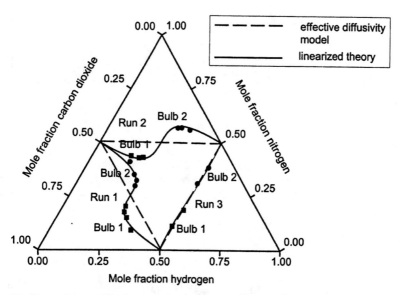

Figure 6.2. Composition profiles in two bulb diffusion cell. Experimental data from Duncan (1960).

Figure 6.1 shows the concentration time history in the diffusion cell for the experiment of Duncan and Toor that was described in detail in Example 5.4.1. The mole fraction of hydrogen predicted by the effective diffusivity model compares well with the experimental data of Duncan (1960). However, the effective diffusivity model suggests that the mole fraction of nitrogen should remain almost constant at approximately 0.5. This is in stark contrast to the experimental data (Fig. 6.1). The results obtained with the effective diffusivity method for nitrogen are completely different from those obtained with the linearized theory. Additional comparisons between the data of Duncan and Toor and the predictions of both the linearized equations and the effective diffusivity models are shown in the triangular diagram in Figure 6.2.

It is worth noting that the almost constant value of the composition of nitrogen would have been predicted with any of the formulas used in Example 6.1.1 to calculate the effective diffusivity. Thus, we have our first demonstration of the inability of the effective diffusivity approach to model multicomponent diffusion processes. ■

6.5 THE LOSCHMIDT TUBE

Our task here is to derive an expression that describes how the composition of a multicomponent mixture changes with time in a Loschmidt diffusion apparatus of the kind described in Section 5.5. The composition profile for a binary system is given by Eqs. 5.5.5 and 5.5.6; the solution to the binarylike multicomponent problem is given by the same expressions on replacing the binary diffusivity in those equations by the effective diffusivity. The average composition in the bottom tube after time t, for example, is given by

$$\frac{(\bar{x}_i - x_{i-})}{(x_{i+} - x_{i-})} = \bar{f}(t, D_{i,\text{eff}}) \tag{6.5.1}$$

where

$$\bar{f} = \left[\frac{1}{2} - \frac{1}{\pi^2} \sum_{k=0}^{\infty} \frac{1}{m^2} \exp\left\{ -\frac{m^2\pi^2}{\ell^2} D_{i,\text{eff}} t \right\} \right] \qquad (6.5.2)$$

with $m = k + \frac{1}{2}$.

Example 6.5.1 Multicomponent Diffusion in The Loschmidt Tube:
Another Test of the Effective Diffusivity

In this example we repeat the calculations of Example 5.5.1, where we used the linearized theory to compute the concentration time history in a Loschmidt tube experiment described by Arnold and Toor (1967). The system they used was methane (1)–argon (2)–hydrogen (3). The diffusion tube had a length of $(\pi/\ell)^2 = 60$ m^{-2}. At the temperature (34°C) and pressure of the experiments (1 atm) the binary diffusion coefficients of the three ternary gas pairs were

$$CH_4-Ar = 21.57 \text{ mm}^2/\text{s}$$

$$CH_4-H_2 = 77.16 \text{ mm}^2/\text{s}$$

$$Ar-H_2 = 83.35 \text{ mm}^2/\text{s}$$

The composition in each tube at the start of one of their experiments was as follows:

$$x_{1-} = 0.0 \qquad x_{2-} = 0.509 \qquad x_{3-} = 0.491$$
$$x_{1+} = 0.585 \qquad x_{2+} = 0.485 \qquad x_{3+} = 0.0$$

SOLUTION   As in the preceding example, none of the simple effective diffusivity formulas are applicable to the situation in the Loschmidt tube. We will proceed with the effective diffusivity formula of Wilke that gives

$$D_{1,\text{eff}} = 28.32 \text{ mm}^2/\text{s}$$

$$D_{2,\text{eff}} = 33.80 \text{ mm}^2/\text{s}$$

These diffusivities were calculated at the average composition

$$x_{1av} = 0.2575 \qquad x_{2av} = 0.4970 \qquad x_{3av} = 0.2455$$

We will illustrate the procedure for computing the composition time history by calculating the mole fractions in the bottom tube after 15 min. This will allow us to compare the results to those obtained in Example 5.5.1. With t equal to 15 min (900 s), we compute the functions $\bar{f}_i$ from Eq. 6.5.2. For present purposes we shall use just the first term of the series in Eq. 6.5.2.

$$\bar{f}_1 = \frac{1}{2} - \frac{4}{\pi^2} \exp\left\{ -\frac{1}{4} \frac{\pi^2}{\ell^2} D_{1,\text{eff}} t \right\}$$

$$= \frac{1}{2} - \frac{4}{\pi^2} \exp\left\{ -\frac{1}{4} \times 60 \times 2.832 \times 10^{-5} \times 900 \right\}$$

$$= 0.2235$$

$$\bar{f}_2 = \frac{1}{2} - \frac{4}{\pi^2} \exp\left\{-\frac{1}{4}\frac{\pi^2}{\ell^2}D_{2,\text{eff}}t\right\}$$

$$= \frac{1}{2} - \frac{4}{\pi^2} \exp\left\{-\frac{1}{4} \times 60 \times 3.380 \times 10^{-5} \times 900\right\}$$

$$= 0.2432$$

The converged values of $\bar{f}_1$ and $\bar{f}_2$ are

$$\bar{f}_1 = 0.2220$$

$$\bar{f}_2 = 0.2425$$

The mole fractions in the bottom tube follow from Eq. 6.5.1 as

$$\bar{x}_1 = 0.1143 \qquad \bar{x}_2 = 0.5032$$

The mole fractions computed with the linearized equations are (from Example 5.5.1)

$$\bar{x}_1 = 0.135 \qquad \bar{x}_2 = 0.553$$

These results are sufficiently different that we ought to be able to determine the better model by comparing the results to the experimental data. The complete concentration time history is shown in Figure 6.3 for this experiment. Note that the mole fraction of methane predicted by the effective diffusivity method is in reasonable agreement with the data (although the results from the linearized equations are better). However, the effective diffusivity method predicts almost no change in the mole fraction of argon; a result that is in marked contrast to both the experimental data and to the predictions of the linearized equations.

Additional data of Arnold and Toor are compared to the predictions of the linearized equations and of the effective diffusivity models in the triangular diagram in Figure 6.4. Clearly, the agreement with the data is very bad indeed. Thus, we have our second demonstration of the inability of the effective diffusivity method to model systems that exhibit strong diffusional interactions. ∎

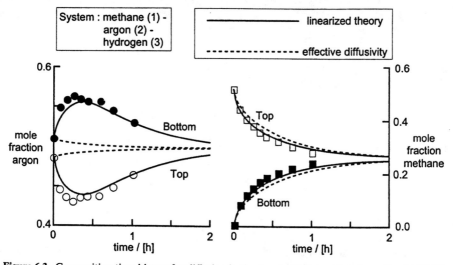

Figure 6.3. Composition time history for diffusion in the Loschmidt tube. Data from Arnold (1965).

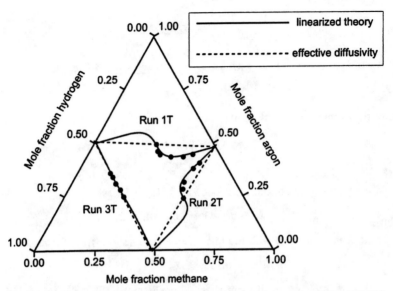

Figure 6.4. Comparison between Loschmidt tube experiments of Arnold and Toor (1967) and the composition trajectories predicted by the linearized theory and effective diffusivity methods.

6.6 DIFFUSION IN A BATCH EXTRACTION CELL

The batch extraction cell experiments of Krishna et al. (1985) were discussed at some length in Section 5.6, where it was shown that the diffusion fluxes could be calculated from

$$J_i = \frac{M_t}{A} \frac{dx_i}{dt} \qquad i = 1, 2 \tag{5.6.8}$$

Here we use an effective diffusivity model for the diffusion fluxes

$$J_i = \frac{c_t D_{i,\,\text{eff}}}{\ell} (x_{il} - x_i) \tag{6.6.1}$$

which may be combined with Eq. 5.6.8 and solved subject to the initial condition $t = 0$, $x_i = x_{i0}$ to give the composition profiles

$$x_{il} - x_i = \exp\{-K_{i,\,\text{eff}} t\} \, \Delta x_{i0} \qquad i = 1, 2 \tag{6.6.2}$$

where $K_{i,\,\text{eff}}$ is a volumetric transfer coefficient [s^{-1}] defined by

$$K_{i,\,\text{eff}} = \frac{c_t A}{\ell M_t} D_{i,\,\text{eff}} \tag{6.6.3}$$

Example 6.6.1   Multicomponent Diffusion in a Batch Extraction Cell

In this example we analyze the composition trajectory in the glycerol rich phase (G) of the system glycerol(1)–water(2)–acetone(3) using a pseudobinary model. The initial composition

G_0 is

$$x_{10} = 0.85 \qquad x_{20} = 0.15 \qquad x_{30} = 0.0$$

The composition of the interface (G_I) is

$$x_{1I} = 0.5480 \qquad x_{2I} = 0.2838 \qquad x_{3I} = 0.1682.$$

DATA The matrix of multicomponent volumetric mass transfer coefficients $[K]$ is given in Example 5.6.1. as

$$[K] = \begin{bmatrix} 0.2327 & -0.3742 \\ -0.1162 & 0.3373 \end{bmatrix} h^{-1}$$

SOLUTION The first calculation step is the determination of the effective volumetric mass transfer coefficients. We may adapt Eqs. 6.1.9 and 6.1.10 for this purpose

$$K_{1,\text{eff}} = K_{11} + K_{12}\frac{\Delta x_{20}}{\Delta x_{10}}$$

$$K_{2,\text{eff}} = K_{22} + K_{21}\frac{\Delta x_{10}}{\Delta x_{20}}$$

The initial composition driving force vector $(\Delta x_0) = (x_I - x_0)$ is

$$(\Delta x_0) = \begin{pmatrix} -0.302 \\ 0.1338 \end{pmatrix}$$

The effective volumetric mass transfer coefficients may now be estimated as

$$K_{1,\text{eff}} = 0.2327 + (-0.3742)[0.1338/(-0.302)]$$
$$= 0.398 \ h^{-1}$$
$$K_{2,\text{eff}} = 0.3373 + (-0.1162)[(-0.302)/0.1338]$$
$$= 0.6 \ h^{-1}$$

The compositions after 1 h are computed as follows:

$$x_1 = x_{1I} - \exp\{-K_{1,\text{eff}}t\}\Delta x_{10}$$
$$= 0.5480 - \exp(-0.398 \times 1) \times (-0.302)$$
$$= 0.7507$$
$$x_2 = x_{2I} - \exp\{-K_{2,\text{eff}}t\}\Delta x_{20}$$
$$= 0.5480 - \exp(-0.6 \times 1) \times (-0.1338)$$
$$= 0.2103$$

The complete concentration time history is shown in Figure 6.5. The composition trajectories do not compare all that favorably with the experimental data of Krishna et al. (1985). The actual observed equilibration paths and the predictions of the linearized equations are more highly curved (see Fig. 5.3). It must also be pointed out that the effective volumetric mass transfer coefficients should take on equal values (cf. discussion in Section 6.1.1 and Krishna et al., 1985) to maintain consistency with the requirement that the mole fractions sum to unity. We conclude that we must reject a pseudobinary mass transfer formulation for the diffusion fluxes in the batch extraction cell. ∎

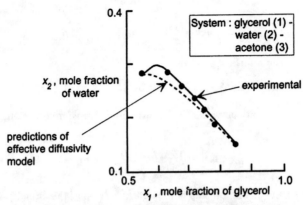

Figure 6.5. Equilibration paths during mass transfer in the system glycerol (1)–water (2)–acetone (3) in a batch extraction cell. Experimental data correspond to Run C of Krishna et al. (1985).

6.7 THE EFFECTIVE DIFFUSIVITY—CLOSING REMARKS

The advantage of the effective diffusivity formulation is its simplicity. The primary disadvantage is that effective diffusivities are not system properties except for the limiting cases noted above. Furthermore, they depend on the fluxes N_i, which are not always known in advance. More complicated variations on the effective diffusivity theme, some requiring iteration on $D_{i,\text{eff}}$, others incorporating a variation of $D_{i,\text{eff}}$ on position or composition have been discussed by Wilke (1950), Toor (1957), Shain (1961) and Hsu and Bird (1960). In view of the many better and, indeed, sometimes simpler methods that have been developed since then, we have not included them here.

The numerical examples in this chapter show the effective diffusivity approach in about as bad a light as possible. This was partly our intention. Strictly speaking, none of the limiting cases presented above applies to the conditions pertaining to the above examples. Nevertheless, it is common practice to use an effective diffusivity in situations where it is not warranted, so these examples provide a small indication of the errors that may be encountered using an effective diffusivity formula in situations it was not designed to handle.

It would be unfair of us to end this chapter with the impression that these results are typical. If conditions are such that the limiting cases do indeed apply we will find that the effective diffusivity approach does much better; Example 8.6.1 is a case in point.

PART II
Interphase Transfer

We have now reached the stage in which the most important problem in mass transfer modeling has to be tackled, namely, the choice of a simplified picture to describe the actual hydrodynamic conditions prevailing in the region of the interface. You may be wondering why we need to use a simplified model? Well, the difficulty is that though the basic governing differential equations are well understood, in many situations the fluid flow is so complicated that the governing equations cannot be solved without gross oversimplifications. In essence, the models used are highly simplified transport phenomena equations; in this sense, they are mathematical models. But we may take another view of the hydrodynamic models and that is to picture them as physical constructs that are more easily analyzed than the actual situation, but that show, in some way, the characteristic behavior of the actual situation. From a pragmatic point of view, the flow field is simplified to the extent that the equations describing the diffusion process can be solved, preferably analytically.

Once the equations have been solved to obtain the composition and temperature profiles, the diffusion fluxes J_i can be calculated and the interfacial transfer rates N_i determined. It is customary to determine, on the basis of the chosen hydrodynamic model, mass transfer coefficients that reflect the overall transfer facility (molecular and turbulent transport) of the phase under consideration.

In Chapter 7 we define mass transfer coefficients for binary and multicomponent systems. In subsequent chapters we develop mass transfer models to determine these coefficients. Many different models have been proposed over the years. The oldest and simplest model is the film model; this is the most useful model for describing multicomponent mass transfer (Chapter 8). Empirical methods are also considered. Following our discussions of film theory, we describe the so-called surface renewal or penetration models of mass transfer (Chapter 9) and go on to develop turbulent eddy diffusivity based models (Chapter 10). Simultaneous mass and energy transport is considered in Chapter 11.

7 Mass Transfer Coefficients

Our knowledge of multicomponent mass transfer coefficients is improving, but this is a slow process. I still occasionally have to pray that my estimate of some coefficient will not be off by more than one order of magnitude.

—J. A. Wesselingh (1992)

7.1 DEFINITION OF MASS TRANSFER COEFFICIENTS

Let us consider the two phase x and y system as shown in Figure 7.1 where typical composition profiles are shown. The interface compositions are x_{iI} on the x side of the interface and y_{iI} on the y side of the interface. The bulk phase mole fractions are denoted by x_{ib} and y_{ib} for the two phases.

We shall adopt a simple model of the interface itself; a surface that offers no resistance to mass transfer and where equilibrium prevails. Thus, the usual equations of phase equilibrium relate the mole fractions y_{iI} and x_{iI}.

The starting point for any analysis of the interphase mass transfer process will be Eqs. 1.3.13 describing the continuity of molar fluxes with respect to the interface. The simpler form, Eq. 1.3.14, will suffice for a majority of the cases in which the interface remains stationary. We proceed with Eq. 1.3.14 with the understanding that if the interface moves, the fluxes N_i must be referred to the interface and not a stationary coordinate reference frame. Furthermore, interphase mass transfer usually takes place in a direction normal to the interface and, therefore, it is sufficient to use the scalar form of Eq. 1.3.14, that is,

$$N_i^x = N_i^y = N_i \qquad i = 1, 2, \ldots, n \tag{7.1.1}$$

where N_i^x is the normal component of N_i and is directed from the bulk phase x to the interface I. The flux N_i^y is directed from the interface I to the bulk phase y.

Equations 7.1.1 may be summed over the n species to yield

$$N_t^x = N_t^y = N_t = \sum_{i=1}^{n} N_i \tag{7.1.2}$$

7.1.1 Binary Mass Transfer Coefficients

The mass transfer coefficient k in phase x for a binary system is best defined in a manner suggested by Bird et al. (1960, p. 639)

$$k_b = \lim_{N_1 \to 0} \frac{N_{1b} - x_{1b}N_t}{c_t(x_{1b} - x_{1I})} = \frac{J_{1b}}{c_t \Delta x_1} \tag{7.1.3}$$

where the driving force for mass transfer Δx_1 is taken to be the difference between the bulk phase mole fraction x_{1b} and the composition of Component 1 at the interface. The diffusion fluxes used in Eq. 7.1.3 are the bulk diffusion fluxes and the mass transfer coefficient obtained in Eq. 7.1.3 are the bulk phase mass transfer coefficients. By using the interface compositions in the calculations of the convective term $x_1 N_t$ in Eq. 7.1.3, we shall obtain the interface transfer coefficients. (In some cases the distinction between k_b and k_I (or equivalently between J_{1b} and J_{1I}) is important; for example, in the description of mass

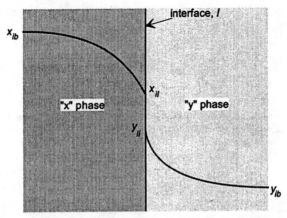

Figure 7.1. Mole fraction profiles in the region close to the interface during interphase mass transfer.

transfer effects in reactors.) We have omitted writing the subscript x (or superscript x) on these last four mentioned quantities because we shall be first considering what happens on the x side of the boundary. When we wish to describe the overall mass transfer behavior of the x–y transfer process we shall distinguish between the quantities on either side of the phase boundary. In proceeding further, it must be remembered that analogous relations will hold for the y phase using the mole fraction difference

$$\Delta y_1 = (y_{1I} - y_{1b}).$$

A word now about the units of k_b. With the fluxes N_i expressed in moles per second per meter squared (mol/s) (m^2 interfacial area), and c_t in the units moles per meter cubed (mol/m^3), the units of k_b are meters per second (m/s). But is it really a velocity? To examine this further, we replace the diffusion flux J_{1b} in Eq. 7.1.3 with $c_t x_{1b}(u_1 - u)$ to obtain (omitting the limit sign)

$$k_b = \frac{(u_1 - u)}{\Delta x_1 / x_{1b}} \qquad (7.1.4)$$

The numerator on the right-hand side of Eq. 7.1.4 is the velocity of transfer of Component 1 with respect to the molar average reference velocity of the mixture u. For a binary system (but not always for a multicomponent system), the quotient in Eq. 7.1.4 is positive and the coefficient k_b is positive. The greatest possible driving force $|\Delta x|$ is unity; thus, with $x_{1b} = 1$ and $x_{1I} = 0$, the maximum value of the denominator of Eq. 7.1.4 is unity, that is, $|\Delta x_1 / x_1| < 1$. Therefore,

$$k_b \geqq (u_1 - u) \qquad (7.1.5)$$

Equation 7.1.5 gives a physical significance to the mass transfer coefficient k_b: It is the maximum velocity (relative to the velocity of the mixture) at which a component can be transferred in the binary system. The actual velocity of Component 1 relative to the mixture velocity is given by

$$(u_1 - u) = \frac{k_b \Delta x_1}{x_{1b}} \qquad (7.1.6)$$

Equation 7.1.6 should be compared to Eq. 3.1.1. It appears that k_b may be related directly to the binary Fick diffusion coefficient D. Indeed, this will be shown to be the case when we examine various specific hydrodynamic models for mass transfer later in this book.

Let us now turn to the more difficult problem of explaining why the limits $N_1 \to 0$, $N_2 \to 0$ appear in Eq. 7.1.3. During the actual mass transfer process itself the composition (and velocity) profiles are distorted by the flow (diffusion) of 1 and 2 across the interface. The mass transfer coefficient defined in Eq. 7.1.3 corresponds to conditions of vanishingly small mass transfer rates, when such distortions are not present. These low-flux or zero-flux coefficients are the ones that are usually available from correlations of mass transfer data. These correlations usually are obtained under conditions where the mass transfer rates are low. For the actual situation under conditions of finite transfer rates, we may write

$$k_b^{\bullet} = \frac{N_1 - x_{1b}N_t}{c_t \Delta x_1} = \frac{J_{1b}}{c_t \Delta x_1} \qquad (7.1.7)$$

The superscript $^{\bullet}$ serves to remind us that the transfer coefficients $k_b^{\bullet}$ correspond to conditions of finite mass transfer rates. For further reading on this point we recommend Bird (1960, Chapter 21).

In order to calculate the flux N_1, we need the finite flux coefficient $k_b^{\bullet}$; this coefficient usually is related to the zero-flux coefficient by a general relation of the form

$$k_b^{\bullet} = k_b \Xi_b \qquad (7.1.8)$$

where Ξ_b is a correction factor to account for the effect of finite fluxes on k_b. The exact form of the correction factor depends, of course, on the composition profiles and, therefore, on the hydrodynamic model chosen to describe the mass transfer process. More about this in subsequent chapters.

7.1.2 Multicomponent Mass Transfer Coefficients

The development for multicomponent mixtures is best carried out by using $n - 1$ dimensional matrix notation. We, therefore, define a matrix of finite flux mass transfer coefficients $[k_b^{\bullet}]$ by

$$(J_b) = (N) - (x_b)N_t = c_t[k_b^{\bullet}](x_b - x_I) = c_t[k_b^{\bullet}](\Delta x) \qquad (7.1.9)$$

The finite flux coefficients are related to the zero-flux or low-flux coefficients by a matrix equation of the form

$$[k_b^{\bullet}] = [k_b][\Xi_b] \qquad (7.1.10)$$

where $[\Xi_b]$ is a matrix of correction factors. The calculation of the mass transfer coefficient matrices and the correction factor matrices for a multicomponent system can be sensitive to the mass transfer model chosen. Even for the simplest mass transfer model—the film model —different approximations lead to different values for the matrices $[k_b]$, $[\Xi_b]$, and $[k_b^{\bullet}]$ (Chapter 8). There is a further problem with the multicomponent mass transfer coefficients, that of nonuniqueness; this is explained below.

In Eqs. 7.1.9 we define $n - 1 \times n - 1$ elements of the mass transfer coefficients with the help of $n - 1$ linear equations. It follows that the elements $k_{bij}^{\bullet}$ are not unique; that is, another set of these coefficients can also lead to the same value of the fluxes N_i. Put another way, making mass transfer measurements in a multicomponent system for the fluxes N_i and Δx_i does not uniquely determine the values of the mass transfer coefficients. A large set of measurements of N_i and Δx_i will be necessary to obtain a set of coefficients. In practice, we proceed in a different manner. We try to predict the values of the multicomponent mass transfer coefficients from binary mass transfer correlations, using as a basis, the generalized Maxwell–Stefan equations or the generalized Fick's law. However, various approximations are necessary before tractable solutions are obtained.

7.1.3 Interaction Effects (Again)

Let us take another look at diffusional interaction effects with the help of Eqs. 7.1.9 and 7.1.10, rewritten for a ternary system as follows:

$$(J) = c_t[k^\bullet](\Delta x) = c_t \begin{bmatrix} k_{11}^\bullet & k_{12}^\bullet \\ k_{21}^\bullet & k_{22}^\bullet \end{bmatrix} \begin{matrix} \Delta x_1 \\ \Delta x_2 \end{matrix} \tag{7.1.11}$$

where

$$[k^\bullet] = [k][\Xi] = \begin{bmatrix} k_{11} & k_{12} \\ k_{21} & k_{22} \end{bmatrix} \begin{bmatrix} \Xi_{11} & \Xi_{12} \\ \Xi_{21} & \Xi_{22} \end{bmatrix}$$

$$= \begin{bmatrix} k_{11}\Xi_{11} + k_{12}\Xi_{21} & k_{11}\Xi_{12} + k_{12}\Xi_{22} \\ k_{21}\Xi_{11} + k_{22}\Xi_{21} & k_{21}\Xi_{12} + k_{22}\Xi_{22} \end{bmatrix} \tag{7.1.12}$$

That is

$$J_1 = c_t k_{11}^\bullet \Delta x_1 + c_t k_{12}^\bullet \Delta x_2$$
$$J_2 = c_t k_{21}^\bullet \Delta x_1 + c_t k_{22}^\bullet \Delta x_2 \tag{7.1.13}$$

Since, in general, $k_{12}^\bullet$, $k_{21}^\bullet$, Δx_1, and Δx_2 can take on any sign, depending on the physical constraints imposed on the system, we could encounter one of the three situations sketched below.

1. Even when its constituent driving force Δx_1 is zero, we could have a nonvanishing flux J_1,

$$J_1 \neq 0 \qquad \Delta x_1 = 0 \tag{7.1.14}$$

This is known as *osmotic diffusion* (Toor, 1957).

2. Under a certain set of operating conditions and system properties the term $k_{12}^\bullet \Delta x_2$ may be of the same magnitude and of opposite sign to $k_{11}^\bullet \Delta x_1$ leading to

$$J_1 = 0 \qquad \Delta x_1 \neq 0 \tag{7.1.15}$$

A *diffusion barrier* is considered to exist for component 1 (Toor, 1957).

3. It is also conceivable that the term $k_{12}^\bullet \Delta x_2$ overshadows $k_{11}^\bullet \Delta x_1$ in magnitude and is of opposite sign giving rise to

$$J_1/\Delta x < 0 \tag{7.1.16}$$

Component 1 experiences *reverse diffusion* in this case (Toor, 1957).

It must be stressed that for a two-component system where

$$J_1 = c_t k^\bullet \Delta x_1 \tag{7.1.17}$$

with $k^\bullet > 0$, none of the three phenomena, sketched above for a ternary mixture, can take place.

The ratio of driving forces $\Delta x_1/\Delta x_2$ plays an important role in enhancing diffusional interaction effects in multicomponent mass transfer. Thus, a small cross-coefficient $k_{12}^\bullet$ may be linked to a large Δx_2, resulting in large interaction effects. The criteria presented above are a little different from those discussed in Section 5.2, where Fick diffusion coefficients and the mole fraction gradients were used. The physical significance is, however, the same.

7.2 THE BOOTSTRAP PROBLEM (AND ITS SOLUTION)

The major part of the next few chapters are devoted to methods of estimating the low flux mass transfer coefficients k and $[k]$ and of calculating the high flux coefficients $k_b^\bullet$ and $[k_b^\bullet]$. In practical applications we will need these coefficients to calculate the diffusion fluxes J_i and the all important molar fluxes N_i. The N_i are needed because it is these fluxes that appear in the material balance equations for particular processes (Chapters 12–14). Thus, even if we know (or have an estimate of) the diffusion fluxes J_k we cannot immediately calculate the molar fluxes N_i because all n of these fluxes are independent, whereas only $n - 1$ of the J_i are independent. We need one other piece of information if we are to calculate the N_i. Usually, the form of this additional relationship is dictated by the context of the particular mass transfer process. The problem of determining the N_i knowing the J_i has been called the bootstrap problem. Here, we consider its solution by considering some particular cases of practical importance.

7.2.1 Equimolar Counterdiffusion

When the total molar flux vanishes

$$N_t = 0 \tag{7.2.1}$$

the component molar fluxes N_i equal the corresponding molar diffusion fluxes J_i for all species of the mixture

$$N_i = J_i \qquad (N_t = 0) \tag{7.2.2}$$

This situation is sometimes referred to as equimolar counterdiffusion (or mass transfer).

In an isobaric closed system like the diffusion cell or Loschmidt tube (see Chapter 5), any movement of the molecules of one (or more) species in one direction must be exactly balanced by a movement of the molecules of other species in the opposite direction and there is no net change in the number of moles $N_t = 0$.

7.2.2 Multicomponent Distillation

The condition $N_t = 0$ is often assumed to be valid in multicomponent distillation calculations. A better approximation is the relationship

$$\sum_{i=1}^{n} N_i \Delta H_{vap,i} = 0 \tag{7.2.3}$$

where the $\Delta H_{vap,i}$ are the molar latent heats of vaporization. It can be seen that if the molar latent heats are equal, the total flux N_t vanishes. However, Eq. 7.2.3 is a special case of a more general relationship between the fluxes that we shall derive in Section 11.5. Nonequimolar effects in distillation are illustrated in Examples 9.2.1 and 11.5.1.

7.2.3 Stefan Diffusion

The general case of mass transfer in a mixture where one component has a zero flux is known as Stefan diffusion. This situation is very common.

- Condensation in the presence of a noncondensing gas is an important operation in many chemical processes. A better known occurrence is the condensation of water vapor on a cold window. The water vapor has a nonzero flux but the air does not condense and so has a zero flux.
- Evaporation is the opposite of condensation in the presence of a noncondensing gas. The most commonplace application is the evaporation into air of a puddle of water lying on the ground. The flux of air is zero.
- Absorption is another very important chemical process where one or more species are removed from a gas stream by absorption into a liquid. One of the components of the gas stream may be insoluble (or is assumed to be insoluble) in the absorbing liquid and, therefore, has a zero flux.

Let us denote that component with zero flux as species n. Thus,

$$N_n = J_n + x_n N_t = 0 \tag{7.2.4}$$

The total molar flux N_t is, therefore, given by

$$N_t = \frac{-J_n}{x_n} \tag{7.2.5}$$

Thus, the relation that allows the calculation of the nonzero N_i from the J_i is

$$N_i = J_i + x_i N_t = J_i - \frac{x_i J_n}{x_n}$$

$$= \left(1 + \frac{x_i}{x_n}\right)J_i + \frac{x_i}{x_n}\sum_{\substack{k=1 \\ k \neq i}}^{n-1} J_k \tag{7.2.6}$$

For a two component mixture Eq. 7.2.6 simplifies to

$$N_1 = (1 + x_1/x_2)J_1 = J_1/x_2 \tag{7.2.7}$$

7.2.4 Flux Ratios Specified

In some situations it is possible to specify the ratio of the component molar flux N_i to the total flux N_t

$$N_i = z_i N_t \tag{7.2.8}$$

where z_i is the specified flux ratio. Equation 7.2.8 may be used to establish the following relationship between N_i and J_i

$$N_i = \frac{J_i}{(1 - x_i/z_i)} \tag{7.2.9}$$

Some situations where the flux ratios may be known are identified below.

- Condensation of mixtures. It is easy to show with the help of a material balance that the composition of the first drop of condensate is determined by the relative rates of condensation. In practice, the condensate composition throughout a horizontal condenser often is assumed to be given by the relative rates of condensation (more on this in Chapter 15).
- For diffusion controlled heterogeneous chemical reactions the reaction stoichiometry dictates the flux ratios.

For example, in the synthesis of ammonia (3) from nitrogen (1) and hydrogen (2) following the reaction

$$N_2 + 3H_2 \rightarrow 2NH_3$$

catalyzed by a porous solid catalyst, the flux ratios are fixed by reaction stoichiometry as

$$N_1 = \tfrac{1}{3}N_2 = -\tfrac{1}{2}N_3 = \tfrac{1}{2}N_t$$

or, in terms of the z_i,

$$z_1 = \tfrac{1}{2} \qquad z_2 = \tfrac{3}{2} \qquad z_3 = -1$$

Note that the z_i sum to unity; however, individual z_i may be positive, zero, or negative. In cases like this one the composition at the surface must be determined in order to satisfy the stoichiometric relations and reaction rate equations (see Examples 8.4.1 and 10.4.1).

7.2.5 The Generalized Bootstrap Problem

The above relationships between the N_i and the J_i are special cases of a more general expression we will derive below. The generalized determinacy condition is written in the form

$$\sum_{i=1}^{n} \nu_i N_i = 0 \tag{7.2.10}$$

where the ν_i can be considered to be determinacy coefficients. To relate the N_i to the J_i we proceed as follows. First, we multiply Eq. 1.2.12 by ν_i to obtain

$$\nu_i N_i = \nu_i J_i + \nu_i x_i N_t \tag{7.2.11}$$

then sum over all species to get, in view of Eq. 7.2.10,

$$\sum_{i=1}^{n} \nu_i J_i + N_t \sum_{i=1}^{n} \nu_i x_i = 0 \tag{7.2.12}$$

The total flux N_t may now be expressed in terms of the diffusion fluxes as

$$N_t = -\left(\sum_{i=1}^{n} \nu_i J_i \bigg/ \sum_{i=1}^{n} \nu_i x_i \right) = -\sum_{k=1}^{n-1} \Lambda_k J_k \qquad (7.2.13)$$

where the coefficients Λ_k are defined by

$$\Lambda_k = (\nu_k - \nu_n) \bigg/ \sum_{j=1}^{n} \nu_j x_j \qquad (7.2.14)$$

Finally, we substitute for N_t in Eq. 1.2.12

$$N_i = \sum_{k=1}^{n-1} \beta_{ik} J_k \qquad (7.2.15)$$

where the β_{ik} are defined by

$$\beta_{ik} \equiv \delta_{ik} - x_i \Lambda_k \qquad (7.2.16)$$

where δ_{ik} is the Kronecker delta.

For equimolar counterdiffusion we make all the ν_i equal.

$$\nu_i = \nu_n \qquad (N_t = 0) \qquad i = 1, 2, \ldots, n-1 \qquad (7.2.17)$$

and the β_{ik} reduce to δ_{ik}.

On the other hand, for nonequimolar distillation the ν_i may be set equal to the molar latent heats of vaporization

$$\nu_i = \Delta H_{\mathrm{vap},i} \qquad i = 1, 2, \ldots, n \qquad (7.2.18)$$

For Stefan diffusion we make all the ν_i zero except one that may take any nonzero value.

$$\nu_i = 0 \qquad \nu_n \neq 0 \qquad (N_n = 0) \qquad (7.2.19)$$

In this case Eq. 7.2.16 simplifies to

$$\beta_{ik} = \delta_{ik} + x_i/x_n \qquad (N_n = 0) \qquad (7.2.20)$$

For situations where the stoichiometry of a chemical reaction controls the flux ratios it is preferable to proceed somewhat differently. Equations 7.2.9 may be combined in the form of Eq. 7.2.15 with β_{ik} given by

$$\beta_{ik} = \delta_{ik}/(1 - x_i/z_i) \qquad (7.2.21)$$

7.2.6 The Bootstrap Matrix

Equations 7.2.13 and 7.2.15 will often be needed in $n - 1$ dimensional matrix form. The required expressions are

$$N_t = -(\Lambda)^T (J) \qquad (7.2.22)$$

and

$$(N) = [\beta](J) \tag{7.2.23}$$

$[\beta]$ is known as the bootstrap matrix and has elements given by Eqs. 7.2.16 or 7.2.21 (Krishna and Standart, 1979). Equation 7.2.22 does not apply if the flux ratios are specified.

For equimolar counterdiffusion we have the simple result that $[\beta] = [I]$. Departures from the identity matrix signify the increasing importance of the convective term $x_i N_t$ in the mass transfer process.

Equations 7.2.9–7.2.21 apply at all points along the diffusion path; in the bulk fluid

$$(N) = [\beta_b](J_b) = c_t[\beta_b][k_b^{\bullet}](\Delta x) \tag{7.2.24}$$

as well as at the interface

$$(N) = [\beta_I](J_I) = c_t[\beta_I][k_I^{\bullet}](\Delta x) \tag{7.2.25}$$

One important difference between the binary and the general multicomponent case is worth recording here. For a binary system all matrices reduce to scalar quantities and we must have

$$\beta_I k_I^{\bullet} = \beta_b k_b^{\bullet} = N_1/c_t \, \Delta x_1 \tag{7.2.26}$$

but we do not have the corresponding equality for the multicomponent system; that is,

$$[\beta_I][k_I^{\bullet}] \neq [\beta_b][k_b^{\bullet}] \tag{7.2.27}$$

because the matrix equation $[A](x) = [B](x)$ does not imply the equality of the matrices $[A]$ and $[B]$. The inequality (Eq. 7.2.27) actually spells out explicitly the problem of nonuniqueness of the multicomponent mass transfer coefficients mentioned earlier. The models of mass transfer that we describe in the next chapters will reveal that the two terms on either side of Eq. 7.2.27 will be unequal except in the case of vanishingly small transfer fluxes. It must be emphasized that the product of the two matrices on either side of inequality (Eq. 7.2.27) with the column matrix of composition driving forces (Δx) will lead to identical values of the molar fluxes N_i.

7.3 INTERPHASE MASS TRANSFER

Consider, once again, mass transfer across the phase boundary in Figure 7.1. We must have continuity of the fluxes N_i across the interface (Eq. 7.1.1). We may express these fluxes in terms of the driving forces for mass transfer on either side of the interface as

$$(N) = c_t^L[k_L^{\bullet}](x^L - x^I) + N_t(x^L) = c_t^L[\beta^L][k_L^{\bullet}](x^L - x^I) \tag{7.3.1}$$

$$(N) = c_t^V[k_V^{\bullet}](y^I - y^V) + N_t(y^V) = c_t^V[\beta^V][k_V^{\bullet}](y^I - y^V) \tag{7.3.2}$$

As noted earlier, we assume that at the interface itself the two phases are in equilibrium with each other. The compositions on either side of the interface are, therefore, related by

$$y_i^I = K_i x_i^I \qquad i = 1, 2, \ldots, n \tag{7.3.3}$$

The K_i are the equilibrium ratios or "K values" (see, e.g., Henley and Seader, 1981; Walas, 1985).

It will sometimes prove useful to linearize the vapor–liquid equilibrium relationship for the interface over the range of compositions obtained in passing from the bulk to the interface conditions

$$(y') = [M](x') + (b) \tag{7.3.4}$$

where $[M]$ is the matrix of equilibrium constants with elements

$$M_{ij} = \partial y_i^* / \partial x_j \qquad i, j = 1, 2, \ldots, n - 1 \tag{7.3.5}$$

where y_i^* is the mole fraction of a vapor in equilibrium with a liquid of composition x_i and (b) is a column matrix of "intercepts."

To evaluate $[M]$ we must differentiate the vapor–liquid equilibrium model with respect to composition. If the vapor liquid equilibrium ratios (K-values) are given by

$$K_i = \gamma_i P_i^s / P \tag{7.3.6}$$

then it can be shown that the matrix $[M]$ is given quite simply by

$$[M] = [K][\Gamma] \tag{7.3.7}$$

where $[K]$ is a diagonal matrix with elements that are the equilibrium ratios (the first $n - 1$ of them) and $[\Gamma]$ is the matrix of thermodynamic factors defined by Eqs. 2.2.5.

A general method of solving Eqs. 7.3.1–7.3.3 is described in Chapter 11, where we complicate matters somewhat by including energy transfer across the phase boundary.

7.3.1 Overall Mass Transfer Coefficients

Summing the resistance in both phases in order to obtain a single expression for computing the fluxes N_i without knowledge of the interface composition is widely discussed in the literature on binary mass transfer. Here we consider the extension to multicomponent systems (Toor, 1964a; Krishna and Standart, 1976b).

We define the composition of a vapor that would be in equilibrium with the bulk liquid with the help of Eq. 7.3.4 as

$$(y^*) = [M](x^L) + (b) \tag{7.3.8}$$

For binary systems we may define an overall mass transfer coefficient, $K_{OV}^\bullet$, by

$$J_1^V = c_t^V K_{OV}^\bullet (y_1^V - y_1^*) \tag{7.3.9}$$

For binary systems all matrices contain $\overset{ONLY}{\text{must}}$ one element and we have no difficulty in deriving the formula for the addition of resistances

$$\frac{1}{c_t^V K_{OV}^\bullet \beta^V} = \frac{1}{c_t^V k_V^\bullet \beta^V} + \frac{M}{c_t^L k_L^\bullet \beta^L} \tag{7.3.10}$$

where, for a binary system, $M = K_1 \Gamma$ (cf. Eq. 7.3.7) with Γ given by Eq. 2.2.12.

For multicomponent systems we may write down the matrix generalization of Eq. 7.3.9 as

$$(J^V) = c_t^V [K_{OV}^\bullet](y^V - y^*) \tag{7.3.11}$$

where the matrix of multicomponent overall mass transfer coefficients, $[K_{OV}^\bullet]$, is defined by a generalization of Eq. 7.3.10

$$[K_{OV}^{\bullet}]^{-1}[\beta^V]^{-1} \equiv [k_V^{\bullet}]^{-1}[\beta^V]^{-1} + \frac{c_t^V}{c_t^L}[M][k_L^{\bullet}]^{-1}[\beta^L]^{-1} \qquad (7.3.12)$$

Unlike the binary case, we cannot define $[K_{OV}^{\bullet}]$ by Eq. 7.3.11 because we have only $n - 1$ independent equations and the matrix $[K_{OV}^{\bullet}]$ contains $(n - 1)^2$ elements (cf. discussion below Eq. 7.2.25).

This cumbersome expression (Eq. 7.3.12) is not particularly useful as it stands. Fortunately, it simplifies considerably in some special cases of interest. First, if the total flux is near zero (as it will often be in distillation), the bootstrap matrices $[\beta]$ reduce to the identity matrix and we have

$$[K_{OV}^{\bullet}]^{-1} = [k_V^{\bullet}]^{-1} + \frac{c_t^V}{c_t^L}[M][k_L^{\bullet}]^{-1} \qquad (7.3.13)$$

Second, in distillation the finite flux mass transfer coefficients will often be well approximated by their low flux limits and we can write a simpler expression for $[K_{OV}]$

$$[K_{OV}]^{-1} = [k^V]^{-1} + \frac{c_t^V}{c_t^L}[M][k^L]^{-1} \qquad (7.3.14)$$

For binary systems Eq. 7.3.14 simplifies to

$$\frac{1}{K_{OV}} = \frac{1}{k^V} + \frac{c_t^V}{c_t^L}\frac{M}{k^L} \qquad (7.3.15)$$

Equations 7.3.11 and 7.3.14 are used in the development of expressions for modeling mass transfer in multicomponent distillation, a topic we consider in Chapter 12. The addition of resistances concept has seen use in distillation models by Krishna et al. (1981a), Burghardt et al. (1983, 1984) and by Gorak and Vogelpohl (1985).

8 Film Theory

Conditions in the immediate region of an interface between phases are hard to explore experimentally. In such situations it is helpful to develop a mathematical model of the process starting with the known basic facts. The result of the analysis is then compared with those experimental measurements which it is possible to make. Good agreement suggests that the model may have been realistic.

—T. K. Sherwood, R. L. Pigford, and C. R. Wilke (1975)

8.1 THE FILM MODEL

In the film model, we imagine that all of the resistance to mass transfer is concentrated in a thin film, or layer, adjacent to the phase boundary. Also that transfer occurs within this film by steady-state molecular diffusion alone and that outside this film, in the bulk fluid, the level of mixing or turbulence is so high that all composition gradients are wiped out, Figure 8.1. Mass transfer occurs through this film essentially in the direction normal to the interface. That is, any constituent molecular diffusion or convection in any flow parallel to the surface due to composition gradients along the interface is negligible in comparison to the mass transfer fluxes normal to the interface. The thickness of this hypothetical film is in the range 0.01–0.1 mm for liquid-phase transport and in the range 0.1–1 mm for gas-phase transport.

Having made the appropriate simplifications to the hydrodynamics, the relevant differential equations describing the molecular diffusion process in the diffusion layer may now be solved. The diffusion process is fully determined by

1. The one-dimensional steady-state of forms of Eqs. 1.3.10 and 1.3.11,

$$\frac{dN_i}{dr} = 0 \qquad \frac{dN_t}{dr} = 0 \qquad (8.1.1)$$

showing that N_i and N_t are r invariant.
2. The constitutive relations (Eqs. 2.2.1 or 3.2.5).
3. The determinacy condition (Eq. 7.2.10).
4. The boundary conditions of the film model.

$$
\begin{aligned}
r = r_0 & \quad x_i = x_{i0} \\
r = r_\delta & \quad x_i = x_{i\delta}
\end{aligned}
\qquad (8.1.2)
$$

We shall develop the solution of these equations for the special case of a two-component mixture before proceeding to the general n-component case.

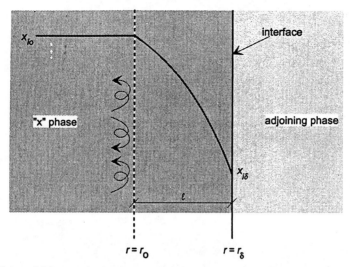

Figure 8.1. Film model for transfer in phase x. Turbulent eddies wipe out composition gradients in the bulk fluid phase. Composition variations are restricted to a layer (film) of thickness ℓ adjacent to the interface. Model due to Lewis and Whitman.

8.2 FILM MODEL FOR BINARY MASS TRANSFER

Let us consider a planar film between the position coordinates $r = r_0$ and $r = r_\delta$. Mass transfer between the two edges of the film occurs purely by molecular diffusion under steady-state conditions. The thickness of the film is $\ell = r_\delta - r_0$. The equation of continuity of moles of species, (Eq. 8.1.1) can be written for species 1 and 2 as

$$\frac{dN_1}{dr} = 0 \qquad \frac{dN_2}{dr} = 0 \qquad \frac{dN_t}{dr} = 0 \qquad (8.2.1)$$

showing that N_1, N_2, and N_t ($= N_1 + N_2$) are r invariant.

The generalized Maxwell–Stefan diffusion equations (Eq. 2.2.1) simplify to

$$\frac{dx_1}{dr} = \frac{x_1 N_2 - x_2 N_1}{c_t D} \qquad (8.2.2)$$

where $D = Ð\Gamma$ is the Fick diffusion coefficient. In the development below we assume that the Fick D is constant. That is equivalent to assuming that the Maxwell–Stefan $Ð$ and the thermodynamic factor Γ are constant. The Fick D is, in fact, constant for ideal gas mixtures at constant temperature and pressure and the assumption of constant D is a fair approximation for small concentration changes in nonideal fluid mixtures.

We rewrite Eq. 8.2.2 by substituting $x_2 = 1 - x_1$, as

$$\frac{dx_1}{dr} = \frac{x_1(N_1 + N_2)}{c_t D} - \frac{N_1}{c_t D} \qquad (8.2.3)$$

It is convenient to define the following parameters

1. A dimensionless distance

$$\eta = \frac{r - r_0}{r_\delta - r_0} = \frac{r - r_0}{\ell} \tag{8.2.4}$$

where ℓ is the thickness of the diffusion layer or "film"

2. A dimensionless mass transfer rate factor

$$\Phi = \frac{(N_1 + N_2)}{c_t D/\ell} = \frac{N_t}{c_t D/\ell} \tag{8.2.5}$$

3. $$\phi = - \frac{N_1}{c_t D/\ell} \tag{8.2.6}$$

With these definitions Eq. 8.2.3 may be written as

$$\frac{dx_1}{d\eta} = \Phi x_1 + \phi \tag{8.2.7}$$

which is to be solved subject to the boundary conditions of a film model

$$\begin{array}{ccc} r = r_0 & \eta = 0 & x_1 = x_{10} \\ r = r_\delta & \eta = 1 & x_1 = x_{1\delta} \end{array} \tag{8.2.8}$$

The linear differential equation (Eq. 8.2.7), can be integrated to give the composition profiles

$$\frac{(x_1 - x_{10})}{(x_{1\delta} - x_{10})} = \frac{(e^{\Phi\eta} - 1)}{(e^\Phi - 1)} \tag{8.2.9}$$

The diffusion flux at $\eta = 0$, J_{10}, is given by

$$J_{10} = - \frac{c_t D}{\ell} \frac{dx_1}{d\eta}\bigg|_{\eta=0} = c_t \frac{D}{\ell} \frac{\Phi}{\exp(\Phi) - 1}(x_{10} - x_{1\delta}) \tag{8.2.10}$$

Similarly the diffusion flux at $\eta = 1$, $J_{1\delta}$ can be obtained

$$J_{1\delta} = - \frac{c_t D}{\ell} \frac{dx_1}{d\eta}\bigg|_{\eta=1} = c_t \frac{D}{\ell} \frac{\Phi \exp(\Phi)}{\exp(\Phi) - 1}(x_{10} - x_{1\delta}) \tag{8.2.11}$$

Comparison of Eqs. 8.2.10 and 8.2.11 with the basic definition of the low flux mass transfer coefficient (Eq. 7.1.3), shows

$$k_0 = k_\delta = k = D/\ell \tag{8.2.12}$$

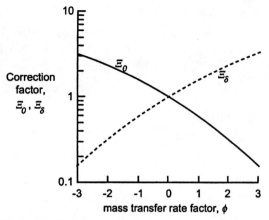

Figure 8.2. High flux correction factor from film theory.

with the correction factors given by

$$\Xi_0 = \frac{\Phi}{\exp(\Phi) - 1} \qquad \Xi_\delta = \frac{\Phi \exp(\Phi)}{\exp(\Phi) - 1} \tag{8.2.13}$$

In the limit of N_t $(= N_1 + N_2)$ tending to zero, $\Phi = 0$ and the correction factors Ξ_0 and Ξ_δ are unity (as can be shown using L'Hopital's rule). If $\Phi < 0$, then the correction factor Ξ_0 is greater than unity (and, therefore, $k_0^\bullet > k_0$). On the other hand, if $\Phi > 0$, then the correction factor, Ξ_0 is less than unity (and, therefore, $k_0^\bullet < k_0$). A graph showing the behavior of the correction factors Ξ_0 and Ξ_δ is shown in Figure 8.2.

The flux N_1 can be calculated by multiplying the diffusion flux by the appropriate bootstrap coefficient

$$N_1 = \beta_0 J_{10} = c_t \beta_0 k \frac{\Phi}{\exp(\Phi) - 1}(x_{10} - x_{1\delta}) \tag{8.2.14}$$

$$= \beta_\delta J_{1\delta} = c_t \beta_\delta k \frac{\Phi \exp(\Phi)}{\exp(\Phi) - 1}(x_{10} - x_{1\delta}) \tag{8.2.15}$$

To compute the flux N_1 from either Eq. 8.2.14 or 8.2.15 requires an iterative procedure since N_1 (and N_2) are involved in the rate factor Φ. Repeated substitution of the fluxes starting from an initial guess calculated with $\Xi = 1$ will usually converge in only a few iterations. It is possible, however, to derive an equation from which the flux N_t (and, hence, N_1 and N_2) may be calculated without iteration. Recognizing that the right-hand members of Eqs. 8.2.14 and 8.2.15 must give the same flux N_1 we have, on cancelling the common terms c_t, k, $\Phi/(\exp(\Phi) - 1)$ and $(x_{10} - x_{1\delta})$,

$$\Phi = N_t/c_t k = \ln\{\beta_0/\beta_\delta\} \tag{8.2.16}$$

Various special cases of this result can now be identified.

8.2.1 Equimolar Counterdiffusion

This situation can arise in, for example, a distillation column when the molar heats of vaporization of the two components are equal to each other or when there is no net change in the total number of moles during diffusion with heterogeneous chemical reaction.

$$N_t = 0 \qquad \beta_0 = \beta_\delta = 1 \qquad \Phi = 0 \qquad \Xi_0 = \Xi_\delta = 1$$

$$N_1 = c_t k (x_{10} - x_{1\delta}) \tag{8.2.17}$$

$$N_2 = -N_1 = c_t k (x_{20} - x_{2\delta})$$

8.2.2 Stefan Diffusion

In this case we have diffusion of component 1 in the presence of an inert or stagnant gas. This situation arises very often during absorption or condensation operations and when reaction takes place in the presence of inerts or diluents. For the case where $N_2 = 0$, we have

$$\beta_0 = 1 + x_{10}/x_{20} = 1/x_{20} \qquad \beta_\delta = 1 + x_{1\delta}/x_{2\delta} = 1/x_{2\delta}$$

$$\Phi = \ln\left\{ \frac{x_{2\delta}}{x_{20}} \right\} = \ln\left\{ \frac{1 - x_{1\delta}}{1 - x_{10}} \right\} = \frac{N_t}{c_t k} = \frac{N_1}{c_t k} \tag{8.2.18}$$

$$N_1 = c_t k \ln\left\{ \frac{1 - x_{1\delta}}{1 - x_{10}} \right\} \tag{8.2.19}$$

8.2.3 Flux Ratios Fixed

When the flux ratios $z_1 = N_1/N_t$ and $z_2 = N_2/N_t$ are fixed by, for example, the stoichiometry of a surface chemical reaction or by the composition of a liquid phase (as in some condensation situations), the β coefficients are given by Eq. 7.2.21 and the mass transfer rate factor reduces to

$$\Phi = \ln\left\{ \frac{1 - x_{1\delta}/z_1}{1 - x_{10}/z_1} \right\} = \frac{N_t}{c_t k} = \frac{N_1/z_1}{c_t k} = \frac{N_2/z_2}{c_t k} \tag{8.2.20}$$

This expression allows calculation of the fluxes N_1 and N_2. It must be noted that in problems of this kind one of the surface mole fractions, x_{10} say, will be unknown and given by an expression for the heterogeneous reaction rate at the surface (e.g., $N_1 = k_s c_t x_{10}$, where k_s is the surface reaction rate constant). The flux ratios in condensation are sometimes set equal to the condensate composition and the vapor composition at the condensate surface is determined from vapor–liquid equilibrium equations. A more detailed discussion of mass (and energy) transfer in condensation is postponed until Chapter 15.

8.2.4 Generalization to Other Geometries

For diffusion in cylindrical or spherical geometries the one-dimensional, steady-state form of Eq. 1.3.10 simplifies to

$$\frac{d r^\alpha N_i}{dr} = 0 \qquad \frac{d r^\alpha N_t}{dr} = 0 \tag{8.2.21}$$

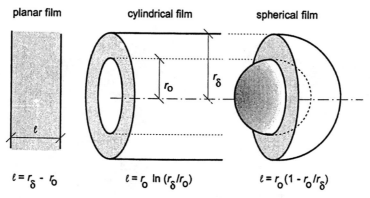

Figure 8.3. Generalization of film model to other geometries by appropriate definition of the characteristic length ℓ.

where α is unity for cylindrical geometries, two for spherical geometries, and zero for plane films. Equation 8.2.21 shows that $r^{\alpha}N_i$ and $r^{\alpha}N_t$ are r invariant.

Equations 8.2.14–8.2.20, derived for the fluxes across a planar film, apply essentially unchanged for diffusion in cylindrical and spherical films. All that needs to be done is to use the appropriate definition of the characteristic length ℓ from Figure 8.3. The flux then calculated from Eqs. 8.2.14 or 8.2.15 would be that at the plane $\eta = 0$.

A very important practical application of the above film model is to determine the external mass transfer resistance to catalyst particles (Example 8.2.2).

Example 8.2.1 Equimolar Distillation of a Binary Mixture

Estimate the rates of mass transfer in the distillation of the system methanol(1)–ethanol(2) under the following conditions:

Temperature = 70°C, pressure = 101.325 kPa.
Composition in the vapor $x_{10} = 0.497$.
Composition at the interface $x_{1\delta} = 0.567$.
The Fick diffusion coefficient $D = 9.1 \times 10^{-6} \text{ m}^2/\text{s}$.
The film thickness $\ell = 1$ mm.

SOLUTION The first step is to calculate the molar density from the ideal gas law

$$c_t = P/RT$$

$$= 101,325/(8.3143 \times 343.15)$$

$$= 35.5 \text{ mol/m}^3$$

The mass transfer coefficient is obtained from Eq. 8.2.12 as

$$k = D/\ell$$

$$= 9.1 \times 10^{-6}/1 \times 10^{-3}$$

$$= 9.1 \times 10^{-3} \text{ m/s}$$

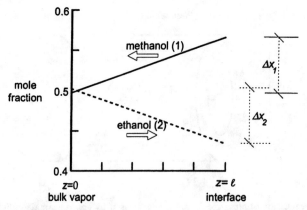

Figure 8.4. Composition profiles in the vapor phase during the distillation of methanol(1) and ethanol(2). Arrows indicate actual directions of mass transfer.

It is common (but not necessarily accurate—see Section 11.5) to assume equimolar counterdiffusion in distillation. Thus, $N_t = 0$ and the flux of methanol is obtained directly from Eq. 8.2.17 as

$$N_1 = c_t k (x_{10} - x_{1\delta})$$

$$= 35.5 \times 9.1 \times 10^{-3} \times (0.497 - 0.567)$$

$$= -0.0226 \text{ mol/m}^2\text{s}$$

The flux of ethanol is just the negative of N_1. Composition profiles are shown in Figure 8.4. The arrows indicate the actual directions of mass transfer. Methanol diffuses from the interface to the bulk vapor, whereas ethanol diffuses in the opposite direction. In other words, the vapor is being enriched in the more volatile methanol. ■

Example 8.2.2   Production of Nickel Carbonyl

Nickel carbonyl is to be produced by passing carbon monoxide over nickel spheres as shown in Figure 8.5. The following reaction takes place at the solid surface:

$$\text{Ni} + 4\text{CO(g)} \rightarrow \text{Ni(CO)}_4\text{(g)}$$

The reaction is very rapid, so that the partial pressure of CO at the metal surface is essentially zero. The carbonyl forms as a gas that diffuses as fast as it forms from the metal surface to the bulk gas stream.

We wish to estimate the rate of mass transfer of CO(1) during the production of nickel carbonyl [Ni(CO)$_4$](2) under the following conditions:

Temperature = 50°C, pressure = 101.325 kPa.
The composition of CO in the gas phase is 50 mol%.
Binary gas diffusivity $D = 2 \times 10^{-5}$ m^2/s.
Catalyst pellets are spherical with diameter $d = 0.0125$ m.
Film thickness = 0.695 mm.

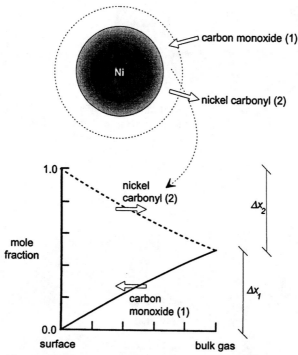

Figure 8.5. Composition profiles in the vapor film surrounding nickel spheres in the production of nickel carbonyl. Arrows indicate actual directions of mass transfer.

SOLUTION For every mole of carbonyl that diffuses to the bulk gas stream, 4 mol of CO diffuse to the metal surface. Thus, the flux ratios are determined as follows:

$$N_1 = -4N_2$$

and

$$N_t = N_1 + N_2 = N_1 - \tfrac{1}{4}N_1 = \tfrac{3}{4}N_1 = -3N_2$$

Hence,

$$z_1 = \tfrac{4}{3} \qquad z_2 = -\tfrac{1}{3}$$

and we may write

$$N_1 = J_1 / \left(1 - x_1 / \tfrac{4}{3} \right) \qquad N_2 = J_2 / \left(1 + x_2 / \tfrac{1}{3} \right)$$

When the flux ratios are fixed, we may evaluate the mass transfer rate factor Φ from Eq. 8.2.20

$$\Phi = \ln \left\{ \frac{1 - x_{1\delta}/z_1}{1 - x_{10}/z_1} \right\}$$

It is convenient to identify the mole fraction x_{10} as the composition at the catalyst surface. The mole fraction in the bulk gas is $x_{1\delta}$ and is 0.5. The mole fraction x_{10} is zero, since carbon monoxide is entirely consumed by the reaction at the catalyst surface. Thus,

Eq. 8.2.20 simplifies to

$$\Phi = \ln(1 - x_{1\delta}/z_1)$$
$$= \ln\left(1 - 0.5/\tfrac{4}{3}\right)$$
$$= -0.4700$$

To evaluate the flux of CO we need the molar density c_t and the mass transfer coefficient $k = D/\ell$. For a spherical film ℓ is obtained form the formula in Figure 8.3; the result is $\ell = 0.625$ mm. The mass transfer coefficient therefore is

$$k = D/\ell$$
$$= 2 \times 10^{-5}/0.625 \times 10^{-3}$$
$$= 0.0320 \text{ m/s}$$

The next step is to calculate the molar density from the ideal gas law

$$c_t = P/RT$$
$$= 37.7 \text{ mol/m}^3$$

The molar flux of CO may now be calculated as

$$N_1 = z_1 c_t k \Phi$$
$$= \tfrac{4}{3} \times 37.7 \times 0.032 \times (-0.47)$$
$$= -0.756 \text{ mol/m}^2\text{s}$$

and N_2 follows as:

$$N_2 = z_2 c_t k \Phi$$
$$= -\tfrac{1}{3} \times 37.7 \times 0.032 \times (-0.47)$$
$$= 0.189 \text{ mol/m}^2\text{s}$$

Composition profiles are shown in Figure 8.5. The directions of mass transfer are shown in the figure by the arrows. ■

Example 8.2.3 Condensation of a Binary Vapor Mixture

A vapor mixture of 40 mol% ethylene dichloride(1) and 60 mol% toluene(2) is fed at 130°C and 101.325 kPa to a condenser. If the composition of the first drop of condensate that forms on the cold surface is 32.5 mol% ethylene dichloride, find the total molar rate of condensation at that position in the condenser.

DATA The interfacial equilibrium relationship is given by

$$y_{1I} = \frac{2.14 x_{1I}}{1 + 1.14 x_{1I}}$$

where x_{1I} and y_{1I} are the interfacial compositions of ethylene dichloride in the liquid and vapor phases, respectively.

The vapor-phase mass transfer coefficient is

$$k = 0.054 \text{ m/s}$$

SOLUTION   The composition of the first drop of condensate is given by the relative rates of condensation

$$x_1 = N_1/(N_1 + N_2)$$

The flux ratio z_1 is, therefore, equal to the liquid mole fraction x_1

$$z_1 = \frac{N_1}{N_1 + N_2} = x_1 = 0.325$$

We will identify the bulk gas with the position $r = 0$ and the interface with $r = r_\delta$. The bulk gas composition y_{10} is 0.4. The interface mole fraction y_{1I} is calculated from the equilibrium relationship as

$$y_{1\delta} = \frac{2.14 \times 0.325}{(1 + 1.14 \times 0.325)}$$

$$= 0.5075$$

Hence, the rate factor Φ is

$$\Phi = \ln\left\{\frac{1 - y_{1\delta}/z_1}{1 - y_{10}/z_1}\right\}$$

$$= \ln\{(1 - 0.5075/0.325)/(1 - 0.4/0.325)\}$$

$$= 0.8893$$

The molar density c_t is evaluated from the ideal gas law as

$$c_t = P/RT$$

$$= 101325/(8.3143 \times 403.15)$$

$$= 30.2 \text{ mol/m}^3$$

The total molar flux follows from Eq. 8.2.16 as

$$N_t = c_t k \Phi$$

$$= 30.2 \times 0.054 \times 0.8893$$

$$= 1.4517 \text{ mol/m}^2\text{s}$$

and the component fluxes are given simply by

$$N_1 = z_1 N_t$$

$$= 0.325 \times 1.4517$$

$$= 0.4718 \text{ mol/m}^2\text{s}$$

$$N_2 = z_2 N_t$$

$$= (1 - 0.325) \times 1.4517$$

$$= 0.9799 \text{ mol/m}^2\text{s}$$

Composition profiles are shown in Figure 8.6.   ■

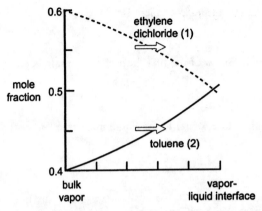

Figure 8.6. Composition profiles in the vapor film adjacent to the condensate.

8.3   EXACT SOLUTIONS OF THE MAXWELL–STEFAN EQUATIONS FOR MULTICOMPONENT MASS TRANSFER IN IDEAL GASES

Let us now turn our attention to n-component mixtures. Exact analytical solutions of the Maxwell–Stefan equations for a film model can be obtained for a mixture of ideal gases for which the binary diffusion coefficients $\mathcal{D}_{ik}$ are independent of composition and identical to the diffusivity of the binary gas $i-k$ pair. Solutions of the Maxwell–Stefan equations for certain special cases involving diffusion in ternary systems have been known for a long time (Gilliland (1937) ($N_3 = 0$); Pratt (1950) ($N_t = 0$); Cichelli et al. (1951); Toor (1957), ($N_t = 0$); Keyes and Pigford (1957); Hsu and Bird (1960); Johns and DeGance (1975)—mass transfer with chemical reaction). A general solution applicable to mixtures with any number of constituents and any relationship between the fluxes seems to have been derived first by Turevskii et al. (1974) and independently by Krishna and Standart (1976a) and is developed in detail below. The relationships between the general solution of Krishna and Standart and the many special case solutions are explored by Taylor (1981a, 1982c).

The starting point for the ideal gas multicomponent film model are the Maxwell–Stefan diffusion equations, written for the y phase, here taken to be gaseous

$$\frac{dy_i}{dr} = \sum_{j=1}^{n} \frac{(y_i N_j - y_j N_i)}{c_t \mathcal{D}_{ij}} \tag{8.3.1}$$

Let us make use of the fact that the mole fractions add up to unity and eliminate y_n from Eq. 8.3.1 and write

$$\frac{dy_i}{d\eta} = \Phi_{ii} y_i + \sum_{\substack{j=1 \\ j \neq i}}^{n-1} \Phi_{ij} y_j + \phi_i \tag{8.3.2}$$

where η is a dimensionless distance defined by

$$\eta = (r - r_0)/\ell \tag{8.3.3}$$

The coefficients Φ_{ij} are defined by

$$\Phi_{ii} = \frac{N_i}{c_t \mathcal{D}_{in}/\ell} + \sum_{\substack{k=1 \\ i \neq k}}^{n} \frac{N_k}{c_t \mathcal{D}_{ik}/\ell} \tag{8.3.4}$$

$$\Phi_{ij} = -N_i \left(\frac{1}{c_t \mathcal{D}_{ij}/\ell} - \frac{1}{c_t \mathcal{D}_{in}/\ell} \right) \tag{8.3.5}$$

and ϕ_i is defined by

$$\phi_i = -\frac{N_i}{c_t \mathcal{D}_{in}/\ell} \qquad i = 1, 2, \ldots, n-1 \tag{8.3.6}$$

Equations 8.3.2 are more conveniently written in $n-1$ dimensional matrix form as

$$\frac{d(y)}{d\eta} = [\Phi](y) + (\phi) \tag{8.3.7}$$

where $[\Phi]$ is a square matrix of mass transfer rate factors of order $n-1$

$$[\Phi] = \begin{bmatrix} \Phi_{11} & \Phi_{12} & \Phi_{13} & \cdots & \Phi_{1,n-1} \\ \Phi_{21} & \Phi_{22} & \Phi_{23} & \cdots & \Phi_{2,n-1} \\ \vdots & & & & \vdots \\ \Phi_{n-1,1} & \Phi_{n-1,2} & \Phi_{n-1,3} & \cdots & \Phi_{n-1,n-1} \end{bmatrix}$$

with elements given by Eqs. 8.3.4 and 8.3.5 and where (ϕ) is a column matrix of order $n-1$

$$(\phi) = \begin{pmatrix} \phi_1 \\ \phi_2 \\ \vdots \\ \phi_{n-1} \end{pmatrix}$$

The boundary conditions of the film model are

$$\begin{aligned} r = r_0 \quad \eta = 0 \quad y_i = y_{i0} \\ r = r_\delta \quad \eta = 1 \quad y_i = y_{i\delta} \end{aligned} \tag{8.3.8}$$

Since c_t, the N_i and the $\mathcal{D}_{ij}$ are constant, Eqs. 8.3.7 represents a set of linear differential equations with constant coefficients. The solution, in $n-1$ dimensional matrix form, is (see Appendix B.2 for derivation)

$$(y) = [\exp[\Phi]\eta](y_0) + [[\exp[\Phi]\eta] - [I]][\Phi]^{-1}(\phi) \tag{8.3.9}$$

which we rewrite by subtracting (y_0) from each side

$$(y - y_0) = [\exp[\Phi]\eta - [I]]((y_0) + [\Phi]^{-1}(\phi)) \tag{8.3.10}$$

Equation 8.3.9 was presented as a solution of the Maxwell–Stefan equations by Turevskii et al. (1974) and by Krishna and Standart (1976a). Krishna and Standart went further, however, by applying the second boundary condition to Eq. 8.3.10

$$(y_\delta - y_0) = [\exp[\Phi] - [I]]((y_0) + [\Phi]^{-1}(\phi)) \tag{8.3.11}$$

This result can be combined with Eq. 8.3.10 to eliminate the column matrix $((y_0) + [\Phi]^{-1}(\phi))$

$$(y - y_0) = [\exp[\Phi]\eta - [I]][\exp[\Phi] - [I]]^{-1}(y_\delta - y_0) \tag{8.3.12}$$

which is an exact matrix analog of Eq. 8.2.9 for binary systems.

The next task is to obtain an expression for the diffusion fluxes J_i. For one-dimensional diffusion the Maxwell–Stefan relations can be written in terms of the diffusion fluxes as (cf. Eq. 2.1.24)

$$(J) = -\frac{c_t}{\ell}[B]^{-1}\frac{d(y)}{d\eta} = -\frac{c_t}{\ell}[D]\frac{d(y)}{d\eta} \tag{8.3.13}$$

where the matrix $[B]$ has elements given by Eqs. 2.1.20 and 2.1.21. The gradients in the mole fractions are obtained from Eq. 8.3.12 as follows:

$$\frac{d(y)}{d\eta} = [\Phi][\exp[\Phi]\eta][\exp[\Phi] - [I]]^{-1}(y_\delta - y_0) \tag{8.3.14}$$

Equations 8.3.13 and 8.3.14 can be used to evaluate the diffusion fluxes at any position η. The diffusion fluxes at $\eta = 0$, J_{i0}, can be evaluated from

$$(J_0) = -\frac{c_t[D_0]}{\ell}\frac{d(y)}{d\eta}\bigg|_{\eta=0}$$

$$= \frac{c_t[D_0]}{\ell}[\Phi][\exp[\Phi] - [I]]^{-1}(y_0 - y_\delta) \tag{8.3.15}$$

where the elements of the matrix $[D_0]$ are obtained from $[D_0] = [B_0]^{-1}$, with the elements B_{ik} evaluated from Eqs. 2.1.21 and 2.1.22 using the mole fractions y_{i0}.

With the finite flux mass transfer coefficients matrix $[k^\bullet]$ defined by Eq. 7.1.9 we have

$$[k_0^\bullet] = \frac{[D_0]}{\ell}[\Phi][\exp[\Phi] - [I]]^{-1} \tag{8.3.16}$$

To calculate the zero flux mass transfer coefficients $[k]$ we take the limit as the N_i (all n of them) go to zero

$$\lim_{N_i \to 0}[\Phi] \to [0] \tag{8.3.17}$$

and

$$\lim_{N_i \to 0}[\Phi][\exp[\Phi] - [I]]^{-1} \to [I] \tag{8.3.18}$$

The proof of this is left as an exercise. Thus, the matrix of zero-flux mass transfer coefficients $[k_0]$ is

$$[k_0] = [D_0]/\ell \tag{8.3.19}$$

and the matrix of correction factors $[\Xi_0]$, is given by

$$[\Xi_0] = [\Phi][\exp[\Phi] - [I]]^{-1} \tag{8.3.20}$$

Alternatively, we may proceed via the diffusion fluxes at $\eta = 1$

$$
\begin{aligned}
(J_\delta) &= -\frac{c_t[D_\delta]}{\ell} \frac{d(y)}{d\eta}\bigg|_{\eta=1} \\
&= \frac{c_t[D_\delta]}{\ell}[\Phi][\exp[\Phi]][\exp[\Phi] - [I]]^{-1}(y_0 - y_\delta)
\end{aligned}
\tag{8.3.21}
$$

giving the zero-flux mass transfer coefficients

$$[k_\delta] = [D_\delta]/\ell \tag{8.3.22}$$

and the matrix of correction factors

$$[\Xi_\delta] = [\Phi]\exp[\Phi][\exp[\Phi] - [I]]^{-1} = [\Xi_0]\exp[\Phi] \tag{8.3.23}$$

By invoking the "bootstrap" solution, the fluxes N_i can be evaluated from one of two equivalent expressions

$$
\begin{aligned}
(N) &= c_t[\beta_0][k_0][\Xi_0](y_0 - y_\delta) = c_t[W_0](y_0 - y_\delta) \\
&= c_t[\beta_\delta][k_\delta][\Xi_\delta](y_0 - y_\delta) = c_t[W_\delta](y_0 - y_\delta)
\end{aligned}
\tag{8.3.24}
$$

where we have defined a matrix of combined mass transfer coefficients as $[W] = [\beta][k][\Xi]$.

8.3.1 Formulation in Terms of Binary Mass Transfer Coefficients

In proceeding with the discussions we define a matrix $[R]$ with elements

$$
\begin{aligned}
R_{ii} &= \frac{y_i}{\kappa_{in}} + \sum_{\substack{k=1 \\ k\neq i}}^{n} \frac{y_k}{\kappa_{ik}} \\
R_{ij} &= -y_i\left(\frac{1}{\kappa_{ij}} - \frac{1}{\kappa_{in}}\right)
\end{aligned}
\tag{8.3.25}
$$

where κ_{ij} is a low flux mass transfer coefficient for the binary i–j pair defined by

$$\kappa_{ij} = Ð_{ij}/\ell \tag{8.3.26}$$

The matrix of low flux mass transfer coefficients may be expressed in terms of the R_{ij} as

$$[k] = [R]^{-1} \tag{8.3.27}$$

We may also write the elements of the rate factor matrix $[\Phi]$ in terms of these binary mass transfer coefficients.

$$\Phi_{ii} = \frac{N_i}{c_t \kappa_{in}} + \sum_{\substack{k=1 \\ k \neq i}}^{n} \frac{N_k}{c_t \kappa_{ik}} \tag{8.3.28}$$

$$\Phi_{ij} = -N_i \left(\frac{1}{c_t \kappa_{ij}} - \frac{1}{c_t \kappa_{in}} \right) \tag{8.3.29}$$

For a ternary system we may carry out the inversion of $[R]$ explicitly and calculate the elements of $[k]$ as follows:

$$
\begin{aligned}
k_{11} &= \kappa_{13} (y_1 \kappa_{23} + (1 - y_1)\kappa_{12})/S \\
k_{12} &= y_1 \kappa_{23} (\kappa_{13} - \kappa_{12})/S \\
k_{21} &= y_2 \kappa_{13} (\kappa_{23} - \kappa_{12})/S \\
k_{22} &= \kappa_{23} (y_2 \kappa_{13} + (1 - y_2)\kappa_{12})/S
\end{aligned}
\tag{8.3.30}
$$

where

$$S = y_1 \kappa_{23} + y_2 \kappa_{13} + y_3 \kappa_{12} \tag{8.3.31}$$

The elements of $[\Phi]$ are given by

$$
\begin{aligned}
\Phi_{11} &= \frac{N_1}{c_t \kappa_{13}} + \frac{N_2}{c_t \kappa_{12}} + \frac{N_3}{c_t \kappa_{13}} \\[2mm]
\Phi_{12} &= -N_1 \left(\frac{1}{c_t \kappa_{12}} - \frac{1}{c_t \kappa_{13}} \right) = -N_1 \alpha_1 \\[2mm]
\Phi_{21} &= -N_2 \left(\frac{1}{c_t \kappa_{12}} - \frac{1}{c_t \kappa_{23}} \right) = -N_2 \alpha_2 \\[2mm]
\Phi_{22} &= \frac{N_1}{c_t \kappa_{12}} + \frac{N_2}{c_t \kappa_{23}} + \frac{N_3}{c_t \kappa_{23}}
\end{aligned}
\tag{8.3.32}
$$

where

$$
\begin{aligned}
\alpha_1 &= \frac{1}{c_t \kappa_{12}} - \frac{1}{c_t \kappa_{13}} \\[2mm]
\alpha_2 &= \frac{1}{c_t \kappa_{12}} - \frac{1}{c_t \kappa_{23}}
\end{aligned}
\tag{8.3.33}
$$

Some special cases of these expressions will be encountered in the illustrative examples we consider below.

8.3.2   Limiting Cases of the General Solution

Let us now examine the structure of the matrices $[k]$, $[\Phi]$, $[\Xi]$, and $[k^\bullet]$ for two limiting cases.

- All binary diffusion (or mass transfer) coefficients equal.
- Species one and two present in very low concentrations.

If all the binary diffusion coefficients have the same value then we see from Eq. 8.3.25 that $[k]$ degenerates to a scalar times the identity matrix

$$[k] = \kappa[I] \tag{8.3.34}$$

That is,

$$k_{ii} = \kappa \qquad k_{ij} = 0 \qquad i \neq j = 1, 2, \ldots, n - 1 \tag{8.3.35}$$

The rate factor matrix $[\Phi]$ also is diagonal with all diagonal elements equal

$$[\Phi] = \Phi[I] \qquad \Phi = N_t/c_t\kappa \tag{8.3.36}$$

The correction factor matrix is particularly simple to calculate since it too is diagonal with all diagonal elements equal

$$[\Xi_0] = \Xi_0[I] \qquad \text{with } \Xi_0 = \Phi/(e^\Phi - 1) \tag{8.3.37}$$

and the general relation for the diffusion fluxes (Eq. 8.3.15) simplifies to

$$(J) = c_t\kappa\Xi_0[I](\Delta y) \tag{8.3.38}$$

That is,

$$J_i = c_t\kappa\frac{\Phi}{\exp \Phi - 1}\Delta y_i \tag{8.3.39}$$

In fact, for this special case it is possible to derive an explicit relation for the molar fluxes N_i (see Section 8.5 for further discussion).

In dilute solutions the concentrations of all but one of the components are so low that we may assume that $y_i \approx 0$ for $i = 1, 2, \ldots, n - 1$ and $y_n \approx 1$. Returning to Eqs. 8.3.25 we find, for this special case

$$k_{ii} = \kappa_{in} \qquad k_{ij} = 0 \qquad i \neq j \tag{8.3.40}$$

The rate factor matrix follows from Eqs. 8.3.28 and 8.3.29

$$\Phi_{ii} = N_t/c_t\kappa_{in} \qquad \Phi_{ij} = 0 \qquad i \neq j \tag{8.3.41}$$

In presenting these last results we make use of the fact that if the mole fractions are vanishingly small, then so are the corresponding driving forces Δy_i.

Since $[k]$ and $[\Phi]$ are diagonal we can calculate the correction factor matrix and the finite flux mass transfer coefficients from

$$k_{0ii}^{\bullet} = \kappa_{in} \frac{\Phi_{ii}}{\exp \Phi_{ii} - 1} \tag{8.3.42}$$

and

$$J_i = c_t k_{ii}^{\bullet}(y_{i0} - y_{i\delta}) \tag{8.3.43}$$

Only in these limiting cases is the computation of the fluxes from an exact solution of the Maxwell–Stefan equations so straightforward. In most cases of practical importance we must make use of the full matrix solution (Eq. 8.3.24).

8.3.3 Computation of the Fluxes

The computation of the fluxes N_i from either of Eqs. 8.3.24 necessarily involves an iterative procedure (except for the special cases discussed above), partly because the N_i themselves are needed for the evaluation of the matrix of correction factors and also because an explicit relation for the matrix $[\Phi]$ cannot be derived as a generalization of Eq. 8.2.16 for binary mass transfer; there is no requirement in matrix algebra for the matrices $[W_0]$ and $[W_\delta]$ to be equal to each other even though the fluxes calculated from both parts of these equations must be equal. Indeed, these two matrices will be equal only in the case of vanishingly small mole fraction differences $(y_0 - y_\delta)$ and vanishingly small mass transfer rates. In almost all cases of interest these two matrices are quite different. An explicit solution was possible for binary systems only because all matrices reduce to scalar quantities.

The method of successive substitution can be a very effective way of computing the N_i from Eqs. 8.3.24 when the mole fractions at both ends of the diffusion path y_{i0} and $y_{i\delta}$, are known. In practice, we start from an initial guess of the fluxes N_i and compute the rate factor matrix $[\Phi]$. The correction factor matrix $[\Xi]$ may be calculated from an application of Sylvester's expansion formula (Eq. A.5.20)

$$[\Xi] = \sum_{i=1}^{m} \hat{\Xi}_i \left\{ \prod_{\substack{j=1 \\ j \neq i}}^{m} \left[[\Phi] - \hat{\Phi}_j[I] \right] \Big/ \prod_{\substack{j=1 \\ j \neq i}}^{m} \left(\hat{\Phi}_i - \hat{\Phi}_j \right) \right\} \tag{8.3.44}$$

where m is the number of distinct eigenvalues of $[\Phi]$ ($m \leqq n - 1$). The eigenvalue functions $\hat{\Xi}_i$ are given by

$$\hat{\Xi}_{i0} = \frac{\hat{\Phi}_i}{\exp \hat{\Phi}_i - 1} \qquad \hat{\Xi}_{i\delta} = \frac{\hat{\Phi}_i \exp \hat{\Phi}_i}{\exp \hat{\Phi}_i - 1} \tag{8.3.45}$$

depending on whether the flux correction factor matrix is to be evaluated at $\eta = 0$ or at $\eta = 1$.

For the ternary case with two distinct eigenvalues $\hat{\Phi}_1 \neq \hat{\Phi}_2$, Eq. 8.3.44 simplifies to

$$[\Xi] = \frac{\hat{\Xi}_1 \left[[\Phi] - \hat{\Phi}_2[I] \right]}{\hat{\Phi}_1 - \hat{\Phi}_2} + \frac{\hat{\Xi}_2 \left[[\Phi] - \hat{\Phi}_1[I] \right]}{\hat{\Phi}_2 - \hat{\Phi}_1} \tag{8.3.46}$$

which can be expanded as follows:

$$\Xi_{11} = \frac{\hat{\Xi}_1(\Phi_{11} - \hat{\Phi}_2)}{\hat{\Phi}_1 - \hat{\Phi}_2} + \frac{\hat{\Xi}_2(\Phi_{11} - \hat{\Phi}_1)}{\hat{\Phi}_2 - \hat{\Phi}_1} \qquad (8.3.46a)$$

$$\Xi_{12} = \frac{(\hat{\Xi}_1 - \hat{\Xi}_2)}{(\hat{\Phi}_1 - \hat{\Phi}_2)}\Phi_{12} \qquad (8.3.46b)$$

$$\Xi_{21} = \frac{(\hat{\Xi}_1 - \hat{\Xi}_2)}{(\hat{\Phi}_1 - \hat{\Phi}_2)}\Phi_{21} \qquad (8.3.46c)$$

$$\Xi_{22} = \frac{\hat{\Xi}_1(\Phi_{22} - \hat{\Phi}_2)}{\hat{\Phi}_1 - \hat{\Phi}_2} + \frac{\hat{\Xi}_2(\Phi_{22} - \hat{\Phi}_1)}{\hat{\Phi}_2 - \hat{\Phi}_1} \qquad (8.3.46d)$$

The two eigenvalues $\hat{\Phi}_1$ and $\hat{\Phi}_2$ are the roots of the quadratic equation

$$\hat{\Phi}^2 - (\Phi_{11} + \Phi_{22})\hat{\Phi} + (\Phi_{11}\Phi_{22} - \Phi_{12}\Phi_{21}) = 0 \qquad (8.3.47)$$

that is,

$$\hat{\Phi}_1 = \tfrac{1}{2}\left\{\mathrm{tr}[\Phi] + \sqrt{\mathrm{disc}[\Phi]}\right\}$$
$$\hat{\Phi}_2 = \tfrac{1}{2}\left\{\mathrm{tr}[\Phi] - \sqrt{\mathrm{disc}[\Phi]}\right\} \qquad (8.3.48)$$

where

$$\mathrm{tr}[\Phi] = \Phi_{11} + \Phi_{22}$$
$$|\Phi| = \Phi_{11}\Phi_{22} - \Phi_{12}\Phi_{21}$$
$$\mathrm{disc}[\Phi] = (\mathrm{tr}[\Phi])^2 - 4|\Phi|$$

The fluxes (N) can then be calculated from either of Eqs. 8.3.24 (having previously calculated $[\beta]$ and $[k]$). The new estimates of the N_i are used to recalculate $[\Phi]$ and the procedure is repeated until convergence is obtained. An initial estimate of the N_i can be computed from Eqs. 8.3.24 with the correction factor matrices set equal to the identity matrix $[I]$. This procedure is summarized in Algorithm 8.1 and illustrated in Example 8.3.1.

Before continuing we note that this procedure will not always converge and, if it does, it may not do so particularly efficiently. In Section 8.3.4 we discuss some of the computational subtleties of these equations and provide a more efficient and robust algorithm for computing the fluxes.

Algorithm 8.1 Algorithm for Calculation of Mass Transfer Rates from an Exact Solution of the Maxwell–Stefan Equations

Given:	$(y_0), (y_\delta), c_t, \kappa_{ij}$.
Step 1:	Compute $[k]$.
Step 2:	Compute $[\beta]$.
Step 3:	Estimate $(N) = c_t[\beta][k](\Delta y)$.
Step 4:	Calculate $[\Phi]$.
Step 5:	Calculate $[\Xi]$.
Step 6:	Calculate $(J) = c_t[k][\Xi](\Delta y)$.
Step 7:	Calculate $(N) = [\beta](J)$.
Step 8:	Check for convergence of the N_i If fluxes have not converged return to Step 4.

Example 8.3.1   Equimolar Counterdiffusion in a Ternary Mixture

Estimate the rates of mass transfer during the distillation of ethanol(1)–*t*-butyl alcohol(2)*–water(3) under the following conditions:

$$\text{Bulk vapor-phase composition} \quad y_{10} = 0.65, \ y_{20} = 0.13$$
$$\text{Interface vapor composition} \quad y_{1\delta} = 0.50, \ y_{2\delta} = 0.14$$

The distillation process may be assumed to be equimolar.

DATA   The diffusivities of the three binary pairs are

$$Ð_{12} = 0.80 \times 10^{-5} \text{ m}^2/\text{s}$$
$$Ð_{13} = 2.1 \times 10^{-5} \text{ m}^2/\text{s}$$
$$Ð_{23} = 1.7 \times 10^{-5} \text{ m}^2/\text{s}$$

Molar density of vapor mixture: $c_t = 30 \text{ mol/m}^3$
Film thickness: $\ell = 1$ mm

SOLUTION   We follow below the procedure outlined as Algorithm 8.1; final results are summarized in Table 8.1.

Step 1: Calculate $[k_0] = \dfrac{[D_0]}{\ell} = \dfrac{[B_0]^{-1}}{\ell} = [R_0]^{-1}$. $[k_0]$ can be calculated directly from Eqs. 8.3.30 using y_0. The results are

$$[k_0] = \begin{bmatrix} 0.01872 & 0.00924 \\ 0.00158 & 0.01060 \end{bmatrix} \text{ m/s}$$

Step 2: Calculate $[\beta]$. For equimolar counterdiffusion $[\beta] = [I]$ and the fluxes are given by

$$(N) = (J_0) = c_t[k_0][\Xi_0](y_0 - y_\delta)$$
$$= (J_\delta) = c_t[k_\delta][\Xi_\delta](y_0 - y_\delta)$$

Step 3: We begin the iterations by assuming that all the fluxes are zero, $N_i^{(0)} = 0$. Hence, $[\Phi] = [0]$, $[\Xi] = [I]$ and we obtain our first real estimate of the fluxes from

$$(N^{(1)}) = c_t[k_0](y_0 - y_\delta)$$
$$= \begin{pmatrix} 8.1450 \times 10^{-2} \\ 3.9347 \times 10^{-3} \end{pmatrix} \text{ mol/m}^2\text{s}$$

Step 4: $[\Phi]$ follows from Eqs. 8.3.32 as

$$[\Phi] = \begin{bmatrix} N_2\alpha_1 & -N_1\alpha_1 \\ -N_2\alpha_2 & N_1\alpha_2 \end{bmatrix}$$

tert-Butyl alcohol is the common name for 2-methyl-2-propanol.

where α_1 and α_2 are defined by Eqs. 8.3.33. The determinant of this matrix is zero and its inverse, defined by Eq. A.3.8, does not exist. {In fact, $[\Phi]$ is always singular if $N_t = 0$ regardless of the number of components (Taylor, 1981a).} One of the eigenvalues of $[\Phi]$, $\hat{\Phi}_2$ say, must be zero. The other eigenvalue is given by

$$\hat{\Phi}_1 = N_1\alpha_2 + N_2\alpha_1$$

which, for this special case, is just the sum of the diagonal elements of $[\Phi]$.

Substituting the current values of the molar fluxes gives

$$[\Phi] = \begin{bmatrix} 0.01015 & -0.2109 \\ -0.00868 & 0.1797 \end{bmatrix}$$

and the eigenvalues are

$$\hat{\Phi}_1 = 0.18982 \qquad \hat{\Phi}_2 = 0.0$$

Step 5: $[\Xi]$ is calculated using Sylvester's theorem. First, the eigenvalues $\hat{\Xi}_{0i}$ from Eqs. 8.3.45

$$\hat{\Xi}_1 = \frac{\hat{\Phi}_1}{\exp \hat{\Phi}_1 - 1}$$

$$\hat{\Xi}_2 = 1$$

The latter expression can be obtained by a straightforward use of L'Hopital's rule. Carrying out the computations gives

$$\hat{\Xi}_1 = 0.90809 \qquad \hat{\Xi}_2 = 1.0$$

For this special case Eqs. 8.3.46 for the correction factor matrix simplify to

$$\Xi_{11} = \frac{\left(\hat{\Xi}_1\Phi_{11} + \Phi_{22}\right)}{\hat{\Phi}_1}$$

$$\Xi_{12} = \frac{\left(\hat{\Xi}_1 - 1\right)}{\hat{\Phi}_1}\Phi_{12}$$

$$\Xi_{21} = \frac{\left(\hat{\Xi}_1 - 1\right)}{\hat{\Phi}_1}\Phi_{21}$$

$$\Xi_{22} = \frac{\left(\hat{\Xi}_1\Phi_{22} + \Phi_{11}\right)}{\hat{\Phi}_1}$$

with the following numerical values (based on the current values of the N_i)

$$[\Xi] = \begin{bmatrix} 0.9951 & 0.10172 \\ 0.0042 & 0.913 \end{bmatrix}$$

Step 6: Although it is more efficient *NOT* to calculate $[k^{\bullet}]$ here, we include the results for illustrative purposes.

$$[k_0^{\bullet}] = [k_0][\Xi_0] = \begin{bmatrix} 0.01866 & 0.01034 \\ 0.00162 & 0.00984 \end{bmatrix} \text{m/s}$$

Now the diffusion fluxes J_i are calculated as

$$(J^{(2)}) = c_t[k_0^{\bullet}](y_0 - y_\delta) = 0.03 \begin{bmatrix} 0.01866 & 0.01034 \\ 0.001618 & 0.00984 \end{bmatrix} \begin{pmatrix} 0.15 \\ -0.01 \end{pmatrix}$$

$$= \begin{pmatrix} 8.0881 \times 10^{-2} \\ 4.3287 \times 10^{-3} \end{pmatrix} \text{mol/m}^2\text{s}$$

Step 7: Since $[\beta] = [I]$, the third estimate of the N_i is, therefore,

$$(N^{(2)}) = \begin{pmatrix} 8.0881 \times 10^{-2} \\ 4.3287 \times 10^{-3} \end{pmatrix} \text{mol/m}^2\text{s}$$

Comparing this result with the last shows that convergence has not been obtained and we must continue for at least one more iteration.

Using the latest values of the N_i we recalculate $[\Phi]$, $[\Xi]$, and $[k^{\bullet}]$ as follows:

$$[\Phi] = \begin{bmatrix} 0.011165 & -0.20862 \\ -0.009548 & 0.17841 \end{bmatrix}$$

$$\hat{\Phi}_1 = 0.18958 \qquad \hat{\Phi}_2 = 0.0$$

$$\hat{\Xi}_1 = 0.9082 \qquad \hat{\Xi}_2 = 1.0$$

$$[\Xi] = \begin{bmatrix} 0.9946 & 0.1010 \\ 0.0046 & 0.9136 \end{bmatrix}$$

$$[k_0^{\bullet}] = \begin{bmatrix} 0.01866 & 0.01034 \\ 0.00162 & 0.00984 \end{bmatrix} \text{m/s}$$

and so we calculate our third estimate of the fluxes from

$$(N) = c_t[k_0^{\bullet}](y_0 - y_\delta) = \begin{pmatrix} 8.0859 \times 10^{-2} \\ 4.3436 \times 10^{-3} \end{pmatrix} \text{mol/m}^2\text{s}$$

The fluxes are now converged to an accuracy of less than 1%. One further iteration will give only marginal improvement ($N_1 = 8.0858 \times 10^{-2}$, $N_2 = 4.3442 \times 10^{-3}$ mol/m^{2}s). The four iterations required here is fairly typical of problems involving equimolar counter transfer $N_t = 0$.

Interestingly, even for equimolar transfer $N_t = 0$, the correction factor matrix $[\Xi]$ does not reduce to $[I]$ (although it is well approximated by $[I]$). This means that the composition profiles (as computed from Eq. 8.3.12 and shown in Fig. 8.7) will not be truly linear (although it is hard to discern this fact from Fig. 8.7). Contrast this with a binary system for which $N_t = 0$ leads to linear composition profiles and Ξ_0 equal to unity.

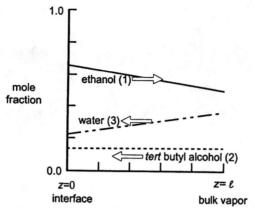

Figure 8.7. Composition profiles for Example 8.3.1. Arrows indicate directions of mass transfer. *tert*-Butyl alcohol is transported towards the interface when its own driving force suggests transfer away from the interface.

The final results are summarized in Table 8.1. It is to be noted that component 2, *t*-butyl alcohol, is being transported towards the interface when its own driving force suggests transfer away from the interface. However, it can clearly be seen in Figure 8.7 that the driving force for *t*-butyl alcohol is rather small and the effects of coupling between species transfers, quantified by the terms $k^{\bullet}_{12} \Delta y_2$ and $k^{\bullet}_{21} \Delta y_1$, are significant here (see Table 8.1) and cannot be ignored. ∎

TABLE 8.1   Equimolar Transfer in Ethanol(1)–*tert*-Butyl alcohol(2)–Water(3)

Position	$r = 0$	$r = \delta$
Composition (y)	$\begin{pmatrix} 0.650 \\ 0.130 \end{pmatrix}$	$\begin{pmatrix} 0.500 \\ 0.140 \end{pmatrix}$
Driving force (Δy)	$\begin{pmatrix} 0.150 \\ -0.010 \end{pmatrix}$	
Matrix of low flux mass transfer coefficients $[k]$ (mm/s)	$\begin{bmatrix} 18.72 & 9.243 \\ 1.581 & 10.60 \end{bmatrix}$	$\begin{bmatrix} 18.33 & 7.716 \\ 1.848 & 11.65 \end{bmatrix}$
Matrix of rate factors $[\Phi]$ (−)	$\begin{bmatrix} 0.0112 & -0.2086 \\ -0.0096 & 0.1784 \end{bmatrix}$	
Matrix of correction factors $[\Xi]$ (−)	$\begin{bmatrix} 0.9946 & 0.1010 \\ 0.0046 & 0.9136 \end{bmatrix}$	$\begin{bmatrix} 1.006 & -0.1076 \\ -0.0049 & 1.092 \end{bmatrix}$
Matrix of high flux mass transfer coefficients $[k^{\bullet}]$ (mm/s)	$\begin{bmatrix} 18.658 & 1.034 \\ 1.622 & 9.845 \end{bmatrix}$	$\begin{bmatrix} 18.399 & 6.454 \\ 1.801 & 12.532 \end{bmatrix}$
$k^{\bullet}_{12} \Delta y_2 / k^{\bullet}_{11} \Delta y_1$ $k^{\bullet}_{21} \Delta y_1 / k^{\bullet}_{22} \Delta y_2$	-0.03694 -2.4713	-0.02338 -2.1557
Molar fluxes (mmol/m^{2}s)	$\begin{pmatrix} 80.86 \\ 4.344 \end{pmatrix}$	

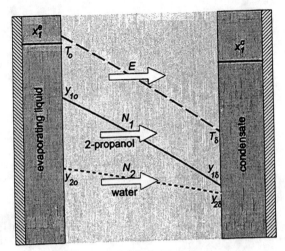

Figure 8.8. Diagram of diffusional distillation process.

Example 8.3.2   *Diffusional Distillation*

Fullarton and Schlünder (1983) investigated the process of "diffusional distillation" for separating liquid mixtures of azeotropic composition. The process is shown schematically in Figure 8.8. A liquid mixture is evaporated at a temperature below its boiling point, diffuses through a vapor space filled with inert gas and condenses at a lower temperature. The inert gas functions as a selective filter that allows preferential passage of those components that diffuse more quickly. Thus, the condensed liquid has a composition different from that of the original mixture.

In modeling the process Fullarton and Schlünder made the following assumptions:

1. The composition of each liquid film and the vapor mixture does not change along the length of the film.
2. The vapor space is regarded as a plane film and the vapor is ideal at constant pressure and temperature.
3. Mass transfer in the vapor space can be described by steady-state molecular diffusion (in other words, by a film model).
4. The flux of inert gas is zero.
5. The liquid films are sufficiently well mixed that the interface composition is the same as the bulk composition. (This is equivalent to assuming that there is no resistance to mass transfer in the liquid film.)
6. The vapor–liquid interfaces are at equilibrium.

In experiment A4 of Fullarton and Schlünder the evaporating liquid mixture of 2-propanol(1) and water(2) has a composition $x_1^e = 0.7625$, $x_2^e = 0.2375$, $T^e = 40°C$. The temperature of the condensation side is 15°C. The total pressure of the system is 101.3 kPa. The gap between the evaporating and condensing films was 6.5 mm. Air (3) was used as the inert gas. Estimate the fluxes under these conditions.

DATA   The diffusivities of the three binary pairs are

$$Đ_{12} = 1.393 \times 10^{-5} \text{ m}^2/\text{s}$$
$$Đ_{13} = 1.046 \times 10^{-5} \text{ m}^2/\text{s}$$
$$Đ_{23} = 2.554 \times 10^{-5} \text{ m}^2/\text{s}$$

Following Fullarton and Schlünder we use the Van Laar equation for the evaluation of $\ln \gamma_i$

$$\ln \gamma_i = A_{ij} A_{ji}^2 x_j^2 / (A_{12} x_1 + A_{21} x_2)^2 \qquad i \neq j = 1, 2$$

The parameters for 2-propanol(1)–water(2) mixtures are

$$A_{12} = 2.3405 \qquad A_{21} = 1.1551$$

The vapor pressures P_i^s may be calculated from the Antoine equation

$$\ln P_i^s = A_i - B_i/(T + C_i)$$

The Antoine constants are

Constant	2-Propanol	Water
A_i	20.44302	18.58488
B_i	4628.956	3984.923
C_i	252.636	233.426

which give P_i^s in millimeters of mercury and T is in degrees celsius.

SOLUTION   Before we can compute the molar fluxes we need to know the composition in the vapor phase at both ends of the diffusion path. At the evaporating side the calculation of y_{i0} is quite simple. From the assumption of interfacial equilibrium we have

$$y_{i0} = \gamma_i P_i^s x_i^e / P$$

The activity coefficients γ_i are calculated at the known temperature T^e and composition x_i^e. Hence, at the azeotropic composition x^e, the activity coefficients are

$$\gamma_1 = 1.04243 \qquad \gamma_2 = 2.38166$$

The vapor pressures P_i^s are calculated from the Antoine equation.
   At the evaporation side we have

$$P_1^s = 101.991 \qquad P_2^s = 55.1929 \text{ mmHg}$$

and the vapor composition at the evaporating side is

$$y_{10} = \gamma_1 P_1^s x_1^e / P = 0.10667$$
$$y_{20} = \gamma_2 P_2^s x_2^e / P = 0.04108$$

The mole fraction of inert gas at the evaporating side is

$$y_{30} = 1 - y_{10} - y_{20} = 0.85225$$

Since $N_3 = 0$, a mass balance on the condensation side yields

$$N_1/(N_1 + N_2) = x_1^c \qquad N_2/(N_1 + N_2) = x_2^c = 1 - x_1^c$$

which, together with the equations representing phase equilibrium at the condensing film interface

$$y_{i\delta} = \gamma_i P_i^s x_i^c / P$$

must be solved simultaneously with Eq. 8.3.24 for the N_i.

One way to solve this set of simultaneous equations is to start by guessing the composition of the condensed film (we need guess only the mole fraction of one component here). Then we may compute the composition of the vapor in equilibrium at the interface from the equilibrium equations and compute the fluxes using Algorithm 8.3.1. The flux ratios may then be checked to see if they equal the estimated liquid-phase composition. If not, then the composition of the condensing film is reestimated and the procedure repeated.

We assume the composition of the condensing film to be

$$x_1^c = 0.5755 \qquad x_2^c = 0.4245$$

The vapor interface composition is computed directly from the equilibrium relations above as

$$y_{1\delta} = 0.02082 \qquad y_{2\delta} = 0.01323$$

The molar density c_t may be calculated from the ideal gas law at the average temperature

$$c_t = 40.53 \text{ mol/m}^3$$

We follow the procedure in Algorithm 8.1 below

Step 1: Calculate $[k_0] = [D_0]/\ell = [B_0]^{-1}/\ell$ where $[k_0]$ can be calculated directly from Eq. 8.3.30. The results are

$$[k_0] = \begin{bmatrix} 1.6245 & -0.0968 \\ 0.0511 & 3.6054 \end{bmatrix} \times 10^{-6} \text{ m/s}$$

Step 2: Calculate $[\beta]$. For Stefan diffusion in a ternary mixture the elements of $[\beta]$ are given by Eq. 7.2.20

$$[\beta] = \begin{bmatrix} 1 + y_1/y_3 & y_1/y_3 \\ y_2/y_3 & 1 + y_2/y_3 \end{bmatrix}$$

$[\beta_0]$, therefore, is

$$[\beta_0] = \begin{bmatrix} 1.1252 & 0.1252 \\ 0.0482 & 1.0482 \end{bmatrix}$$

Step 3: The first estimates of the diffusion fluxes are calculated from

$$(J_0) = c_t[k_0](y_0 - y_\delta)$$

$$= \begin{pmatrix} 5.543 \times 10^{-3} \\ 4.247 \times 10^{-3} \end{pmatrix} \text{mol/m}^2\text{s}$$

and the N_i are computed from

$$(N) = [\beta_0](J_0)$$

$$= \begin{pmatrix} 6.768 \times 10^{-3} \\ 4.719 \times 10^{-3} \end{pmatrix} \text{mol/m}^2\text{s}$$

Step 4: $[\Phi]$ follows from Eqs. 8.3.32 (setting $N_3 = 0$) as

$$[\Phi] = \begin{bmatrix} 0.1581 & 0.0259 \\ 0.0247 & 0.1076 \end{bmatrix}$$

The eigenvalues of this matrix may be computed from the general Eqs. 8.3.48. However, for Stefan diffusion in a ternary mixture, it is possible to derive simple analytical expressions for the eigenvalues as

$$\hat{\Phi}_1 = \frac{N_1}{c_t \kappa_{13}} + \frac{N_2}{c_t \kappa_{23}}$$

$$\hat{\Phi}_2 = \frac{N_1 + N_2}{c_t \kappa_{12}}$$

The numerical values are

$$\hat{\Phi}_1 = 0.1334 \qquad \hat{\Phi}_2 = 0.1323$$

Step 5: Next, the eigenvalues $\hat{\Xi}_{0i}$ are obtained from Eqs. 8.3.45

$$\hat{\Xi}_1 = 0.9348 \qquad \hat{\Xi}_2 = 0.9353$$

$[\Xi]$ follows from Eqs. 8.3.46 as

$$[\Xi] = \begin{bmatrix} 0.9230 & -0.0124 \\ 0.0118 & 0.9471 \end{bmatrix}$$

Step 6: The next estimate of the diffusion fluxes is computed from

$$(J_0) = c_t[k_0][\Xi_0](\Delta y)$$

$$= \begin{pmatrix} 5.087 \times 10^{-3} \\ 4.166 \times 10^{-3} \end{pmatrix} \text{mol/m}^2\text{s}$$

The N_i follow from the bootstrap relation as before

$$(N) = [\beta_0](J_0) = \begin{pmatrix} 6.245 \times 10^{-3} \\ 4.611 \times 10^{-3} \end{pmatrix} \text{mol/m}^2\text{s}$$

Convergence has not yet been obtained and so we need to carry out a few more iterations. Successive estimates of the N_i are tabulated below.

Iteration	$N_1 \times 10^3$	$N_2 \times 10^3$
	[mol/m^{2}s]	
1	6.768	4.719
2	6.245	4.611
3	6.275	4.624
4	6.273	4.624

The ratio of the fluxes is

$$N_1/(N_1 + N_2) = 0.5757$$

which is quite close to our "guess" of the condensed film composition x_1^c. If it had not been, then we would have had to reestimate the condensed film composition. Any one-dimensional search method could be used for this purpose. An alternative and probably more efficient way (in this particular case) is to solve all of the independent equations simultaneously using a multidimensional Newton–Raphson procedure as described in Appendix C. Final results for the matrices of mass transfer coefficients at both

TABLE 8.2 Diffusional Distillation

Position	$r = 0$	$r = \delta$
Composition (y)	$\begin{pmatrix} 0.1067 \\ 0.0411 \end{pmatrix}$	$\begin{pmatrix} 0.0208 \\ 0.0132 \end{pmatrix}$
Driving force (Δy)	$\begin{pmatrix} 0.08587 \\ 0.02788 \end{pmatrix}$	
Bootstrap matrix $[\beta]$	$\begin{bmatrix} 1.125 & 0.125 \\ 0.048 & 1.048 \end{bmatrix}$	$\begin{bmatrix} 1.022 & 0.022 \\ 0.014 & 1.014 \end{bmatrix}$
Matrix of low flux mass transfer coefficients $[k]$ (mm/s)	$\begin{bmatrix} 1.625 & -0.097 \\ 0.051 & 3.605 \end{bmatrix}$	$\begin{bmatrix} 1.614 & -0.020 \\ 0.018 & 3.862 \end{bmatrix}$
Matrix of rate factors $[\Phi]$ (−)	$\begin{bmatrix} 0.1495 & 0.0240 \\ -0.0242 & 0.1013 \end{bmatrix}$	
Matrix of correction factors $[\Xi]$ (−)	$\begin{bmatrix} 0.927 & -0.011 \\ 0.011 & 0.950 \end{bmatrix}$	$\begin{bmatrix} 1.077 & 0.013 \\ -0.013 & 1.052 \end{bmatrix}$
Matrix of high flux mass transfer coefficients $[k^\bullet]$ (mm/s)	$\begin{bmatrix} 1.505 & -0.111 \\ 0.0892 & 3.425 \end{bmatrix}$	$\begin{bmatrix} 1.738 & 0.000 \\ -0.030 & 4.061 \end{bmatrix}$
Diffusion fluxes (J) (mmol/m^2 s)	$\begin{pmatrix} 5.114 \\ 4.182 \end{pmatrix}$	$\begin{pmatrix} 6.050 \\ 4.486 \end{pmatrix}$
Molar fluxes (N) (mmol/m^2 s)	$\begin{pmatrix} 6.273 \\ 4.624 \end{pmatrix}$	

ends of the diffusion path are shown in Table 8.2. Composition profiles, computed from Eq. 8.3.12 using the converged values of the fluxes, are shown in Figure 8.8.

The essence of the diffusional distillation process is that the lighter component (in this case the water vapor) condenses at a faster rate through air than the heavier 2-propanol. This results in a condensate composition of $x_1^c = 0.5757$, lower than the azeotropic composition of $x_1^e = 0.7625$. McDowell and Davis (1988) extended the analysis of Fullarton and Schlünder by accounting for concentration changes along the length of the diffusion distillation column (see Assumption 1 above) and simultaneous heat transfer. Their numerical simulations are in good agreement with the data (in better agreement than the computations of Fullarton and Schlünder). ∎

8.3.4 Advanced Computational Strategies

Now that we have worked through some typical problems it will be clear that the amount of computation involved is not trivial. Clearly, efficient computer based solution procedures are desirable (and that used in the above example is not optimal except for some problems involving only three components). The success or failure of any iterative scheme is very closely related to the goodness of the initial guess of the independent variables (the N_i here) (Step 2 in Algorithm 8.1). Krishna and Standart (1979) suggested taking the N_i as zero in Step 2, in which case $[\Phi]$ reduces to the null matrix and $[\Xi]$ to the identity matrix $[I]$, as shown above. Direct substitution of the N_i calculated in Step 8 back into Step 4 may lead to very slow convergence from a null starting guess. Instead, we recommend that damped substitution be employed (Taylor and Webb, 1981). Convergence will generally be obtained within about 10 iterations. In Section 8.5 we will discuss a method of obtaining much better initial estimates of the fluxes; estimates from which convergence can usually be obtained in two or three iterations even in "difficult" cases.

Taylor and Webb (1980a) found that the second member of Eq. 8.3.24 converges most rapidly if the eigenvalues of $[\Phi]$ are positive in sign (this will be the case if the N_i are positive), whereas the fourth member converges most rapidly if the eigenvalues of $[\Phi]$ are negative in sign (corresponding to negative fluxes). In fact, the first part of Eq. 8.3.24 may never converge if, in addition to being negative in sign, the eigenvalues $\hat{\Phi}_i$ are "large" in absolute value ("large" means > 7). Conversely, the latter part may never converge if the $\hat{\Phi}_i$ are large and positive. The explanation for this behavior is given by Taylor and Webb and not repeated here. It should be noted that large $\hat{\Phi}_i$ are characteristic of problems involving mass transfer at very high rates. The possible occurrence of large $\hat{\Phi}_i$ must be allowed for when drawing up a general purpose algorithm, such as that summarized in Algorithm 8.2.

The more appropriate member of Eq 8.3.24 is determined from the sign of the dominant eigenvalue of $[\Phi]$ (the eigenvalue with the largest absolute value). Computing eigenvalues can be quite time consuming and, in this case, can be avoided altogether if we recognize that the trace of $[\Phi]$ (i.e., the sum of the diagonal elements Φ_{ii}) is equal to the sum of the eigenvalues. Thus, the sign of Φ_{av}, which is easily calculated in Step 4 of Algorithm 8.2, suffices to determine which member of Eq. 8.3.24 should be used. In cases where the eigenvalues are of mixed sign, they are generally low in value and both members could equally well be used.

For the evaluation of $[\Xi]$ (Step 5 of Algorithm 8.1) Sylvester's expansion formula should be used only if the number of components is small (3 or 4 say). For larger problems the use of power series as discussed in Appendix A.6 is recommended (Taylor and Webb, 1981). The power series expansions (Eqs. A.6.4 and A.6.7) may be used if the eigenvalues of $[\Phi]$ are repeated or are complex with no special treatment and should be the default methods in any computer program that performs the relevant computations.

Algorithm 8.2   Improved Algorithm for Calculation of Mass Transfer Rates from an Exact Solution of the Maxwell–Stefan Equations

Given:	$(y_0), (y_\delta), c_t, \kappa_{ij}$
Step 1:	Estimate the N_i (see text).
Step 2:	Compute $[R]$ using y_0.
Step 3:	Compute (Λ) using y_0.
Step 4:	Compute $[\Phi]$.
Step 5:	Compute trace of $[\Phi]$: $\mathrm{tr}[\Phi]$.
Step 6:	Calculate $\Phi_{av} = \mathrm{tr}[\Phi]/(n-1)$.
	If $\Phi_{av} < -5$, recalculate $[R]$ and (Λ) using y_δ.
Step 7:	Compute correction factor matrices:
	If $N_t = 0$, calculate $[\Xi]^{-1}$ from a power series (Appendix A.6)
	Do *NOT* invert $[\Xi]^{-1}$.
	If $N_t \neq 0$, compute $\exp[\Phi]$ from a power series.
	Subtract the identity matrix to obtain $[\exp[\Phi] - [I]]$.
Step 8:	Compute the diffusion fluxes as follows:
	If $N_t = 0$: Solve the linear system

$$[\Xi]^{-1}[R](J) = c_t(\Delta y)$$

for (J).
If $N_t \neq 0$: Solve the linear system

$$[[\exp[\Phi] - [I]]](z) = (\Delta y)$$

for (z).
Compute $(z') = [\Phi](z)$.
If $\Phi_{av} < -5$: Compute $(z') = \exp[\Phi](z')$.
Solve the linear system:

$$[R](J) = c_t(z')$$

	for (J).
Step 9:	Calculate $N_t = -(\Lambda)^{\mathrm{T}}(J)$.
	If $\Phi_{av} < -5$: $N_i = J_i + y_{i\delta}N_t$.
	Otherwise: $N_i = J_i + y_{i0}N_t$.
Step 10:	Check for convergence of the N_i. If fluxes have converged stop.
Step 11:	If number of iterations exceed five, then set N_i equal to the average of the last two estimates of the N_i.
Step 12:	Compute $[\Phi]$.
	Return to Step 7.

In problems where the flux ratios are known (e.g., condensation and heterogeneous reacting systems where the reaction rate is controlled by diffusion) the mole fractions at the interface are not known in advance and it is necessary to solve the mass transfer rate equations simultaneously with additional equations (these may be phase equilibrium and/or reaction rate equations). For these cases it is possible to embed Algorithms 8.1 or 8.2 within another iterative procedure that solves the additional equations (as was done in Example 8.3.2). However, we suggest that a better procedure is to solve the mass transfer rate equations simultaneously with the additional equations using Newton's method. This approach will be developed below for cases where the mole fractions at both ends of the film are known. Later we will extend the method to allow straightforward solution of more complicated problems (see Examples 9.4.1, 11.5.2, 11.5.3, and others).

The molar fluxes in a ternary system are calculated from

$$N_1 = J_{10} + y_{10}(N_1 + N_2 + N_3)$$
$$N_2 = J_{20} + y_{20}(N_1 + N_2 + N_3)$$

with the molar diffusion fluxes J_{10} and J_{20} given by

$$J_{10} = c_t^V k_{11}^\bullet (y_{10} - y_{1\delta}) + c_t^V k_{12}^\bullet (y_{20} - y_{2\delta})$$
$$J_{20} = c_t^V k_{21}^\bullet (y_{10} - y_{1\delta}) + c_t^V k_{22}^\bullet (y_{20} - y_{2\delta})$$

Given the binary mass transfer coefficients and the mole fractions y_{i0} and $y_{i\delta}$, there are three unknown quantities in these equations; the molar fluxes, N_1, N_2, N_3. However, there are only two independent mass transfer rate equations. Thus, one more equation is needed; this will be the bootstrap relation:

$$\nu_1 N_1 + \nu_2 N_2 + \nu_3 N_3 = 0$$

These three equations can be solved simultaneously using Newton's method as described in Appendix C.2. First, we write the independent equations in the form $F(\chi) = 0$

$$F_1 \equiv J_{10} + y_{10}(N_1 + N_2 + N_3) - N_1 = 0$$
$$F_2 \equiv J_{20} + y_{20}(N_1 + N_2 + N_3) - N_2 = 0$$
$$F_3 \equiv \nu_1 N_1 + \nu_2 N_2 + \nu_3 N_3 = 0$$

The first step is to estimate the molar fluxes. This can be done as described above and elsewhere (Section 8.5). The mass transfer coefficients are calculated and the values of the discrepancy functions F_i evaluated. To reestimate the molar fluxes we must evaluate the Jacobian matrix [J]. The elements of this matrix are obtained by differentiating the above equations with respect to the independent variables. These derivatives may be approximated by

$$\partial F_1 / \partial N_1 = y_{10} - 1$$
$$\partial F_1 / \partial N_2 = y_{10}$$
$$\partial F_1 / \partial N_3 = y_{10}$$
$$\partial F_2 / \partial N_1 = y_{20}$$
$$\partial F_2 / \partial N_2 = y_{20} - 1$$
$$\partial F_2 / \partial N_3 = y_{20}$$
$$\partial F_3 / \partial N_1 = \nu_1$$
$$\partial F_3 / \partial N_2 = \nu_2$$
$$\partial F_3 / \partial N_3 = \nu_3$$

In deriving these expressions we ignored the fact that the elements of $[k^\bullet]$ are complicated functions of the mole fractions and of the molar fluxes. It would not be at all straightforward to allow for this dependence. In practice, it is found that the rate of convergence of Newton's method is not seriously impaired by this simplification.

Following the calculation of [J] we solve the linear system (Eq. C.2.5) and obtain new estimates of the independent variables. The entire procedure, summarized in Algorithm 8.3, is repeated until convergence. We do not particularly recommend Algorithm 8.3 for solving this class of mass transfer problem; for this category of problem Algorithm 8.2 (and often Algorithm 8.1) will prove distinctly superior. However, Algorithm 8.3 is much more easily

Algorithm 8.3   Algorithm for Calculation of Mass Transfer Rates from an Exact Solution of the Maxwell–Stefan Equations Using Newton's Method

Given:	$(y_0), (y_\delta), c_t, \kappa_{ij}$
Step 1:	Estimate the N_i.
Step 2:	Calculate $[k]$.
Step 3:	Calculate $[\Phi]$.
Step 4:	Calculate $[\Xi]$.
Step 5:	Calculate $(J) = c_t[k][\Xi](\Delta y)$.
Step 6:	Evaluate discrepancy functions:

$$F_i = J_i + y_{i0}N_t - N_i \qquad (i = 1, \ldots, n-1)$$

$$F_n = \sum_{i=1}^{n} \nu_i N_i$$

Step 7:	Check values of the F_i. If converged, stop.
Step 8:	Compute the Jacobian matrix.
Step 9:	Solve linear system (Eq. C.2.5) for new values of the fluxes. Return to Step 3.

adapted to those cases where the mole fractions at one or more film boundary must be determined simultaneously with the fluxes. This, after all, will nearly always be the case.

8.3.5   An Alternative Formulation

An exact solution of the Maxwell–Stefan equations may be obtained in a somewhat different way as shown below. Equations 1.2.12 are written in $n - 1$ dimensional matrix form

$$(N) = (J) + N_t(y) \tag{8.3.49}$$

which is combined with Eq. 8.3.13 to give

$$\frac{d(y - z)}{d\eta} = [\Psi]((y) - (z)) \tag{8.3.50}$$

where z_i is the (constant) flux ratio N_i/N_t. The matrix $[\Psi]$ is defined by

$$[\Psi] = \frac{N_t \ell}{c_t}[D]^{-1} = \frac{N_t \ell}{c_t}[B] \tag{8.3.51}$$

In so far as $[B]$ (or, more generally, $[D]^{-1}$) depends on the composition of the mixture and as the composition is, in turn, a function of position η, we may regard $[\Psi]$ as a function of η. Thus, Eq. 8.3.50 is a first-order matrix differential equation of order $n - 1$ with a variable coefficient matrix $[\Psi(\eta)]$. This equation may be solved by the method of repeated substitution as shown in Appendix B.2. The solution is

$$(y - z) = [\Omega_0^\eta(\Psi)](y_0 - z) \tag{8.3.52}$$

where $[\Omega_0^\eta(\Psi)]$ is the matrizant of $[\Psi]$ defined by Eq. B.2.16. Since the matrizant at $\eta = 0$ is the identity matrix $[I]$, we may eliminate the column matrix (z) from the left-hand side of

Eq. 8.3.52 as follows:

$$(y - y_0) = [[\Omega_0^\eta(\Psi)] - [I]](y_0 - z) \tag{8.3.53}$$

Now, applying the second boundary condition Eq. 8.3.8 gives

$$(y_\delta - y_0) = [[\Omega_0^1(\Psi)] - [I]](y_0 - z) \tag{8.3.54}$$

which allows us to express the composition profiles in the form

$$(y - y_0) = [[\Omega_0^\eta(\Psi)] - [I]][[\Omega_0^1(\Psi)] - [I]]^{-1}(y_\delta - y_0) \tag{8.3.55}$$

The diffusion fluxes at $\eta = 0$ are evaluated from Eq. 8.3.13 with the composition gradient obtained by differentiating Eq. 8.3.55 (cf., Eq. B.2.18).

$$\frac{d(y)}{d\eta} = [\Psi][\Omega_0^\eta(\Psi)][[\Omega_0^1(\Psi)] - [I]]^{-1}(y_\delta - y_0) \tag{8.3.56}$$

Thus, the diffusion fluxes at $\eta = 0$ (where $[\Omega_0^0(\Psi)] = [I]$) are given by

$$(J_0) = \frac{c_t[D_0]}{\ell}[\Psi_0][[\Omega_0^1(\Psi)] - [I]]^{-1}(y_0 - y_\delta) \tag{8.3.57}$$

and the diffusion fluxes at $\eta = 1$ are given by

$$(J_\delta) = \frac{c_t[D_\delta]}{\ell}[\Psi_\delta][\Omega_0^1(\Psi)][[\Omega_0^1(\Psi)] - [I]]^{-1}(y_0 - y_\delta) \tag{8.3.58}$$

With the finite flux mass transfer coefficients matrix $[k^\bullet]$ defined by Eq. 7.1.9 we have

$$[k_0^\bullet] = \frac{[D_0]}{\ell}[\Psi_0][[\Omega_0^1(\Psi)] - [I]]^{-1} \tag{8.3.59}$$

$$[k_\delta^\bullet] = \frac{[D_\delta]}{\ell}[\Psi_\delta][\Omega_0^1(\Psi)][[\Omega_0^1(\Psi)] - [I]]^{-1} \tag{8.3.60}$$

In the limit of the total flux N_t going to zero we have

$$(J_0) = -(c_t[\overline{D}]/\ell)(y_0 - y_\delta) \tag{8.3.61}$$

where $[\overline{D}]$ is defined by

$$[\overline{D}] = \int_0^1 [D(\eta)]\, d\eta \tag{8.3.62}$$

Equations 8.3.52–8.3.57 were presented as an exact solution of the Maxwell–Stefan equations for diffusion in ideal gas mixtures by Burghardt (1984). Equations 8.3.52–8.3.57 are somewhat less useful than Eqs. 8.3.15–8.3.24 because we need to know the composition profiles in order to evaluate the matrizant. Even if the profiles are known, the computation of the fluxes from either of Eqs. 8.3.62 or 8.3.63 is not straightforward and not recommended. It is with the development in Section 8.4 in mind that we have included these results here.

8.4 MULTICOMPONENT FILM MODEL BASED ON THE ASSUMPTION OF CONSTANT [D] MATRIX: THE LINEARIZED THEORY OF TOOR, STEWART, AND PROBER

In 1964 Toor and Stewart and Prober independently put forward a general approach to the solution of multicomponent diffusion problems. Their method, which was discussed in detail in Chapter 5, relies on the assumption of constancy of the Fick matrix $[D]$ along the diffusion path. The so-called "linearized theory" of Toor, Stewart, and Prober is not limited to describing steady-state, one-dimensional diffusion in ideal gas mixtures (as we have already demonstrated in Chapter 5); however, for this particular situation Eq. 5.3.5, with $[D]$ given by Eq. 4.2.2, simplifies to

$$(J) = -c_t[B]^{-1}\frac{d(y)}{dr} = -c_t[D]\frac{d(y)}{dr} \tag{8.4.1}$$

We emphasize the difference between the Toor, Stewart, and Prober approach and the exact method considered above by using the subscript av; thus, $[D_{av}]$. For practical purposes, this means that $[D_{av}]$ has to be evaluated by employing suitable average mole fractions, $y_{i,av}$, in the definition of the B_{ik} (Eqs. 2.1.21 and 2.1.22). The arithmetic average mole fraction $y_{i,av} = \frac{1}{2}(y_{i0} + y_{i\delta})$, normally is recommended for calculation of $[B_{av}]$ (Stewart and Prober, 1964; Arnold and Toor, 1967; Smith and Taylor, 1983). Thus, for gas mixtures

$$[D_{av}] = [B_{av}]^{-1} \tag{8.4.2}$$

is assumed constant over the diffusion path.

The solution to the film model diffusion equations for constant $[D]$ can be obtained directly from Eq. 8.3.55. For constant $[D]$ (and, hence, $[\Psi]$), the matrizant simplifies to an exponential matrix as shown in Section B.2. The composition profiles, for example, are given by (Taylor and Webb, 1980b, Taylor, 1982b; Burghardt, 1984)

$$(y - y_0) = [\exp[\Psi]\eta - [I]][\exp[\Psi] - [I]]^{-1}(y_\delta - y_0) \tag{8.4.3}$$

where

$$[\Psi] = \frac{N_t\ell}{c_t}[D_{av}]^{-1} = \frac{N_t\ell}{c_t}[B_{av}] \tag{8.4.4}$$

is the (now constant) matrix of mass transfer rate factors. The diffusion fluxes at $\eta = 0$ are evaluated from

$$(J_0) = \frac{c_t[D_{av}]}{\ell}[\Psi][\exp[\Psi] - [I]]^{-1}(y_0 - y_\delta) \tag{8.4.5}$$

and the diffusion fluxes at $\eta = 1$ are given by

$$(J_\delta) = \frac{c_t[D_{av}]}{\ell}[\Psi]\exp[\Psi][\exp[\Psi] - [I]]^{-1}(y_0 - y_\delta) \tag{8.4.6}$$

With the finite flux mass transfer coefficients matrix $[k^\bullet]$ defined by Eq. 7.1.9 we have

$$[k_0^\bullet] = \frac{[D_{av}]}{\ell}[\Psi][\exp[\Psi] - [I]]^{-1} \tag{8.4.7}$$

$$[k_\delta^\bullet] = \frac{[D_{av}]}{\ell}[\Psi]\exp[\Psi][\exp[\Psi] - [I]]^{-1} \tag{8.4.8}$$

To obtain the low flux mass transfer coefficients we take the limit as the total flux N_t goes to zero

$$\lim_{N_t \to 0} [\Psi] \to [0] \tag{8.4.9}$$

and

$$\lim_{N_t \to 0} [\Psi][\exp[\Psi] - [I]]^{-1} \to [I] \tag{8.4.10}$$

$$\lim_{N_t \to 0} [\Psi]\exp[\Psi][\exp[\Psi] - [I]]^{-1} \to [I] \tag{8.4.11}$$

The matrices of zero-flux mass transfer coefficients $[k_0]$ and $[k_\delta]$ are, therefore, equal

$$[k_0] = \frac{[D_{av}]}{\ell} = [k_\delta] = [k_{av}] \tag{8.4.12}$$

The matrices of correction factors $[\Xi_0]$ and $[\Xi_\delta]$, are given by

$$[\Xi_0] = [\Psi][\exp[\Psi] - [I]]^{-1} \tag{8.4.13}$$

$$[\Xi_\delta] = [\Psi]\exp[\Psi][\exp[\Psi] - [I]]^{-1} \tag{8.4.14}$$

where the matrix $[\Psi]$ is given by

$$[\Psi] = N_t[k_{av}]^{-1}/c_t \tag{8.4.15}$$

Finally, the fluxes N_i can be evaluated from

$$\begin{aligned}(N) &= c_t[\beta_0][k_{av}][\Xi_0](y_0 - y_\delta)\\ &= c_t[\beta_\delta][k_{av}][\Xi_\delta](y_0 - y_\delta)\end{aligned} \tag{8.4.16}$$

8.4.1 Comparison With Exact Method

Let us now take a closer look at the relationship between the exact solution of the Maxwell–Stefan equations and the solution of the linearized equations developed above. If we compare Eqs. 8.4.5 and 8.4.6 with Eqs. 8.3.15 and 8.3.21, their counterparts in the exact solution, it can be seen that $[B_{av}]$, $[k_{av}]$ and $[\Psi]$ correspond exactly with $[B_0]$ (or $[B_\delta]$), $[k_0]$ (or $[k_\delta]$), and $[\Phi]$. Furthermore, the similarity between the two methods extends right down to the calculation of the individual elements of these matrices as is emphasized in Table 8.3.

TABLE 8.3   Matrices in the Exact Solution and the Approximate Solution of the Maxwell–Stefan Equations[a]

Matrix $[M]$	Meaning	m_i
Exact Solution due to Krishna and Standart		
$[B_0] = [D_0]^{-1}$	Inverted matrix of Fick diffusion coefficients evaluated at the origin	y_{i0}
$[k_0]^{-1}$	Inverted matrix of low flux mass transfer coefficients	$y_{i0}\ell$
$[\Phi]$	Matrix of mass transfer rate factors	$N_i\ell/c_t$
Linearized Method of Toor, Stewart, and Prober		
$[B_{av}] = [D_{av}]^{-1}$	Inverted matrix of Fick diffusion coefficients evaluated at the average composition	$y_{i\,av}$
$[k_{av}]^{-1}$	Inverted matrix of low flux mass transfer coefficients	$y_{i\,av}\ell$
$[\Psi]$	Matrix of mass transfer rate factors	$y_{i\,av}N_t\ell/c_t$

[a] A matrix $[M]$ has elements

$$M_{ii}^{-1} = \frac{m_i}{\mathit{Ð}_{in}} + \sum_{\substack{k=1 \\ k \neq i}}^{n} \frac{m_k}{\mathit{Ð}_{ik}} \qquad M_{ij}^{-1} = -m_i\left(\frac{1}{\mathit{Ð}_{ij}} - \frac{1}{\mathit{Ð}_{in}}\right)$$

We see that the only differences between the exact solution and the approximate method described above are

1. The use of average mole fractions rather than the boundary conditions in the calculation of $[k]$.
2. The average convective flux $y_{i,\,av}N_t$ replaces the molar flux N_i in the calculation of the rate factor matrix $[\Psi]$.

Despite these differences both solutions of the multicomponent diffusion equations will give identical results if

1. All the binary diffusivities are equal; that is,

$$\mathit{Ð}_{ij} = \mathit{Ð} \tag{8.4.17}$$

Then $[k_0]$ and $[k_{av}]$ are equal (see Table 8.3) with

$$k_{0ii} = k_{av\,ii} = \mathit{Ð}/\ell \qquad k_{0ij} = k_{av\,ij} = 0 \qquad i \neq j \tag{8.4.18}$$

The rate factor matrices $[\Psi]$ and $[\Phi]$ are also equal with elements

$$\Psi_{ii} = \Phi_{ii} = N_t\ell/c_t\mathit{Ð} \qquad \Psi_{ij} = \Phi_{ij} = 0 \qquad i \neq j \tag{8.4.19}$$

The correction factor matrices will, therefore, be equal. In this case, however, the diffusion equations are not coupled, and the ith diffusion flux is given by

$$J_{0i} = c_t\frac{\mathit{Ð}}{\ell}\frac{\Phi}{\exp\Phi - 1}(x_{i0} - x_{i\delta}) \tag{8.4.20}$$

and the matrix formulations discussed above are not required.

2. The $n - 1$ species are present in vanishingly low concentrations ($y_{0i} \approx 0$, $i = 1, 2, \ldots, n - 1$; $y_{0n} \approx y_{\delta n} = 1$). In this case the matrices $[k]$ become diagonal (see Table 8.3); that is,

$$k_{0ii} = k_{\text{av } ii} = Ð_{in}/\ell \qquad k_{0ij} = k_{\text{av } ij} = 0 \qquad i \neq j \qquad (8.4.21)$$

The matrices $[\Psi]$ and $[\Phi]$ also become diagonal with elements given by

$$\Psi_{ii} = \Phi_{ii} = N_t \ell/c_t Ð_{in} \qquad \Psi_{ij} = \Phi_{ij} = 0 \qquad i \neq j \qquad (8.4.22)$$

The validity of the approximation in Eq. 8.4.22 for the exact solution depends on the observation that if $y_{0i} \approx y_{\delta i} \approx 0$, where $i = 1, 2, \ldots, n - 1$; then the first $n - 1 N_i$ will be vanishingly low. Under these conditions Eq. 8.4.20 and 8.4.21 will be valid whatever variation may exist between the $Ð_{ij}$.

We have outlined above, situations where the two methods may be expected to be in close agreement. It would seem that large differences between the methods are particularly likely in mixtures of constituents with quite different $Ð_{ij}$ at high molar concentrations. Nevertheless, even in such cases the errors caused by linearizing the equations are not usually large.

The goodness of the assumption of constant $[D]$ for describing steady-state, one-dimensional diffusion has been fairly thoroughly tested by several investigators (cf. the discussion in Chapter 5). The most complete computational comparison of the approximate method with the exact method described in the preceding section has been carried out by Smith and Taylor (1983) who solved more than 10,000 problems for each of 23 real ternary gas systems covering a wide range of ratios of diffusion coefficients and determinacy coefficients (the ν_i). In general, the solution of the linearized equations can be expected always to provide excellent estimates of the total flux and very good estimates of the individual fluxes unless one (or more) of the components has a very small flux relative to the fluxes of the other species. The largest discrepancies between the two methods occur when inert species are present. We have no reservations about recommending the use of solutions derived on the basis of constant $[D]$ when a more exact method is not available or too difficult to use (as, e.g., in Section 10.3).

8.4.2 The Toor–Stewart–Prober Formulation

We have developed the solution to the linearized equations as a special case of an exact solution in Section 8.3.5 in order to emphasize the close relationship that exists between the two methods. It should be noted, however, that this is not the way in which Toor or Stewart and Prober obtained their results. Indeed, these equations are not to be found in this form in the papers that first presented the linearized theory. Both Toor and Stewart and Prober obtained their results using the procedure described in Chapter 5; that is, by diagonalizing the matrix $[D]$ and solving sets of uncoupled equations. The final result is

$$(J) = c_t [k_{\text{av}}^\bullet](y_0 - y_\delta) \qquad (8.4.23)$$

where $[k_{\text{av}}^\bullet]$ is given by

$$[k_{\text{av}}^\bullet] = [P][\hat{k}^\bullet][P]^{-1} \qquad (8.4.24)$$

where $[\hat{k}^\bullet]$ is a diagonal matrix whose nonzero elements are the eigenvalues of $[k^\bullet]$ and $[P]$ is the modal matrix, of $[D]$. The eigenvalues of $[k_{\text{av}}^\bullet]$ are related to the eigenvalues of $[D]$ as follows:

$$\hat{k}_i^\bullet = \hat{k}_i \hat{\Xi}_i \qquad (8.4.25)$$

where

$$\hat{k}_i = \hat{D}_i/\ell \qquad (8.4.26)$$

$$\hat{\Psi}_i = N_t/c_t\hat{k}_i \qquad (8.4.27)$$

$$\hat{\Xi}_{0i} = \hat{\Psi}_i/(\exp\hat{\Psi}_i - 1) \qquad (8.4.28)$$

$$\hat{\Xi}_{\delta i} = \hat{\Psi}_i \exp\hat{\Psi}_i/(\exp\hat{\Psi}_i - 1) \qquad (8.4.29)$$

are the eigenvalues of $[k_{av}]$, $[\Psi]$, $[\Xi_0]$, and $[\Xi_\delta]$, respectively.

It is important to recognize that this representation of $[k^\bullet]$ is possible only because the matrix $[D]$ was assumed constant along the diffusion path. The matrix $[P]$, which diagonalizes $[D]$, is constant along the diffusion path and the transformation that reduces $[D]$ to a diagonal matrix

$$[D] = [P]^{-1}[\hat{D}][P] \qquad (8.4.30)$$

also diagonalizes $[k]$, $[\Psi]$, and $[\Xi]$. Thus, the matrix $[k^\bullet]$ is really a function of only one matrix, $[D]$. In the exact solution the matrix $[k^\bullet]$ is a function of two matrices $[B]$ and $[\Phi]$ that cannot be diagonalized by the same modal matrix (except for the limiting cases identified above). The matrix $[D]$ possesses a complete set of eigenvectors regardless of the multiplicity of the eigenvalues of $[D]$ (Cullinan, 1965). This means that the transformation (Eq. 8.4.30) is always possible and Eq. 8.4.24 may always be used to evaluate the matrix of mass transfer coefficients.

We may also use Sylvester's expansion formula (Eq. A.5.20) to compute $[k^\bullet]$

$$[k^\bullet] = \sum_{i=1}^{m} \hat{k}_i^\bullet \left\{ \prod_{\substack{j=1 \\ j \neq i}}^{m} [[D] - \hat{D}_j[I]] \bigg/ \prod_{\substack{j=1 \\ j \neq i}}^{m} (\hat{D}_i - \hat{D}_j) \right\} \qquad (8.4.31)$$

where m is the number of distinct eigenvalues of the Fick matrix $[D]$ ($m \leq n - 1$). The eigenvalues $\hat{k}_i^\bullet$ are given by Eq. 8.4.25.

For the ternary case with two distinct eigenvalues, we have

$$[k^\bullet] = \frac{\hat{k}_1^\bullet[[D] - \hat{D}_2[I]]}{\hat{D}_1 - \hat{D}_2} + \frac{\hat{k}_2^\bullet[[D] - \hat{D}_1[I]]}{\hat{D}_2 - \hat{D}_1} \qquad (8.4.32)$$

which may be expanded as follows:

$$k_{11}^\bullet = \frac{\hat{k}_1^\bullet(D_{11} - \hat{D}_2)}{\hat{D}_1 - \hat{D}_2} + \frac{\hat{k}_2^\bullet(D_{11} - \hat{D}_1)}{\hat{D}_2 - \hat{D}_1} \qquad (8.4.32a)$$

$$k_{12}^\bullet = \frac{(\hat{k}_1^\bullet - \hat{k}_2^\bullet)}{(\hat{D}_1 - \hat{D}_2)} D_{12} \qquad (8.4.32b)$$

$$k_{21}^\bullet = \frac{(\hat{k}_1^\bullet - \hat{k}_2^\bullet)}{(\hat{D}_1 - \hat{D}_2)} D_{21} \qquad (8.4.32c)$$

$$k_{22}^\bullet = \frac{\hat{k}_1^\bullet(D_{22} - \hat{D}_2)}{\hat{D}_1 - \hat{D}_2} + \frac{\hat{k}_2^\bullet(D_{22} - \hat{D}_1)}{\hat{D}_2 - \hat{D}_1} \qquad (8.4.32d)$$

8.4.3   Computation of the Fluxes

The computation of the fluxes N_i from either of Eqs. 8.4.16 involves an iterative procedure only if the total flux N_t is not specified. If it is zero, then no iterations are required since the correction factor $[\Xi]$ reduces to the identity matrix. If it is not zero but can be specified in advance (only rarely is this possible) then the calculations again involve no iteration but the flux correction factors will have to be evaluated. The formal similarity between the exact solution and the Toor, Stewart, and Prober method means that we may use essentially the same algorithm for computing the fluxes; the only changes to the procedure in Algorithms 8.1–8.3 are that $[k]$ is calculated at the average composition and $[\Psi]$ is used in place of $[\Phi]$. Thus, given a computer program for calculating the fluxes from the exact solution, it is a simple exercise to modify it so that the solution of the linearized film model equations may be used instead. It is worth noting that the linearized equations are much more stable than the exact solution; it is much harder to find a problem that causes divergence with the linearized equations than it is with the exact method.

An alternative algorithm based on Eqs. 8.4.23 and 8.4.24 is presented in Algorithm 8.4. Unlike modified versions of Algorithms 8.1 or 8.2, Algorithm 8.4 requires us to evaluate the eigenvalues of $[D]$ and the modal matrix $[P]$ (but not its inverse). Fortunately, $[P]$ need be evaluated only once; thereafter, only two square matrix–column vector multiplications (and one of the square matrices is a diagonal matrix) are required in each iteration. This is very much less than the number of matrix–matrix multiplications required if Eq. 8.4.31 is used. Algorithm 8.4 may also be used for other models of mass transfer (see Sections 9.3 and 10.4). Initial estimates of N_t are best calculated from Eq. 8.4.33 below with $\hat{\Xi}_i = 1$ and with the Λ coefficients calculated at the arithmetic average mole fraction (Vickery et al., 1984).

An alternative method is summarized in Algorithm 8.5 based on the fact that we really only need to determine the total flux N_t in order to completely determine the matrix of mass transfer coefficients and, therefore, the molar fluxes N_i. We start from Eq. 7.2.22 and substitute for (J) using Eqs. 8.4.23 and 8.4.24 to get

$$N_t = -c_t(\Lambda)^T[P][\hat{k}][\hat{\Xi}][P]^{-1}(\Delta y) \tag{8.4.33}$$

Algorithm 8.4   Algorithm Based on Repeated Substitution for Calculation of Mass Transfer Rates from Solutions of the Linearized Equations

Given:	$(y_0), (y_\delta), c_t, [D]$
Step 1:	Obtain eigenvalues and eigenvectors of $[D]$: $\hat{D}$ and $[P]$.
Step 2:	Solve $[P](\Delta \hat{y}) = (\Delta y)$ for $(\Delta \hat{y})$.
Step 3:	Compute $\hat{k}_i$.
Step 4:	Compute (Λ) at average composition.
Step 5:	Estimate $N_t = -c_t(\Lambda)^T[P][\hat{k}](\Delta \hat{y})$.
Step 6:	Recalculate (Λ) at y_0.
Step 7:	Calculate $\hat{\Psi}_i$.
Step 8:	Calculate $\hat{\Xi}_i$.
Step 9:	Calculate $\hat{k}_i^\bullet$.
Step 10:	Calculate $(J) = c_t[P][\hat{k}^\bullet](\Delta \hat{y})$.
Step 11:	Calculate $N_t = -(\Lambda)^T(J)$.
Step 12:	Check for convergence on N_t. If N_t has not converged, return to Step 7. Otherwise, continue with Step 13.
Step 13:	Compute $N_i = J_i + y_i N_t$.
Step 14:	Compute $[k^\bullet] = [P][\hat{k}^\bullet][P]^{-1}$ only if needed.

where $[\hat{\Xi}]$ is a diagonal matrix whose elements are the eigenvalues of $[\Xi]$. The function of N_t whose root we seek is obtained by dividing Eq. 8.4.33 by N_t to give

$$F(N_t) \equiv -(\Lambda)^T[P]\big[\hat{f}\big][P]^{-1}(\Delta y) - 1 = 0 \qquad (8.4.34)$$

where

$$\hat{f}_i = \frac{1}{\exp \hat{\Phi}_i - 1} \qquad (8.4.35)$$

This last step, which involves division by N_t, is possible only if $N_t \neq 0$. For $N_t = 0$, the linearized equations are explicit with $\Xi_{ii} = 1$.

This equation in the single unknown N_t is easily and rapidly solved using Newton's method. The relevant calculations are shown in Algorithm 8.5. First derivatives of F with respect to N_t (needed in Step 6) are obtained from

$$F'(N_t) = -(\Lambda)^T[P]\big[\hat{f}'\big][P]^{-1}(\Delta y) \qquad (8.4.36)$$

where

$$\hat{f}'_i = \frac{-\exp \hat{\Phi}_i}{c_t \hat{k}_i \big(\exp \hat{\Phi}_i - 1\big)^2} \qquad (8.4.37)$$

The function $F(N_t)$ possesses many desirable properties that make very rapid convergence a virtual certainty; it is a highly unusual problem that needs more than three iterations (Vickery et al., 1984).

Algorithm 8.5   Algorithm Based on Newton's Method for Calculation of Mass Transfer Rates from Solutions of the Linearized Equations

Given:	$(y_0), (y_\delta), c_t, [D]$
Step 1:	Obtain eigenvalues and eigenvectors of $[D]$: $\hat{D}$ and $[P]$.
Step 2:	Solve $[P](\Delta \hat{y}) = (\Delta y)$ for $(\Delta \hat{y})$.
Step 3:	Compute $\hat{k}_i$.
Step 4:	Compute (Λ) at average composition.
Step 5:	Estimate $N_t = -c_t(\Lambda)^T[P][\hat{k}](\Delta \hat{y})$.
Step 6:	Recalculate (Λ) at boundary composition.
Step 7:	Calculate $\hat{f}_i, \hat{f}'_i$.
Step 8:	Calculate $F(N_t)$ and $F'(N_t)$.
Step 9:	Reestimate $N_t = N_t - F/F'$.
Step 10:	Check for convergence on N_t. If N_t has not converged return to Step 6.

Remaining calculations done only as needed

Step 11:	Calculate $(J) = c_t[P][\hat{k}^\bullet](\Delta \hat{y})$.
Step 12:	Calculate $(N) = (J) + N_t(y_0)$.
Step 13:	Compute $[k^\bullet] = [P][\hat{k}^\bullet][P]^{-1}$.

Example 8.4.1   Vapor-Phase Dehydrogenation of Ethanol

Consider the vapor-phase dehydrogenation of ethanol(1):

$$\text{Ethanol} \rightarrow \text{acetaldehyde} + \text{hydrogen}$$
$$\textbf{1} \qquad\qquad \textbf{2} \qquad\qquad \textbf{3}$$

carried out at a temperature of 548 K and 101.3 kPa (see Froment and Bischoff (1979, p. 151), from whose work this example has been adapted, for discussion of other aspects of this process). The reaction is assumed to take place at a catalyst surface with a reaction rate that is first order in ethanol mole fraction

$$\text{Rate} = -k_r y_1^I \text{ mol ethanol produced}/(\text{m}^2 \text{ surfaced area})(\text{s})$$

where k_r is the reaction rate coefficient and where y_1^I is the mole fraction of ethanol at the catalyst surface.

Estimate the overall rate of reaction when the bulk phase gas-phase composition is

$$y_1^V = 0.6 \qquad y_2^V = 0.2 \qquad y_3^V = 0.2$$

DATA

1. Reaction rate constant:

$$k_r = 10 \text{ mol}/(\text{m}^2 \text{ s mol fraction})$$

2. The Maxwell–Stefan diffusion coefficients are

$$Đ_{12} = 7.2 \times 10^{-5} \text{ m}^2/\text{s}$$
$$Đ_{13} = 23.0 \times 10^{-5} \text{ m}^2/\text{s}$$
$$Đ_{23} = 23.0 \times 10^{-5} \text{ m}^2/\text{s}$$

3. The film thickness: $\ell = 1$ mm

ANALYSIS   The flux of ethanol is determined by the rate of chemical reaction at the surface

$$N_1 = -k_r y_1^I$$

Furthermore, for every mole of ethanol that diffuses to the catalyst surface, 1 mol of acetaldehyde and 1 mol of hydrogen are produced and these components diffuse away from the catalyst surface. That is,

$$N_2 = N_3 = -N_1$$

The total flux, therefore, is

$$N_t = N_1 + N_2 + N_3 = -N_1$$

The ratios of the fluxes are fixed at

$$N_1/N_t = -1 \qquad N_2/N_t = 1 \qquad N_3/N_t = 1$$

The molar fluxes in the vapor phase will be calculated from

$$N_1 = J_1 + y_1^I N_t$$
$$N_2 = J_2 + y_2^I N_t$$

where N_t is the total molar flux

$$N_t = N_1 + N_2 + N_3$$

and the molar diffusion fluxes at the interface J_1 and J_2 are given by

$$J_1 = c_t^V k_{11}^\bullet (y_1^I - y_1^V) + c_t^V k_{12}^\bullet (y_2^I - y_2^V)$$
$$J_2 = c_t^V k_{21}^\bullet (y_1^I - y_1^V) + c_t^V k_{22}^\bullet (y_2^I - y_2^V)$$

We have identified the interface mole fractions y^I with y_0 and the bulk vapor mole fractions y^V with y_δ.

There are five unknown quantities in these equations: N_1, N_2, N_3, y_1^I, and y_2^I. The mole fractions at the interface are included in the list of unknowns because they were not specified in the problem statement. Indeed, we are not free to specify the interface composition unless we relax the reaction stoichiometry relations and that is not possible. Thus, the interface mole fractions must be determined simultaneously with the molar fluxes. We suggest that this is best done with the help of Newton's method in a way similar to that in Algorithm 8.3.

The stoichiometric relations allow us to eliminate N_2 and N_3 from the mass transfer rate equations that we rewrite below together with the reaction rate equation in the form $F(\chi) = 0$

$$F_1 \equiv J_1 - y_1^I N_1 - N_1 = 0$$
$$F_2 \equiv J_2 - y_2^I N_1 + N_1 = 0$$
$$F_3 \equiv -k_r y_1^I - N_1 = 0$$

These three equations depend on just three variables: N_1, y_1^I, and y_2^I. We may solve these equations using Newton's method as described in Appendix C.2.

SOLUTION   The total molar concentration c_t, in the gas phase (assumed to be isothermal) is estimated from the ideal gas law as 22.2 mol/m^3.

We begin the iterative calculations by making an initial estimate of the three unknown variables. We shall take the following values

$$N_1 = -0.1 \text{ mol/m}^2\text{s}$$
$$y_1^I = 0.01$$
$$y_2^I = 0.6$$

The interface mole fraction of component 3, therefore, is

$$y_3^I = 0.39$$

As a basis for choosing these values we note that since ethanol is consumed by reaction at the surface, its mole fraction is expected to be low. On the other hand, acetaldehyde is produced at the surface, 1 mol of aldehyde for every 1 mol of ethanol consumed; we, therefore, estimate its interface mole fraction as equal to the bulk mole fraction of ethanol. The flux N_1 is estimated directly from the reaction rate equation.

The fluxes N_2 and N_3 may be computed from the stoichiometric relations as

$$N_2 = N_3 = -N_1 = 0.1 \text{ mol/m}^2\text{s}$$

and the total flux is estimated as

$$N_t = N_2 = 0.1 \text{ mol/m}^2\text{s}$$

In order to evaluate the discrepancy functions F_1–F_3 we must compute the matrix of finite flux mass transfer coefficients. The procedure is illustrated below.

The first step is to determine the matrix of Fick diffusion coefficients $[D]$. The arithmetic average mole fractions will be needed in the evaluation of the $[D]$; these average mole fractions are

$$y_{1_{av}} = 0.305 \qquad y_{2_{av}} = 0.4 \qquad y_{3_{av}} = 0.295$$

The matrix of Fick diffusivities is calculated from Eqs. 4.2.7 as

$$[D] = \begin{bmatrix} 15.073 & 6.044 \\ 7.926 & 16.956 \end{bmatrix} \times 10^{-5} \text{ m}^2/\text{s}$$

The eigenvalues of this matrix are

$$\hat{D}_1 = 23.000 \times 10^{-5} \text{ m}^2/\text{s} \qquad \hat{D}_2 = 9.030 \times 10^{-5} \text{ m}^2/\text{s}$$

The eigenvalues of the matrix of low flux mass transfer coefficients are computed from Eq. 8.4.26 as

$$\begin{aligned}
\hat{k}_1 &= \hat{D}_1/\ell \\
&= 23.0 \times 10^{-5}/1.0 \times 10^{-3} \\
&= 0.230 \text{ m/s} \\
\hat{k}_2 &= 0.0903 \text{ m/s}
\end{aligned}$$

The next step is to compute the mass transfer rate factors from Eq. 8.4.27

$$\begin{aligned}
\hat{\Psi}_1 &= N_t/c_t\hat{k}_1 \\
&= 0.1/(22.2 \times 0.230) \\
&= 0.01955 \\
\hat{\Psi}_2 &= N_t/c_t\hat{k}_2 \\
&= 0.0498
\end{aligned}$$

The eigenvalues of the correction factor matrix follow from Eq. 8.4.28 as

$$\begin{aligned}
\hat{\Xi}_1 &= \hat{\Psi}_1/\left(\exp \hat{\Psi}_1 - 1\right) \\
&= 0.01955/(\exp(0.01955) - 1) \\
&= 0.9902 \\
\hat{\Xi}_2 &= \hat{\Psi}_2/\left(\exp \hat{\Psi}_2 - 1\right) \\
&= 0.9753
\end{aligned}$$

The eigenvalues of the matrix of finite flux mass transfer coefficients is given by Eqs. 8.4.25 as

$$\hat{k}_1^\bullet = \hat{k}_1 \hat{\Xi}_1$$
$$= 0.230 \times 0.9902$$
$$= 0.2277 \text{ m/s}$$
$$\hat{k}_2^\bullet = \hat{k}_2 \hat{\Xi}_2$$
$$= 0.08807 \text{ m/s}$$

We may now compute the matrix of high flux mass transfer coefficients from Eq. 8.4.32 as

$$[k^\bullet] = \begin{bmatrix} 0.1485 & 0.0604 \\ 0.0793 & 0.1673 \end{bmatrix} \text{ m/s}$$

With the molar density c_t given above we may compute the diffusion fluxes as

$$J_1 = -1.411 \text{ mol/m}^2\text{s} \qquad J_2 = 0.4485 \text{ mol/m}^2\text{s}$$

Finally, we may evaluate the discrepancy functions F_1–F_3 as follows:

$$F_1 = J_1 - y_1^I N_1 - N_1$$
$$= -1.411 - 0.01 \times (-0.1) - (-0.1)$$
$$= -1.31$$
$$F_2 = J_2 - y_2^I N_1 \overparen{(-N_2)} = +N_1$$
$$= 0.4485 - 0.6 \times (-0.1) - (-0.1)$$
$$= 0.4085$$
$$F_3 = -k_r y_1^I - N_1$$
$$= -10 \times 0.01 - (-0.1)$$
$$= 0$$

Since these values of F_i clearly do not represent a converged solution it will be necessary to reestimate the flux N_1 and the two independent interface mole fractions. This requires the evaluation of the Jacobian matrix [J]. The elements of this matrix are obtained by differentiating the above equations with respect to the independent variables. The nonzero derivatives of F_1 and F_2 are as follows:

$$\partial F_1/\partial N_1 = -y_1^I - 1$$
$$\partial F_1/\partial y_1^I = c_t^V k_{11}^\bullet - N_1$$
$$\partial F_1/\partial y_2^I = c_t^V k_{12}^\bullet$$
$$\partial F_2/\partial N_1 = -y_2^I + 1$$
$$\partial F_2/\partial y_1^I = c_t^V k_{21}^\bullet$$
$$\partial F_2/\partial y_2^I = c_t^V k_{22}^\bullet - N_1$$

As suggested in the discussion surrounding Algorithm 8.3, we have ignored the fact that the elements of $[k^\bullet]$ are complicated functions of the mole fractions and of the molar fluxes in working out the above partial derivatives.

The partial derivatives of the reaction rate equation are

$$\partial F_3/\partial N_1 = -1$$

$$\partial F_3/\partial y_1^I = -k_r$$

$$\partial F_3/\partial y_2^I = 0$$

Following the calculation of [J] we solve the linear system (Eq. C.2.5) and obtain the following new estimates of the independent variables

$$N_1 = -1.117 \, \text{mol/m}^2\text{s}$$

$$y_1^I = 0.1117 \qquad y_2^I = 0.5526$$

We may now recalculate the functions F_1–F_3 with the following results:

$$F_1 = 0.3253 \qquad F_2 = -0.2038 \qquad F_3 \approx 0$$

The next computed set of independent variables is

$$N_1 = -0.8715 \, \text{mol/m}^2\text{s}$$

$$y_1^I = 0.08715 \qquad y_1^I = 0.5845$$

We must now reevaluate the discrepancy functions F_1–F_3 using these new values of the unknown variables. All of the calculations illustrated above must be repeated, including that of the matrix of Fick diffusion coefficients (because the average mole fractions have changed), the matrix of finite flux mass transfer coefficients (because $[D]$ and the fluxes have changed) and the diffusion fluxes (because $[k^\bullet]$ has changed). Repeating the above calculation steps with the most recent values of the unknown variables gives the following values for the functions

$$F_1 = -0.0368 \qquad F_2 = 0.0203 \qquad F_3 \approx 0$$

This set of function values does not correspond to a converged solution either so we must compute the Jacobian matrix a second time and solve the linear system (Eq. C.2.5) in order to obtain a third estimate of the solution

$$N_1 = -0.8997 \, \text{mol/m}^2\text{s}$$

$$y_1^I = 0.08997 \qquad y_2^I = 0.5812$$

Three more iterations yields the converged values

$$N_1 = -0.8960 \, \text{mol/m}^2\text{s}$$

$$y_1^I = 0.08960 \qquad y_2^I = 0.5818$$

The values of the various matrices at these converged values are listed in Table 8.4.

TABLE 8.4   Dehydrogenation of Ethanol at 548 K and 1.0135 bar.

Position	$r = 0$	$r = \delta$
Composition (y)	$\begin{pmatrix} 0.08960 \\ 0.5818 \end{pmatrix}$	$\begin{pmatrix} 0.6 \\ 0.2 \end{pmatrix}$
Matrix of low flux mass transfer coefficients $[k]$ (m/s)	$\begin{bmatrix} 0.155 & 0.067 \\ 0.0755 & 0.163 \end{bmatrix}$	
Eigenvalues of rate factor matrix	0.1752 0.4579	
Eigenvalues of correction factor matrix	0.9149 0.7884	
Matrix of high flux mass transfer coefficients $[k^{\bullet}]$ (m/s)	$\begin{bmatrix} 0.1355 & 0.066 \\ 0.0750 & 0.144 \end{bmatrix}$	
Diffusion fluxes (J) $(\text{mol/m}^2\text{s})$	$\begin{pmatrix} -0.9763 \\ 0.3747 \end{pmatrix}$	
Molar flux N_1 $(\text{mol/m}^2\text{s})$	-0.8960	

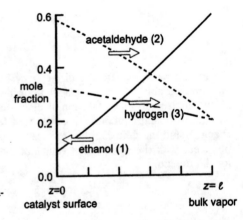

Figure 8.9. Composition profiles in dehydrogenation of ethanol.

The composition profiles, calculated from Eq. 8.4.3 are shown in Figure 8.9. Notice the quite large change in the mole fraction of ethanol. Despite this large change, the profiles themselves are not highly curved; this being the result of the relatively low total molar flux. ∎

8.5   SIMPLIFIED EXPLICIT METHODS

Both the exact solution and the Toor, Stewart, and Prober methods discussed above require an iterative approach to the calculation of the fluxes. In addition, the calculations are somewhat time consuming, especially when done by hand. It would be nice to have a method of calculating the fluxes that involved no iterations and yet was sufficiently accurate (when compared to these more rigorous methods) to be useful in engineering calculations.

There are two such methods; one is due to Krishna (1979d, 1981b), the other to Burghardt and Krupiczka (1975) and its generalization by Taylor and Smith (1982).

Our starting point for the development of these explicit methods is the diffusion equation

$$(J) = -c_t[B]^{-1}\frac{d(y)}{dr} = -c_t[D]\frac{d(y)}{dr} \tag{8.5.1}$$

and the bootstrap solution

$$(N) = [\beta](J) \tag{8.5.2}$$

We combine these two equations as follows:

$$(N) = -c_t[\beta][B]^{-1}\frac{d(y)}{dr} \tag{8.5.3}$$

Equations 8.5.1.3 are quite general; they involve no assumptions regarding the constancy of particular matrices; they apply to mixtures with any number of components and for any relationship between the fluxes. It is at this point where any assumptions necessary to solve Eqs. (8.5.1–8.5.3) must be made. In the three methods to be discussed below we proceed in exactly the same way as we did when deriving the exact solution and the solution to the linearized equations; first obtain the composition profiles, then differentiate to obtain the gradients at the film boundary, and combine the result with Eq. 8.5.3 to obtain the working flux equations.

8.5.1 Method of Krishna

The simplest solution to Eqs. 8.5.3 is obtained by assuming that the matrix $[\beta][B]^{-1}$ can be considered constant over the film. Thus, since the N_i are also invariant, Eqs. 8.5.3 can be integrated directly to give linear composition profiles

$$\frac{(y_i - y_{i0})}{(y_{i\delta} - y_{i0})} = \eta \tag{8.5.4}$$

The molar fluxes N_i are then given by Eq. 8.5.3 with $[\beta][B]^{-1}$ evaluated at some average composition (the arithmetic average of y_0 and y_δ is that used in practice) (Krishna, 1979d, 1981b)

$$(N) = -\frac{c_t[\beta_{av}][D_{av}]}{\ell}\frac{d(y)}{d\eta}\bigg|_{\eta=0} \tag{8.5.5}$$

With the mole fraction gradients obtained by differentiating Eq. 8.5.4 we have

$$(N) = \frac{c_t}{\ell}[\beta_{av}][D_{av}](y_0 - y_\delta) \tag{8.5.6}$$

8.5.2 Method of Burghardt and Krupiczka

Burghardt and Krupiczka (1975) developed an approximate explicit solution to Eqs. 8.5.3 for the special case of diffusion of m species through $n - m$ stagnant gases. We develop their method below for the case of diffusion in the presence of a single stagnant gas identified as species n. While we are more interested in general solutions to the multicomponent diffusion equations, a derivation of their method will help us to understand the generalization of it to be considered afterwards.

For the special case of Stefan diffusion we may relate the $n - 1$ independent molar fluxes N_i to the $n - 1$ independent composition gradients as follows:

$$(N) = -c_t[A]^{-1}\frac{d(y)}{dr} \tag{8.5.7}$$

where the matrix $[A]$ has elements

$$A_{ii} = \sum_{\substack{k=1 \\ i \neq k}}^{n} \frac{y_k}{Ð_{ik}} \tag{8.5.8}$$

$$A_{ij} = -\frac{y_i}{Ð_{ij}} \tag{8.5.9}$$

Burghardt and Krupiczka (1975) did not assume $[A]$ to be constant. Instead, they assumed the matrix $[[A]/y_n]$ was constant over the film. To develop the solution of Burghardt and Krupiczka we must, therefore, rewrite Eq. 8.5.7 as

$$(N) = -\frac{c_t[A]^{-1}}{\ell}\frac{y_n}{y_n}\frac{d(y)}{d\eta} \tag{8.5.10}$$

Thus, on separating variables and integrating Eq. 8.5.10 we find

$$(N)\int y_n d\eta = -c_t\left[[A_{av}]^{-1}y_{n,av}\right]\int d(y) \tag{8.5.11}$$

The subscript av serves to remind us that $[A]$ and y_n are evaluated at the arithmetic average composition. To solve Eq. 8.5.11 we need an expression for the variation of y_n over the film. The required equation is provided by the last of the set of n Maxwell–Stefan relations (Eq. 8.3.2), which simplifies to give

$$\frac{dy_n}{d\eta} = \Phi y_n \tag{8.5.12}$$

where the mass transfer rate factor Φ is given by

$$\Phi = \sum_{k=1}^{n-1} \frac{N_k}{c_t Ð_{kn}/\ell} \tag{8.5.13}$$

In view of the assumptions underlying the film model, Φ is constant. Equation 8.5.12 is easily integrated to give the variation through the film of the composition of species n

$$y_n/y_{n0} = \exp(\Phi\eta) \tag{8.5.14}$$

If we integrate over the film from $\eta = 0$ to $\eta = 1$ we obtain an explicit relation for the rate factor Φ

$$\Phi = \ln(y_{n\delta}/y_{n0}) \tag{8.5.15}$$

Note that Eqs. 8.5.12–8.5.15 are exact. Now, if we substitute for y_n in Eq. 8.5.11 and carry out the required integrations we find that the composition profiles for all components are identical

$$\frac{(y_i - y_{i0})}{(y_{i\delta} - y_{i0})} = \frac{(e^{\Phi \eta} - 1)}{(e^{\Phi} - 1)} \tag{8.5.16}$$

The molar fluxes are given by

$$(N) = -\frac{c_t [A_{av}]^{-1}}{\ell} \frac{y_{n,av}}{y_{n0}} \frac{d(y)}{d\eta}\bigg|_{\eta=0} \tag{8.5.17}$$

Note that all nonconstant terms in Eq. 8.5.17 must be evaluated at the boundary at which the gradient is calculated. With the derivatives of composition at $\eta = 0$ obtained from Eq. 8.5.16 we find

$$(N) = \frac{c_t}{\ell} [A_{av}]^{-1} \Xi (y_0 - y_\delta) \tag{8.5.18}$$

where Ξ is given by

$$\Xi = \tfrac{1}{2} \Phi \frac{(\exp \Phi + 1)}{(\exp \Phi - 1)} \tag{8.5.19}$$

The parameter Ξ in this expression accounts for the nonlinearity of the composition profiles; it is, in fact, a high flux correction factor as is $[\Xi]$ in the matrix methods described above. Equation 8.5.18 involves no iteration because the rate factor Φ and the correction factor Ξ can be calculated from Eqs. 8.5.15; all we need to know are the boundary conditions y_{i0} and $y_{i\delta}$.

For diffusion of m species into $n - m$ stagnant gases $[A]$ is of order m (the elements are still given by Eqs. 8.5.8 and 8.5.9). The parameter Φ is calculated from Eq. 8.5.15 but y_n is replaced by the sum of the mole fractions of the $n-m$ stagnant gases. This method is illustrated in Example 8.5.1.

8.5.3 Method of Taylor and Smith

The method of Taylor and Smith (1982) is a generalization of the method of Burghardt and Krupiczka for Stefan diffusion. We use the determinacy condition (Eq. 7.2.10) to eliminate the nth flux from the Maxwell–Stefan relations (Eq. 2.1.16) and combine the first $n-1$ equations in matrix form as

$$(N) = -c_t [A]^{-1} \frac{d(y)}{dr} \tag{8.5.20}$$

where the matrix $[A]$ has elements

$$A_{ii} = \frac{y_i (\nu_i / \nu_n)}{\mathcal{D}_{in}} + \sum_{\substack{k=1 \\ i \neq k}}^{n} \frac{y_k}{\mathcal{D}_{ik}} \tag{8.5.21}$$

$$A_{ij} = -y_i \left(\frac{1}{\mathcal{D}_{ij}} - \frac{(\nu_j / \nu_n)}{\mathcal{D}_{in}} \right) \tag{8.5.22}$$

For the special case of Stefan diffusion $N_n = 0$, Eqs. 8.5.20 reduce to Eqs. 8.5.7.

It can be shown by inverting the bootstrap matrix $[\beta]$ using the Sherman–Morrison formula (see Ortega and Rheinbolt, 1970, p. 50) that, in fact

$$[A] = [B][\beta]^{-1} \quad \text{or} \quad [A]^{-1} = [\beta][B]^{-1} \quad (8.5.23)$$

That is, the matrices $[A]^{-1}$ and $[B][\beta]^{-1}$ are equal and not simply equivalent. Thus, the explicit method of Krishna could equally well be written in the form

$$(N) = \frac{c_t}{\ell}[A_{av}]^{-1}(y_0 - y_\delta) \quad (8.5.24)$$

The generalization of the method of Burghardt and Krupiczka is based on the assumption that the matrix $[A]$ divided by the mole fraction weighted sum of the ν_i

$$\bar{\nu} = \sum_{j=1}^{n} \nu_j y_j$$

{i.e., $\bar{\nu}[A]^{-1}$} can be considered constant over the diffusion path. For the special case of Stefan diffusion $\bar{\nu} = y_n$. The derivation of the final working equations follows the procedure used above to obtain Eq. 8.5.18. In fact, all of Eqs. 8.5.7–8.5.19 apply with y_n replaced by $\bar{\nu}$ and with the rate factor Φ given by the (now approximate) relation

$$\Phi = \ln(\bar{\nu}_\delta/\bar{\nu}_0) \quad (8.5.25)$$

The only difference between Eq. 8.5.6 obtained by Krishna and Eq. 8.5.18 obtained by Burghardt and Krupiczka (1975) and its generalization by Taylor and Smith (1982) is the inclusion of the scalar correction factor Ξ. For equimolar countertransfer ($N_t = 0$; $\nu_i = \nu_n$; $\Phi = 0$; $\Xi = 1$), the methods of Krishna and of Taylor and Smith are equal and, indeed, equal to the limiting form of Eq. 8.4.16 obtained from the linearized theory. In all other instances, the two explicit methods give results that differ only by the scalar factor Ξ. However, this correction factor can result in a clear improvement in the predicted fluxes.

8.5.4 Computation of the Fluxes

Equation 8.5.23 allows us to rewrite Eq. 8.5.18 as

$$(N) = c_t[\beta_{av}][k_{av}]\Xi(y_0 - y_\delta) \quad (8.5.26)$$

$[k_{av}]$ may be evaluated from Eq. 8.3.25 as $[k_{av}] = [R_{av}]^{-1}$.

For computational purposes it is better to break Eq. 8.5.26 into an expression for the average diffusion fluxes $J_{i,av}$

$$[R_{av}](J_{av}) = c_t\Xi(y_0 - y_\delta) \quad (8.5.27)$$

which is solved for the $J_{i,av}$. The molar fluxes N_i may then be computed from

$$N_i = J_{i,av} + y_{i,av}N_t \quad (8.5.28)$$

with the total flux N_t given by

$$N_t = -(\Lambda)^T(J_{av}) \quad (8.5.29)$$

The complete procedure for computing the fluxes from one of the explicit methods is given in Algorithm 8.6.

Equation 8.5.24 may also be written in the form

$$(N) = c_t [W_{av}] \Xi (y_0 - y_\delta) \tag{8.5.30}$$

where

$$[W_{av}]^{-1} = [\beta_{av}][k_{av}] = [\beta_{av}][R_{av}]^{-1} \tag{8.5.31}$$

The matrix $[W]$ has elements

$$W_{ii}^{-1} = \frac{y_i(\nu_i/\nu_n)}{\kappa_{in}} + \sum_{\substack{k=1 \\ i \neq k}}^{n} \frac{y_k}{k_{ik}} \tag{8.5.32}$$

$$W_{ij}^{-1} = -y_i \left(\frac{1}{\kappa_{ij}} - \frac{(\nu_j/\nu_n)}{\kappa_{in}} \right) \tag{8.5.33}$$

where the κ_{ij} are defined by Eq. 8.3.26. The parameter $[W_{av}]$ is evaluated from Eqs. 8.5.32 and 8.5.33 using the average mole fractions, $y_{i,av}$.

Equation 8.5.27 is a convenient starting point on which to base an algorithm for computing the fluxes.

Algorithm 8.6   Algorithm for Calculation of Mass Transfer Rates from Explicit Solutions of the Maxwell–Stefan Equations

Given:	Mole fractions y_{i0}, $y_{i\delta}$.
	Binary mass transfer coefficients k_{ij}.
	Molar density: c_t.
Step 1:	Compute average composition.
Step 2:	Compute $[R]$ using $y_{i,av}$ and k_{ij}.
Step 3:	Compute Ξ.
Step 4:	Solve Eq. 8.5.27 for (J_{av}).
Step 5:	Compute (Λ_{av}).
Step 6:	Compute N_t from Eq. 8.5.29.
Step 7:	Calculate N_i from Eq. 8.5.28.

Example 8.5.1   Evaporation into Two Inert Gases

Estimate the rate of evaporation of ethyl propionate(1) into mixtures of air(2) and hydrogen(3) using the method of Burghardt and Krupiczka. This problem is based on experiments conducted by Fairbanks and Wilke (1950) with a view to assessing the validity of Wilke's effective diffusivity formula (Eq. 6.1.14).

DATA

Temperature = 29.9°C, pressure = 101.325 kPa.
Composition of gas phase:

$$y_{10} = 0.0, \ y_{20} = 0.3, \ y_{30} = 0.7$$

Composition at interface

$$y_{1\delta} = 0.0634, \ y_{2\delta} = 0.2594, \ y_{3\delta} = 0.6772$$

The Maxwell–Stefan diffusion coefficients:

$$Đ_{12} = 8.5 \text{ mm}^2/\text{s}$$

$$Đ_{13} = 37.4 \text{ mm}^2/\text{s}$$

Film thickness $\ell = 1$ mm.

SOLUTION For diffusion of a single species into a mixture of inert gases all matrices in the Burghardt–Krupiczka method are scalars and the flux of species 1 is given by

$$N_1 = c_t \Xi (y_{10} - y_{1\delta})/(A_{11}\ell) \qquad (8.5.34)$$

where A_{11} is given by Eq. 8.5.8. For the ternary mixture in this problem this gives

$$A_{11} = y_2/Đ_{12} + y_3/Đ_{13}$$

$$= 51{,}318 \text{ s/m}^2$$

Arithmetic average mole fractions were used in the computation of A_{11}.

The mass transfer rate factor is computed from Eq. 8.5.15 using the total mole fraction of inert species

$$\Phi = \ln\{(1 - y_{1\delta})/(1 - y_{10})\}$$

$$= \ln\{(1 - 0.0634)/(1 - 0.0)\}$$

$$= -0.0655$$

The correction factor Ξ follows from Eq. 8.5.19 as

$$\Xi = \frac{1}{2}\Phi\frac{(\exp(\Phi) + 1)}{(\exp(\Phi) - 1)}$$

$$= \frac{1}{2}(-0.0655)\frac{(\exp(-0.0655) + 1)}{(\exp(-0.0655) - 1)}$$

$$= 1.00036$$

Clearly, the flux correction is rather low and there was no real need to compute Ξ in this problem.

The molar density is computed from the ideal gas law

$$c_t = P/RT$$

$$= 101{,}325/(8.3143 \times 303.05)$$

$$= 40.21 \text{ mol/m}^3$$

The numerical value of the flux of ethyl propionate therefore is

$$N_1 = c_t \Xi (y_{10} - y_{1\delta})/(A_{11}\ell)$$

$$= 40.21 \times 1.00036 \times (0.0 - 0.0634)/(51{,}318 \times 1 \times 10^{-3})$$

$$= -0.0497 \text{ mol/m}^2\text{s}$$

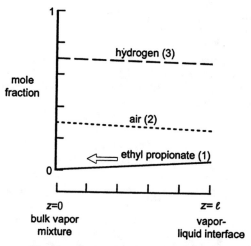

Figure 8.10. Composition profiles for evaporation of ethyl propionate into a mixture of air and hydrogen.

There is, in fact, a little more to this problem than might be indicated by the above computation. Since two of the fluxes are specified ($N_2 = N_3 = 0$), we are not at liberty to specify the mole fractions of two species at the interface. The mole fraction of ethyl propionate in the vapor at the liquid surface is determined from vapor–liquid equilibrium relations. Since the liquid is pure ethyl propionate this means that $y_{1\delta}$ is the ratio of the vapor pressure of ethyl propionate at the system temperature divided to the total pressure. With the vapor pressure given by the Antoine equation with constants taken from Reid et al. (1977), this calculation gives the value specified above. Ethyl propionate is not present in the bulk gas phase (at least initially), so $y_{1\delta} = 0.0$. The relative mole fractions of air and hydrogen in the experiment of Fairbanks and Wilke therefore become the actual mole fractions.

The mole fractions of air and hydrogen at the interface are given by exact solutions of the Maxwell–Stefan equations for those components

$$N_1 \ell / c_t D_{in} = \ln(y_{i\delta}/y_{i0}) \qquad i \neq 1 \qquad (8.5.35)$$

together with an equation that forces the mole fractions of all species to sum to unity. Not all of these equations are independent, of course, and in our solution Eq. 8.5.35 for $i = 3$ was not used. The values of the interface compositions specified in the problem statement are those that satisfy Eqs. 8.5.35. The values were found by solving Eq. 8.5.34 simultaneously with Eqs. 8.5.35. The resulting composition profiles are shown in Figure 8.10. ∎

8.5.5 Comparison With Exact and Linearized Solutions

The great advantage of the methods described in this section over those described earlier is, of course, rapidity in computation. This gain in computational simplicity is, however, at the expense of theoretical rigor. It is, therefore, important to establish the accuracy of the methods described above using the exact method of Section 8.3 as a basis for comparison. The extensive numerical computations made by Smith and Taylor (1983) showed that the explicit method of Taylor and Smith ranked second overall among seven approximate methods tested (the linearized method of Section 8.4 was best). For some determinacy conditions the explicit method of Taylor and Smith was actually the better of the approximate methods although the advantage over the linearized theory is negligibly small. The method of Burghardt and Krupiczka was second best when stagnant components are in the mixture. The average discrepancy in the predicted total flux is comparable to the discrepanceis in this flux obtained from the linearized equations.

The explicit method of Krishna (1979d, 1981b) is most successful if the ν_i are close together and, therefore (or for other reasons), the total flux is low. At high rates of mass transfer, the assumption of constant $[A]^{-1}$ (or of $[\beta][B]^{-1}$) is a poor one, particularly in cases involving an inert species.

A clue to the success of the explicit methods is provided by a comparison of the composition profiles predicted by the explicit methods to the profiles obtained from the exact and linearized methods. Krishna's explicit method always yields linear profiles and is unable to account for large deviations from linearity. The methods of Burghardt and Krupiczka and of Taylor and Smith yield identical dimensionless profiles for all species of the correct (exponential) form. Another reason for the better performance of the Burghardt–Krupiczka/Taylor–Smith method is that Eq. 8.5.24 represents an exact solution of the Maxwell–Stefan equations if all the $Ð_{ij}$ are equal (Burghardt, 1984; Taylor, 1984). Krishna's explicit method is exact only if all the $Ð_{ij}$ are equal and the total flux is zero (the ν_i are equal). There is, however, more to it. The scalar correction factor Ξ does a pretty good job of correcting what we might consider as the "low-flux" estimates obtained from Eq. 8.5.6. In fact, the rate factor Φ defined in Eq. 8.5.25 is an exact eigenvalue of the matrix $[\Phi]$ for three cases:

- All diffusivities are equal (regardless of the values of the ν_i).
- Stefan diffusion ($N_n = 0$) (regardless of the values of $Ð_{ij}$ and ν_i).
- Equimolar countertransfer ($N_t = 0$, $\Phi = 0$) (regardless of the values of the $Ð_{ij}$ and ν_i).

It is the eigenvalues (literally; "characteristic values") of $[\Phi]$ that characterize the correction factor matrix $[\Xi]$. Thus, the scalar rate factor Φ and correction factor Ξ when multiplied by identity matrices frequently are quite good models for the behavior of the complete matrices $[\Phi]$ (or $[\Psi]$) and $[\Xi]$ in the exact and linearized methods.

One especially good use for the Taylor–Smith/Burghardt–Krupiczka method is to generate initial estimates of the fluxes for use with the Krishna–Standart or Toor–Stewart–Prober methods. It is a very rare problem that requires more than two or three iterations if Eq. 8.5.26 is used to generate initial estimates of the fluxes (Step 3 in Algorithm 8.2) (Krishnamurthy and Taylor, 1982).

8.6 EFFECTIVE DIFFUSIVITY METHODS

The oldest, simplest, and still widely used methods, pioneered by Hougen and Watson (1947) and by Wilke (1950), employ the concept of an effective diffusion coefficient. The effective diffusivity concept was discussed in detail in Chapter 6; here we show how the effective diffusivity can be used to calculate mass transfer rates.

The starting point for this analysis is the one-dimensional form of Eq. 6.1.1

$$J_i = -c_t D_{i,\,\text{eff}} \frac{dy_i}{dr} \tag{8.6.1}$$

with the effective diffusivity, $D_{i,\text{eff}}$, given by Eqs. 6.1.7 or 6.1.8 or the special cases and approximate relations Eqs. 6.1.13–6.1.15.

If $D_{i,\text{eff}}$ can be assumed constant at some suitably averaged composition then the composition profiles are easily obtained as

$$\frac{(y_i - y_{i0})}{(y_{i\delta} - y_{i0})} = \frac{(\exp(\Phi_{i,\text{eff}}\eta) - 1)}{(\exp(\Phi_{i,\text{eff}}) - 1)} \tag{8.6.2}$$

where

$$\Phi_{i,\text{eff}} = N_t \ell / c_t D_{i,\text{eff}} \qquad i = 1, 2, \ldots, n - 1 \tag{8.6.3}$$

The diffusion fluxes at $\eta = 0$ are calculated from

$$J_i = -\frac{c_t D_{i,\text{eff}}}{\ell} \frac{d(y)}{d\eta}\bigg|_{\eta = 0} \tag{8.6.4}$$

With the composition derivatives obtained from Eq. 8.6.2 we may define an effective mass transfer coefficient by

$$J_{i0} = c_t k_{i,\text{eff}}^{\bullet}(y_{i0} - y_{i\delta}) \tag{8.6.5}$$

with the high flux mass transfer coefficient $k_{i,\text{eff}}^{\bullet}$ expressed as

$$k_{i,\text{eff}}^{\bullet} = k_{i,\text{eff}}\Xi_{i,\text{eff}} \tag{8.6.6}$$

where we have defined a low-flux mass transfer coefficient $k_{i,\text{eff}}$

$$k_{i,\text{eff}} = D_{i,\text{eff}}/\ell \tag{8.6.7}$$

and high flux correction factor $\Xi_{i,\text{eff}}$

$$\Xi_{i,\text{eff}} = \Phi_{i,\text{eff}}/(\exp(\Phi_{i,\text{eff}}) - 1) \tag{8.6.8}$$

In the limit that $N_t \to 0$, $\Phi_{i,\text{eff}} \to 0$, $\Xi_{i,\text{eff}} \to 1$ and the diffusion fluxes are obtained directly from

$$J_i = c_t k_{i,\text{eff}}(y_{i0} - y_{i\delta}) \qquad [N_t = 0] \tag{8.6.9}$$

The molar fluxes are obtained from the appropriate bootstrap relation as discussed in Chapter 7.

If the total flux N_t is not specified, an iterative approach is required for evaluation of the molar fluxes N_i. A procedure based on repeated substitution is provided in Algorithm 8.7. A still more efficient procedure can be devised using Newton's method (cf. Algorithm 8.5) (Krishna and Taylor, 1986).

**Algorithm 8.7   Algorithm Based on Repeated Substitution
for Calculation of Mass Transfer Rates
from an Effective Diffusivity Method**

Given:	$(y_0), (y_\delta), c_t, Ð_{ij}$
Step 1:	Compute effective diffusivities
	$D_{i,\text{eff}}$ $(i = 1, \ldots, n - 1)$
Step 2:	Compute effective mass transfer coefficients, $k_{i,\text{eff}}$.
Step 3:	Compute $[\beta]$.
Step 4:	Estimate N_i.
Step 5:	Calculate $\Phi_{i,\text{eff}}$.
Step 6:	Calculate $\Xi_{i,\text{eff}}$.
Step 7:	Calculate $J_i = c_t k_{i,\text{eff}} \Xi_{i,\text{eff}} \Delta y_i$.
Step 8:	Calculate $(N) = [\beta](J)$.
Step 9:	Check for convergence on N_t
	If N_t has not converged return to Step 4.

In the other category of this class of methods, the effective diffusion coefficient is defined with respect to the molar flux N_i. That is,

$$N_i = -c_t D_{i,\text{eff}} \frac{dy_i}{dr} \tag{8.6.10}$$

with $D_{i,\text{eff}}$ given by Eq. 6.1.2. If the flux ratios are known or can be approximated in some way (as is the case when we have diffusion controlled chemical reactions taking place on catalyst surfaces) then the $D_{i,\text{eff}}$ defined in Eqs. 6.1.2 may be calculated.

Now, if $D_{i,\text{eff}}$ can be considered constant at some average value, then integration of Eq. 8.6.10 yields linear composition profiles and a simple expression for calculating the fluxes without iteration

$$N_i = \frac{c_t}{\ell} D_{i,\text{eff}}(y_0 - y_\delta) \tag{8.6.11}$$

Examples illustrating the application of this method are given by Kubota et al. (1969) and by Geankoplis (1972).

The only advantage of the effective diffusivity definitions is their simplicity in computation; no matrix functions need be evaluated. The primary disadvantage of the use of $D_{i,\text{eff}}$ is that these parameters are not, in general, system properties except for the limiting cases noted in Chapter 6.

Example 8.6.1   Diffusion in a Stefan Tube

In Example 2.1.1 we described the experiments of Carty and Schrodt (1975) who evaporated a binary liquid mixture of acetone(1) and methanol(2) in a Stefan tube. Air(3) was used as the carrier gas. Using an effective diffusivity method calculate the composition profiles.

DATA

Composition at the interface: $y_1 = 0.319$, $y_2 = 0.528$.
Composition at the top of the tube: $y_1 = 0.0$, $y_2 = 0.0$.
Pressure = 99.4 kPa.
Temperature = 328.5 K.
The length of the diffusion path is 238 mm.

The Maxwell–Stefan diffusion coefficients are

$$Ð_{12} = 8.48 \text{ mm}^2/\text{s}$$

$$Ð_{13} = 13.72 \text{ mm}^2/\text{s}$$

$$Ð_{23} = 19.91 \text{ mm}^2/\text{s}$$

SOLUTION   The composition profiles in the Stefan tube are given by Eq. 8.6.2. Before we can compute the profiles we must determine the rates of evaporation of acetone and methanol. Since the evaporating species are present in low concentrations at the top of the tube (although not at the bottom) we shall use the dilute solution limit for the effective diffusivities

$$D_{1,\text{eff}} = Ð_{13} = 13.72 \text{ mm}^2/\text{s}$$

$$D_{2,\text{eff}} = Ð_{23} = 19.91 \text{ mm}^2/\text{s}$$

The effective mass transfer coefficients therefore are

$$k_{1,\text{eff}} = D_{1,\text{eff}}/\ell$$
$$= 13.72/238$$
$$= 0.05765 \text{ mm}/\text{s}$$
$$k_{2,\text{eff}} = D_{2,\text{eff}}/\ell$$
$$= 19.91/238$$
$$= 0.08366 \text{ mm}/\text{s}$$

Solution of Eq. 8.6.5 together with the determinacy condition $N_3 = 0$ yields the following values for the fluxes:

$$N_1 = 1.781 \times 10^{-3} \qquad N_2 = 3.309 \times 10^{-3} \text{ mol}/\text{m}^2\text{s}$$

results that are in excellent agreement with the exact values (from Example 2.1.1)

$$N_1 = 1.783 \times 10^{-3} \qquad N_2 = 3.127 \times 10^{-3} \text{ mol}/\text{m}^2\text{s}$$

It is worth noting that simple repeated substitution of the fluxes is not effective for solving this particular problem if the calculations are started with $\Xi_{i,\text{eff}} = 1$ (corresponding to a null estimate of the fluxes). The oscillations in the fluxes that result with simple repeated substitution can be avoided by using an average of the last two computed estimates of the fluxes in the evaluation of the mass transfer rate factors. In this case, however, we used Newton's method to solve a single function of the total flux (cf. Algorithm 8.5).

Using the converged set of fluxes, the rate factors have the values

$$\Phi_{1,\text{eff}} = 2.427 \qquad \Phi_{2,\text{eff}} = 1.672$$

and these values may be used directly in Eq. 8.6.2 to compute the profiles shown in Figure 8.11. The effective diffusivity profiles are in rather good agreement with profiles given by an exact solution and with the experimental data of Carty and Schrodt (1975). The good results obtained here must be considered to be somewhat fortuitous; the fluxes computed with

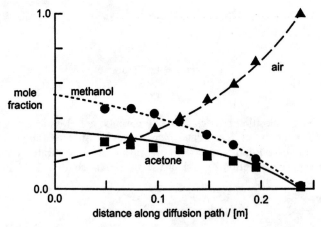

Figure 8.11. Composition profiles in a Stefan tube. Lines are computed from effective diffusivity model. Data of Carty and Schrodt (1975).

other effective diffusivity formulas are not as good as those given above (Krishna and Taylor, 1986). ∎

8.6.1 Comparison With the Matrix Methods

The simple effective diffusivity method (Eq. 8.6.5), represents an exact solution of the multicomponent diffusion equations if conditions are such that the appropriate limiting forms of Eqs. 6.1.10 and 6.1.11 apply. Thus, Eq. 8.6.5 can be used with confidence for systems where the binary $Đ_{ij}$ display little or no variation or in mixtures where one component is present in very large excess. However, the identification of these conditions requires some a priori knowledge that none of the matrix methods require. "To knowingly, or unknowingly, use these formulas in situations where these limiting cases do not apply would stand about as much chance of successfully predicting the fluxes as the throw of a dice to predict the magnitude and the toss of a coin to determine the sign" (Smith and Taylor, 1983). Two problems that could be placed in this category were presented in Chapter 5.

If the nonzero fluxes have the same sign (i.e., they are all in the same direction), then effective diffusivity methods are more likely to give reasonable results. This is nearly always the case in condensation and absorption processes and this goes some way at least to explaining why effective diffusivity methods usually give good estimates of the total amount condensed and the total heat load even if the individual condensation rates are not so well predicted. Webb et al. (1981) discussed in detail the conditions that must apply for an effective diffusivity method to be a useful model in multicomponent condensation.

The effective diffusivity formula of Stewart (Eq. 6.1.8) is by far the best of this class of methods. This should not come as a surprise since this method is capable of correctly identifying the various interaction phenomena possible in multicomponent systems. Indeed, for equimolar countertransfer, this effective diffusivity method is equivalent to the linearized theory and to both explicit methods discussed above. In fact, for some systems Stewart's effective diffusivity method is superior to Krishna's explicit method (Smith and Taylor, 1983). However, since the explicit methods are actually simpler to use than Stewart's effective diffusivity method (all methods require the same basic data) and, in general provide such superior results we see no reason for any further use of the effective diffusivity for describing diffusion in multicomponent ideal gas mixtures.

8.7 MULTICOMPONENT FILM MODEL FOR MASS TRANSFER IN NONIDEAL FLUID SYSTEMS

In this section we briefly describe how the solutions developed above for mass transfer in ideal gases can be extended to cover nonideal fluids. The starting point for the analysis of mass transfer in nonideal fluid mixtures is the set of generalized Maxwell–Stefan Eqs. 2.2.1, which for one-dimensional mass transfer, may be written as

$$d_i = -\sum_{j=1}^{n} \frac{x_i x_j (u_i - u_j)}{Ð_{ij}}$$

$$= \sum_{j=1}^{n} \frac{(x_i N_j - x_j N_i)}{c_t Ð_{ij}} \tag{8.7.1}$$

where

$$d_i = \frac{x_i}{RT} \frac{d\mu_i}{dr} \tag{8.7.2}$$

and where the thermodynamic factors Γ_{ik} are defined by Eq. 2.2.5. The Γ_{ik} and $Ð_{ik}$ are, in general, composition dependent.

8.7.1 Exact Solutions

An exact solution to Eqs. 8.7.1 and 8.7.2 may be obtained by generalizing the analysis of Section 8.3.5. In fact, the entire development of Section 8.3.5 holds for nonideal fluids as long as we use $[B]^{-1}[\Gamma]$ for the matrix $[D]$.

Alternatively, we may cast Eqs. 8.7.1 and 8.7.2 in $n - 1$ dimensional matrix form as

$$[\Gamma]\frac{d(x)}{d\eta} = [\Phi](x) + (\phi) \tag{8.7.3}$$

where η, $[\Phi]$, and (ϕ) are as defined before for the ideal gas case. Equations 8.7.3 may be solved by the method of repeated solution (Exercise 8.7.1).

The results of both approaches are expressed in terms of a matrizant and are not at all straightforward to use. For this reason, we recommend the approximate methods discussed below for use in applications.

8.7.2 Approximate Methods

Krishna (1977) presented an approximate solution of Eqs. 8.7.3 by assuming that the coefficients Γ_{ik} and $Ð_{ik}$ could be considered constant along the diffusion path. With these assumptions Eq. 8.7.3 represents a linear matrix differential equation, the solution of which can be written down in a manner exactly analogous to the ideal gas case. Thus, the composition profiles are given by

$$(x - x_0) = [\exp[\Theta]\eta - [I]][\exp[\Theta] - [I]]^{-1}(x_\delta - x_0) \tag{8.7.4}$$

where $[\Theta]$ is the augmented matrix of rate factors

$$[\Theta] = [\Gamma_{av}]^{-1}[\Phi].$$ (8.7.5)

The molar fluxes N_i can be calculated from

$$\begin{aligned}(N) &= c_t[\beta_0][k_0][\Xi_0](x_0 - x_\delta) \\ &= c_t[\beta_\delta][k_\delta][\Xi_\delta](x_0 - x_\delta)\end{aligned}$$ (8.7.6)

where the matrices of mass transfer coefficients are given by

$$\begin{aligned}[k_0] &= [D_0]/\ell = [B_0]^{-1}[\Gamma_{av}]/\ell \\ [k_\delta] &= [D_\delta]/\ell = [B_\delta]^{-1}[\Gamma_{av}]/\ell\end{aligned}$$ (8.7.7)

The subscripts 0 and δ serve as reminders that the appropriate compositions, x_{i0} and $x_{i\delta}$, respectively, have to be used in the defining equations for B_{ik}, (Eqs. 2.1.20 and 2.1.21). The subscript av on $[\Gamma]$ emphasizes the assumption that the matrix of thermodynamic factors evaluated at the average mole fraction and is assumed constant along the diffusion path.

The correction factors $[\Xi_0]$ and $[\Xi_\delta]$ are given by

$$\begin{aligned}[\Xi_0] &= [\Theta][\exp[\Theta] - [I]]^{-1} \\ [\Xi_\delta] &= [\Theta]\exp[\Theta][\exp[\Theta] - [I]]^{-1}\end{aligned}$$ (8.7.8)

It is interesting to note the direct influence of the thermodynamic factors Γ_{ik} on the mass transfer correction factors $[\Xi]$.

Equations 8.7.4–8.7.8 may also be obtained as limiting cases of the exact (matrizant) solution to Eqs. 8.7.3 (Exercise 8.7.1).

The interaction phenomena discussed earlier for the ideal gas case will also be possible for nonideal fluid mixtures, for which $[\Gamma]$ contribute to the matrix $[k^\bullet]$ by means of its separate influence on $[k]$, the zero flux matrix, and $[\Xi]$, the correction factor matrix.

The linearized theory of Toor (1964) and of Stewart and Prober (1964) discussed in Section 8.4 can be extended to nonideal fluids simply by using the appropriate relation for the matrix of multicomponent diffusion coefficients. For nonideal mixtures the matrix $[D]$ is evaluated as

$$[D_{av}] = [B_{av}]^{-1}[\Gamma_{av}]$$ (8.7.9)

The explicit methods developed in Section 8.5 can be generalized in similar ways. Krishna (1979a, 1981b), for example, assumes constancy of the matrix product $[\beta][B]^{-1}[\Gamma]$ and obtains

$$(N) = \frac{c_t}{\ell}[\beta_{av}][B_{av}]^{-1}[\Gamma_{av}](x_0 - x_\delta)$$ (8.7.10)

Extensions of the method of Taylor and Smith (1982) are described by Kubaczka and Bandrowski (1990) and by Taylor (1991).

For liquid mixtures, effective diffusion coefficients may be defined using the generalized Maxwell–Stefan equations (Lightfoot and Scattergood, 1965).

The computational methods discussed above for use with the ideal solution simplifications of the general relations presented could also be employed here. We recommend the use of repeated substitution to calculate the N_i from Eqs. 8.7.4 and Newton's method for calculating N_t from the linearized equations. The explicit methods are useful for generating initial estimates of the N_i.

Example 8.7.1   Mass Transfer in a Nonideal Fluid Mixture

To illustrate the application of the film model for nonideal fluid mixtures we consider steady-state diffusion in the system glycerol(1)–water(2)–acetone(3). This system is partially miscible (see Krishna et al., 1985). Determine the fluxes N_1, N_2, and N_3 in the glycerol-rich phase if the bulk liquid composition is

$$x_{1\delta} = 0.7824 \qquad x_{2\delta} = 0.1877 \qquad x_{3\delta} = 0.0299$$

and the interface composition is maintained constant at

$$x_{10} = 0.5480 \qquad x_{20} = 0.2838 \qquad x_{30} = 0.1682$$

The flux ratio, $z_1 = N_1/N_t$ is constant during the diffusion process and given by $z_1 = \frac{1}{12}$.

DATA

1. The generalized Maxwell–Stefan diffusion coefficients (to be assumed constant over the diffusion path) are $[10^{-10} \text{ m}^2/\text{s}]$

$$Đ_{12} = 11 \qquad Đ_{13} = 18 \qquad Đ_{23} = 56$$

2. The matrix of thermodynamic factors, evaluated at the average composition is

$$[\Gamma] = \begin{bmatrix} 0.95934 & -1.2392 \\ -0.3586 & 1.8739 \end{bmatrix}$$

3. The effective film thickness is $\ell = 10^{-5}$ m.
4. The mixture density c_t is 15 kmol/m^3.

SOLUTION   Either of Algorithms 8.1 or 8.2 may be adapted for computing the fluxes in a nonideal solution. We give only the final results of the iteration process below.

$[k_0]$ is obtained from Eq. 8.7.7 as

$$[k_0] = [B_0]^{-1}[\Gamma_{av}]/\ell$$

The results of this computation are

$$[k_0] = \begin{bmatrix} 143 & -104 \\ -11 & 286 \end{bmatrix} \times 10^{-6} \text{ m/s}$$

The total flux is found with the help of Eq. 1.1.12 as

$$N_t = J_1/(z_1 - x_{10})$$

where $z_1 = \frac{1}{12}$.

The converged values for the diffusion fluxes are

$$(J_0) = \begin{pmatrix} -0.527 \\ 0.378 \end{pmatrix} \text{ mol/m}^2\text{s}$$

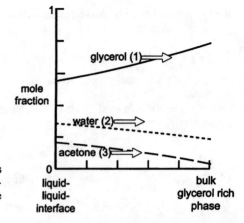

Figure 8.12. Composition profiles during mass transfer in the system glycerol(1)–water(2)–acetone(3). Note that despite a large negative driving force for glycerol(1), its flux is positive.

and for the N_i

$$(N) = \begin{pmatrix} 0.0945 \\ 0.700 \end{pmatrix} \text{mol/m}^2\text{s}$$

with $N_3 = 0.340 \text{ mol/m}^2\text{s}$.

At these values for the fluxes the various matrices are given below

$$[\Theta] = \begin{bmatrix} 0.4984 & 0.1351 \\ -0.0867 & 0.1226 \end{bmatrix}$$

$$[\Xi] = \begin{bmatrix} 0.7705 & -0.0606 \\ 0.0389 & 0.9390 \end{bmatrix}$$

$$[k_0^\bullet] = \begin{bmatrix} 106.2 & -106.6 \\ 2.89 & 269.5 \end{bmatrix} \times 10^{-6} \text{ m/s}$$

Composition profiles are shown in Figure 8.12. Arrows denote the actual directions of mass transfer. It is interesting to note that despite a large negative driving force for glycerol(1), $\Delta x_1 = -0.2344$, the flux N_1 is positive, that is, in a direction opposite to its driving force.   ∎

8.8 ESTIMATION OF MASS TRANSFER COEFFICIENTS FROM EMPIRICAL CORRELATIONS

In many cases, a priori estimates of the film thickness ℓ cannot be made, and we resort to empirical methods of estimating the mass transfer coefficients. Most published experimental works have concentrated on two component systems and there are no correlations for the multicomponent $[k]$. The need to estimate multicomponent mass transfer coefficients is very real, however. The question is How can we estimate multicomponent mass transfer coefficients when all we have to go on are binary correlations? In this section we look at the various methods that have been proposed to answer this question.

8.8.1 Estimation of Binary Mass Transfer Coefficients

Binary mass transfer data usually are correlated in terms of dimensionless groups, such as the Sherwood number,

$$\text{Sh} = kd/D \tag{8.8.1}$$

the Stanton number

$$\text{St} = k/\bar{u} \tag{8.8.2}$$

and the Chilton–Colburn j factor

$$j_D = \text{St Sc}^{2/3} \tag{8.8.3}$$

where d is some characteristic dimension of the mass transfer equipment, $\bar{u}$ is the mean velocity for flow, D is the Fick diffusion coefficient, and Sc is the Schmidt number ν/D.

Dimensional analysis suggests that the Sherwood number be a function of the Reynolds number ($\text{Re} = \bar{u}d/\nu$) and Schmidt number, Sc (Sherwood et al., 1975)

$$\text{Sh} = f(\text{Re}, \text{Sc}) \tag{8.8.4}$$

There are a great many correlations available in the literature for estimating binary mass transfer coefficients. It is beyond the scope of this book to review these correlations in detail [the reader is referred to the text by Sherwood et al. (1975) for more information]. For present purposes it suffices to cite only a couple of examples of useful empirical expressions. Other correlations are discussed in Sections 12.1.5 and 12.3.3.

The Gilliland–Sherwood correlation for gas-phase binary mass transfer in a wetted wall column is

$$\text{Sh} = 0.023\,\text{Re}^{0.83}\text{Sc}^{0.44} \tag{8.8.5}$$

An apparent weakness of the film model is that it suggests that the mass transfer coefficient is directly proportional to the diffusion coefficient raised to the first power. This result is in conflict with most experimental data, as well as with more elaborate models of mass transfer [surface renewal theory considered in the next chapter, e.g., or boundary layer theory (Bird et al., 1960)]. However, if we substitute the film theory expression for the mass transfer coefficient (Eq. 8.2.12) into Eq. 8.8.1 for the Sherwood number we find

$$\text{Sh} = d/\ell \tag{8.8.6}$$

and we see that the inverse Sherwood number Sh^{-1} may be regarded as a dimensionless film thickness. The film thickness obtained from a Sherwood number correlation, such as Eq. 8.8.5, will be a function of the flow conditions, system geometry, and fluid properties like viscosity and density. In addition, ℓ is proportional to the Fick diffusion coefficient D raised to some fraction of unity. The observation that ℓ may be a function of D removes the objection that film theory does not predict the correct dependance of k on D.

A widely used expression for estimating mass transfer coefficients is the Chilton–Colburn analogy

$$j_D = f/2 \quad \text{or} \quad \text{St} = \tfrac{1}{2}f\text{Sc}^{-2/3} \tag{8.8.7}$$

where f is the Fanning friction factor. Further discussion of the Chilton–Colburn analogy can be found in the books by Bird et al. (1960) and Sherwood et al. (1975), and in Chapters 10 and 11.

The Sherwood number Sh for spherical bodies is defined by

$$\text{Sh} \equiv k2r_0/D \tag{8.8.8}$$

The classic result Sh = 2 is obtained for the case where a spherical body is immersed in an infinite fluid medium ($r_\delta \gg r_0$) and for vanishingly small total molar flux N_t. For finite slip between the spherical body and the surrounding fluid the Sherwood number is usually obtained from correlations of the form (see Sherwood et al., 1975):

$$\text{Sh} = 2 + f(\text{Re}, \text{Sc}) \tag{8.8.9}$$

where $f(\text{Re}, \text{Sc})$ is some function of the Reynolds and Schmidt numbers.

The binary mass transfer coefficients estimated from these correlations and analogies are the low flux coefficients and, therefore, need to be corrected for the effects of finite transfer rates before use in design calculations.

In some correlations it is the binary mass transfer coefficient—interfacial area product that is correlated. In this case, then k should be considered to be this product and the N_i that are calculated from Eqs. 8.2.14 et seq are the mass transfer rates themselves with units moles per second (mol/s) (or equivalent).

8.8.2 Estimation of Multicomponent Mass Transfer Coefficients: The Method of Toor, Stewart, and Prober

In their original development of the linearized theory Toor (1964) and Stewart and Prober (1964) proposed that correlations of the type given by Eqs. 8.8.5 and 8.8.7 could be generalized by replacing the Fick diffusivity D by the charactersitic diffusion coefficients of the multicomponent system; that is, by the eigenvalues of the Fick matrix $[D]$. The mass transfer coefficient calculated from such a substitution would be a characteristic mass transfer coefficient; an eigenvalue of $[k]$. For example, the Gilliland–Sherwood correlation (Eq. 8.8.5) would be modified as follows:

$$\widehat{\text{Sh}}_i = 0.023 \, \text{Re}^{0.83} \, \widehat{\text{Sc}}_i^{0.44} \tag{8.8.10}$$

and the Chilton–Colburn analogy (Eq. 8.8.6), becomes

$$\widehat{\text{St}}_i = \tfrac{1}{2} f \, \widehat{\text{Sc}}_i^{2/3} \tag{8.8.11}$$

where

$$\widehat{\text{Sc}}_i \equiv \nu/\hat{D}_i$$

$$\widehat{\text{Sh}}_i \equiv \hat{k}_i d/\hat{D}_i$$

$$\widehat{\text{St}}_i \equiv \hat{k}_i/\overline{u}$$

are Schmidt, Sherwood, and Stanton numbers in terms of the eigenvalues of $[D]$. The $\hat{k}_i$ computed from these (or other similar) relations are used together with (Eqs. 8.4.23–8.4.32) and Algorithms 8.4 and 8.5 to compute the matrices of multicomponent mass transfer coefficients and the molar fluxes N_i.

This approach is, in fact, equivalent to replacing the binary diffusivity D by the matrix of multicomponent diffusion coefficients $[D]$ and the binary mass transfer coefficient with the matrix of multicomponents $[k]$. Thus, we have the matrix generalization of Eq. 8.8.5

$$[\text{Sh}] = [k]d[D]^{-1} = 0.023 \, \text{Re}^{0.83}[\text{Sc}]^{0.44} \tag{8.8.12}$$

and the matrix generalization of Eq. 8.8.2 is

$$[\text{St}] = [k]/u = \tfrac{1}{2}f[\text{Sc}]^{-2/3} \tag{8.8.13}$$

where [Sc] is a matrix of Schmidt numbers $[\text{Sc}] = \nu[D]^{-1}$.

Burghardt and Krupizcka suggested a variation on the Toor–Stewart–Prober approach for use with their own explicit method for Stefan diffusion developed in Section 8.5.2 [see Taylor (1984) for further discussion of their method].

8.8.3 Estimation of Multicomponent Mass Transfer Coefficients for Gas Mixtures from Binary Mass Transfer Coefficients

Krishna and Standart suggested that, in situations where the film thickness ℓ is not known the matrix of low flux mass transfer coefficients be calculated directly from Eq. 8.3.27

$$[k] = [R]^{-1} \tag{8.3.27}$$

where $[R]$ is calculated from Eqs. 8.3.25

$$R_{ii} = \frac{y_i}{\kappa_{in}} + \sum_{\substack{k=1 \\ k \neq i}}^{n} \frac{y_k}{\kappa_{ik}}$$

$$R_{ij} = -y_i \left(\frac{1}{\kappa_{ij}} - \frac{1}{\kappa_{in}} \right) \tag{8.3.25}$$

The binary κ_{ij} may be calculated as a function of the appropriate Maxwell–Stefan diffusion coefficient from a suitable correlation or physical model (e.g., the surface renewal models of Chapter 10). These binary κ_{ij} must also be used directly in the calculation of the rate factor matrix $[\Phi]$ (cf. Eqs. 8.3.28 and 8.3.29).

The same approach may be used in conjunction with the linearized equations and explicit methods of Section 8.5. Basically, $[k]$ is calculated from Eq. 8.3.25 using the average mole fractions $y_{i,\text{av}}$, and the binary $i - j$ pair mass transfer coefficients κ_{ij} in the evaluation of $[R]$.

The approach of Toor, Stewart, and Prober and that of Krishna and Standart are equivalent for the two limiting cases discussed in Section 8.3.1: ideal gas mixtures in which the binary $Đ_{ij}$ are equal or very dilute mixtures. In our experience the two approaches almost always give virtually identical results [although Young and Stewart (1986) might disagree]. We, therefore, recommend the use of $[k_{\text{av}}] = [R_{\text{av}}]^{-1}$ in view of the ease of computation and because the results are sufficiently accurate. In the Toor–Stewart–Prober approach we need to evaluate fractional powers of matrices. This calculation can be done using Sylvester's expansion formula or the modal matrix transformation approach but either way is rather more involved than the direct use of binary κ_{ij} in the calculation of $[k]$ (or equivalent matrix).

8.8.4 Estimation of Mass Transfer Coefficients for Nonideal Multicomponent Systems

Comparison of Eqs. 8.3.19 and 8.7.5 for the low flux mass transfer coefficients for ideal and nonideal systems, respectively, suggests that in cases where the film thickness is not known we may estimate the low flux mass transfer coefficient matrix for nonideal systems from

$$[k] = [R]^{-1}[\Gamma] \tag{8.8.14}$$

where the elements of the matrix $[R]$ are given by Eqs. 8.3.25 with the liquid-phase mole fractions x_i replacing the y_i.

For a two component system Eq. 8.8.14 simplifies to

$$k = \kappa\Gamma \tag{8.8.15}$$

The κ_{ij} may be estimated using an empirical correlation or alternative physical model (e.g., surface renewal theory) with the Maxwell–Stefan diffusivity of the appropriate i–j pair $Ð_{ij}$ replacing the binary Fick D. Since most published correlations were developed with data obtained with nearly ideal or dilute systems where Γ is approximately unity, we expect this separation of diffusive and thermodynamic contributions to k to work quite well. We may formally define the Maxwell–Stefan mass transfer coefficient κ_{ij} as (Krishna, 1979a)

$$\kappa_{ij} = \mathop{\substack{\text{limit} \\ \text{all } N_i \to 0 \\ \Gamma_{ij} \to \delta_{ij}}} \frac{J_i}{c_t \Delta x_i} \tag{8.8.16}$$

Example 8.8.1 *Ternary Distillation in a Wetted Wall Column*

Dribicka and Sandall (1979) distilled ternary mixtures of benzene (1), toluene (2), and ethylbenzene (3) at total reflux in a wetted wall column made from stainless steel and of 2.21-cm inside diameter (d). A schematic diagram of the column is shown in Figure 8.13. Samples of the vapor and liquid phases were taken from various points along the column. During one of their experiments the vapor phase at a height of 300 mm from the bottom of the column had the composition

$$y_{10} = 0.7471 \qquad y_{20} = 0.2072 \qquad y_{30} = 0.0457$$

Estimate the rates of mass transfer assuming the vapor composition at the interface is

$$y_{1\delta} = 0.8906 \qquad y_{2\delta} = 0.0995 \qquad y_{3\delta} = 0.0099$$

DATA

Viscosity of vapor mixture: $\mu = 8.819 \times 10^{-6}$ Pa s.

Vapor mass density: $\rho_t = 2.810$ kg/m^3.

Molar density: $c_t = 34.14$ mol/m^3.

The vapor phase diffusivities of the three binary pairs are $Ð_{12} = 2.228 \times 10^{-6}$ m^2/s; $Ð_{13} = 2.065 \times 10^{-6}$ m^2/s; and $Ð_{23} = 1.832 \times 10^{-6}$ m^2/s.

Vapor mass velocity: $G = 8.68$ kg/m^{2}s.

SOLUTION In this illustration we adopt the suggestion of Toor, Stewart, and Prober and compute $[k]$ from Eq. 8.8.12. The first step must, therefore, be to compute the matrix of

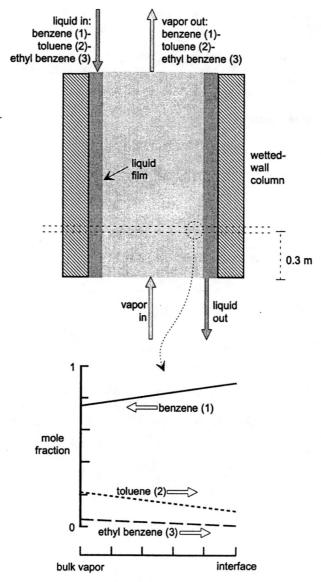

Figure 8.13. Schematic diagram of wetted wall column. Inset shows composition profiles in vapor film at the interface.

multicomponent diffusion coefficients $[D]$, which follows directly from Eqs. 4.2.7 as

$$[D] = \begin{bmatrix} 2.093 & -0.130 \\ -0.067 & 2.148 \end{bmatrix} \times 10^{-6} \text{ m}^2/\text{s}$$

The arithmetic average mole fractions were used in the computation of $[D]$. The eigenvalues of this matrix are

$$\hat{D}_1 = 2.218 \times 10^{-6} \text{ m}^2/\text{s} \qquad \hat{D}_2 = 2.0231 \times 10^{-6} \text{ m}^2/\text{s}$$

The Gilliland–Sherwood correlation in the form of Eq. 8.8.12 may be used to estimate the mass transfer coefficients in the vapor phase. The Reynolds number is found as follows:

$$\text{Re} = Gd/\mu$$
$$= 8.68 \times 0.0221/(8.819 \times 10^{-6})$$
$$= 21,752$$

The Schmidt numbers are computed next as

$$\widehat{\text{Sc}}_1 = \mu/\left(\rho_t \hat{D}_1\right)$$
$$= 8.819 \times 10^{-6}/(2.810 \times 2.218 \times 10^{-6})$$
$$= 1.4153$$
$$\widehat{\text{Sc}}_2 = \mu/\left(\rho_t \hat{D}_2\right)$$
$$= 1.5513$$

The Sherwood numbers are obtained from Eqs. 8.8.10

$$\widehat{\text{Sh}}_1 = 0.023\,\text{Re}^{0.83}\,\widehat{\text{Sc}}_1^{0.44}$$
$$= 0.023 \times 21,752^{0.83} \times 1.4153^{0.44}$$
$$= 106.71$$
$$\widehat{\text{Sh}}_2 = 0.023\,\text{Re}^{0.83}\,\widehat{\text{Sc}}_2^{0.44}$$
$$= 111.11$$

The eigenvalues of the matrix of mass transfer coefficients are computed next as

$$\hat{k}_1 = \widehat{\text{Sh}}_1 \hat{D}_1/d$$
$$= 106.71 \times 2.218 \times 10^{-6}/0.0221$$
$$= 1.071 \times 10^{-2} \text{ m/s}$$
$$\hat{k}_2 = \widehat{\text{Sh}}_2 \hat{D}_2/d$$
$$= 1.017 \times 10^{-2} \text{ m/s}$$

The matrix of mass transfer coefficients may now be computed using Eqs. 8.4.31

$$[k] = \begin{bmatrix} 1.036 & -0.0359 \\ -0.0184 & 1.0516 \end{bmatrix} \times 10^{-2} \text{ m/s}$$

The total molar flux may safely be assumed to be zero so the diffusion and molar fluxes are equal.

$$(N) = \begin{pmatrix} -5.209 \times 10^{-2} \\ 3.957 \times 10^{-2} \end{pmatrix} \text{mol/m}^2\text{s}$$

The flux of ethylbenzene N_3 is

$$N_3 = -N_1 - N_2$$
$$= 1.252 \times 10^{-2}$$

No iterations are required since the correction factors are unity in this case.

Since the experiments were carried out at total reflux, the bulk liquid at the same point in the column must have the same composition as the bulk vapor. Furthermore, if the liquid film is assumed to be well mixed, the liquid interface composition is the same as the bulk liquid composition. Thus, a simple bubble point computation on y_b gives the vapor composition at the interface. The values we found were reported above for use in this illustration.

Composition profiles are shown in Figure 8.13. Diffusional interactions are quite limited in this system due to the similar nature of all three components. As a result, all three components diffuse "normally" (Fig. 8.13). ■

8.8.5 Estimation of Overall Mass Transfer Coefficients: A Simplified Result

We may use the results of Sections 8.8.3 and 8.8.4 to develop a simple method for estimating overall mass transfer coefficients. The starting point for this development is Eq. 7.3.14 for $[K_{OV}]$

$$[K_{OV}]^{-1} = [k^V]^{-1} + \frac{c_t^V}{c_t^L}[M][k^L]^{-1} \qquad (7.3.14)$$

For the vapor phase—here assumed ideal—$[k^V]$ may be estimated from Eq. 8.3.27

$$[k^V] = [R^V]^{-1} \qquad (8.8.17)$$

and $[k^L]$ may be obtained from (cf. Eq. 8.8.14)

$$[k^L] = [R^L]^{-1}[\Gamma] \qquad (8.8.18)$$

The matrices $[R^V]$ and $[R^L]$ have elements defined by Eqs. 8.3.25 using the appropriate mole fractions and "ideal" mass transfer coefficients of the binary i–j pair k_{ij} for the appropriate phase. If we substitute Eqs. 8.8.17 and 8.8.18, together with Eq. 7.3.7 for the linearized equilibrium matrix, into Eq. 7.3.16 we find

$$[R^{OV}] = [R^V] + \frac{c_t^V}{c_t^L}[K][R^L] \qquad (8.8.19)$$

where $[R^{OV}]$ is a matrix of overall resistances to mass transfer. The matrix of overall mass transfer coefficients is the inverse of $[R_{OV}]$

$$[K_{OV}] = [R^{OV}]^{-1} \qquad (8.8.20)$$

For two component systems Eqs. 8.8.19 and 8.8.20 simplify to

$$\frac{1}{K_{OV}} = \frac{1}{k^V} + \frac{c_t^V}{c_t^L}\frac{K}{k^L} \qquad (8.8.21)$$

It is interesting to note that the thermodynamic factors cancel out of Eqs. 8.8.19 and 8.8.20. The elimination of the thermodynamic factors will prove particularly useful in the estimation of transfer efficiencies in multicomponent distillation (Section 12.3).

9 Unsteady-State Mass Transfer Models

> ... my earliest work, done at a time when I had no experimental facilities and was much taken up with the transient processes which occur when a gas comes into contact with a semi-infinite liquid (particularly one containing a reagent) and was also somewhat fascinated by the possibilities of mathematics (to which I had recently been introduced after a lapse of many years).
> —P. V. Danckwerts (1981)

9.1 SURFACE RENEWAL MODELS

In the penetration or surface renewal models fluid elements (or eddies) are pictured as arriving at the interface from the bulk fluid phase and residing at the interface for a period of time t_e (the exposure time). During the time t_e that the fluid element resides at the interface, mass exchange takes place with the adjoining phase by a process of unsteady-state diffusion. The fluid element is quiescent during this exposure period at the interface and the diffusion process is purely molecular. The element may, however, move in plug flow along the interface. After exposure, and consequent mass transfer, the fluid elements return to the bulk fluid phase and are replaced by fresh eddies. A pictorial representation of this model, adapted from Scriven (1968, 1969), is given in Figure 9.1.

The governing differential equations for the unsteady-state diffusion process experienced by the fluid element during its residence at the interface is Eq. 1.3.10 for each species. For one-dimensional, unsteady-state diffusion in a planar coordinate system these equations may be written as

$$c_t \frac{\partial x_i}{\partial t} + \frac{\partial N_i}{\partial z} = 0 \tag{9.1.1}$$

where z represents the direction coordinate for diffusion. Summing Eqs. 9.1.1 for all species in the mixture gives

$$\frac{\partial N_t}{\partial z} = 0 \tag{9.1.2}$$

from which we conclude that the mixture flux N_t is not a function of position z and depends only on time. If we substitute Eq. 1.2.12 for the molar fluxes N_i into Eq. 9.1.1 we obtain

$$c_t \frac{\partial x_i}{\partial t} + N_t \frac{\partial x_i}{\partial z} = \frac{\partial J_{iz}}{\partial z} \tag{9.1.3}$$

The molar diffusion flux J_{iz} is given by Eq. 3.1.1 for a binary system and either Eq. 2.1.25 or Eq. 3.2.5 for a multicomponent system.

The assumptions of the model are incorporated into the initial and boundary conditions. During the diffusion process the interface has the composition x_{i0}. This composition usually

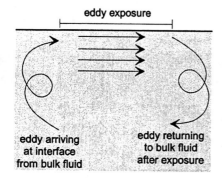

eddy exposure

eddy arriving
at interface
from bulk fluid

eddy returning
to bulk fluid
after exposure

Figure 9.1. General representation of the surface renewal model. An eddy arrives at the interface and resides there for randomly varying periods of time. During this period, there is plug flow of fluid elements. The bulk fluid is considered to be located at an infinite distance from the interface. Pictorial representation adapted from Scriven (1968, 1969).

is assumed to be constant (this assumption is not essential to the penetration model), and we have the boundary condition

$$z = 0 \qquad t > 0 \qquad x_i = x_{i0} \qquad (9.1.4)$$

Before the start of the diffusion process, the compositions are everywhere uniform in the phase under consideration, and equal to the bulk fluid composition $x_{i\infty}$. Thus we have the initial condition

$$z \geqq 0 \qquad t = 0 \qquad x_i = x_{i\infty} \qquad (9.1.5)$$

Finally, we have the last boundary condition that is valid for short contact times, that is,

$$z \to \infty \qquad t > 0 \qquad x_i = x_{i\infty} \qquad (9.1.6)$$

which essentially states that the diffusing component has not penetrated into the bulk fluid phase.

The classic penetration model of Higbie (1935) is based on the assumption that all the fluid elements reside at the interface for the same length of time. The surface age distribution for this model is

$$\psi(t) = 1/t_e \qquad (9.1.7)$$

for all $t \leq t_e$ and $\psi(t) = 0$ for $t > t_e$ and is shown in Figure 9.2.

The basis for the Danckwerts (1951) surface renewal model is the idea that the chance of an element of surface being replaced with fresh liquid from the bulk is independent of the

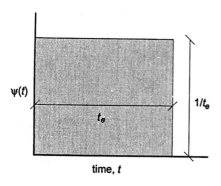

$\psi(t)$

t_e

$1/t_e$

time, t

Figure 9.2. Surface age distribution function $\psi(t)$ in which each fluid element stays the same period of time at the interface before being replenished from the bulk fluid (Higbie model).

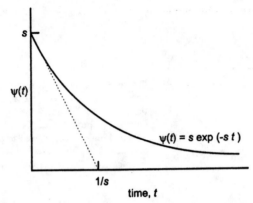

Figure 9.3. Surface age distribution according to Danckwerts; the elements undergo random surface renewal at frequency s.

length of time for which it has been exposed. The age distribution function assumed is

$$\psi(t) = s\exp(-st) \tag{9.1.8}$$

and is depicted in Figure 9.3. Here, s is the fraction of the area of surface that is replaced with fresh liquid in unit time.

In the remainder of this chapter we present solutions to this set of equations. As in Chapter 8 on film theory we begin with the binary case and then go on to consider multicomponent systems.

9.2 UNSTEADY-STATE DIFFUSION IN BINARY SYSTEMS

For a binary system with no convection perpendicular to the interface, Eq. 9.1.1, combined with Eq. 3.1.1 for J_1, simplifies to

$$c_t\frac{\partial x_1}{\partial t} + N_t\frac{\partial x_1}{\partial z} = c_t D\frac{\partial^2 x_1}{\partial z} \tag{9.2.1}$$

We have assumed c_t and D to be constant in presenting Eq. 9.2.1.

We use the method of combination of variables, with the combined variable $\zeta = z/\sqrt{4t}$. This allows us to rewrite Eq. 9.2.1 as

$$D\frac{d^2x_1}{d\zeta^2} + 2(\zeta - \phi)\frac{dx_1}{d\zeta} = 0 \tag{9.2.2}$$

where ϕ is defined by

$$\phi = (N_t/c_t)\sqrt{t} \tag{9.2.3}$$

The parameter ϕ is not a function of ζ (see Section 9.3 and Bird et al. (1960) for justification). In terms of ζ, the initial and boundary conditions become

$$
\begin{array}{ll}
\zeta = 0 & x_1 = x_{10} \\
\zeta = \infty & x_1 = x_{1\infty}
\end{array}
\tag{9.2.4}
$$

Equation 9.2.2 can be integrated to give the concentration profiles as (Arnold, 1944; Bird et al., 1960)

$$
\frac{(x_1 - x_{10})}{(x_{1\infty} - x_{10})} = \frac{1 - \operatorname{erf}((\zeta - \phi)/\sqrt{D})}{1 + \operatorname{erf}(\phi/\sqrt{D})}
\tag{9.2.5}
$$

The molar diffusion flux at the interface $z = 0$ is obtained from the one-dimensional form of Eq. 3.1.1

$$
J_{10} = -c_t D \frac{\partial x_1}{\partial z}\bigg|_{z=0}
\tag{9.2.6}
$$

The composition derivative $\partial x_1/\partial x|_{z=0}$ is obtained by differentiating Eq. 9.2.5, setting $z = 0$, and the resulting expression combined with Eq. 9.2.6 to give

$$
J_{10} = c_t\sqrt{D/\pi t}\,\frac{\exp(-\phi^2/D)}{\left(1 + \operatorname{erf}(\phi/\sqrt{D})\right)}(x_{10} - x_{1\infty})
\tag{9.2.7}
$$

In the limit that N_t goes to zero, Eq. 9.2.7 simplifies to

$$
J_{10} = c_t\sqrt{D/\pi t}\,(x_{10} - x_{1\infty}) \qquad (N_t = 0)
\tag{9.2.8}
$$

which shows that the instantaneous value of the low flux mass transfer coefficient k, defined by Eq. 7.1.4, is (Bird et al., 1960)

$$
k(t) = \sqrt{D/\pi t}
\tag{9.2.9}
$$

The average mass transfer coefficient over the total exposure period t_e, is given by

$$
k = \int_0^t k(t)\psi(t)\,dt
\tag{9.2.10}
$$

where $\psi(t)$ is a surface age distribution function representing the fraction of elements having ages between t and $t + dt$ at the surface. To evaluate this integral we need a model for $\psi(t)$.

With the age distribution function for the classic Higbie model (Eq. 9.1.5), the average mass transfer coefficient is

$$
k = 2\sqrt{D/\pi t_e}
\tag{9.2.11}
$$

The Danckwerts surface age distribution (Eq. 9.1.6) leads to an average value of the mass transfer coefficient given by

$$
k = \sqrt{Ds}
\tag{9.2.12}
$$

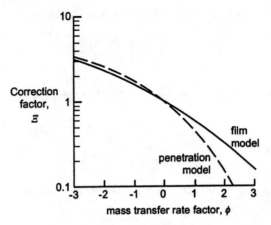

Figure 9.4. High flux correction factor from film and penetration models.

The correction factor for finite mass transfer rates Ξ is given by [see Bird, et al. 1960)]

$$\Xi = \frac{\exp\{-\Phi^2/\pi\}}{\{1 + \mathrm{erf}(\Phi/\sqrt{\pi}\,)\}} \tag{9.2.13}$$

in which the mass transfer rate factor Φ is defined as

$$\Phi = N_t/c_t k \tag{9.2.14}$$

A graphical comparison of the film and penetration model correction factors is provided by Figure 9.4. It can be seen that for a given Φ the film model and penetration model predictions of Ξ are close to each other. For this reason we recommend the use of the film model Ξ in design calculations because of the relative simplicity in computations; to use the penetration model Ξ, we need to evaluate error functions that are more time consuming.

The calculation of k using Eqs. 9.2.11 and 9.2.12 requires a priori estimation of the exposure time t_e or the surface renewal rate s. In some cases this is possible. For bubbles rising in a liquid the exposure time is the time the bubble takes to rise its own diameter. In other words, the jacket of the bubble is renewed every time it moves a diameter. If we consider the flow of a liquid over a packing, when the liquid film is mixed at the junction between the packing elements, then t_e is the time for the liquid to flow over a packing element. For flow of liquid in laminar jets and in thin films, the exposure time is known but in these cases it may be important to take into account the distribution of velocities along the interface. In the penetration model, this velocity profile is assumed to be flat (i.e., plug flow). For gas–liquid mass transfer in stirred vessels, the renewal frequency in the Danckwerts model s may be related to the speed of rotation (see Sherwood et al. 1975).

The molar flux at the interface N_{10} can be calculated by multiplying the diffusion flux J_{10} by the appropriate bootstrap coefficient β evaluated at the interface composition

$$N_{10} = \beta_0 J_{10} = c_t \beta_0 k \Xi (x_{10} - x_{1\delta}) \tag{9.2.15}$$

It is necessary to use an iterative method to compute the flux N_{10} from Eq. 9.2.15. Repeated substitution of the fluxes, starting from an initial guess calculated with $\Xi = 1$, will usually converge in only a few iterations.

Example 9.2.1 Regeneration of Triethylene Glycol

Triethylene glycol (TEG) is used in the drying of natural gas. After the drying operation the aqueous solution of TEG is regenerated by stripping off the water. This stripping operation is usually carried out in a distillation tray column and the process is illustrated schematically in Figure 9.5. We shall assume that the vapor rises through the liquid on the tray in the form of bubbles of 5-mm diameter. The bubbles rise with a velocity of 0.25 m/s. The interphase mass transfer process is largely controlled by the liquid-phase resistance. Furthermore, there is a substantial difference between the molar latent heats of vaporization of TEG and water, 70 and 40.5 kJ/mol, respectively. Estimate the mass transfer coefficient in the liquid phase and determine the mass transfer rates under the following conditions (component 1 is TEG and component 2 is water).

Composition in the bulk liquid: $x_{1\infty} = 0.6$.
Composition at the interface: $x_{10} = 0.99$.
The Fick diffusion coefficient: $D = 2 \times 10^{-10}$ m²/s.
Liquid density: $c_t = 6000$ mol/m³.

SOLUTION The contact time is the time required for a bubble to rise one diameter.

$$t_e = d/U$$

$$= 5 \times 10^{-3}/0.25$$

$$= 0.02 \text{ s}$$

The low flux mass transfer coefficient may now be calculated from Eq. 9.2.11 as

$$k = 2\sqrt{D/\pi t_e}$$

$$= 2 \times \left(2 \times 10^{-10}/\pi \times 0.02\right)^{1/2}$$

$$= 1.128 \times 10^{-4} \text{ m/s}$$

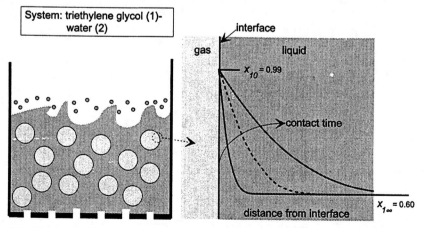

Figure 9.5. Schematic diagram of the froth on a distillation tray used for regeneration of triethylene glycol. Inset shows composition profiles in the liquid phase.

The first estimate of J_{10} is calculated from Eq. 9.2.8 as

$$J_{10} = c_t k(x_{10} - x_{1\infty})$$
$$= 6000 \times 1.128 \times 10^{-4} \times (0.99 - 0.6)$$
$$= 0.2640 \text{ mol/m}^2\text{s}$$

The molar flux at the interface follows from the bootstrap solution (Eqs. 7.2.15, 7.2.16, and 7.2.18)

$$N_{10} = \beta_0 J_{10}$$

where

$$\beta_0 = 1 - x_{10}\Lambda_1$$

with

$$\Lambda_1 = \frac{\Delta H_{vap1} - \Delta H_{vap2}}{x_{10} \Delta H_{vap1} + x_{20} \Delta H_{vap2}}$$

Hence,

$$\Lambda_1 = \frac{70 - 40.5}{0.99 \times 70 + 0.01 \times 40.5}$$
$$= 0.4232$$
$$\beta_0 = 1 - 0.99 \times 0.4232$$
$$= 0.5810$$

Thus, the first estimate of N_{10} is

$$N_{10} = \beta_0 J_{10}$$
$$= 0.5810 \times 0.2640$$
$$= 0.1534 \text{ mol/m}^2\text{s}$$

The flux of water is given by

$$N_{20} = -\Delta H_{vap1} N_{10}/\Delta H_{vap2}$$
$$= -70 \times 0.1534/40.5$$
$$= -0.2652 \text{ mol/m}^2\text{s}$$

The mass transfer rate factor is obtained from Eq. 9.2.14 as

$$\Phi = (N_{10} + N_{20})/c_t k$$
$$= (0.1534 - 0.2652)/(6000 \times 1.128 \times 10^{-4})$$
$$= -0.1651$$

and the high flux correction factor follows from Eq. 9.2.13

$$\Xi = \frac{\exp\{-\Phi^2/\pi\}}{\{1 + \text{erf}(\Phi/\sqrt{\pi})\}}$$

$$= \frac{\exp\{-(-0.1649)^2/\pi\}}{1 + \text{erf}\{(-0.1649)/\sqrt{\pi}\}}$$

$$= 1.108$$

The high flux mass transfer coefficient is

$$k^\bullet = k\Xi$$

$$= 1.128 \times 10^{-4} \times 1.108$$

$$= 1.250 \times 10^{-4} \text{ m/s}$$

We may now reestimate the molar diffusion flux from

$$J_{10} = c_t k^\bullet (x_{10} - x_{1\infty})$$

$$= 6000 \times 1.250 \times 10^{-4} \times (0.99 - 0.6)$$

$$= 0.2924 \text{ mol/m}^2\text{s}$$

The molar fluxes are calculated as before with the results

$$N_{10} = 0.1699 \text{ mol/m}^2\text{s} \qquad N_{20} = -0.2936 \text{ mol/m}^2\text{s}$$

We may continue the iterative procedure as outlined above. After five iterations we obtained the following converged results:

$$\Phi = -0.1849$$

$$\Xi = 1.1206$$

$$k^\bullet = 1.265 \times 10^{-4} \text{ m/s}$$

$$J_{10} = 0.2959 \text{ mol/m}^2\text{s}$$

$$N_{10} = 0.1719 \text{ mol/m}^2\text{s} \qquad N_{20} = -0.2972 \text{ mol/m}^2\text{s}$$

We can see from the results that the high flux correction is significant here. Also important is the effect of the unequal molar latent heats of vaporization resulting in a net mixture flux of

$$N_t = -0.1252 \text{ mol/m}^2\text{s}$$

If we had assumed equimolar counterdiffusion the molar fluxes would be equal to our first estimate of the diffusion fluxes

$$N_{10} = 0.2640 \text{ mol/m}^2\text{s} \qquad N_{20} = -0.2640 \text{ mol/m}^2\text{s}$$

values that are substantially different from the correct results. ∎

9.3   UNSTEADY-STATE DIFFUSION IN MULTICOMPONENT SYSTEMS

Let us now turn our attention to multicomponent systems. An exact analytical solution of the multicomponent penetration model for ideal gas mixtures has been presented by Olivera-Fuentes and Pasquel-Guerra (1987). Their analysis, which, in many ways, is similar to the film model analysis of Section 8.3.5, is generalized below to any system described by constitutive relations of the form of Eqs. 2.2.9 or 3.2.5.

9.3.1   An Exact Solution of the Multicomponent Penetration Model

In our analysis of the multicomponent penetration model we used the combined variable $\zeta = z/\sqrt{4t}$.

Following Olivera-Fuentes and Pasquel-Guerra (1987) we introduce transformed molar fluxes ν_i

$$\nu_i = N_i\sqrt{t} \qquad \nu_t = N_t\sqrt{t} \tag{9.3.1}$$

The fluxes ν_i are assumed to be functions only of ζ: $\nu_i = \nu_i(\zeta)$. In terms of ζ the mixture conservation Eq. 9.1.2 becomes

$$d\nu_t/d\zeta = 0 \tag{9.3.2}$$

from which we conclude that ν_t is constant. The individual species conservation Eqs. 9.1.1 become

$$-(c_t\zeta - \nu_t)\frac{dx_i}{d\zeta} = \frac{d\chi_i}{d\zeta} \tag{9.3.3}$$

where we have defined transformed diffusion fluxes χ_i by

$$\chi_i = J_i\sqrt{t} = \nu_i - \nu_t x_i \tag{9.3.4}$$

We must also write the one-dimensional form of Eq. 5.1.5 in terms of the χ_i and ζ as

$$c_t\frac{d(x)}{d\zeta} = -2[D]^{-1}(\chi) \tag{9.3.5}$$

which we use to eliminate the mole fraction gradients from the left-hand side of Eqs. 9.3.3

$$\frac{d(\chi)}{d\zeta} = -2(\zeta - \phi)[D]^{-1}(\chi) \tag{9.3.6}$$

where $\phi = \nu_t/c_t$ and is a constant.

In so far as $[D]$ depends on the composition of the mixture and as the composition is, in turn, a function of ζ we may regard $[D]$ as a function of ζ. Thus, Eq. 9.3.6 is a first-order matrix differential equation of order $n - 1$ in terms of (χ) with a variable coefficient matrix $[A(\zeta)]$

$$\frac{d(\chi)}{d\zeta} = [A(\zeta)](\chi) \tag{9.3.7}$$

where

$$[A(\zeta)] = -2(\zeta - \phi)[D]^{-1} \tag{9.3.8}$$

In terms of ζ the initial and boundary conditions become

$$\zeta = 0 \qquad x_i = x_{i0} \tag{9.3.9}$$

$$\zeta = \infty \qquad x_i = x_{i\infty} \tag{9.3.10}$$

Equations 9.3.7, subject to the transformed initial condition (Eq. 9.3.9) may be solved by the method of repeated substitution as described in Appendix B.2. The solution is (Eq. B.2.15)

$$(x) = \left[\Omega_0^\zeta(A)\right](x_0) \tag{9.3.11}$$

where $[\Omega_0^\zeta(A)]$ is the matrizant of $[A]$ defined by Eq. B.2.16. The column matrix (χ_0) is the (unknown) matrix of transformed diffusion fluxes at $\zeta = 0$. We now substitute the right-hand side of Eq. 9.3.11 into Eq. 9.3.5 to obtain

$$c_t \frac{d(x)}{d\zeta} = -2[D]^{-1}\left[\Omega_0^\zeta(A)\right](\chi_0) \tag{9.3.12}$$

This equation may be integrated to give

$$c_t(x - x_0) = -2\left[\int_0^\zeta [D]^{-1}\left[\Omega_0^\zeta(A)\right] d\zeta\right](\chi_0) \tag{9.3.13}$$

with, in particular

$$c_t(x_\infty - x_0) = -2\left[\int_0^\infty [D]^{-1}\left[\Omega_0^\zeta(A)\right] d\zeta\right](\chi_0) \tag{9.3.14}$$

which allows us to evaluate the diffusion fluxes at the interface as

$$(J_0) = \frac{c_t}{\sqrt{4t}}\left[\int_0^\infty [D]^{-1}\left[\Omega_0^\zeta(A)\right] d\zeta\right]^{-1}(x_0 - x_\infty) \tag{9.3.15}$$

The composition profiles are obtained by combining Eqs. 9.3.13 and 9.3.14.

$$(x - x_0) = \left[\int_0^\zeta [D]^{-1}\left[\Omega_0^\zeta(A)\right] d\zeta\right]\left[\int_0^\infty [D]^{-1}\left[\Omega_0^\zeta(A)\right]\zeta\right]^{-1}(x_\infty - x_0) \tag{9.3.16}$$

Equations 9.3.15 and 9.3.16 represent an exact analytical solution of the multicomponent penetration model. For two component systems, these results reduce to Eqs. 9.2.7. Unfortunately, the above results are of little practical use for computing the diffusion fluxes because they require an a priori knowledge of the composition profiles (cf. Section 8.3.5). Thus, a degree of trial and error over and above that normally encountered in multicomponent mass transfer calculations enters into their use. Indeed, Olivera-Fuentes and Pasquel-Guerra did not perform any numerical computations with this method and resorted to a numerical integration technique.

9.3.2 Multicomponent Penetration Model Based on the Assumption of Constant [D] Matrix

The only really practical approach is to use the Toor–Stewart–Prober approximation of constant $[D]$. The starting point for our analysis is the matrix generalization of Eq. 9.1.1

$$c_t \frac{\partial(x)}{\partial t} + N_t \frac{\partial(x)}{\partial z} = c_t [D] \frac{\partial^2(x)}{\partial z} \tag{9.3.17}$$

in which we have assumed the $[D]$ matrix to be constant (we will drop the subscript av in this section; it is understood that $[D]$, if calculated as described in Section 4.3, is evaluated at an average composition). Equation 9.3.17 may be solved in a variety of ways; one possible approach is to use the procedure devised by Toor and by Stewart and Prober, uncoupling the equations using a similarity transformation and solving equivalent binary-type problems as described in Section 5.3. An alternative derivation using the matrizant formulation is given by Taylor (1982c).

The solution of the linearized Eqs. 9.3.17 may also be obtained as a special case of the exact Eqs. 9.3.15 and 9.3.16. If we take $[D]$ to be independent of composition (and, hence, of ζ), Eq. 9.3.15 for the diffusion fluxes simplifies to (cf. Taylor, 1982c)

$$(J_0) = \frac{c_t}{\sqrt{4t}} [D] \left[\int_0^\infty [\Omega_0^\zeta(A)] \, d\zeta \right]^{-1} (x_0 - x_\infty) \tag{9.3.18}$$

and Eq. 9.3.16 for the composition profiles becomes

$$(x - x_0) = \left[\int_0^\zeta [\Omega_0^\zeta(A)] \, d\zeta \right] \left[\int_0^\infty [\Omega_0^\zeta(A)] \, d\zeta \right]^{-1} (x_\infty - x_0) \tag{9.3.19}$$

For the case of constant $[D]$ we may carry out the integrations required by Eq. B.2.16 to evaluate the matrizant. The result is (Taylor, 1982c)

$$\begin{aligned}
[\Omega_0^\zeta(A)] &= [I] + \sum_{k=1}^\infty \frac{(-1)^k}{k!} (\eta^2 - 2\phi\eta)^k [D]^{-k} \\
&= \exp\left[-(\eta^2 - 2\phi\eta)[D]^{-1} \right] \\
&= \exp\left[-((\eta - \phi)^2 - \phi^2)[D]^{-1} \right] \\
&= \exp\left[\phi^2 [D]^{-1} \right] \exp\left[-(\eta - \phi)^2 [D]^{-1} \right]
\end{aligned} \tag{9.3.20}$$

The last part of Eq. 9.3.20 follows because functions of one matrix ($[D]$) commute.

The integral of the matrizant can now be expressed as

$$\begin{aligned}
\int_0^\zeta [\Omega_0^\zeta(A)] \, d\zeta &= \exp\left[\phi^2 [D]^{-1} \right] \int_0^\zeta \exp\left[-(\eta - \phi)^2)[D]^{-1} \right] d\zeta \\
&= \exp\left[\phi^2 [D]^{-1} \right] \int_{-\phi}^{\zeta-\phi} \exp\left[-\xi^2 [D]^{-1} \right] d\xi \\
&= \exp\left[\phi^2 [D]^{-1} \right] \left[\int_0^{\zeta-\phi} \exp\left[-\xi^2 [D]^{-1} \right] d\xi \right. \\
&\qquad\qquad \left. - \int_0^{-\phi} \exp\left[-\xi^2 [D]^{-1} \right] d\xi \right]
\end{aligned} \tag{9.3.21}$$

where $\xi = \eta - \phi$.

Consider the integral

$$\frac{2}{\sqrt{\pi}} \int_0^{\zeta} \exp\left[-\xi^2[M]^2\right] d\xi$$

where $[M]$ is any arbitrary constant matrix. Expanding the exponential matrix in a power series and integrating term by term yields

$$\frac{2}{\sqrt{\pi}} \int_0^{\zeta} \exp\left[-\xi^2[M]^2\right] d\xi = [M]^{-1}\text{erf}[\zeta[M]] \tag{9.3.22}$$

where $\text{erf}[\zeta[M]]$ is the matrix error function defined by

$$\text{erf}[\zeta[M]] = \frac{2}{\sqrt{\pi}}\left[\zeta[M] - \frac{\zeta^3}{3 \times 1!}[M]^3 + \frac{\zeta^5}{5 \times 2!}[M]^5 - \frac{\zeta^7}{7 \times 3!}[M]^7 + \cdots\right] \tag{9.3.23}$$

which, if $[M]$ is positive definite and diagonalizable, has the properties

$$\text{erf}[\infty[M]] = [I] \qquad \text{erf}[-\zeta[M]] = -\text{erf}[\zeta[M]] \tag{9.3.24}$$

In view of Eq. 9.3.24, Eq. 9.3.21 becomes

$$\int_0^{\zeta}\left[\Omega_0^{\zeta}(A)\right] d\zeta = \frac{1}{2}\sqrt{\pi}[D]^{1/2}\exp\left[\phi^2[D]^{-1}\right]\left[\text{erf}\left[(\eta - \phi)[D]^{-1/2}\right]\right]$$
$$+ \text{erf}\left[\phi[D]^{-1/2}\right]\right] \tag{9.3.25}$$

Substituting Eq. 9.3.25 into Eq. 9.3.19, subtracting (x_∞) from both sides and after some manipulation we obtain the composition profiles

$$(x - x_\infty) = \left[[I] - \text{erf}\left[(\eta - \phi)[D]^{-1/2}\right]\right]$$
$$\times \left[[I] + \text{erf}\left[\phi[D]^{-1/2}\right]\right]^{-1}(x_0 - x_\infty) \tag{9.3.26}$$

where we have again made use of the fact that functions of one matrix commute.

The diffusion fluxes at the interface are obtained from Eq. 9.3.18 with the integral of the matrizant given by Eq. 9.3.25 as

$$(J_0) = \frac{c_t}{\sqrt{\pi t}}[D]^{1/2}\exp\left[\phi[D]^{-1/2}\right]\left[[I] + \text{erf}\left[\phi[D]^{-1/2}\right]\right]^{-1}(x_0 - x_\infty) \tag{9.3.27}$$

The preceding development holds for the special case of $N_t = 0$; if we set $\phi = 0$ in Eq. 9.3.27 the composition profiles become

$$(x - x_0) = \left[[I] - \text{erf}\left[z[D]^{-1/2}/2\sqrt{t}\right]\right](x_0 - x_\infty) \qquad [N_t = 0] \tag{9.3.28}$$

and the diffusion fluxes are obtained from

$$(J_0) = \frac{c_t}{\sqrt{\pi t}}[D]^{1/2}(x_0 - x_\infty) \qquad [N_t = 0] \tag{9.3.29}$$

When we compare Eq. 9.3.29 to Eq. 7.1.9 we see that the matrix of "low flux" mass transfer coefficients is given by

$$[k] = \frac{[D]^{1/2}}{\sqrt{\pi t}} \qquad (9.3.30)$$

and the matrix of correction factors $[\Xi]$ has the form

$$[\Xi] = \exp\left[-[\Phi]^2/\sqrt{\pi}\right]\left[[I] + \text{erf}[\Phi]/\pi\right]^{-1} \qquad (9.3.31)$$

$[\Phi]$ is the matrix generalization of Φ in Eq. 9.2.14.

$$[\Phi] = u[D]^{-1/2}\sqrt{\pi t} = N_t[k]^{-1}/c_t \qquad (9.3.32)$$

The matrix $[k]$ in Eq. 9.3.30 is the matrix of mass transfer coefficients at any time t. The matrix of time-averaged zero-flux mass transfer coefficients is given for the Higbie model by

$$[k] = 2\frac{[D]^{1/2}}{\sqrt{\pi t}} \qquad (9.3.33)$$

and for the Danckwerts model by

$$[k] = \sqrt{s}\,[D]^{1/2} \qquad (9.3.34)$$

With the above equations for $[k]$ and $[\Xi]$, the diffusion fluxes at the interface ($z = 0$) are obtained from

$$(J_0) = c_t[k][\Xi](x_0 - x_\infty) = c_t[k^\bullet](x_0 - x_\infty) \qquad (9.3.35)$$

and the molar fluxes at the interface follow from

$$(N_0) = c_t[\beta_0][k^\bullet](x_0 - x_\infty) \qquad (9.3.36)$$

Notice that it is not possible to derive a similar expression for the bulk fluid ($z \to \infty$) where all of the composition gradients vanish.

9.3.3 Toor–Stewart–Prober Formulation

Toor (1964) and Stewart and Prober (1964) did not use the method presented above; they used the method described in Chapter 5. For the multicomponent penetration model, the following expression for the matrix of mass transfer coefficients is obtained (cf. Section 8.4.2):

$$[k^\bullet] = [P][\hat{k}^\bullet][P]^{-1} \qquad (8.4.24a)$$

where $[\hat{k}^\bullet]$ is a diagonal matrix whose nonzero elements are the eigenvalues of $[k^\bullet]$

$$\hat{k}_i^\bullet = \hat{k}_i\hat{\Xi}_i \qquad (8.4.25)$$

and where

$$\hat{k}_i = 2\sqrt{(\hat{D}_i/\pi t)} \tag{9.3.37}$$

$$\hat{\Phi}_i = N_t/c_t\hat{k}_i \tag{9.3.38}$$

$$\hat{\Xi}_i = \exp(-\hat{\Phi}_i^2/\pi)/(1 + \text{erf}(\hat{\Phi}_i/\sqrt{\pi})) \tag{9.3.39}$$

These are the eigenvalues of $[k]$, $[\Phi]$, and $[\Xi]$, respectively.

Equation 8.4.24a serves as a starting point for computing the mass transfer coefficients and, hence, the molar fluxes. As is the case with the film theory result a trial and error procedure is needed if the total flux is not specified. We do not particularly recommend computing the matrix functions needed in the penetration models from power series expansions (even though this is possible; Taylor, 1982c). We prefer to use the algorithms developed for computing the fluxes from the solution of the linearized film model equations (Algorithms 8.4 and 8.5).

Example 9.3.1   Mass Transfer in a Stirred Cell

In this example we reexamine the mass transfer in the system glycerol (1)–water (2)–acetone (3) considered in Example 8.7.1. A large amount of a glycerol-rich phase of bulk composition

$$x_{1\infty} = 0.7824 \qquad x_{2\infty} = 0.1877 \qquad x_{3\infty} = 0.0299$$

is brought into contact with another (immiscible) phase such that the interface composition is maintained at

$$x_{10} = 0.5480 \qquad x_{20} = 0.2838 \qquad x_{30} = 0.1682.$$

The contactor is a stirred cell and was depicted in Figure 5.9. The stirrer is rotated at such a speed that the surface renewal frequency of the phase under consideration is

$$s = 25 \text{ s}^{-1}$$

Using the same physical property data as in Example 8.7.1, calculate the fluxes N_i

SOLUTION   With the Maxwell–Stefan diffusion coefficients and thermodynamic factors as given in Example 8.7.1, the matrix of Fick diffusivities $[D]$ can be evaluated from Eqs. 4.2.12 using the average mole fractions with the following results:

$$[D] = \begin{bmatrix} 14.40 & -9.97 \\ -1.66 & 25.62 \end{bmatrix} \times 10^{-10} \text{ m}^2/\text{s}$$

The eigenvalues of this matrix are

$$\hat{D}_1 = 2.694 \times 10^{-9} \text{ m}^2/\text{s} \qquad \hat{D}_2 = 1.31 \times 10^{-9} \text{ m}^2/\text{s}$$

The eigenvalues of $[k]$ are calculated from Eq. 9.3.37 as

$$\hat{k}_1 = 2.59 \times 10^{-4} \text{ m/s} \qquad \hat{k}_2 = 1.81 \times 10^{-4} \text{ m/s}$$

The matrix of low flux mass transfer coefficients is found directly from Eqs. 8.4.31 or 8.4.24 as

$$[k] = \begin{bmatrix} 1.88 & -0.566 \\ -0.094 & 2.52 \end{bmatrix} \times 10^{-4} \text{ m/s}$$

The first estimate of the diffusion fluxes is calculated from Eq. 9.3.35 with the correction factor matrix taken to be the identity matrix

$$(J) = \begin{pmatrix} -0.744 \\ 0.396 \end{pmatrix} \text{ mol/m}^2\text{s}$$

The total flux is found from

$$N_t = J_1/(z_1 - x_{10})$$

with $z_1 = \frac{1}{12}$. This gives

$$N_t = 1.6 \text{ mol/m}^2\text{s}$$

The N_i follow from Eq. 1.1.12 as

$$N_1 = 0.133 \qquad N_2 = 0.851 \qquad N_3 = 0.616 \text{ mol/m}^2\text{s}$$

These values are used in the calculation of the mass transfer rate factors form Eqs. 9.3.38, the high flux correction factors from Eqs. 9.3.39 and, hence, new values of the high flux mass transfer coefficients. The cycle of flux-coefficient calculations is repeated and after 10 iterations we obtain the following converged values:

$$\hat{\Phi}_1 = 0.316 \qquad \hat{\Phi}_2 = 0.453$$

$$\hat{k}_1^\bullet = 2.097 \times 10^{-4} \text{ m/s} \qquad \hat{k}_2^\bullet = 1.321 \times 10^{-4} \text{ m/s}$$

$$[k^\bullet] = \begin{bmatrix} 1.398 & -0.558 \\ -0.093 & 2.02 \end{bmatrix} \times 10^{-4} \text{ m/s}$$

The diffusion fluxes are

$$(J) = \begin{pmatrix} -0.571 \\ 0.324 \end{pmatrix} \text{ mol/m}^2\text{s}$$

The total flux is

$$N_t = 1.23 \text{ mol/m}^2\text{s}$$

and the N_i are

$$N_1 = 0.102 \qquad N_2 = 0.673 \qquad N_3 = 0.453 \text{ mol/m}^2\text{s}$$

The above computations were carried out with Eqs. 9.3.39 for the eigenvalues of the correction factor matrix. As noted earlier, this involves the computation of the error function that is significantly more time consuming than the exponential function needed for the film model correction factor. With the eigenvalues of $[\Xi]$ given by the film model

Eq. 8.4.28 we obtain the following converged values of the fluxes

$$N_1 = 0.108 \qquad N_2 = 0.705 \qquad N_3 = 0.483 \, \text{mol}/\text{m}^2\text{s}$$

These values are within 5% of the values calculated with the penetration theory correction factor matrix and support our earlier suggestion that it is sufficient to use the simpler film model correction factor matrix in multicomponent mass transfer calculations at high mass transfer rates. ∎

9.4 DIFFUSION IN BUBBLES, DROPS, AND JETS

One very important restriction in the development of the surface renewal models is the assumption that the penetrating, or diffusing, component does not "see" the bulk fluid, which, to all intents and purposes, is located at an infinite distance from the interface. This assumption is implicit in the boundary condition (Eq. 9.1.6) and is strictly true only for short Fourier times

$$\text{Fo} = Dt/\delta^2 < 0.20 \tag{9.4.1}$$

where δ is the distance from the interface to the "core" of the fluid phase. For example, if we consider diffusion inside rigid droplets, the distance δ corresponds to the radius of the drop.

For long contact times and/or short distances between the interface and the "core" δ, the solution given above for the zero-flux coefficient does not apply. This situation may arise for mass transfer inside liquid droplets that stay sufficiently long in contact with the surrounding gas or liquid. For long contact times, the diffusing species will penetrate deep into the heart of the bubble (or drop), and it is important in such cases to define the mass transfer coefficient in terms of the driving forces $\Delta x_i = x_{iI} - \langle x_i \rangle$, where x_{iI} represents the interface composition and $\langle x_i \rangle$ is the cup-mixing composition of the spherical dispersed phase.

9.4.1 Binary Mass Transfer in Spherical and Cylindrical Geometries

In this section we present expressions for the mass transfer coefficients for diffusion in spherical and cylindrical geometries. The results presented here are useful in the modeling of mass transfer in, for example, gas bubbles in a liquid, liquid droplets in a gas, or gas jets in a liquid as shown in Figure 9.6.

For a binary system, under conditions of small mass transfer fluxes, the unsteady-state diffusion equations may be solved to give the fractional approach to equilibrium F defined by (see Clift et al., 1978)

$$F \equiv \frac{(x_{10} - \langle x_1 \rangle)}{(x_{10} - x_{1I})} \tag{9.4.2}$$

where x_{10} is the initial composition (at $t = 0$) within the particle and x_{1I} is the composition at the surface of the particle (held constant for the duration of the diffusion process).

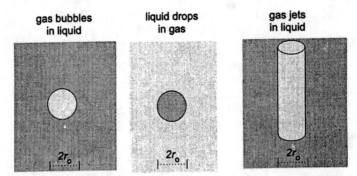

Figure 9.6. Idealized view of (a) spherical gas bubbles in a liquid, (b) liquid droplets in a gas, and (c) cylindrical gas jets in a liquid. Diffusion in bubbles, drops, and jets may be modeled by solving the diffusion equations for cylindrical and spherical coordinates.

The Sherwood number for a spherical particle, $\text{Sh} \equiv k \cdot 2r_0/D$, at time t, defined by taking the driving force to be $(x_{1I} - \langle x_1 \rangle)$, may be expressed in terms of F as (Clift et al., 1978, p. 58)

$$\text{Sh} = \frac{2}{3(1 - F)} \frac{\partial F}{\partial \text{Fo}} \tag{9.4.3}$$

where Fo is the Fourier number: Dt/r_0^2; r_0 is the radius of the particle. The time averaged Sherwood number is

$$\overline{\text{Sh}} = -2\ln(1 - F)/3\text{Fo} \tag{9.4.4}$$

The time averaged mass transfer coefficient $\bar{k}$ may be extracted from Eq. 9.4.4 as

$$\bar{k} = -\ln(1 - F)/a't \tag{9.4.5}$$

where a' is the surface area per unit volume of particle $a' = 3/r_0$ and t is the contact time.
For a rigid spherical particle (bubble or droplet) F is given by (see Clift et al., 1978)

$$F = 1 - \frac{6}{\pi^2} \sum_{m=1}^{\infty} \frac{1}{m^2} \exp\{-m^2\pi^2 \text{Fo}\} \tag{9.4.6}$$

We see from Eqs. 9.4.2 and 9.4.6 that when $t \to \infty$ equilibrium is attained, and the average composition $\langle x_1 \rangle$ will equal the surface composition x_{1I}. The time averaged Sherwood number and mass transfer coefficients for a rigid spherical particle may be obtained directly from Eqs. 9.4.4 and 9.4.5 with F given by Eq. 9.4.6 above. The Sherwood number at time t may be found using Eqs. 9.4.3 and 9.4.6 as

$$\text{Sh} = \frac{2}{3}\pi^2 \left(\sum_{m=1}^{\infty} \exp\{-m^2\pi^2 \text{Fo}\} \Big/ \sum_{m=1}^{\infty} \frac{1}{m^2} \exp\{-m^2\pi^2 \text{Fo}\} \right) \tag{9.4.7}$$

Figure 9.7 shows the variation of Sh with Fo. For large values of Fo, Sh approaches the

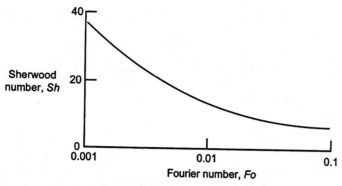

Figure 9.7. Sherwood number for mass transfer within a rigid spherical particle of radius r_0 as a function of the Fourier number $Fo = Dt/r_0^2$.

asymptotic value $\frac{2}{3}\pi^2 = 6.58$. For this steady-state limit, the zero-flux mass transfer coefficient is

$$k = \pi^2 D/3r_0 \tag{9.4.8}$$

showing, as for the film model discussed in Chapter 8, a unity-power dependence on the Fick coefficient D. This result is to be contrasted with the square-root dependence for small values of Fo; see the variation in the curvature in Figure 9.7.

When the size of the bubble (or droplet) exceeds a certain limit the dispersed phase may begin to circulate or oscillate. The Kronig–Brink model for circulation within the dispersed phase gives the following expression for the fractional approach to equilibrium

$$F = 1 - \frac{3}{8} \sum_{m=1}^{\infty} A_m^2 \exp\{-16\lambda_m \, Fo\} \tag{9.4.9}$$

and the Sherwood number at time t is

$$Sh = \frac{1}{3}32\left(\sum_{m=1}^{\infty} A_m^2 \lambda_m \exp\{-16\lambda_m \, Fo\} \middle/ \sum_{m=1}^{\infty} A_m^2 \exp\{-16\lambda_m \, Fo\} \right) \tag{9.4.10}$$

The eigenvalues A_n and λ_n have been tabulated by Sideman and Shabtai (1964). For $Fo \to \infty$, the asymptotic value of Sh is

$$Sh = 32\lambda_1/3 = 17.66 \tag{9.4.11}$$

which is 2.7 times the corresponding limit for noncirculating particles, demonstrating the enormous influence of the system hydrodynamics on the mass transfer behavior.

Another situation that is of practical importance is radial diffusion inside a cylindrical jet of gas or liquid. The fractional approach to equilibrium is given by

$$F = 1 - 4 \sum_{m=1}^{\infty} \frac{\exp\{-j_m^2 \, Fo\}}{j_m^2} \tag{9.4.12}$$

where the j_m are the roots of the zero-order Bessel function $J_0(j_m) = 0$. For this geometry the time averaged Sherwood number and mass transfer coefficient are given by

$$\overline{Sh} = -\ln(1 - F)/Fo \qquad (9.4.13)$$

$$\overline{k} = -\ln(1 - F)/a't \qquad (9.4.14)$$

where a' is the surface area per unit volume of particle $a' = 2/r_0$.

9.4.2   Transport in Multicomponent Drops and Bubbles

The analyses in Section 9.4.1 for binary systems can be extended to multicomponent systems by using the Toor–Stewart–Prober approximation of constant $[D]$. We will not go through the details of the derivations here, our readers can verify these results for themselves.

The fractional approach to equilibrium in a multicomponent system is given by the $n - 1$ dimensional matrix analog of Eq. 9.4.2.

$$(x_0 - \langle x \rangle) = [F](x_0 - x_I) \qquad (9.4.15)$$

For mass transfer in a rigid spherical drop the matrix $[F]$ is given by the $n - 1$ dimensional matrix generalization of Eq. 9.4.6

$$[F] = \left[[I] - \frac{6}{\pi^2} \sum_{m=1}^{\infty} \frac{1}{m^2} \exp\left[-m^2\pi^2 \, Fo_{ref}[D'] \right] \right] \qquad (9.4.16)$$

where we have introduced an $n - 1 \times n - 1$ matrix of normalized Fick diffusion coefficients

$$[D'] = [D]/D_{ref} \qquad (9.4.17)$$

and a reference Fourier number of $Fo_{ref} \equiv D_{ref}t/r_0^2$.

The matrix of Sherwood numbers at time t

$$[Sh] \equiv [k] \cdot 2r_0[D]^{-1} \qquad (9.4.18)$$

is given by

$$[Sh] = \frac{2}{3}\pi^2 \left[\sum_{m=1}^{\infty} \exp\left[-m^2\pi^2 \, Fo_{ref}[D'] \right] \right] \left[\sum_{m=1}^{\infty} \frac{1}{m^2} \exp\left[-m^2\pi^2 \, Fo_{ref}[D'] \right] \right]^{-1} \qquad (9.4.19)$$

In the limit $Fo_{ref} \to \infty$, $[Sh]$ approaches the asymptotic limit

$$[Sh] = \tfrac{2}{3}\pi^2[I] \qquad (9.4.20)$$

or

$$[k] = \tfrac{1}{3}\pi^2[D]/r_0 \qquad (9.4.21)$$

A very interesting difference between the short contact time value: $[k] = 2[D]^{1/2}/\sqrt{\pi t}$ and the long contact time steady-state value: $[k] = \tfrac{1}{3}\pi^2[D]/r_0$ is the variation in the influence of diffusional coupling. The influence of molecular diffusional coupling will be maximum when $[k]$ is proportional to $[D]$ as is the case for the long-time asymptote. This influence is reduced as the contact time is reduced, and it is at its lowest for the short-contact time square-root dependence (see Example 9.4.1).

The Kronig–Brink model for circulation within the dispersed phase can be generalized to n-component systems to give the following expression for the matrix $[F]$

$$[F] = \left[[I] - \frac{3}{8} \sum_{m=1}^{\infty} A_m^2 \exp\left[-16\lambda_m \, \text{Fo}_{\text{ref}}[D'] \right] \right] \tag{9.4.22}$$

and the matrix of Sherwood numbers is

$$[\text{Sh}] = \frac{32}{3} \left[\sum_{m=1}^{\infty} \exp\left[A_m^2 \lambda_m \, \text{Fo}_{\text{ref}}[D'] \right] \right] \left[\sum_{m=1}^{\infty} \exp\left[A_m^2 \, \text{Fo}_{\text{ref}}[D'] \right] \right]^{-1} \tag{9.4.23}$$

For $\text{Fo}_{\text{ref}} \to \infty$, the asymptotic value of $[\text{Sh}]$ is

$$[\text{Sh}] = 32\lambda_1[I]/3 = 17.66[I] \tag{9.4.24}$$

For radial, unsteady diffusion in a cylindrical geometry the matrix $[F]$ is given by

$$[F] = \left[[I] - 4 \sum_{m=1}^{\infty} \frac{\exp\left[-j_m^2 \, \text{Fo}_{\text{ref}}[D'] \right]}{j_m^2} \right] \tag{9.4.25}$$

For calculating $[\text{Sh}]$ from these models we may make use of Sylvester's formula as follows:

$$[\text{Sh}] = \sum_{i=1}^{s} \widehat{\text{Sh}}_i \left\{ \prod_{\substack{j=1 \\ j \neq i}}^{s} \left[[D] - \hat{D}_j[I] \right] \Big/ \prod_{\substack{j=1 \\ j \neq i}}^{s} \left(\hat{D}_i - \hat{D}_j \right) \right\} \tag{9.4.26}$$

where $\widehat{\text{Sh}}_i$ is the ith eigenvalue of $[\text{Sh}]$ and s is the number of such eigenvalues that are distinct ($s \leq n - 1$). The $\widehat{\text{Sh}}_i$ are obtained from the equations in Section 9.2 with $\hat{D}_i$, the corresponding eigenvalue of $[D]$, replacing the binary diffusivity in the Fourier number Fo. A value for the reference diffusivity is not needed in the computation of $[\text{Sh}]$ (or $[k]$), since D_{ref} cancels out of the calculations.

Alternatively, we may use the similarity transformation

$$[\text{Sh}] = [P][\widehat{\text{Sh}}][P]^{-1} \tag{9.4.27}$$

where $[\widehat{\text{Sh}}]$ is a diagonal matrix whose nonzero elements are the eigenvalues of $[\text{Sh}]$. The matrix $[P]$ is the modal matrix of the Fick matrix $[D]$.

To calculate $[k]$, which is the value needed to compute mass transfer fluxes, we may avoid computing $[\text{Sh}]$. Equations 8.4.23 or 8.4.30 and 8.4.31 may be used directly as written to compute the multicomponent mass transfer coefficients with the eigenvalues of $[k]$ computed from the appropriate expression in Section 9.2 as described above.

We illustrate the use of these models in Example 9.4.1.

Example 9.4.1   Diffusion in a Multicomponent Drop

A droplet 5 mm in diameter containing a mixture of acetone(1)–benzene(2)–methanol(3) is brought into contact with a surrounding vapor phase that maintains the surface of the droplet at a fixed composition. Investigate the influence of the contact time between vapor and liquid on the ratios k_{12}/k_{11} and k_{21}/k_{22} of the matrix of multicomponent mass transfer coefficients $[k]$. Assume the droplet initially has the composition $x_1 = 0.52$, $x_2 = 0.28$, and $x_3 = 0.2$. What are the directions of mass transfer if the ratio of driving force of acetone to the driving force for benzene is

$$\Delta x_2/\Delta x_1 = -5$$

DATA   The matrix of Fick diffusion coefficients is

$$[D] = \begin{bmatrix} 4.280 & 1.040 \\ -0.67 & 2.260 \end{bmatrix} \times 10^{-9} \text{ m}^2/\text{s}$$

SOLUTION   For the purposes of this calculation we shall assume the drop to be noncirculating. Thus, the matrix of multicomponent mass transfer coefficients $[k]$ may be computed from Eqs. 9.4.18 and 9.4.19 with the help of Sylvester's expansion formula or the modal transformation. At both the long and short contact time limits, however, we may calculate the ratios of mass transfer coefficients k_{12}/k_{11} and k_{21}/k_{22} without evaluating the series expansions needed in Eq. 9.4.19.

For short contact times, $[k]$ is proportional to $[D]^{1/2}$, which may be computed using Sylvester's expansion formula, or by the modal transformation method. Both methods require the eigenvalues of $[D]$, which are

$$\hat{D}_1 = 3.839 \times 10^{-9} \text{ m}^2/\text{s} \qquad \hat{D}_2 = 2.701 \times 10^{-9} \text{ m}^2/\text{s}$$

$[D]^{1/2}$ is found to be

$$[D]^{1/2} = \begin{bmatrix} 6.583 & 0.913 \\ -0.588 & 4.810 \end{bmatrix} \times 10^{-5} \text{ m/s}^{1/2}$$

Thus,

$$k_{12}/k_{11} = 0.913/6.583$$
$$= 0.139$$
$$k_{21}/k_{22} = -0.588/4.810$$
$$= -0.122.$$

In the other extreme of long contact times, $[k]$ is directly proportional to $[D]$, and so

$$k_{12}/k_{11} = D_{12}/D_{11} = 0.243 \qquad k_{21}/k_{22} = D_{21}/D_{22} = -0.295.$$

The influence of contact time is illustrated in Figure 9.8, where k_{12}/k_{11} and $-k_{21}/k_{11}$ are plotted against $\text{Fo}_{\text{ref}} = D_{\text{ref}}t/r_0^2$, where the reference diffusivity D_{ref} is taken to be D_{11} in the calculations.

Under conditions of low transfer fluxes (no correction factor to take into account) for the steady-state asymptotic limit

$$(k_{11}\Delta x_1 + k_{12}\Delta x_2)/k_{11}\Delta x_1 = 1 + \frac{k_{12}}{k_{11}}\frac{\Delta x_2}{\Delta x_1}$$
$$= 1 + 0.243 \times (-5)$$
$$= -0.215$$

that is, acetone will experience reverse mass transfer; while in the other limit of short

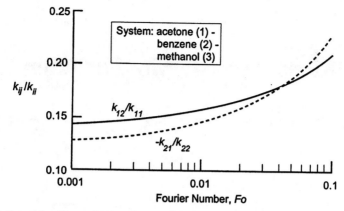

Figure 9.8. Ratios of k_{12}/k_{11} and $-k_{21}/k_{22}$ for transfer inside a spherical rigid droplet containing a mixture of acetone(1)–benzene(2)–methanol(3). Variation with the Fourier number $\text{Fo}_{\text{ref}} = D_{\text{ref}}t/r_0^2$. The parameter D_{ref} is taken equal to D_{11} in the calculations.

contact times

$$(k_{11}\,\Delta x_1 + k_{12}\,\Delta x_2)/k_{11}\,\Delta x_1 = 1 + \frac{k_{12}}{k_{11}}\frac{\Delta x_2}{\Delta x_1}$$

$$= 1 + 0.139 \times (-5)$$

$$= 0.307$$

and we see that though coupling effects still are significant, they are insufficient to drive acetone against its driving force. Thus, the effect of contact time could be to alter the *direction* of mass transfer. This effect was first pointed out by Krishna (1978a) who showed that the film and penetration models could predict different directions of mass transfer.

From the calculations presented above, it should be clear that coupling effects in multicomponent mass transfer will be influenced not only by the structure of the Fick matrix $[D]$ but also by the hydrodynamics of two-phase contacting, which influences both the contact time between the phases and the distribution of sizes of droplets or bubbles. Very small bubbles (or drops) may approach the steady-state limit (largest influence of coupling), while larger bubbles will transfer mass in the "short-contact regime" (least influence of coupling). ■

10 Mass Transfer in Turbulent Flow

Chemical engineers cannot escape dealing with mass and heat transfer across fluid interfaces that are in more or less chaotic motion, chaos that goes under name of turbulence.

—L. E. Scriven (1968, 1969)

Until now we have considered mass transfer by molecular diffusion. If turbulent conditions prevail there will be an additional transport contribution by the turbulent eddies. The understanding of turbulent (eddy) momentum transport is a prerequisite to the understanding of turbulent mass transport. There is an astronomical amount of literature on turbulence modeling. For a chemical engineer who is interested in learning something about the state of affairs here, we can recommend the book by Launder and Spalding (1972). An excellent review of turbulent heat and mass transfer at interfaces is given by Sideman and Pinczewski (1975). This latter work reviews turbulent heat and mass transfer from the viewpoint of a chemical engineer.

This chapter describes models of mass transfer in turbulent conditions. Beginning with a brief survey of turbulent eddy diffusivity models we develop solutions to the binary mass transport equations at length before presenting the corresponding multicomponent results.

10.1 BALANCE AND CONSTITUTIVE RELATIONS FOR TURBULENT MASS TRANSPORT

If the conditions prevailing in the phase under consideration are turbulent, then it will be necessary to time average the conservation equations. The time averaging procedure is discussed by, for example, Bird et al. (1960). Time averaging the component material balances (Eqs. 1.3.6) gives

$$\frac{\partial \bar{\omega}_i}{\partial t} + \bar{v} \cdot \nabla \bar{\omega}_i = -\frac{1}{\rho_t}\nabla \cdot (j_i + j_{i,\text{turb}}) \tag{10.1.1}$$

where $j_{i,\text{turb}}$ is the turbulent diffusion flux caused by the turbulent eddies present in the system. The overbars in Eq. 10.1.1 denote time-averaged quantities. In subsequent discussions we shall omit writing the overbars and take it as understood that time-smoothed variables are considered.

A time-averaged velocity profile in fully developed turbulent flow is shown in Figure 10.1 where it can be seen that the profile is considered in three sections.

1. A wall region or viscous sublayer where the flow is laminar.
2. A buffer zone that acts as a transition between the laminar sublayer.
3. A fully developed turbulent core.

In the core of the bulk fluid phases in turbulent flow, the turbulent diffusion flux $j_{i,\text{turb}}$ predominates over the molecular diffusion flux j_i. Close to the interface the turbulence is

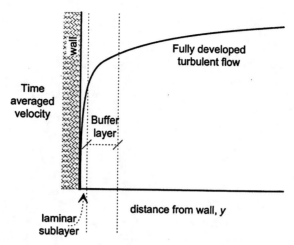

Figure 10.1. Time-averaged velocity profiles in fully developed turbulent flow identifying the presence of a laminar sublayer, a buffer layer, and the turbulent core of the flow. Velocity profile calculated from Eqs. 10.2.14–10.2.16.

damped and the molecular diffusion flux j_i predominates. It is generally only possible to use the simplest description of turbulent diffusion, namely, Boussinesq's hypothesis

$$j_{i,\text{turb}} = -\rho_t D_{\text{turb}} \nabla \omega_i \qquad i = 1, 2, \ldots, n \qquad (10.1.2)$$

where D_{turb} is the turbulent eddy diffusivity of mass. There are no coupling effects in turbulent diffusion because turbulent eddy mass transport is not species specific. In other words, all the components are transported by the same mechanism. We use mass units and the mass average reference velocity frame because the eddy diffusivity approach requires simultaneous consideration of the equations of motion.

Even with this simple constitutive relationship (Eq. 10.1.2) for the turbulent diffusion flux, the problem remains as to the value of the turbulent diffusivity D_{turb}. The most usual procedure for the prediction of D_{turb} is to proceed through a knowledge of ν_{turb}, the turbulent kinematic viscosity. We define a turbulent Schmidt number

$$\text{Sc}_{\text{turb}} = \nu_{\text{turb}}/D_{\text{turb}} \qquad (10.1.3)$$

If we accept the analogy between heat and mass transfer, Sc_{turb} must be equal to Pr_{turb}, the turbulent Prandtl number defined by

$$\text{Pr}_{\text{turb}} = C_p \mu_{\text{turb}}/\lambda_{\text{turb}} \qquad (10.1.4)$$

where λ_{turb} is the turbulent eddy thermal conductivity. Experimental values of Pr_{turb} show a marked dependence on the molecular Prandtl number and vary between 0.2 and 2.5 as indicated by Figure 10.2 (adapted from a thesis by Blom, 1970). Not surprisingly, there is a variety of models available for Pr_{turb} (see Sideman and Pinczewski, 1975) but a choice of the "best" model cannot easily be made and for most practical design purposes we are forced to assume $\text{Sc}_{\text{turb}} = \text{Pr}_{\text{turb}} = 1$, for want of more reliable information.

By defining Sc_{turb}, we have replaced the problem of estimating D_{turb} with the problem of estimating ν_{turb}, the turbulent kinematic viscosity. To estimate ν_{turb}, we need to know the velocity profiles between the interface and the bulk fluid phase. For simple flow situations,

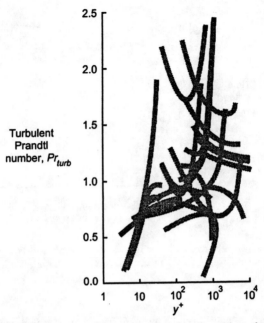

Figure 10.2. Compilation of experimental values for the turbulent Prandtl number Pr_{turb}. The horizontal axis is a dimensionless distance from the wall defined by Eq. 10.2.6. Adapted from the thesis by Blom (1970).

such as flow over flat plates and inside circular tubes, sufficient information is available concerning the velocity profiles to allow estimation of the turbulent eddy diffusivities and, hence, calculation of the mass and heat transfer coefficients and fluxes between the "wall" and the flowing stream. We discuss this issue in Section 10.2.

10.2 TURBULENT EDDY DIFFUSIVITY MODELS

For definiteness, we consider the transfer processes between a cylindrical wall and a turbulently flowing n-component fluid mixture. For condensation of vapor mixtures flowing inside a vertical tube, for example, the "wall" can be considered to be the surface of the liquid condensate film. We examine the phenomena occurring at any axial position in the tube, assuming that fully developed flow conditions are attained. For steady-state conditions, the equations of continuity of mass of component i (assuming no chemical reactions), Eqs. 1.3.7 take the form

$$\frac{d(rn_{ir})}{dr} = 0 \tag{10.2.1}$$

where r represents the radial coordinate and n_{ir} is the mass flux of component i. Equation 10.2.1 shows that rn_{ir} is r invariant. It will prove to be more convenient to work in a coordinate system measuring the distance from the wall y

$$y = R_w - r \tag{10.2.2}$$

where R_w represents the radius of the circular tube. We consider n_{iy} to be positive if the flux is directed in the positive y direction, that is, from the wall towards the flowing fluid mixture. The mass flux n_{iy} can be written in terms of the diffusive mass flux j_{iy} with respect to the mass average mixture velocity, and a bulk-flow contribution $\omega_i n_{ty}$

$$n_{iy} = j_{iy} + \omega_i n_{ty}, \qquad i = 1, 2, \ldots, n \tag{10.2.3}$$

where n_{ty} is the total mixture mass flux. Since the conditions inside the tube are considered to be turbulent, we use time-smoothed fluxes and compositions in Eq. 10.2.3; j_{iy} is the sum of the molecular and turbulent contributions to diffusion.

The boundary conditions for this model are

$$
\begin{aligned}
y &= 0 & \omega &= \omega_0 & &\text{(the interface)} \\
y &= y_b & \omega &= \omega_b & &\text{(bulk fluid)}
\end{aligned}
\tag{10.2.4}
$$

where y_b is a distance from the wall beyond which we may safely assume that the turbulence level is high enough to wipe out any further radial composition variations.

Before proceeding further, it is convenient to define the following parameters and variables that incorporate information on the flow

1. Friction velocity u^*

$$u^* = \sqrt{\tau_0/\rho_t} = \sqrt{f/2}\,\bar{u} \tag{10.2.5}$$

where τ_0 is the shear stress at the interface or wall, and $\bar{u}$ is the average velocity of flow of the multicomponent fluid mixture inside the tube. The parameter f is the Fanning friction factor.

2. A dimensionless distance from the wall.

$$y^+ = yu^*\rho_t/\mu = yu^*/\nu \tag{10.2.6}$$

3. A dimensionless velocity.

$$u^+ = u/u^* = \sqrt{2/f}\,u/\bar{u} \tag{10.2.7}$$

4. A dimensionless tube radius.

$$R_W^+ = R_W u^*/\nu = \text{Re}\,\sqrt{f/8} \tag{10.2.8}$$

where Re is the Reynolds number for flow $2R_w\bar{u}/\nu$.

In terms of the reduced distance from the wall the boundary conditions are

$$
\begin{aligned}
y^+ &= 0, & \omega_i &= \omega_{i0} & &\text{(the interface)} \\
y^+ &= y_b^+ & \omega_i &= \omega_{ib} & &\text{(bulk fluid)}
\end{aligned}
\tag{10.2.9}
$$

The first difficulty we encounter is that the position y_b^+, at which the bulk fluid-phase composition ω_b is reached, is not known precisely. To overcome this shortcoming in our knowledge, we proceed to divide the region $0 - y_b^+$ as follows:

1. A region $0 - y_1^+$ in which both molecular and turbulent eddy contributions to mass transfer are important.
2. A turbulent core from y_1^+ to y_b^+, where the contribution of turbulent eddy transport is much larger than the molecular contribution

$$(\nu_{\text{turb}}/\nu) \gg 1 \tag{10.2.10}$$

10.2.1 Estimation of the Turbulent Eddy Viscosity

We now turn our attention to the estimation of the ratio ν_{turb}/ν. The starting point for our analysis is the equation of motion, which for steady-state conditions gives the shear stress profile as (see, e.g., Bird et al., 1960)

$$(\tau_r + \tau_{r,\text{turb}}) = (r/R_w)\tau_w \tag{10.2.11}$$

Transforming to the y-coordinate system and introducing the constitutive relations for the shear stresses due to the molecular contribution ($\tau_y = -\mu \, du/dy$) and due to the turbulent eddy contribution ($\tau_{\text{turb}} = -\mu_{\text{turb}} \, du/dy$) we can write Eq. 10.2.11 in terms of reduced parameters as follows:

$$\left(1 + \frac{\nu_{\text{turb}}}{\nu}\right)\frac{du^+}{dy^+} = 1 - \frac{y^+}{R_w^+} \tag{10.2.12}$$

The ratio y^+/R_w^+ is negligibly small for problems in which the major concern is the estimation of heat and mass transfer rates. This can be seen from the fact that most of the resistance to mass and heat transfer is concentrated in a thin zone of thickness $y^+ = 5$. The term R_w^+ has a value exceeding 300 for Re = 10,000. We shall, therefore, approximate the term $(1 - y^+/R_w^+)$ by unity in the ensuing analysis. With the right-hand side of Eq. 10.2.12 set to unity we may rewrite it as

$$\frac{\nu_{\text{turb}}}{\nu} = \frac{1}{du^+/dy^+} - 1 \tag{10.2.13}$$

This ratio can be estimated if the functional form $u^+(y^+)$ is known.

The von Karman velocity profile for turbulent flow is given by three equations.

1. For the wall region or viscous sublayer.

$$u^+ = y^+ \qquad 0 \leq y^+ < 5 \tag{10.2.14}$$

2. For a buffer zone.

$$u^+ = 5.0 \ln y^+ - 3.05 \qquad 5 \leq y^+ < 30 \tag{10.2.15}$$

3. For the turbulent core.

$$u^+ = 2.5 \ln y^+ + 5.0 \qquad 30 \leq y^+ \tag{10.2.16}$$

The von Karman velocity profile is illustrated in Figure 10.3. Substitution of Eqs. 10.2.14–10.2.16 into Eq. 10.2.13 yields the following expressions for ν_{turb}/ν

$$\frac{\nu_{\text{turb}}}{\nu} = 0 \qquad 0 < y^+ < 5 \tag{10.2.17}$$

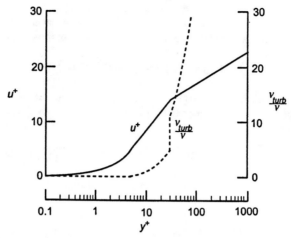

Figure 10.3. von Karman model for the velocity profile in turbulent flow $u^+(y^+)$. Also shown is the ratio of turbulent to molecular kinematic viscosities ν_{turb}/ν that results from the von Karman velocity profile.

for the viscous sublayer, and

$$\frac{\nu_{\text{turb}}}{\nu} = \frac{y^+}{5} - 1 \qquad 5 \leqq y^+ < 30 \qquad (10.2.18)$$

for the buffer zone in which both molecular and turbulent contributions play a role. The ratio ν_{turb}/ν is shown as a function of y^+ in Figure 10.3. The parameter y_1^+ is taken to be 30 in the von Karman development. Beyond this distance from the wall the transport is purely turbulent.

For another class of models, the mixing length models, a different approach is used. Here, the turbulent eddy viscosity is assumed to be of the form

$$\nu_{\text{turb}} = \ell^2 \left| \frac{du}{dy} \right| \qquad (10.2.19)$$

where u is the velocity in the direction of mean flow. The parameter ℓ is known as the mixing length and is analogous to the mean free path in the kinetic theory of gases. The physical interpretation of ℓ is that it is the distance over which a turbulent eddy retains its identity. The absolute value of du/dy is required in Eq. 10.2.19 to ensure that the shear stress changes when the flow field changes direction. The problem of estimating ν_{turb} now rests with a method for calculating the mixing length ℓ. From physical considerations, ℓ must be a function of distance from the wall. The simplest model for ℓ is to take it to be proportional to the distance from the wall, as hypothesized by Prandtl

$$\ell^+ = \lambda y^+ \qquad (10.2.20)$$

where we have defined a reduced mixing length

$$\ell^+ = \ell u^*/\nu \qquad (10.2.21)$$

The constant λ in Eq. 10.2.20 is the von Karman constant, equal to 0.4. Though the Prandtl mixing length hypothesis (Eq. 10.2.20) works for conditions in the turbulent core

($y^+ > 30$), it greatly overestimates the values of ℓ^+ closer to the wall, where the solid surface hinders the mixing mechanisms. An important modification of Prandtl's development was introduced by van Driest (1956) who introduced a damping factor

$$\ell^+ = \lambda y^+ (1 - \exp(-y^+/A^+)) \tag{10.2.22}$$

where A^+ is a damping length constant that is interpreted as the distance from the wall beyond which viscous effects are negligible. In the notation used here, A^+ corresponds to y_1^+. van Driest empirically determined the value of $A^+ = y_1^+ = 26$. Introducing Eq. 10.2.19 into Eq. 10.2.13 we find a quadratic expression for du^+/dy^+

$$(\ell^+)^2 \left(\frac{du^+}{dy^+} \right)^2 + \frac{du^+}{dy^+} - 1 = 0 \tag{10.2.23}$$

and so du^+/dy^+ is obtained explicitly as

$$\frac{du^+}{dy^+} = \frac{-1 + \sqrt{1 + 4(\ell^+)^2}}{2(\ell^+)^2} \tag{10.2.24}$$

Equation 10.2.24, in combination with Eq. 10.2.13, allows ν_{turb}/ν to be estimated.

10.3   TURBULENT MASS TRANSFER IN A BINARY FLUID

10.3.1   Solution of the Diffusion Equations

The constitutive relation for j_{iy}, taking account of the molecular diffusion and turbulent eddy contributions, is

$$j_y = -\rho_t (D + D_{\text{turb}}) \frac{d\omega}{dr} \tag{10.3.1}$$

where D is the Fick diffusion coefficient and D_{turb} is the turbulent eddy diffusivity of mass.
With the help of definitions (Eqs. 10.2.6–10.2.8), we may combine Eqs. 10.3.1 and 10.2.3 and write the following expression for the mass flux of component 1 at the wall n_0

$$n_0 = -\rho_t u^* \left(\text{Sc}^{-1} + \text{Sc}_{\text{turb}}^{-1} \frac{\nu_{\text{turb}}}{\nu} \right) \left(1 - \frac{y^+}{R_w^+} \right) \frac{d\omega}{dr} + \omega n_{t0} \tag{10.3.2}$$

For the reasons discussed earlier, we shall assume that the term $(1 - y^+/R_w^+)$ in Eq. 10.3.2 is unity. With this simplification we rewrite Eq. 10.3.2 as follows:

$$\frac{d\omega}{dy^+} = A(y^+)(\omega + \xi) \tag{10.3.3}$$

where $A(y^+)$ is defined by

$$A(y^+) = \frac{n_{t0}}{\rho_t u^*} \left(\text{Sc}^{-1} + \text{Sc}_{\text{turb}}^{-1} \frac{\nu_{\text{turb}}}{\nu} \right)^{-1} \tag{10.3.4}$$

and where ξ is the flux ratio $\xi = n_0/n_{t0}$. Since ξ is a constant we may write

$$\frac{d(\omega + \xi)}{dy^+} = A(y^+)(\omega + \xi) \tag{10.3.5}$$

The solution of this first-order differential equation may be obtained through use of an integrating factor. The solution is

$$(\omega + \xi) = \exp\{-B(y^+)\}C \tag{10.3.6}$$

where

$$B(y^+) = \int_0^{y^+} A(y^+) \, dy^+ \tag{10.3.7}$$

and where C is a constant of integration. To determine C and eliminate ξ from Eq. 10.3.6 we make use of the boundary conditions (Eq. 10.2.9) to obtain the composition profiles as

$$\frac{(\omega - \omega_0)}{(\omega_b - \omega_0)} = \frac{(\exp \Psi - 1)}{(\exp \Phi - 1)} \tag{10.3.8}$$

where the factors Ψ and Φ are defined by

$$\Psi = B(y^+) = \int_0^{y^+} A(y^+) \, dy^+ \tag{10.3.9}$$

$$\Phi = B(y_b^+) = \int_0^{y_b^+} A(y^+) \, dy^+ \tag{10.3.10}$$

Equation 10.3.8 is identical to the corresponding film theory result (Eq. 8.2.9), with Ψ and Φ taking the place of $\Phi\eta$ and Φ in the film model.

The diffusion flux at the wall ($y = 0$) is given by Eq. 10.3.1 with $\nu_{\text{turb}} = 0$ (the turbulent eddy diffusivity dies out near fluid interfaces and is zero at the interface)

$$j_0 = -\rho_t D \left. \frac{d\omega}{dy} \right|_{y=0} \tag{10.3.11}$$

which we rewrite in terms of y^+ as

$$j_0 = -\frac{\rho_t u^* D}{\nu} \left. \frac{d\omega}{dy^+} \right|_{y^+=0} \tag{10.3.12}$$

The composition gradient is obtained by differentiating Eq. 10.3.8 (using the Leibnitz rule for the derivative of an integral)

$$\left. \frac{d\omega}{dy^+} \right|_{y^+=0} = \frac{A(y^+=0)}{(\exp \Phi - 1)} (\omega_b - \omega_0) \tag{10.3.13}$$

Hence, with $A(0) = n_{t0} \, Sc/\rho_t u^*$ (from Eq. 10.3.4) with $\nu_{\text{turb}} = 0$), the diffusion flux is given by

$$j_0 = -n_{t0} \frac{1}{(\exp \Phi - 1)} (\omega_b - \omega_0) \tag{10.3.14}$$

Now, with the mass transfer coefficient $k^\bullet$ defined by

$$j_0 = \rho_t k^\bullet (\omega_0 - \omega_b) \tag{10.3.15}$$

we see that

$$k^\bullet = \frac{\rho_t}{n_{t0}} \{\exp \Phi - 1\}^{-1} \tag{10.3.16}$$

To obtain the low flux mass transfer coefficient k we take the limit as n_{t0} goes to zero

$$k = n_{t0}/\rho_t \Phi \tag{10.3.17}$$

and substitute for Φ using Eqs. 10.3.10 and 10.3.4 to get

$$k^{-1} = \int_0^{y_b^+} \frac{1}{u^*} \left(Sc^{-1} + Sc_{turb}^{-1} \frac{\nu_{turb}}{\nu} \right)^{-1} dy^+ \tag{10.3.18}$$

where Sc is the Schmidt number $Sc = \nu/D$ and Sc_{turb} is the turbulent Schmidt number defined by Eq. 10.1.3. Note that k is not a function of the total flux n_{t0}. The high flux correction factor Ξ given by $k^\bullet = k\Xi$, is

$$\Xi = \Phi/(\exp \Phi - 1) \tag{10.3.19}$$

which is formally identical to the corresponding film theory result (Eq. 8.2.13). The molar flux of component 1 is obtained from the appropriate from the bootstrap solution

$$n_0 = \rho_t \beta_0 k^\bullet (\omega_0 - \omega_b) \tag{10.3.20}$$

β is given by Eq. 7.2.16 with the mole fractions replaced by mass fractions.

10.3.2 Mass Transfer Coefficients

Having completed the formal development of an expression for the evaluation of the interfacial fluxes n_{i0}, we turn to the actual evaluation of the zero-flux mass transfer coefficient k for some specific models of turbulence. It is usual in such developments to define the Stanton number

$$St = k/\bar{u} = \sqrt{f/2}\, k/u^* \tag{10.3.21}$$

We see from Eqs. 10.3.13, 10.3.18, and 10.3.21) that the inverse Stanton number is given by the following expression:

$$St^{-1} = \sqrt{2/f} \int_0^{y_b^+} \left(Sc^{-1} + Sc_{turb}^{-1} \frac{\nu_{turb}}{\nu} \right)^{-1} dy^+ \tag{10.3.22}$$

We now write Eq. 10.3.22 as a sum of two integrals

$$St^{-1} = \sqrt{2/f} \int_0^{y_i^+} \left(Sc^{-1} + Sc_{turb}^{-1} \frac{\nu_{turb}}{\nu} \right)^{-1} dy^+ + \sqrt{2/f} \int_{y_i^+}^{y_b^+} \left(\frac{\nu_{turb}}{\nu} \right)^{-1} dy^+ \tag{10.3.23}$$

where we have made use of the approximation (Eq. 10.2.10), which states that molecular diffusion is of no importance in the turbulent core. We must eliminate the upper limit on the second integral in Eq. 10.3.23 as this distance is not known precisely. To this end we integrate Eq. 10.2.12 (with the right-hand side set to unity) as follows:

$$\int_0^{y_b^+} du^+ = u_b^+$$

$$= \int_0^{y_i^+} \left(1 + \frac{\nu_{turb}}{\nu}\right)^{-1} dy^+ + \int_{y_i^+}^{y_b^+} \left(\frac{\nu_{turb}}{\nu}\right)^{-1} dy^+ \qquad (10.3.24)$$

The above equation can be used to eliminate the second integral on the right-hand side of Eq. 10.3.23. With this substitution and using $u^+ = \sqrt{2/f}$, the reduced bulk flow velocity, we obtain the final working expression for the estimation of St

$$St^{-1} = 2/f + \sqrt{2/f}\int_0^{y_i^+} \left\{ \left(Sc^{-1} + Sc_{turb}^{-1}\frac{\nu_{turb}}{\nu}\right)^{-1} - \left(1 + \frac{\nu_{turb}}{\nu}\right)^{-1}\right\} dy^+ \quad (10.3.25)$$

In order to carry out the integrations required by Eq. 10.3.25 we need to know how Sc_{turb} and the ratio ν_{turb}/ν depend on y^+ (the Schmidt number Sc is assumed constant).

Some special cases may now be derived from the general expression for St (Eq. 10.3.25). When $Sc_{turb} = 1$ and the Schmidt number is assumed to be equal to unity, that is, $Sc = 1$, then the integrand vanishes and we have

$$St = f/2 \qquad (10.3.26)$$

which is Reynolds analogy for mass transfer.

With the help of Eqs. 10.2.17 and 10.2.18, derived from the von Karman universal velocity profile, and assuming $Sc_{turb} = 1$ we may carry out the integrations required by Eq. 10.3.25 to obtain

$$St^{-1} = (2/f) + 5\sqrt{2/f}\left\{Sc - 1 + \ln\left(1 + \tfrac{5}{6}(Sc - 1)\right)\right\} \qquad (10.3.27)$$

an expression sometimes referred to as the von Karman analogy. The variation of Stanton number with Schmidt number is shown in Figure 10.4. Several other analogies have been

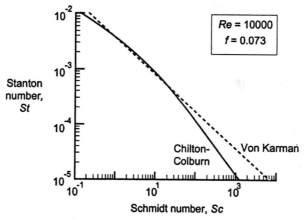

Figure 10.4. Stanton number, (St) as a function of Schmidt number (Sc) from the von Karman velocity profile and the Chilton–Colburn analogy. The friction factor is maintained at 0.073 for this illustration.

proposed based on different models for $u^+(y^+)$ (see the review of Sideman and Pinczewski, 1975).

Insertion of a mixing length model (e.g., the one due to van Driest, Eq. 10.2.22) for ℓ^+ into Eq. 10.2.24 and combining with Eqs. 10.2.13 and 10.3.25 allows calculation of St. In this case, however, numerical integration is required.

A modification of Reynolds analogy to take account of fluids whose Schmidt number is not unity is the Chilton–Colburn analogy

$$\text{St} = \tfrac{1}{2}f \, \text{Sc}^{-2/3} \qquad (10.3.28)$$

which is compared to the von Karman model (Eq. 10.3.27) in Figure 10.4.

Provided f is based on shear friction and not on total drag, Eq. 10.3.28 has been found to hold remarkably well for many types of flow systems and geometries (Sherwood et al., 1975). As originally presented, the Chilton–Colburn analogy was purely empirical. However, if the turbulent eddy viscosity is taken to vary according to (Vieth et al., 1963)

$$\nu_{\text{turb}}/\nu = 1.77\left(\frac{f}{2}\right)^{2/3}(y^+)^3 \qquad (10.3.29)$$

then the Chilton–Colburn analogy may be derived from Eq. 10.3.25.

A clue to the success of the Chilton–Colburn analogy can be found in an analysis carried out by Fletcher et al. (1982). Using a modified van Driest mixing length model, Fletcher et al. showed that for the Prandtl (Schmidt) number greater than unity, this model predicted a $-\tfrac{2}{3}$ power dependence of the Stanton number on the Prandtl (Schmidt) number, in agreement with the empirical Chilton–Colburn assumption. It is, however, interesting to note that the studies of Fletcher et al. shows that for $\text{Pr(Sc)} < 0.1$, the Stanton number varies inversely as the Prandtl (Schmidt) number for the van Driest model. The applicability of the Chilton–Colburn analogy to systems with high molecular diffusivities of a particular binary pair requires careful investigation.

Even the film theory discussed in Chapter 8 falls neatly into the framework provided by Eq. 10.3.22. If we assume that the mass transfer process is governed by molecular transport within an "effective" film of thickness y_b and that the level of turbulence is such as to wash out completely all composition gradients beyond this distance, then we see that

$$\text{St}^{-1} = \sqrt{2/f} \, \text{Sc} \, y_b^+ = \bar{u}y_b/D \qquad (10.3.30)$$

where we have introduced $y_b^+ = y_b\sqrt{f/2}\,\bar{u}/\nu$. The zero-flux mass transfer coefficients $k = D/y_b$, the classic film theory result.

Example 10.3.1   Thin-Film Sulfonation of Dodecyl Benzene

Sulfonation of dodecyl benzene (DDB) is an industrially important process for detergent manufacture. A gas mixture containing SO_3 (1) and N_2 (2) is brought into contact with a thin "film" of DDB inside a cooled tube as shown in Figure 10.5. The gas and liquid phases flow in a cocurrent downward manner. The reaction between SO_3 and DDB occurs instantaneously and, therefore, the bulk overall reaction rate is governed by the mass transfer of SO_3 from the bulk gas phase to the gas–liquid interface. Calculate the flux of SO_3 at the entrance to the tubular reactor.

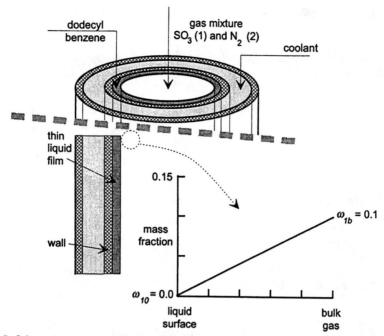

Figure 10.5. Schematic diagram of a thin-film sulfonator. Inset shows composition profiles (mass fraction units) in the region close to the "wall," the surface of the liquid film of dodecyl benzene.

DATA

Tube diameter: $d = 25$ mm.

Gas/vapor temperature at the reactor inlet: $T = 50°C$.

Pressure: $P = 130$ kPa.

Composition of entering gas mixture (mass fraction):

$$\omega_{1b} = 0.1 \qquad \omega_{2b} = 0.9$$

Average molar mass of gas phase: $M = 0.029$ kg/mol.

Gas velocity at the reactor inlet: $u = 30$ m/s.

Gas viscosity: $\mu = 19\ \mu$Pa s.

Diffusivity of SO_3 in N_2 at 50°C and 130 kPa; $D = 12$ mm²/s.

The friction factor inside the tube with the falling liquid film may be calculated from

$$f/2 = 0.023/Re^{0.2}$$

SOLUTION   Since the reaction between SO_3 and DDB is instantaneous, the mass fraction of SO_3 at the gas–liquid interface will be zero, that is,

$$\omega_{10} = 0.0 \qquad \omega_{20} = 1.0$$

We proceed by evaluating the mass density of the gas mixture from

$$\rho_t = MP/RT$$
$$= 0.029 \times 130 \times 10^3/(8.3144 \times 323.15)$$
$$= 1.403 \text{ kg/m}^3$$

The Reynolds number is

$$\text{Re} = \rho_t ud/\mu$$
$$= 0.025 \times 30 \times 1.403/(19 \times 10^{-6})$$
$$= 55,388$$

and we see that the flow is completely turbulent. The friction factor may now be calculated as

$$f = 2 \times 0.023/(55,388)^{0.2}$$
$$= 0.005176$$

The Schmidt number of the gas mixture is

$$\text{Sc} = \mu/\rho_t D$$
$$= 19 \times 10^{-6}/(1.403 \times 12 \times 10^{-6})$$
$$= 1.128$$

The Stanton number may be evaluated from Eq. 10.3.27 as

$$\text{St} = 0.00244$$

and so the low flux mass transfer coefficient is

$$k = u \times \text{St}$$
$$= 30 \times 0.00244 = 0.0734 \text{ m/s}$$

Since the flux of nitrogen is zero we may evaluate the mass transfer rate factor Φ directly as (cf. Eq. 8.2.18)

$$\Phi = \ln(\omega_{2b}/\omega_{20})$$
$$= \ln(0.9/1.0)$$
$$= -0.1054$$

and the high flux mass transfer coefficient is

$$k^\bullet = k\Phi/(e^\Phi - 1)$$
$$= 0.0773 \text{ m/s}$$

The mass flux of SO_3 may now be calculated from

$$n_1 = \rho_t \beta k^\bullet(\omega_{10} - \omega_{1b})$$

where β is the bootstrap coefficient

$$\beta = 1/\omega_{20} = 1$$

and so

$$n_1 = 1.403 \times 1 \times 0.0773 \times (0.0 - 0.1)$$

$$= -0.0108 \text{ kg/m}^2\text{s}$$

The composition profiles in the vapor adjacent to the liquid surfaces are shown in Figure 10.5. ■

10.4 TURBULENT EDDY TRANSPORT IN MULTICOMPONENT MIXTURES

10.4.1 Solution of the Multicomponent Diffusion Equations

The analysis of turbulent eddy transport in binary systems given above is generalized here for multicomponent systems. The constitutive relation for j_{i_y} in multicomponent mixtures taking account of the molecular diffusion and turbulent eddy contributions, is given by the matrix generalization of Eq. 10.3.1

$$(j_y) = -\rho_t[[D] + [D_{\text{turb}}]]\frac{d(\omega)}{dr} \tag{10.4.1}$$

where $[D]$ is the matrix of Fick diffusion coefficients in the mass average velocity reference frame and $[D_{\text{turb}}]$ is the matrix of turbulent eddy diffusivities of mass. Now, since eddy diffusion is not specifies specific, $[D_{\text{turb}}]$ reduces to the form of a scalar times the identity matrix (Toor, 1960; Stewart, 1973)

$$[D_{\text{turb}}] = D_{\text{turb}}[I] \tag{10.4.2}$$

With the help of Eqs. 10.4.1 and 10.4.2 and definitions (Eqs. 10.4.6–10.4.11), we may derive the following expression for the mass fluxes at the wall:

$$(n_0) = -\rho_t u^*\left[[Sc]^{-1} + Sc_{\text{turb}}^{-1}\frac{\nu_{\text{turb}}}{\nu}[I]\right]\left(1 - \frac{y^+}{R_w^+}\right)\frac{d(\omega)}{dr} + (\omega)n_{t0} \tag{10.4.3}$$

where we have introduced a matrix of Schmidt numbers $[Sc] = \nu[D]^{-1}$.

Let us proceed to integrate the differential Eq. 10.4.3. As in our analysis of binary mass transfer we shall approximate the term $(1 - y^+/R_w^+)$ by unity. With this simplification we may rewrite Eq. 10.4.3 as

$$\frac{d(\omega)}{dy^+} = [A(y^+)](\omega + \xi) \tag{10.4.4}$$

where the matrix $[A(y^+)]$ is defined by

$$[A(y^+)] = \frac{n_{t0}}{\rho_t u^*}\left[[Sc]^{-1} + Sc_{\text{turb}}^{-1}\frac{\nu_{\text{turb}}}{\nu}[I]\right]^{-1} \tag{10.4.5}$$

(ξ) is a column matrix of flux ratios $\xi_i = n_{i0}/n_{t0}$. The ξ_i are constants and so we may write

$$\frac{d(\omega + \xi)}{dy^+} = [A(y^+)](\omega + \xi) \tag{10.4.6}$$

The solution to the matrix differential Eq. 10.4.6 can be found using the method of successive substitution (Appendix B.2). Here we follow closely the treatment by Taylor (1981b) (see, also Krishna, 1982). The solution to Eq. 10.4.6 can be written as

$$(\omega + \xi) = \left[\Omega_0^{y^+}(A)\right](\omega_0 + \xi) \tag{10.4.7}$$

where $[\Omega_0^{y^+}(A)]$ is the matrizant defined by Eq. B.2.16. Substituting $\omega = \omega_0$ when $y^+ = 0$ in Eq. 10.4.7 we obtain, after noting that $[\Omega_0^0(A)] = [I]$, the identity matrix,

$$(\omega - \omega_0) = \left[\left[\Omega_0^{y^+}(A)\right] - [I]\right](\omega_0 + \xi) \tag{10.4.8}$$

At the distance $y^+ = y_b^+$, $\omega = \omega_b$ and we have (from Eq. 10.4.8).

$$(\omega_b - \omega_0) = \left[\left[\Omega_0^{y_b^+}(A)\right] - [I]\right](\omega_0 + \xi) \tag{10.4.9}$$

which may be used to eliminate $(\omega_0 + \xi)$ to give

$$(\omega - \omega_0) = \left[\left[\Omega_0^{y^+}(A)\right] - [I]\right]\left[\left[\Omega_0^{y_b^+}(A)\right] - [I]\right]^{-1}(\omega_b - \omega_0) \tag{10.4.10}$$

Let us now consider the evaluation of the matrizant $[\Omega_0^{y_b^+}(A)]$. The matrix $[A(y^+)]$ is given by Eq. 10.4.5 and it is easy to see that the inverse matrix $[A(y^+)]^{-1}$ exhibits a very simple dependence on the position coordinate y^+ ([Sc] is assumed to be constant in our model): only the diagonal elements of $[A(y^+)]^{-1}$ are position dependent. Furthermore, the position dependence is the same for all the principal diagonal elements because ν_{turb}/ν is not species dependent. The matrices $[Sc]^{-1}$, $[[Sc]^{-1} + Sc_{turb}^{-1}(\nu_{turb}/\nu)[I]]$, $[A(y^+)]^{-1}$, and $[A(y^+)]$ all commute with each other and with $\int[A(y^+)]\,dy^+$. All this means is that the integrations required by Eq. B.2.16 can be carried out by parts to give

$$\left[\Omega_0^{y^+}(A)\right] = \exp[\Psi] \tag{10.4.11}$$

where

$$[\Psi] = \int_0^{y^+} [A(y^+)]\,dy^+ \tag{10.4.12}$$

Taking the upper limit of this integral to be y_b^+, the reduced distance from the wall at which the bulk compositions (ω_b) are attained, we define a matrix of rate factors $[\Phi]$ by

$$[\Phi] = \int_0^{y_b^+} [A(y^+)]\,dy^+ \tag{10.4.13}$$

The matrices $[\Psi]$ and $[\Phi]$ assume the roles of Ψ and Φ used earlier. We may write Eq. 10.4.10 in terms of $[\Phi]$ and $[\Psi]$ as follows:

$$(\omega - \omega_0) = [\exp[\Psi] - [I]][\exp[\Phi] - [I]]^{-1}(\omega_b - \omega_0) \tag{10.4.14}$$

where (ω_0) and (ω_b) are the compositions at the wall $(y^+ = 0)$ and in the bulk fluid $(y^+ = y_b^+)$.

The diffusion fluxes at the wall $y = 0$ are given by Eq. 10.4.5 with $\nu_{turb} = 0$

$$(j_0) = -\rho_t[D]\frac{d(\omega)}{dy}\bigg|_{y=0} \tag{10.4.15}$$

which we rewrite in terms of y^+ as

$$(j_0) = -\frac{\rho_t u^*[D]}{\nu}\frac{d(\omega)}{dy^+}\Bigg|_{y^+=0} \tag{10.4.16}$$

The composition gradients are obtained by differentiating Eq. 10.4.14

$$\frac{d(\omega)}{dy^+}\Bigg|_{y^+=0} = [A(y^+=0)][\exp[\Phi]-[I]]^{-1}(\omega_b-\omega_0) \tag{10.4.17}$$

Thus, with $[A(0)] = n_{t0}[\text{Sc}]/\rho_t u^*$ (from Eq. 10.4.5 with $\nu_{\text{turb}} = 0$), the diffusion flux is given by

$$(j_0) = -n_{t0}[\exp[\Phi]-[I]]^{-1}(\omega_b-\omega_0) \tag{10.4.18}$$

The matrix of multicomponent mass transfer coefficients defined by

$$(j_0) = \rho_t[k^\bullet](\omega_0-\omega_b) \tag{10.4.19}$$

and we see that $[k^\bullet]$ is given by

$$[k^\bullet] = \frac{\rho_t}{n_{t0}}[\exp[\Phi]-[I]]^{-1} \tag{10.4.20}$$

Equation 10.4.20 is the matrix generalization of Eq. 10.3.17. To obtain the matrix of low flux mass transfer coefficients $[k]$ we take the limit as n_t goes to zero

$$[k]^{-1} = \rho_t[\Phi]/n_{t0} \tag{10.4.21}$$

and substitute for $[\Phi]$ using Eq. 10.4.13 to get

$$[k]^{-1} = \int_0^{y_b^+}\frac{1}{u^*}\left[[\text{Sc}]^{-1} + \text{Sc}_{\text{turb}}^{-1}\frac{\nu_{\text{turb}}}{\nu}[I]\right]^{-1} dy^+ \tag{10.4.22}$$

The high flux correction factor matrix $[\Xi]$ given by $[k^\bullet] = [k][\Xi]$, is

$$[\Xi] = [\Phi][\exp[\Phi]-[I]]^{-1} \tag{10.4.23}$$

The molar fluxes are obtained from the appropriate form of the bootstrap solution

$$(n_0) = \rho_t[\beta_0][k^\bullet](\omega_0-\omega_b) \tag{10.4.24}$$

with the elements of $[\beta]$ given by Eq. 7.2.16 where the mole fractions are replaced by mass fractions.

10.4.2 Multicomponent Mass Transfer Coefficients

In proceeding with the development of an expression for the evaluation of $[k]$ we define a matrix of Stanton numbers by

$$[\text{St}] = [k]/\bar{u} = \sqrt{f/2}[k]/u^* \tag{10.4.25}$$

Thus, the inverse of [St] is given by the following expression:

$$[St]^{-1} = \sqrt{2/f} \int_0^{y_b^+} \left[[Sc]^{-1} + Sc_{turb}^{-1} \frac{\nu_{turb}}{\nu} [I] \right]^{-1} dy^+ \qquad (10.4.26)$$

As in our development of Eq. 10.3.27 for binary mass transfer, we divide the region $0 - y_b^+$ into two parts, make use of Eq. 10.3.24, and obtain

$$[St]^{-1} = (2/f)[I] + \sqrt{2/f} \int_0^{y_b^+} \left\{ \left[[Sc]^{-1} + Sc_{turb}^{-1} \frac{\nu_{turb}}{\nu} [I] \right]^{-1} - \left[[I] + \frac{\nu_{turb}}{\nu} [I] \right]^{-1} \right\} dy^+ \qquad (10.4.27)$$

When $Sc_{turb} = 1$ and the matrix of Schmidt numbers is assumed to be equal to the identity matrix, that is, $[Sc] = [I]$, then we have the Reynolds analogy for multicomponent mass transfer.

$$[St] = (f/2)[I] \qquad (10.4.28)$$

The requirement $[Sc] = [I]$ for a multicomponent system is a much more special case than for a corresponding binary system for it requires that all binary pair diffusivities in the multicomponent system be equal to one another and, furthermore, that $\nu/D = 1$, a situation realizable only for ideal mixtures made up of species of similar size and nature.

The matrix generalization of the Chilton–Colburn analogy is

$$[St] = \tfrac{1}{2} f [Sc]^{-2/3} \qquad (10.4.29)$$

With the universal velocity profile Eqs. 10.2.14–10.2.16, we obtain the multicomponent form of the von Karman analogy

$$[St]^{-1} = \frac{2}{f}[I] + 5\sqrt{2/f} \left[[Sc] - [I] + \ln\{[I] + \tfrac{5}{6}[Sc - I]\} \right] \qquad (10.4.30)$$

10.4.3 Computational Issues

For calculating [St] from Eq. 10.4.30 we may make use of Sylvester's formula as follows:

$$[St] = \sum_{i=1}^m \widehat{St}_i \left\{ \prod_{\substack{j=1 \\ j \neq i}}^m [[Sc] - \widehat{Sc}_j[I]] \Big/ \prod_{\substack{j=1 \\ j \neq i}}^m \left(\widehat{Sc}_i - \widehat{Sc}_j \right) \right\} \qquad (10.4.31)$$

where m is the number of distinct eigenvalues of [Sc] ($m \leq n - 1$). The eigenvalues of [St] are given by Eq. 10.3.27 with $\widehat{Sc}_i$, the corresponding eigenvalue of [Sc], replacing the binary Schmidt number.

Alternatively, we may use the familiar similarity transformation

$$[St] = [P][\widehat{St}][P]^{-1} \qquad (10.4.32)$$

where $[\widehat{St}]$ is a diagonal matrix whose nonzero elements are the eigenvalues of [St]. The matrix $[P]$ is the modal matrix of [Sc] *and* of [D].

To calculate $[k]$ we would normally not bother to compute the matrix $[St]$, only its eigenvalues. Equations 8.4.24 or 8.4.31 may be used directly as written to compute the multicomponent mass transfer coefficients with

$$\hat{k}_i^{\bullet} = \hat{k}_i \hat{\Xi}_i \tag{8.4.25}$$

where

$$\hat{k}_i = \widehat{St}_i \bar{u} \tag{10.4.33}$$

$$\hat{\Phi}_i = n_t / \rho_t \hat{k}_i \tag{10.4.34}$$

$$\hat{\Xi}_i = \hat{\Phi}_i / \left(\exp \hat{\Phi}_i - 1 \right) \tag{10.4.35}$$

are the eigenvalues of $[k]$, $[\Phi]$, and $[\Xi]$, respectively.

An alternative technique may sometimes be useful for computing $[St]$ from Eq. 10.4.30 based on a series expansion of the natural logarithm of a matrix. The method is especially useful if no method of computing the eigenvectors is readily available.

The function $\ln(1 + x)$ may be evaluated from the series

$$\ln(1 + x) = x - \tfrac{1}{2}x^2 + \tfrac{1}{3}x^3 - \tfrac{1}{4}x^4 + \cdots \tag{10.4.36}$$

which is convergent as long as x lies between -1 and $+1$. The matrix generalization of the series representation of $\ln(1 + x)$

$$\ln[[I] + [A]] = [A] - \tfrac{1}{2}[A]^2 + \tfrac{1}{3}[A]^3 - \tfrac{1}{4}[A]^4 + \cdots \tag{10.4.37}$$

may be used as long as the absolute value of all the eigenvalues of $[A]$ is less than or equal to unity.

We may identify $[A]$ as the matrix $\tfrac{5}{6}[[Sc] - [I]]$. If the matrix $\tfrac{5}{6}[[Sc] - [I]]$ will have eigenvalues lying between -1 and $+1$ then the term $\ln[[I] + \tfrac{5}{6}[[Sc] - [I]]]$ may be evaluated from the convergent series (Eq. 10.4.37). The matrix $[St]^{-1}$ may then be computed from Eq. 10.4.30 using only elementary matrix computations. For gas mixtures the eigenvalues of $[Sc]$ are of order 1 and this procedure can, therefore, be used.

For computation of the fluxes themselves, Algorithms 8.1, 8.4, and 8.5 may be used more or less as written; simply replace all quantities in molar units by the corresponding quantities in the turbulent eddy diffusivity model.

Example 10.4.1   Methanation in a Tube Wall Reactor

The methanation reaction

$$CO(1) + 3H_2(2) \rightarrow CH_4(3) + H_2O(4)$$

takes place in a circular tube with a coating of the catalyst on the inside wall as illustrated in Figure 10.6. The pressure at the entrance to the reactor is 2.1 MPa and the temperature is 658 K. The reaction between CO and H_2 may be assumed to be instantaneous at the prevailing operating conditions. Estimate the rate of production of methane at the entrance of the tubular reactor where the bulk gas mole fractions are

$$y_{1b} = 0.10 \qquad y_{2b} = 0.82 \qquad y_{3b} = 0.04 \qquad y_{4b} = 0.04$$

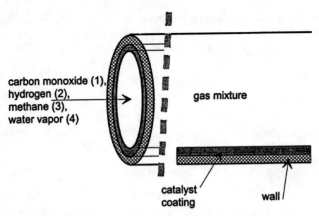

Figure 10.6. Schematic diagram of methanation reactor showing catalyst coating on the inside of the reactor tube wall.

DATA

Tube diameter: $d = 0.05$ m.

Gas velocity at the reactor inlet: $u = 1.5$ m/s.

Gas viscosity: $\mu = 1 \times 10^{-5}$ Pa s.

The Fanning friction factor: $f = 0.006$.

Maxwell–Stefan diffusivities of the binary pairs

$$\mathcal{D}_{12} = \mathcal{D}_{23} = \mathcal{D}_{24} = 13.5 \times 10^{-6} \text{ m}^2/\text{s}$$

$$\mathcal{D}_{13} = \mathcal{D}_{14} = \mathcal{D}_{34} = 4.0 \times 10^{-6} \text{ m}^2/\text{s}$$

ANALYSIS   We are asked to determine the rate of production of methane. We shall use the turbulent eddy diffusivity model to represent the transport processes in the gas phase. The mass fluxes are given by Eq. 10.4.24

$$(n_0) = \rho_t[\beta_0][k^\bullet](\omega_0 - \omega_b) \tag{10.4.24}$$

with the elements of $[\beta]$ given by Eq. 7.2.16 where the mole fractions are replaced by mass fractions. The high flux mass transfer coefficients are functions of the multicomponent Fick diffusion coefficients, the system hydrodynamics, and the total mass flux. The Fick diffusion coefficients are a function of the mixture composition and the Maxwell–Stefan diffusion coefficients.

Since the reaction is assumed to be instantaneous and hydrogen is present in excess of the stoichiometric requirements (stoichiometric ratio of $H_2 : CO$ is $3 : 1$ but the bulk gas contains the same components in the ratio $8.2 : 1$), the mole fraction of carbon monoxide must be zero at the tube wall catalytic surface:

$$y_{10} = 0.0$$

The four molar fluxes are related through the reaction stoichiometry as follows:

$$N_1 = -N_3 \qquad N_2 = -3N_3 \qquad N_4 = N_3$$

The total molar flux is the sum of the N_i and can be expressed as

$$N_t = N_1 + N_2 + N_3 + N_4$$
$$= -N_3 - 3N_3 + N_3 + N_3$$
$$= -2N_3$$

Thus, there is only one flux to be determined; we shall take that to be the flux of methane. We may make use of the following relationship between mass and molar fluxes

$$n_i = N_i M_i$$

and the stoichiometric relations discussed above to express the total mass flux as follows:

$$n_t = n_1 + n_2 + n_3 + n_4$$
$$= N_1 M_1 + N_2 M_2 + N_3 M_3 + N_4 M_4$$
$$= -N_3 M_1 - 3N_3 M_2 + N_3 M_3 + N_3 M_4$$
$$= N_3(M_3 + M_4 - M_1 - 3M_2)$$

The term $(M_3 + M_4 - M_1 - 3M_2)$ is identically zero; mass is conserved in a chemical reaction even though, as in this case, moles may not be. Thus, the total mass flux is zero. This means that the high flux correction factor matrix and the bootstrap matrix simplify to identity matrices and the mass fluxes are given by the simpler expression

$$(n_0) = \rho_t[k](\omega_0 - \omega_b)$$

With the matrix of low flux mass transfer coefficients related to the matrix of Stanton numbers by Eq. 10.4.25 we may write

$$(n_0) = \rho_t u[\text{St}](\omega_0 - \omega_b)$$

The unknown variables to be determined are the mass flux of methane and the mass fractions at the catalyst surface

$$n_3, \omega_{20}, \omega_{30}, \text{ and } \omega_{40}$$

The mass fractions at the interface are constrained by the summation equation

$$\omega_{10} + \omega_{20} + \omega_{30} + \omega_{40} = 1$$

with $\omega_{10} = 0$.

The method of solution adopted in this case was to search for the value of the mass flux of methane that makes $\omega_{10} = 0$. Thus, instead of computing fluxes from the rate equations knowing the mass fractions in the bulk and at the interface, we solve the linear system

$$[\text{St}]^{-1}(n_0) = \rho_t u(\omega_0 - \omega_b)$$

for the mass fractions at the interface knowing only the bulk composition and having guessed the mass flux of methane. The correct value of the mass flux of methane has been obtained when we compute $\omega_{10} = 0$ (or as close to zero as our chosen convergence tolerance will permit). This procedure is fairly simple to converge because the total mass flux is zero.

SOLUTION   The first step is to convert the bulk phase mole fractions to mass fractions. This conversion is done with the help of the relations in Table 1.1 with the result

$$\omega_{1b} = 0.4816 \qquad \omega_{2b} = 0.2842 \qquad \omega_{3b} = 0.1103 \qquad \omega_{4b} = 0.12389$$

In order to calculate the composition at the interface we need to estimate the multicomponent Fick diffusion coefficients. This is usually done at the average composition for which the composition at the interface is required. We shall begin our illustration with the following values

$$\omega_{10} = 0 \qquad \omega_{20} = 0.2112 \qquad \omega_{30} = 0.3716 \qquad \omega_{40} = 0.4172$$

which allows us to determine the average mass fractions as

$$\omega_1 = 0.2408 \qquad \omega_2 = 0.2477 \qquad \omega_3 = 0.2409 \qquad \omega_4 = 0.2706$$

The average mole fractions are

$$y_1 = 0.0532 \qquad y_2 = 0.7608 \qquad y_3 = 0.0930 \qquad y_4 = 0.0930$$

The average mole fractions were obtained by converting the average mass fractions and not by averaging the mole fractions at the interface and in the bulk fluid.

The average molecular weight of the gas mixture is 0.00619 kg/mol. The mass density of the gas mixture is estimated from the ideal gas law as

$$\rho_t = MP/RT$$
$$= 0.00619 \times 2.1 \times 10^6/(8.3143 \times 658)$$
$$= 2.3766 \text{ kg/m}^3$$

The procedure for computing the multicomponent Fick diffusion coefficients in the mass average velocity reference frame was illustrated in Example 4.2.5 and the steps shown there have been repeated for this example with the result

$$[D] = \begin{bmatrix} 8.4596 & -1.5968 & 0.0515 \\ 0.4673 & 13.598 & -0.1610 \\ -0.1497 & -1.5978 & 8.6607 \end{bmatrix} \times 10^{-6} \text{ m}^2/\text{s}$$

The eigenvalues of this matrix are found to be

$$\hat{D}_1 = 8.609 \times 10^{-6} \text{ m}^2/\text{s}$$
$$\hat{D}_2 = 13.50 \times 10^{-6} \text{ m}^2/\text{s}$$
$$\hat{D}_3 = 8.609 \times 10^{-6} \text{ m}^2/\text{s}$$

The matrix of multicomponent mass transfer coefficients may be evaluated from Eq. 10.4.30 as

$$[St]^{-1} = \frac{2}{f}[I] + 5\sqrt{2/f}\left[[Sc] - [I] + \ln\{[I] + \tfrac{5}{6}[Sc - I]\}\right] \qquad (10.4.30)$$

where [Sc], the matrix of Schmidt numbers defined by $[Sc] = \nu[D]^{-1}$, has values

$$[Sc] = \begin{bmatrix} 0.4942 & 0.0578 & -0.0019 \\ -0.0169 & 0.3081 & 0.0058 \\ 0.0054 & 0.0578 & 0.4869 \end{bmatrix}$$

It is possible to check that the matrix $\frac{5}{6}[[Sc] - [I]]$ has eigenvalues that lie between 0 and +1 and the term $\ln[[I] + \frac{5}{6}[[Sc] - [I]]]$ may, therefore, be evaluated from the convergent series (Eq. 10.4.37) as

$$\ln\left[[I] + \tfrac{5}{6}[[Sc] - [I]]\right] = \begin{bmatrix} -0.5461 & 0.0970 & -0.0031 \\ -0.0284 & -0.8583 & 0.0098 \\ 0.0091 & 0.0971 & -0.5584 \end{bmatrix}$$

Seventeen terms of the series were required to converge all elements of the matrix to five significant digits.

The inverse of the Stanton number matrix may now be computed from Eq. 10.4.30 as

$$[St]^{-1} = \begin{bmatrix} 237.302 & 14.134 & -0.456 \\ -4.136 & 191.820 & 1.425 \\ 1.324 & 14.143 & 235.522 \end{bmatrix}$$

The matrix of mass transfer coefficients may be computed by inverting this matrix and multiplying by the velocity. Although we do not actually need to carry out this computation, we have included its result for illustrative purposes

$$[k] = \begin{bmatrix} 6.3128 & -0.4663 & 0.0150 \\ 0.13645 & 7.8133 & -0.0470 \\ -0.0437 & -0.4666 & 6.3716 \end{bmatrix} \times 10^{-3} \text{ m/s}$$

In order to compute the interface composition we need an estimate of the fluxes. We will take the flux n_3 to be

$$n_3 = 4.086556 \times 10^{-3} \text{ kg/m}^2\text{s}$$

from which we may compute the remaining mass fluxes using the stoichiometric relations discussed above as

$$n_1 = -0.00713 \qquad n_2 = -0.00154 \qquad n_4 = 0.00459 \text{ kg/m}^2\text{s}$$

Armed with these values we solve the matrix equation

$$[St]^{-1}(n_0) = \rho_t u(\omega_0 - \omega_b)$$

for the interface mass fractions. This calculation gives

$$\omega_{10} = 0.0 \qquad \omega_{20} = 0.2112 \qquad \omega_{30} = 0.3716 \qquad \omega_{40} = 0.4172$$

which, in fact, is the estimate that we started with.

The interface composition in mole fraction units is

$$y_{10} = 0.0 \qquad y_{20} = 0.69345 \qquad y_{30} = 0.15328 \qquad y_{40} = 0.15328$$

and the rate of production (molar flux) of methane is

$$N_3 = 0.25373 \text{ mol/m}^2\text{s}$$

The composition profiles in the gas close to the catalyst surface are shown in Figure 10.7. The profiles are linear because the total mass flux n_t is zero.

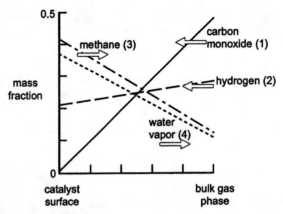

Figure 10.7. Composition profiles (mass fraction units) in the gas close to the catalyst surface. The profiles are linear because the total mass flux n_t is zero.

We have illustrated the calculations necessary to solve this problem by providing first estimates of the flux of methane and of the interface mass fractions that were the converged values. Thus, all the numerical results given above represent the final values of the corresponding quantities. When we solved this problem with no knowledge of the interface composition, our first step was to set the average mass fractions equal to the bulk mass fractions in order to estimate the Fick diffusion coefficients. The first computed values of the interface mass fractions are quite close to the final solution and determining the mass flux of methane necessary to zero the interface mass fraction of CO is quite simple, since ω_1 is almost a linear function of n_3. ∎

10.4.4 Comparison of the Chilton–Colburn Analogy with Turbulent Eddy Diffusivity Based Models

A fundamental shortcoming of the Chilton–Colburn approach for multicomponent mass transfer calculations is that the assumed dependence of $[k]$ on $[Sc]$ takes no account of the variations in the level of turbulence, embodied by ν_{turb}/ν, with variations in the flow conditions. The reduced distance y^+ is a function of the Reynolds number: $y^+ = (y/R_w)(f/8)^{1/2}$ Re; consequently, Re affects the reduced mixing length ℓ^+ defined by Eq. 10.2.21. An increase in the turbulence intensity should be reflected in a relative decrease in the influence of the molecular transport processes. So, for a given multicomponent mixture the increase in the Reynolds number should have the direct effect of reducing the effect of the phenomena of molecular diffusional coupling. That is, the ratios of mass transfer coefficients k_{12}/k_{11} and k_{21}/k_{22} should decrease as Re increases.

In Figure 10.8 we have plotted the variation of the ratios of mass transfer coefficients k_{12}/k_{11} and k_{21}/k_{22} for an acetone–benzene–helium system considered in Example 11.5.3. The Chilton–Colburn analogy predicts that these ratios would be independent of Re, as shown by the horizontal lines in Figure 10.8. The von Karman turbulent model, on the other hand, predicts that the influence of coupling should decrease with increase in Re. The latter trend is in accord with our physical intuition. Depending on the driving forces for mass transfer, the Chilton–Colburn and the von Karman turbulent models could predict different directions of transfer of acetone (see, e.g., Krishna, 1982).

The influence of turbulence intensity on the extent of diffusional coupling, as portrayed in Figure 10.8, is exactly analogous to the influence of the Fourier number Fo on the extent

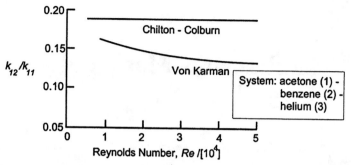

Figure 10.8. Ratio k_{12}/k_{11}, which are elements of the zero-flux matrix of mass transfer coefficients $[k]$, as a function of the gas-phase Reynolds number. Mass transfer between a gaseous mixture of acetone (1)–benzene (2)–helium (3) and a liquid film containing acetone and benzene. Calculations by Krishna (1982) based on the von Karman turbulent film model and the Chilton–Colburn analogy.

of diffusional coupling for transfer inside a rigid droplet, considered earlier (Example 9.4.1). We conclude that the precise description of the mechanism of mass transfer is important in the estimation of mass transfer rates in multicomponent systems for systems in which the $[D]$ matrix has sizeable off-diagonal elements. The calculation of the flux of a component with a relatively small driving force is particularly sensitive to the choice of the model for describing the multicomponent mass transfer process.

11 Simultaneous Mass and Energy Transfer

> Accurate prediction of heat and mass transfer rates in applications of practical interest is still a formidable problem.
>
> —E. N. Lightfoot (1969)

Perfectly isothermal systems are rare in chemical engineering practice and many processes, such as distillation, gas absorption, stripping, condensation, and evaporation, involve the simultaneous transfer of mass and energy across fluid–fluid interfaces. Representative temperature profiles in some nonisothermal processes are shown in Figure 11.1. The temperature profile also has a large influence in chemically reacting systems. For non-isothermal systems it is important to consider simultaneous heat transfer even though we are primarily interested in the mass transfer process.

The purpose of this chapter is to present a general framework for dealing with the effect of mass transfer on heat transfer and the effect of heat transfer on mass transfer. Applications to distillation operations are included in this chapter; mass and energy transfer in multicomponent condensation is considered in Chapter 15.

11.1 BALANCE EQUATIONS FOR SIMULTANEOUS HEAT AND MASS TRANSFER

For transfer in either fluid phase of the two-phase system considered in Figure 1.1, the differential energy balance relation in Table 1.5 provides the additional physical law necessary to determine the temperature profiles and energy fluxes. This balance relationship may be rewritten in several alternative, equivalent, forms (see Bird et al., 1960). Two useful forms of the energy balance relation, assuming mechanical equilibrium, are in terms of the partial molar enthalpies $\bar{H}_i$.

$$\frac{\partial \sum\limits_{i=1}^{n} c_i \bar{H}_i}{\partial t} = -\nabla \cdot \left\{ q + \sum_{i=1}^{n} \bar{H}_i N_i \right\} \tag{11.1.1}$$

or

$$\frac{\partial \{c_t C_p T\}}{\partial t} = -\nabla \cdot q - \nabla \cdot \{c_t C_p T u\} - \sum_{i=1}^{n} J_i \cdot \nabla \bar{H}_i \tag{11.1.2}$$

where q is the conductive heat flux.

In addition to the differential balance relationships (Eqs. 11.1.1 and 11.1.2) that are valid in the bulk phases, we have the jump balance condition at the interface "I"

$$\xi \cdot E^x = \xi \cdot E^y \tag{11.1.3}$$

where E^x and E^y are the energy fluxes in the adjoining phases at the interface. Equation 11.1.3 states that the normal component of the energy fluxes must be continuous across the

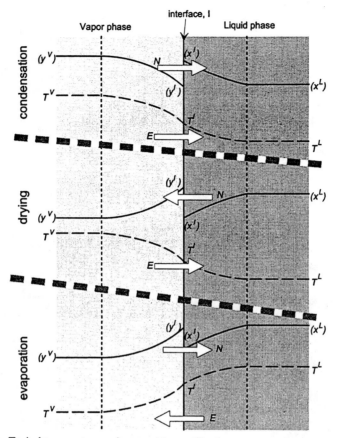

Figure 11.1. Typical temperature and composition profiles in some nonisothermal processes.

phase boundary I and is entirely analogous to Eq. 1.3.14 for mass transfer. In either fluid phase the energy flux takes the form

$$E = q + \sum_{i=1}^{n} \bar{H}_i N_i \tag{11.1.4}$$

where E plays a role analogous to the molar fluxes N_i in the interphase energy-transfer process. The conductive heat flux q plays a role analogous to the molar diffusion fluxes J_i. We shall be making further use of these analogies later on.

For turbulent flow conditions, on time averaging the differential energy balance relations, we note that the time smoothed heat flux q (caused by molecular transport processes) is augmented by a turbulent contribution q_{turb}.

11.2 CONSTITUTIVE RELATIONS FOR SIMULTANEOUS HEAT AND MASS TRANSFER

The most appropriate starting point for setting up the constitutive relation for q is the theory of irreversible thermodynamics and the expression for the rate of production of entropy due to mass and energy-transfer processes (see Hirschfelder et al., 1964;

Standart et al., 1979).

$$\sigma = q \cdot \nabla\left(\frac{1}{T}\right) - \frac{c_t R}{n} \sum_{i=1}^{n} \sum_{j=1}^{n} d_i \cdot (u_i - u_j) \tag{11.2.1}$$

For diffusion in isothermal multicomponent systems the generalized driving force d_i was written as a linear function of the relative velocities $(u_i - u_j)$. In the general case, we must allow for coupling between the processes of heat and mass transfer and write constitutive relations for d_i and q in terms of the $(u_i - u_j)$ and $\nabla(1/T)$. With this allowance, the complete expression for the conductive heat flux is

$$q = -\lambda\nabla T + c_t RT \sum_{\substack{i=1}}^{n} \sum_{\substack{j=1 \\ j\neq i}}^{n} \frac{x_i x_j}{\mathcal{D}_{ij}}\left(\frac{D_i^T}{\rho_i}\right)(u_i - u_j)$$

$$= -\lambda\nabla T + \frac{1}{2}c_t RT \sum_{\substack{i=1}}^{n} \sum_{\substack{j=1 \\ j\neq i}}^{n} \frac{x_i x_j}{\mathcal{D}_{ij}}\left(\frac{D_i^T}{\rho_i} - \frac{D_j^T}{\rho_j}\right)(u_i - u_j) \tag{11.2.2}$$

where λ is the thermal conductivity of the mixture and D_i^T is the thermal diffusion coefficient of component i. It is convenient to define multicomponent thermal diffusion factors

$$\alpha_{ij} = \frac{1}{\mathcal{D}_{ij}}\left(\frac{D_i^T}{\rho_i} - \frac{D_j^T}{\rho_j}\right) \qquad i \neq j = 1, 2, \ldots, n \tag{11.2.3}$$

which have the antisymmetric property

$$\alpha_{ij} = -\alpha_{ji} \qquad i \neq j = 1, 2, \ldots, n \tag{11.2.4}$$

where α_{ii} is undefined. With this definition of α_{ij}, we may write

$$q = -\lambda\nabla T + \frac{1}{2}c_t RT \sum_{\substack{i=1}}^{n} \sum_{\substack{j=1 \\ j\neq i}}^{n} x_i x_j \alpha_{ij}(u_i - u_j) \tag{11.2.5}$$

The second term on the right-hand side of Eq. 11.2.5 gives the contribution of the mass fluxes to the heat flux q; this contribution is commonly referred to as the Dufour effect.

The inverse of the Dufour effect is the production of mass fluxes due to temperature gradients; this is referred to as thermal diffusion or the Soret effect. To account for this effect, we need to augment the generalized Maxwell–Stefan diffusion equations in the following manner:

$$d_i = -\sum_{\substack{j=1 \\ j\neq i}}^{n} \frac{x_i x_j(u_i - u_j)}{\mathcal{D}_{ij}} - \sum_{\substack{j=1 \\ j\neq i}}^{n} x_i x_j \alpha_{ij}\frac{\nabla T}{T} \tag{11.2.6}$$

where the generalized driving force d_i is defined by Eq. 2.3.12

$$c_t RT d_i \equiv c_i \nabla_{T,P}\mu_i + (\phi_i - \omega_i)\nabla P - \left(c_i F_i - \omega_i \sum_{j=1}^{n} c_j F_j\right) \tag{2.3.12}$$

If we substitute these expressions into Eq. 11.2.1 for the rate of entropy production per unit volume we obtain (after some manipulation) the remarkably simple expression

$$\sigma_{\text{diff}} = \lambda \left| \frac{\nabla T}{T} \right|^2 + \frac{1}{2} c_t R \sum_{i=1}^{n} \sum_{j=1}^{n} \frac{x_i x_j}{\mathcal{D}_{ij}} \left| (u_i - u_j) \right|^2 \geq 0 \qquad (11.2.7)$$

This leads us to conclude that $\lambda \geq 0$, in addition to the constraint (Eq. 2.3.21) for the positive definiteness of the Maxwell–Stefan diffusion coefficients at infinite dilution $\mathcal{D}_{ij}^{\circ}$.

Equations 11.2.5 and 11.2.6 are the complete forms of the constitutive relations for simultaneous mass and energy transport. The reader is referred to the treatise by Hirschfelder et al. (1964) and to the papers by Merk (1960) and Standart et al. (1979) for further background to these derivations.

The interactions between thermal and mass fluxes have been recognized for a long time. A great deal of effort has been devoted to the study of thermal diffusion and a number of reviews are available on the subject (Grew and Ibbs, 1962; Grew, 1969; Mason et al., 1966). Its practical application has been highlighted by the successful application of a thermal diffusion process to the separation of isotopes (Glasstone, 1958). Strong coupling effects may be found in processes involving the transport of "heavy" species in very steep temperature gradients; conditions encountered, for example, in ablation cooling during space vehicle reentry, chemical vapor deposition (CVD) processes, and aerosol capture applications. Rosner (1980) discusses the influence of thermal diffusion (Soret) effects on interfacial mass transport rates in some of these (and other) cases. The effect of thermal diffusion in nonisothermal gas absorption has been discussed by DeLancey and Chiang (1968, 1970a, b) (see, also, DeLancey, 1972). It must be said that the coupling between thermal and mass fluxes is of little significance in the classical unit operations, such as distillation and condensation, since temperature gradients are rarely high enough, or sustained long enough, to make the thermal contributions to the mass fluxes worth considering. The Dufour and Soret effects, quantified by the members on the right of Eqs. 11.2.5 and 11.2.6, will, henceforth, be ignored. With this simplification, Eq. 11.2.6 reduces to Eq. 2.3.17 and Eq. 11.2.5 reduces to

$$q = -\lambda \nabla T \qquad (11.2.8)$$

which is Fourier's law of heat conduction.

For the turbulent heat flux q_{turb} we may write

$$q_{\text{turb}} = -\lambda_{\text{turb}} \nabla T \qquad (11.2.9)$$

The estimation procedure for λ_{turb} is analogous to that of D_{turb} and the approach is to proceed via the turbulent Prandtl number

$$\text{Pr}_{\text{turb}} = \frac{C_p \mu_{\text{turb}}}{\lambda_{\text{turb}}} \qquad (11.2.10)$$

For lack of better information, we must take $\text{Pr}_{\text{turb}} = \text{Sc}_{\text{turb}} = 1$.

11.3   DEFINITION OF HEAT TRANSFER COEFFICIENTS

By analogy with Eq. 7.1.3, we may define heat transfer coefficients, in either fluid phase, by

$$h = \lim_{N_i \to 0} \left[\frac{E - \sum_{i=1}^{n} \bar{H}_i N_i}{c_t C_p (T_b - T_I)} = \frac{q}{c_t C_p \, \Delta T} \right] \qquad (11.3.1)$$

where h is a heat transfer coefficient. We have defied convention in the above definition for one very good reason: the heat transfer coefficient h has the dimensions of velocity, just as for the mass transfer coefficient. The physical significance of h is simple and straightforward. It represents the maximum speed at which heat can be transferred in the phase under consideration.

It is more conventional to define the heat transfer coefficients as

$$h = \lim_{N_i \to 0} \left(\frac{q}{\Delta T} \right) \qquad (11.3.2)$$

but with this definition the parallelism with the mass transfer coefficient definition is somewhat lost. We shall proceed further with the conventional definition because it is firmly enshrined in all textbooks. The coefficient defined by Eq. 11.3.2 corresponds to the heat transfer coefficient under conditions of negligible mass fluxes, that is, zero flux heat transfer coefficients.

Just as finite mass fluxes distort the composition profiles during the interphase mass transfer process, they exert a similar effect on the temperature profiles and the interfacial heat fluxes. Witness the presence of the term

$$\sum_{i=1}^{n} N_i \overline{H}_i$$

in the differential energy balance relation (Eq. 11.1.1).

For the actual conditions of finite mass transfer rates, we have

$$h^{\bullet} = \frac{q}{\Delta T} \qquad (11.3.3)$$

where the superscript $\bullet$ serves to remind us that the transfer coefficient corresponds to conditions of finite mass transfer rates. The finite-flux heat transfer coefficients are related to the zero-flux coefficients by a relationship of the form

$$h^{\bullet} = h \Xi_H \qquad (11.3.4)$$

where Ξ_H is the correction factor to account for the effect of finite mass fluxes on the heat transfer coefficient h. Both h and Ξ_H depend on the temperature and composition profiles in the region adjacent to the interface which, in turn, are influenced by the hydrodynamics prevalent in the phase.

11.4 MODELS FOR SIMULTANEOUS MASS AND ENERGY TRANSFER

The presence of temperature gradients in a multicomponent system introduces an additional complication in the analysis of the mass transfer process; such gradients influence the values of physical, thermodynamic, and transport properties, such as the diffusion coefficients. These property variations may be taken care of by introducing temperature dependent property functions or by using average values of the properties (as is done here). The consequence of this simplification is that the basic mass transfer analysis remains essentially unchanged from those in Chapters 8–10 and we need only consider the effect of mass transfer on the heat transfer process.

11.4.1 The Film Model

In the film model we assume that all the resistance to mass and heat transfer is concentrated in a thin film and that transfer occurs within this film by steady-state diffusion and

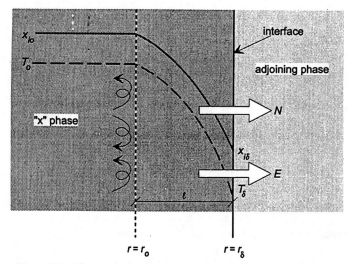

Figure 11.2. Film model for simultaneous mass and energy transfer.

heat conduction. In the bulk fluid all concentration and temperature gradients are wiped out. Figure 11.2 illustrates the model.

For steady-state heat transfer within a planar film, the energy balance relation (Eq. 11.1.1) simplifies to

$$\frac{dE}{dr} = 0 \tag{11.4.1}$$

that is, the energy flux E is constant through the film

$$E = E_0 = E_\delta = \text{constant} \tag{11.4.2}$$

(Recall our earlier analysis of the mass transfer process in Section 8.1 when we found that the N_i were constant through the film.) The energy flux E is related to the conductive heat flux by Eq. 11.1.4 which, when combined with Eq. 11.2.8, simplifies to

$$E = -\lambda \frac{dT}{dr} + \sum_{i=1}^{n} N_i \bar{H}_i \tag{11.4.3}$$

If the reference state for the calculation of the partial molar enthalpies $\bar{H}_i$ is taken to be the pure component at temperature T_{ref}, then we may write

$$\bar{H}_i = C_{pi}(T - T_{\text{ref}}) \tag{11.4.4}$$

where C_{pi} is the molar heat capacity of the species i and is assumed independent of temperature.

Equation 11.4.3 is to be solved subject to the boundary conditions of a film model

$$\begin{array}{lll} r = r_0 & \eta = 0 & T = T_0 \\ r = r_\delta & \eta = 1 & T = T_\delta \end{array} \tag{11.4.5}$$

where η is a dimensionless distance defined as in Section 8.2.

In proceeding further, it is convenient to define a heat transfer rate factor

$$\Phi_H = \frac{\sum\limits_{i=1}^{n} N_i C_{pi}}{\lambda/\ell} \tag{11.4.6}$$

where ℓ is the characteristic diffusion path length, which is equal to the film thickness for a plane film.

With definitions (Eqs. 11.4.4 and 11.4.6) we may write Eq. 11.4.3 in the form

$$\frac{d(T - T_{\text{ref}})}{d\eta} = \Phi_H(T - T_{\text{ref}}) - \frac{E}{\lambda/\ell} \tag{11.4.7}$$

Assuming constant thermal conductivity and heat capacities (in practice this means evaluating them at the average temperature), the differential Eq. 11.4.7 can be solved with the boundary conditions (Eqs. 11.4.5) to yield the temperature profile

$$\frac{(T - T_0)}{(T_\delta - T_0)} = \frac{\exp(\Phi_H \eta) - 1}{\exp(\Phi_H) - 1} \tag{11.4.8}$$

The conductive heat flux at $\eta = 0$, q_0 may be evaluated from

$$q_0 = -\frac{\lambda}{\ell}\frac{dT}{d\eta}\bigg|_{\eta=0} \tag{11.4.9}$$

and, with the temperature gradient determined by differentiating Eq. 11.4.8, we may identify the low flux heat transfer coefficient h and high flux correction factor Ξ_H, defined by

$$q_0 = h\Xi_H(T_0 - T_\delta) = h^\bullet(T_0 - T_\delta) \tag{11.4.10}$$

as

$$h = \lambda/\ell \tag{11.4.11}$$

and

$$\Xi_H = \Phi_H/(\exp \Phi_H - 1) \tag{11.4.12}$$

where Ξ_H is called the Ackermann correction factor (Ackermann, 1937). Note that the correction factor for heat transfer behaves in exactly the same way that the corresponding correction factor for mass transfer behaves (see Eqs. 8.2.13 and Figure 8.2). Thus, for vanishingly small mass transfer fluxes, that is, $N_i \to 0$, Ξ_H reduces to unity, leading to linear temperature profiles; for $\Phi_H > 0$, Ξ_H is less than unity, whereas for $\Phi_H < 0$, Ξ_H is greater than unity. At very high mass transfer rates (as sometimes encountered in, e.g., spray vaporization, or coal gasification processes when small droplets or particles are introduced to a high temperature environment) the high flux heat transfer correction factor may be an order of magnitude different from the low flux heat transfer coefficient.

The energy flux E can be calculated from

$$E = q_0 + \sum_{i=1}^{n} N_i H_i$$

$$= h^\bullet(T_0 - T_\delta) + h\Phi_H(T_0 - T_{\text{ref}}) \tag{11.4.13}$$

which shows that the value of E depends on the arbitrary choice of the reference temperature T_{ref} for the calculation of enthalpies. This is no drawback because, in practice, we use the interfacial energy balance $E^y = E^x$, and the choice of T_{ref} is of no importance. The film model for estimation of the conductive flux q must actually be seen in the context of our model independent analysis for the determination of N_i and E in Section 11.5.

The above analysis was first carried out by Ackermann (1937) and by Colburn and Drew (1937) and forms the basis for, among other things, the design of condensers, a topic we shall address in more detail in Chapter 15.

Example 11.4.1   Heat Transfer in Diffusional Distillation

In Example 8.3.2 we determined the rates of mass transfer in diffusional distillation, a process described by Fullarton and Schlünder (1983) for separating liquid mixtures of azeotropic composition. Estimate the heat flux through the gas/vapor mixture under the conditions prevailing in the experiment described in Example 8.3.2.

DATA

Temperature of evaporating liquid = 40°C.
Temperature of condensing liquid = 15°C.
Annular gap (film thickness): ℓ = 6.5 mm.
Molar fluxes of isopropanol (1) and water (2) in air (3) were found in Example 8.3.2 to be

$$N_1 = 6.280 \text{ mmol/m}^2\text{s}$$

$$N_2 = 4.633 \text{ mmol/m}^2\text{s}$$

Thermal conductivity of gas–vapor mixture: λ = 0.025 W/m K.
Molar heat capacity of 2-propanol: C_{p1} = 89.5 J/mol K.
Molar heat capacity of water: C_{p2} = 33.6 J/mol K.

SOLUTION   A schematic diagram of the temperature and composition profiles in this process was given in Example 8.3.2, where the mass transfer part of the problem was solved. All we have to do here is compute the heat flux from the film theory model in Section 11.4.1.

The low flux heat transfer coefficient is computed as follows:

$$h = \lambda/\ell$$
$$= 0.025/6.5 \times 10^{-3}$$
$$= 3.85 \text{ W/m}^2 \text{ K}$$

The heat transfer rate factor follows from Eq. 11.4.6 as

$$\Phi_H = (N_1 C_{p1} + N_2 C_{p2})/h$$
$$= (6.280 \times 10^{-3} \times 89.5 + 4.633 \times 10^{-3} \times 33.6)/3.85$$
$$= 0.187$$

The film theory correction factor is calculated from Eq. 11.4.12 as

$$\Xi_H = \Phi_H/(\exp \Phi_H - 1)$$
$$= 0.187/(\exp(0.187) - 1)$$
$$= 0.909$$

and so the high flux heat transfer coefficient is

$$h^{\bullet} = h\Xi_H$$
$$= 3.85 \times 0.909$$
$$= 3.5 \text{ W/m}^2 \text{ K}$$

Finally, the heat flux q_0 is given by Eq. 11.4.10 as

$$q_0 = h^{\bullet}(T_0 - T_\delta)$$
$$= 3.5 \times (40 - 15)$$
$$= 87.4 \text{ W/m}^2$$

If there were no mass transfer, the heat flux would be almost 10% higher. The effect of mass transfer on heat transfer, although not all that large here, is clearly too large to ignore altogether. ■

11.4.2 The Penetration Model

Our discussions on the film model for heat transfer showed an exact parallel with the corresponding mass transfer problem. The same parallel holds for unsteady-state transfer. Thus, for $\text{Fo} = \lambda t/\rho C_p r_0^2 \to 0$ (short contact time), the time-averaged heat transfer coefficient is given by

$$h = 2\rho C_p \sqrt{\lambda/(\rho C_p \pi t_e)} \tag{11.4.14}$$

The asymptotic, steady-state limit for heat transfer coefficient within a rigid spherical body is

$$h = \frac{\pi^2 \lambda}{3 r_0} \tag{11.4.15}$$

The correction factor for high fluxes is given by Eq. 9.2.13 and Figure 9.4 in which we use the heat transfer rate factor

$$\Phi_H = \frac{\displaystyle\sum_{i=1}^{n} N_i C_{pi}}{h} \tag{11.4.16}$$

in place of the mass transfer rate factor Φ. For chemical engineering design purposes, we recommend the use of the film theory correction factor given by Eq. 11.4.12.

11.4.3 Turbulent Eddy Diffusivity Model

We now take up the problem of estimating the heat transfer coefficients and the energy flux E in turbulent flow in a tube. As in our analysis of the corresponding mass transfer problem (Chapter 10), we consider the transfer processes between a cylindrical wall and a turbulently flowing n-component fluid mixture. We examine the phenomena occurring at any axial position in the tube, assuming that fully developed flow conditions are attained. For steady-state conditions, the differential energy balance (Eqs. 11.1.1 and 11.1.2) takes the form

$$\frac{d(rE_r)}{dr} = 0 \tag{11.4.17}$$

where r represents the radial coordinate and E_r is the energy flux. Equation 11.4.17 shows that rE_r is r invariant.

In Chapter 10, we chose to describe the mass transfer process using mass fluxes and the mass average reference velocity frame because of the need later to solve the equations of continuity of mass in conjunction with the equations of motion. In terms of these mass fluxes the energy flux is given by

$$E_r = q_r + \sum_{i=1}^{n} n_{ir}\overline{H}_i \tag{11.4.18}$$

where $\overline{H}_i$ now represents the partial specific enthalpy with units of joules per kilogram (J/kg). All fluxes in Eqs. 11.4.17 and 11.4.18 are time smoothed quantities; thus, q_r is the sum of molecular and turbulent contributions given by Eqs. 11.2.8 and 11.2.9, respectively.

As discussed in Section 10.2, the description of the turbulent transport processes is much more conveniently carried out in terms of a coordinate system measuring the distance from the wall y

$$y = R_w - r \tag{10.2.2}$$

where R_w represents the radius of the circular tube. We consider E_y to be positive if the flux is directed in the positive y direction, that is, from the wall towards the flowing fluid mixture.

The boundary conditions for this model are

$$\begin{aligned} y = 0 \quad & T = T_0 \quad \text{(the interface)} \\ y = y_b \quad & T = T_b \quad \text{(bulk fluid)} \end{aligned} \tag{11.4.19}$$

where y_b is a distance from the wall beyond which we may safely assume that the turbulence level is high enough to wipe out any further radial variations in temperature.

We now reintroduce a number of parameters that were useful in the corresponding mass transfer problem

1. Friction velocity u^*.

$$u^* = \sqrt{\tau_0/\rho_t} = \sqrt{f/2}\,\overline{u} \tag{10.2.5}$$

where τ_0 is the shear stress at the interface or wall and $\overline{u}$ is the average velocity of flow of the multicomponent fluid mixture inside the tube. The parameter f is the Fanning friction factor.

2. A dimensionless distance from the wall.

$$y^+ = yu^*\rho_t/\mu = yu^*/\nu \tag{10.2.6}$$

3. A dimensionless velocity.

$$u^+ = u/u^* = \sqrt{2/f}\,u/\overline{u} \tag{10.2.7}$$

4. A dimensionless tube radius

$$R_W^+ = R_W u^*/\nu = \text{Re}\sqrt{f/8} \tag{10.2.8}$$

where Re is the Reynolds number for flow $2R_w\overline{u}/\nu$.

With the help of these quantities, we may write the following expression for the energy flux at the wall E_0

$$E_0 = -\rho_t C_p u^* \left(\Pr^{-1} + \Pr_{\text{turb}}^{-1} \frac{\nu_{\text{turb}}}{\nu} \right) \left(1 - \frac{y^+}{R_w^+} \right) \frac{d(T - T_{\text{ref}})}{dr} + \sum_{i=1}^{n} n_{i0} \overline{H}_i \quad (11.4.20)$$

where $\overline{H}_i$ is the partial specific enthalpy of component i and is calculated from

$$\overline{H}_i = C_{pi}(T - T_{\text{ref}}) \quad (11.4.21)$$

with C_{pi} being the specific heat capacity for component i.

The boundary conditions that apply to Eq. 11.4.20, expressed here in terms of y^+, are

$$y^+ = 0 \qquad T = T_0 \qquad \text{(the interface)}$$
$$y^+ = y_b^+ \qquad T = T_b \qquad \text{(bulk fluid)}$$

Taking $(1 - y^+/R_w^+)$ as unity, the energy balance relation can be solved in a manner analogous to the corresponding Eq. 10.3.2 for the continuity of mass. Thus, we obtain the temperature profiles as

$$\frac{(T - T_0)}{(T_b - T_0)} = \frac{(\exp \Psi_H - 1)}{(\exp \Phi_H - 1)} \quad (11.4.22)$$

The heat transfer rate factors Ψ_H and Φ_H are defined by

$$\Psi_H = \int_0^{y^+} A_H(y^+) \, dy^+ \quad (11.4.23)$$

$$\Phi_H = \int_0^{y_b^+} A_H(y^+) \, dy^+ \quad (11.4.24)$$

where $A_H(y^+)$ is defined by

$$A_H(y^+) = \frac{\sum_{i=1}^{n} n_i C_{pi}}{\rho_t C_p u^*} \left(\Pr^{-1} + \Pr_{\text{turb}}^{-1} \frac{\nu_{\text{turb}}}{\nu} \right)^{-1} \quad (11.4.25)$$

The integral Φ_H is a heat transfer rate factor which, for purely molecular heat conduction, reduces to Φ_H defined in Eq. 11.4.6 (with $\ell = y_{bH}$). The integral Ψ_H is equivalent to $\Phi_H \eta$ in Eq. 11.4.8.

Equation 11.4.22 can be differentiated to obtain q_0, the conductive heat flux at the wall

$$q_0 = h^*(T_0 - T_b) = \frac{\sum_{i=1}^{n} n_i C_{pi}}{\exp \Phi_H - 1} (T_0 - T_b) \quad (11.4.26)$$

Thus, the high flux heat transfer coefficient is given by

$$h^* = \frac{\sum_{i=1}^{n} n_i C_{pi}}{\Phi_H} \quad (11.4.27)$$

and the correction factor for high mass fluxes is

$$\Xi_H = \frac{\Phi_H}{\exp \Phi_H - 1} = \frac{h^\bullet}{h} \tag{11.4.28}$$

which is formally identical to the Ackermann film model correction factor derived earlier.

Proceeding in an exactly analogous manner to the corresponding mass transfer problem, we may derive the following explicit expression for the Stanton number for heat transfer $\text{St}_H = h/\rho_t C_p u$

$$\text{St}_H^{-1} = 2/f + \sqrt{2/f} \int_0^{y_i^+} \left\{ \left(\text{Pr}^{-1} + \text{Pr}_{\text{turb}}^{-1} \frac{\nu_{\text{turb}}}{\nu} \right)^{-1} - \left(1 + \frac{\nu_{\text{turb}}}{\nu} \right)^{-1} \right\} dy^+ \tag{11.4.29}$$

The previously discussed turbulence model can be used for the evaluation of the integral in Eq. 11.4.21. For example, with the von Karman velocity profile (Eqs. 10.2.14–10.2.16) we obtain

$$\text{St}_H^{-1} = (2/f) + 5\sqrt{2/f} \left\{ \text{Pr} - 1 + \ln \left(1 + \tfrac{5}{6}(\text{Pr} - 1) \right) \right\} \tag{11.4.30}$$

Equations 11.4.26–11.4.30 allow the calculation of the conductive heat flux q_0; the total energy flux E_0 is the sum of the conductive and the bulk flow enthalpy contributions

$$E_0 = q_0 + \sum_{i=1}^{n} n_i \bar{H}_i \tag{11.4.31}$$

with the mass fluxes calculated from Eqs. 10.3.20 or 10.4.24.

Example 11.4.2   Estimation of the Heat Transfer Coefficient for a Thin-Film Sulfonator

In Example 10.3.1 we considered the calculation of the mass transfer coefficient in the gas phase of a thin-film sulfonator. A schematic diagram of a sulfonation reactor was provided by Figure 10.5. Now, in the modeling of the reactor, the estimation of the temperature profiles along the reactor tube is very important. An important parameter in the determination of the temperature profiles is the gas-phase heat transfer coefficient. Estimate this heat transfer coefficient at the entrance to the reactor for the same set of operating conditions as specified in Example 10.3.1.

DATA   Physical property and system data are given in Example 10.3.1. Additional physical property data needed here are as follows:

Thermal conductivity of gas mixture: $\lambda = 0.026$ W/mK.
Heat capacity of SO_3 at bulk gas temperature: $C_{p1} = 4932$ J/kg K.
Heat capacity of N_2 at bulk gas temperature: $C_{p2} = 801.7$ J/kg K.

SOLUTION   The average heat capacity of the gas mixture is calculated at the mean composition between the interface and bulk gas phases

$$\bar{C}_p = \bar{\omega}_1 C_{p1} + \bar{\omega}_1 C_{p2}$$

$$= 1008 \text{ J/kg K}$$

The Prandtl number of the gas mixture, therefore, is

$$Pr = \overline{C}_p \mu / \lambda$$
$$= 1008 \times 19 \times 10^{-6} / 0.026$$
$$= 0.737$$

The Reynolds number Re and friction factor f were calculated in Example 10.3.1 as 55388 and 0.005176, respectively. The Stanton number for heat transfer may now be evaluated from Eq. 11.4.29 as

$$St_H = 0.002975$$

Hence, the low flux heat transfer coefficient is

$$h = u\rho_t \overline{C}_p St_H$$
$$= 30 \times 1.403 \times 1008 \times 0.002975$$
$$= 126 \text{ W/m}^2\text{K}$$

The mass flux n_1 was calculated in Example 10.3.1.
    The heat transfer rate factor is

$$\Phi_H = n_1 C_{p1} / h$$
$$= (-0.0108) \times 4932 / 126$$
$$= -0.424$$

and the high flux heat transfer coefficient is

$$h^\bullet = h\Phi_H / (\exp(\Phi_H) - 1)$$
$$= 155 \text{ W/m}^2\text{K}$$

Since the heat capacity of SO_3 is very high there is a substantial correction to the heat transfer coefficient caused by simultaneous mass transfer. Simultaneous mass and energy transfer needs to be properly taken into consideration in the design of industrial thin-film sulfonation reactors. ∎

11.4.4   Empirical Methods

Heat transfer data usually are correlated by use of dimensionless groups, such as the Nusselt number (analogous to the Sherwood number for mass transfer)

$$Nu = hd / \lambda \tag{11.4.32}$$

the Stanton number for heat transfer

$$St_H = h / \rho_t C_p u \tag{11.4.33}$$

and the Chilton–Colburn j factor for heat transfer

$$j_H = St_H \, Pr^{2/3} \tag{11.4.34}$$

where d is some characteristic dimension of the equipment, u is the mean velocity for flow, and Pr is the Prandtl number $C_p \mu / \lambda$.

There are scores of expressions available in the literature for estimating heat transfer coefficients; we mention only one here, the well-known Chilton–Colburn analogy

$$j_D = j_H = f/2 \qquad (11.4.35)$$

or

$$\mathrm{St}_H \, \mathrm{Pr}^{2/3} = \mathrm{St} \, \mathrm{Sc}^{2/3} = f/2 \qquad (11.4.36)$$

where f is the Fanning friction factor. Equations 11.4.35 and 11.4.36 express the full form of the Chilton–Colburn analogy. As discussed in Section 10.4.4, the Chilton–Colburn analogy may also be derived from the turbulent eddy diffusivity models of the previous subsection. The Chilton–Colburn analogy is used for estimating mass transfer coefficients in Example 15.1.2.

The heat transfer coefficients estimated from correlations or analogies are the low flux coefficients and, therefore, need to be corrected for the effects of finite transfer rates before use in design calculations. We recommend the film theory correction factor given by Eq. 11.4.12.

11.5 INTERPHASE MASS AND ENERGY TRANSFER

We will often be faced with the problem of determining the rates of mass and energy transfer across a phase boundary. It is these fluxes that appear in the equations that model processes, such as distillation, gas absorption, condensation, and so on. Here we present a summary of the relevant equations and suggest a procedure for determining the required fluxes.

Consider transport across the phase boundary shown in Figure 11.3. We shall denote the two bulk phases by L and V and the interface by I. Though the analysis below is developed for liquid–vapor interphase transport the formalism is generally valid for all two-phase systems. Therefore, what follows applies equally to distillation, stripping, and absorption operations. With a few modifications (to be described later), the analysis below may be used in the determination of rates of condensation, evaporation, vaporization, and boiling.

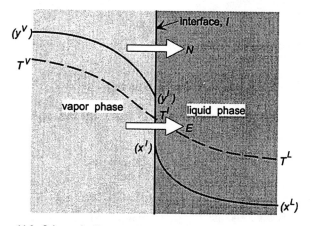

Figure 11.3. Schematic diagram of transport processes at a phase boundary.

At the vapor–liquid interface we have continuity of the component molar fluxes

$$N_i^L = N_i = N_i^V \tag{11.5.1}$$

and of the total molar fluxes

$$N_t^L = N_t = N_t^V \tag{11.5.2}$$

where N_i^L and N_i^V are the normal components of the molar flux N_i at the interface. These fluxes are made up of diffusive and convective contributions as

$$N_i^L = J_i^L + x_i N_t^L = N_i = J_i^V + y_i N_t^V = N_i^V \tag{11.5.3}$$

Using the definitions of the mass transfer coefficients we may write, for the diffusion fluxes

$$(J^L) = c_t^L [k_L^\bullet](x^I - x^L) \tag{11.5.4}$$

$$(J^V) = c_t^V [k_V^\bullet](y^V - y^I) \tag{11.5.5}$$

where, for definiteness, we consider transfer from the "V" phase to the "L" phase as leading to a positive flux. Any of the models described in Chapters 8–10, at least in principle, may be used for the determination of the matrices of mass transfer coefficients $[k^\bullet]$.

We also have continuity of the energy flux across the V–L interface

$$E^V = E^L = E \tag{11.5.6}$$

where E^V and E^L are the normal components of the energy flux at the interface. With the energy flux defined by Eq. 11.1.4 we may rewrite Eq. 11.5.6 as

$$q^V + \sum_{i=1}^n N_i^V \overline{H}_i^V(T^V) = q^L + \sum_{i=1}^n N_i^L \overline{H}_i^L(T^L) \tag{11.5.7}$$

The heat fluxes in the two phases are given by

$$q^V = h_V^\bullet(T^V - T^I) \tag{11.5.8}$$

$$q^L = h_L^\bullet(T^I - T^L) \tag{11.5.9}$$

The finite flux heat transfer coefficient in the vapor phase $h_V^\bullet$ is related to the zero-flux heat transfer coefficient h^V by

$$h_V^\bullet = h^V(\Phi_H^V)/(\exp \Phi_H^V - 1) \tag{11.5.10}$$

where we have used the film model correction factor defined by Eq. 11.4.12, with Φ_H^V obtained from

$$\Phi_H = \sum_{i=1}^n N_i C_{pi}/h \tag{11.4.16}$$

The high flux correction to h^L is not usually needed; the value of the liquid-phase heat transfer coefficient is high enough to ensure that any such correction would be close to unity.

To complete the model of interphase transport, we must say something about the interface. It is usual to assume that equilibrium prevails at the interface, and relate the mole fraction y_i^I and x_i^I by

$$y_i^I = K_i x_i^I \qquad i = 1, 2, \ldots, n \qquad (11.5.11)$$

The K values are to be evaluated at the temperature, pressure, and compositions at the interface.

11.5.1 The Bootstrap Problem Revisited

There is a close relationship between the bootstrap problem discussed in Chapter 7 and the interphase energy balance Eq. 11.5.7. To demonstrate this relationship we rewrite Eq. 11.5.7 as

$$
\begin{aligned}
q^L - q^V &= \sum_{i=1}^{n} \left(\overline{H}_i^V - \overline{H}_i^L \right) N_i = \sum_{i=1}^{n} \lambda_i N_i \\
&= \sum_{i=1}^{n} \lambda_i J_i^V + \sum_{i=1}^{n} \lambda_i y_i N_t \\
&= \sum_{i=1}^{n} \lambda_i J_i^L + \sum_{i=1}^{n} \lambda_i x_i N_t \\
&= \sum_{i=1}^{n-1} (\lambda_i - \lambda_n) J_i^V + \lambda_y N_t \\
&= \sum_{i=1}^{n-1} (\lambda_i - \lambda_n) J_i^L + \lambda_x N_t
\end{aligned}
\qquad (11.5.12)
$$

where we have defined

$$\lambda_y = \sum_{i=1}^{n} \lambda_i y_i \qquad \lambda_x = \sum_{i=1}^{n} \lambda_i x_i \qquad \lambda_i = \overline{H}_i^V - \overline{H}_i^L \qquad (11.5.13)$$

Equations 11.5.12 and 11.5.1–11.5.3 may be combined to give $n - 1$ independent relations for the total flux N_t

$$-\frac{J_i^L - J_i^V}{x_i - y_i} = \frac{q^L - q^V}{\lambda_y} - \frac{\displaystyle\sum_{i=1}^{n} (\lambda_k - \lambda_n) J_k^V}{\lambda_y} = N_t \qquad (11.5.14)$$

In Eqs. 11.5.14 we use the vapor-phase diffusion fluxes. Alternatively, we may use the liquid-phase diffusion fluxes to obtain

$$-\frac{J_i^L - J_i^V}{x_i - y_i} = \frac{q^L - q^V}{\lambda_x} - \frac{\displaystyle\sum_{i=1}^{n} (\lambda_k - \lambda_n) J_k^L}{\lambda_x} = N_t \qquad (11.5.15)$$

We now eliminate N_t from Eqs. 11.5.3 with Eqs. 11.5.15 to give

$$N_i = J_i^V - y_i \sum_{k=1}^{n-1} \Lambda_k J_k^V + y_i \frac{\Delta q}{\lambda_y} \qquad (11.5.16)$$

where we have defined the parameters

$$\Lambda_k = (\lambda_k - \lambda_n)/\lambda_y \qquad \Delta q = q^L - q^V \tag{11.5.17}$$

Equation 11.5.16 may be rewritten in matrix notation ($n - 1$ dimensional) as

$$(N) = [\beta^V](J^V) + (y)\frac{\Delta q}{\lambda_y} \tag{11.5.18}$$

where the matrix $[\beta^V]$ has the elements (cf. Eq. 7.2.16)

$$\beta_{ik}^V = \delta_{ik} - y_i\Lambda_k \qquad i, k = 1, 2, \ldots, n - 1 \tag{11.5.19}$$

An expression analogous to Eq. 11.5.18 may be written in terms of the liquid-phase diffusion fluxes

$$(N) = [\beta^L](J^L) + (x)\frac{\Delta q}{\lambda_x} \tag{11.5.20}$$

Equation 11.5.18 or 11.5.20 may be used to determine the $n - 1$ fluxes N_i where $i = 1, 2, \ldots, n - 1$, given the $n - 1$ diffusion fluxes J_i. The nth flux is determined from

$$N_n = N_t - \sum_{i=1}^{n} N_i \tag{11.5.21}$$

and the total energy flux follows from Eq. 11.1.4.

11.5.2   Nonequimolar Effects in Multicomponent Distillation

In the treatment of transport processes during distillation, most textbooks (e.g., Sherwood et al., 1975) assume that conditions of equimolar countertransfer hold. That is,

$$N_t = 0 \tag{11.5.22}$$

We can see from Eqs. 11.5.15 that the total flux will vanish only if the following two conditions are satisfied

$$\Delta q = q^L - q^V = 0 \tag{11.5.23}$$

$$\lambda_i = \lambda_n \qquad i = 1, 2, \ldots, n - 1 \tag{11.5.24}$$

The first requirement may be written as follows:

$$\Delta q = h_L^\bullet(T^I - T^L) - h_V^\bullet(T^V - T^I) = 0 \tag{11.5.25}$$

This requirement is often met in practice because the heat transfer coefficients in the vapor and liquid phases are such that they will wipe out any temperature gradients locally, say on a tray. In any event, the contribution of the Δq term to the total flux is likely to be small because it is small in comparison to the average latent heat terms λ_y or λ_x.

The second requirement (Eq. 11.5.25) is that the molar latent heats of vaporization of the constituent species be identical. Now, molar latent heats of many compounds are close to one another, but the differences are not zero. Let us examine the effect of such small differences in the latent heats on the interfacial rates of transfer. Example 11.5.1 demonstrates the importance of taking into account nonequimolar effects in distillation.

Example 11.5.1 Nonequimolar Effects in Ternary Distillation

Determine the fluxes N_1, N_2, and N_3 in the vapor phase during the distillation of the mixture isopentane* (1)–ethanol (2)–water (3). The bulk vapor composition is

$$y_1^V = 0.60 \qquad y_2^V = 0.10 \qquad y_3^V = 0.30$$

and the interface composition is

$$y_1^I = 0.62 \qquad y_2^I = 0.16 \qquad y_3^I = 0.22$$

DATA

Binary mass transfer coefficients.

$$\kappa_{12} = 0.036 \qquad \kappa_{13} = 0.072 \qquad \kappa_{23} = 0.105 \text{ m/s}$$

Temperature = 346 K and pressure = 100 kPa.
Partial molar enthalpies are in kilojoules per mole [kJ/mol].

$$\bar{H}_1^V = 38.0 \qquad \bar{H}_2^V = 50.6 \qquad \bar{H}_3^V = 47.0$$

$$\bar{H}_1^L = 15.5 \qquad \bar{H}_2^L = 10.1 \qquad \bar{H}_3^L = 5.0$$

SOLUTION We shall use the method of Krishna and Standart described in Section 8.3 to compute the molar fluxes. Algorithms 8.3.1 or 8.3.2 may be used to determine the molar fluxes. Convergence is very rapid in this example. No more than two iterations are needed. Only the final results of the relevant computations are summarized below.

The interface is taken to be the plane at $\eta = 0$. Thus, $y_0 = y^I$ and $y_\delta = y^V$. The matrix of low flux mass transfer coefficients is calculated from Eq. 8.3.27 as $[k] = [R]^{-1}$ with the elements of $[R]$ given by Eq. 8.3.25. The matrix $[R_0]$ is found to be

$$[R_0] = \begin{bmatrix} 16.11 & -8.6 \\ -2.92 & 20.84 \end{bmatrix} \text{s/m}$$

where we have used the interface composition y_0 in calculating $[R_0]$. The matrix $[k]$ is found by inverting this matrix as

$$[k_0] = \begin{bmatrix} 0.0671 & 0.0277 \\ 0.0094 & 0.0519 \end{bmatrix} \text{m/s}$$

The total flux is computed from Eq. 11.5.14 (cf. Eq. 7.2.13) with the Δq term set to zero. For the record, the bootstrap matrix $[\beta_0]$, although not used directly in the computation of the molar fluxes, is

$$[\beta_0] = \begin{bmatrix} 1.4075 & 0.0313 \\ 0.1052 & 1.0081 \end{bmatrix}$$

---

*Isopentane is the common name for 2-methylbutane.

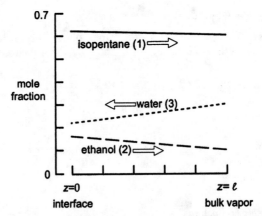

Figure 11.4. Composition profiles in distillation of isopentane (1)–ethanol (2)–water (3). Arrows indicate actual directions of mass transfer.

The converged values for the molar fluxes N_i are

$$N_1 = 0.1506 \qquad N_2 = 0.1228 \qquad N_3 = -0.1991 \text{ mol/m}^2\text{s}$$

At these values for the fluxes the various matrices are given below

$$[\Phi] = \begin{bmatrix} -0.0788 & 0.0602 \\ 0.0645 & -0.0995 \end{bmatrix}$$

$$[\Xi] = \begin{bmatrix} 0.9615 & 0.0292 \\ 0.0313 & 0.9514 \end{bmatrix}$$

$$[k_0^\bullet] = \begin{bmatrix} 0.0654 & 0.0283 \\ 0.0107 & 0.0496 \end{bmatrix} \text{ m/s}$$

Composition profiles are shown in Figure 11.4.

Diffusional interaction effects are moderately important in this system. Evidence for this statement is provided by the values of the off-diagonal coefficients relative to the values of the diagonal elements of the matrices of mass transfer coefficients. These interaction effects will have the effect of making the individual component transfer efficiencies significantly different.

If we repeat the computations but use equal values for the molar heats of vaporization we then find the molar fluxes have the following values

$$N_1 = 0.1053 \qquad N_2 = 0.1131 \qquad N_3 = -0.2184 \text{ mol/m}^2\text{s}$$

showing that the flux of isopentane, assuming equimolar transfer, is significantly lower than that calculated using the proper interfacial energy balance relations. This result is expected, since the molar latent heat of isopentane is much lower than those of the other two species. Water has the highest molar latent heat and the assumption of equimolar transfer overestimates its transfer rate. In dehydration processes by azeotropic distillation, the transfer efficiency of water will be lower than that calculated assuming equimolar distillation. ∎

11.5.3 Computation of the Fluxes

The number of independent equations that model the interphase mass and heat transfer process is $3n + 1$. $3n - 1$ of these equations are

- $n - 1$ rate equations for the vapor phase (Eqs. 11.5.5).
- $n - 1$ rate equations for the liquid phase (Eqs. 11.5.4).
- n equilibrium relations for the interface (Eqs. 11.5.11).
- 1 interfacial energy balance (Eq. 11.5.7).

The set of equations is completed by expressions that force the mole fractions at the interface to sum to unity

$$S^V = \sum_{i=1}^{n} y_i^I - 1 = 0 \qquad (11.5.26)$$

$$S^L = \sum_{i=1}^{n} x_i^I - 1 = 0 \qquad (11.5.27)$$

The $3n + 1$ variables that may be determined by solving this set of equations include

- $2n$ mole fractions on each side of the interface y_i^I and x_i^I.
- n molar fluxes N_i.
- The interface temperature T^I.

All other quantities appearing in Eqs. 11.5.3–11.5.9 must be specified. The specified variables normally will be the bulk phase compositions x_i^L and y_i^V, and the bulk temperatures T^L and T^V, and the system pressure. In practice, the bulk conditions are determined by process material and energy balances (see, e.g., Chapters 14 and 15). Transport, physical, and thermodynamic properties (K values and enthalpies), and the mass and heat transfer coefficients (the low flux ones) can be evaluated from appropriate models in terms of the composition, temperature, and pressure of the appropriate phase.

Given the bulk fluid conditions (mole fractions and temperatures), the system pressures, the low flux mass and heat transfer coefficients (or methods to evaluate them), and appropriate equilibrium models, what is the most effective means of obtaining the rates of mass and heat transfer?

One possible approach is to take the interfacial mole fractions x_i^I as the independent variables. The remaining interfacial parameters (y_i^I and T^I) may then be obtained from a bubble point calculation (Henley and Seader, 1981). Knowing the compositions in the bulk fluid phase, as well as at the interface, allows us to determine the fluxes in each phase separately using the algorithms presented in Chapters 8–10. If the fluxes for the vapor and liquid phases do not agree (as required by the continuity Eq. 11.5.1) then the interfacial composition must be reestimated somehow and the procedure repeated. From one point of view this is an attractive and easily implemented method. It is built from what we may consider to be very well tried algorithms: bubble point calculations for the interfacial state and the multicomponent mass transfer algorithms of Chapters 8–10. A possible drawback is that all three steps require iteration. The net result is that too much time is wasted converging mass transfer rate calculations for estimates of the interfacial state that may be nowhere near the true solution.

We believe that a better way to solve the set of Eqs. 11.5.1–11.5.10 is to solve all the equations simultaneously using Newton's method (Algorithm C.2). This approach avoids the nested iterations of the foregoing procedure and keeps both thermodynamic property evaluations and multicomponent mass transfer coefficient matrix computations to a minimum.

For use with Newton's method, the $3n + 1$ independent equations are rewritten in the form $(F(\chi)) = (0)$.

1. $n - 1$ mass transfer rate equations in the vapor phase.

$$(R^V) \equiv c_t^V[k_V^{\bullet}](y^V - y^I) + N_t(y^V) - (N) = (0) \qquad (11.5.28)$$

2. $n - 1$ mass transfer rate equation is in the liquid phase.

$$(R^L) \equiv c_t^L[k_L^{\bullet}](x^I - x^L) + N_t(x^L) - (N) = (0) \qquad (11.5.29)$$

3. n interfacial equilibrium equations.

$$Q_i^I \equiv K_i x_i^I - y_i^I = 0 \qquad i = 1, 2, \ldots, n \qquad (11.5.30)$$

4. One energy continuity equation $(E^V - E^L = 0)$.

$$E^I \equiv q^V - q^L + \sum_{i=1}^{n} N_i(\overline{H}_i^V - \overline{H}_i^L) = 0 \qquad (11.5.31)$$

The independent equations are ordered into a vector of functions (F) as follows:

$$(F)^T = \left(R_1^V, R_2^V, \ldots, R_{n-1}^V, E^I, R_1^L, R_2^L, \ldots, R_{n-1}^L, \right.$$
$$\left. Q_1^I, Q_2^I, \ldots, Q_{n-1}^I, Q_n^I, S^V, S^L \right)$$

The unknown variables corresponding to this set of equations are ordered into a vector (χ) as follows:

$$(\chi)^T = \left(N_1, N_2, \ldots, N_{n-1}, N_n, x_1^I, x_2^I, \ldots, x_n^I, \right.$$
$$\left. y_1^I, y_2^I, \ldots, y_n^I, T^I \right)$$

We wish to find the vector of variables (χ) that gives $(F) = (0)$. The steps of an algorithm to accomplish this are given in Algorithm 11.1.

Algorithm 11.1 Procedure for Determination of Mass and Energy Fluxes at a Phase Boundary	
Given:	Bulk phase conditions.
	Mass and heat transfer coefficient models.
	Thermodynamic property models.
Step 1:	Generate initial estimates of all unknown variables.
	Set interface temperature to the average of vapor and liquid temperatures.
	Set interface mole fractions equal to bulk values.
	Set molar fluxes to small, nonzero, values.
Step 2:	Compute:
	Transport properties.
	Thermodynamic properties (K values, enthalpies).
	Mass transfer coefficient matrix $[k]$.
	Heat transfer coefficients.
	Vector of discrepancy function (F).
Step 3:	Compute elements of Jacobian matrix [J].
Step 4:	Check for convergence, if not obtained continue with Step 5.
Step 5:	Compute new set of unknown variables by solving Eq. C.2.5. Return to Step 2.

Example 11.5.2 Distillation of a Binary Mixture

Methanol (1) and water (2) are being distilled in a tray column operating at 101.3 kPa. On one of the trays in the column the mole fractions of methanol in the bulk vapor and liquid phases are $y_1^V = 0.5776$ and $x_1^L = 0.3105$. Estimate the rates of interphase mass transfer.

DATA

Vapor-phase mass transfer coefficient: $k^V = 8 \times 10^{-2}$ m/s.
Liquid-phase mass transfer coefficient: $k^L = 2 \times 10^{-4}$ m/s.
Liquid-phase molar density: $c_t^L = 36.58$ kmol/m^3.
Latent heat of vaporization of methanol: $\Delta H_{vap1} = 36$ kJ/mol.
Latent heat of vaporization of water: $\Delta H_{vap2} = 43$ kJ/mol.

The vapor–liquid equilibrium behavior of the methanol–water system at 101.3 kPa may be represented by

$$K_1 = \gamma_1 P_1^s / P \qquad K_2 = \gamma_2 P_2^s / P$$

The pure component vapor pressures may be estimated from the Antoine equation as follows:

$$\ln P_1^s = 23.402 - 3593.4/(T - 34.29)$$

$$\ln P_2^s = 23.196 - 3816.4/(T - 46.13)$$

where P_1^s and P_2^s are in pascals (Pa) and T is in kelvin (K).

The activity coefficients may be estimated from the Margules equation (Table D.2) with parameters from Gmehling and Onken (1977ff, Vol. 1/1a, p. 53):

$$A_{12} = 0.8517 \qquad A_{21} = 0.4648$$

ANALYSIS Since there are only two components, the number of equations we must solve is just 7. We list these equations below to assist us in taking the derivatives that we need if we are to use Newton's method.

$$F_1 \equiv c_t^V k^V \Xi^V \left(y_1^V - y_1^I \right) + y_1^V (N_1 + N_2) - N_1$$

$$F_2 \equiv N_1 \, \Delta H_{vap1} + N_2 \, \Delta H_{vap2}$$

$$F_3 \equiv c_t^L k^L \Xi^L \left(x_1^I - x_1^L \right) + x_1^L (N_1 + N_2) - N_1$$

$$F_4 \equiv K_1 x_1^I - y_1^I$$

$$F_5 \equiv K_2 x_2^I - y_2^I$$

$$F_6 \equiv y_1^I + y_2^I - 1$$

$$F_7 \equiv x_1^I + x_2^I - 1$$

The discrepancy functions F_1–F_7 are zero at the solution. We have ignored the Δq term in the interfacial energy balance F_2. The temperature of the two bulk phases will, therefore, be assumed to be equal to the (as yet unknown) interface temperature.

During the solution of these nonlinear equations we will need the partial derivatives of these functions with respect to each of the independent variables: N_1, N_2, y_1^I, y_2^I, x_1^I, x_2^I, and T^I. Many of the required partial derivatives are zero. Examples include the partial derivatives of the summation equations F_6 and F_7 with respect to everything except the two mole fractions that appear in those expressions. The nonzero elements of the Jacobian matrix are

$$\partial F_1/\partial N_1 = y_1^V - 1$$
$$\partial F_1/\partial N_2 = y_1^V$$
$$\partial F_1/\partial y_1^I = -c_t^V k^V \Xi^V$$
$$\partial F_2/\partial N_1 = \Delta H_{\text{vap1}}$$
$$\partial F_2/\partial N_2 = \Delta H_{\text{vap2}}$$
$$\partial F_3/\partial N_1 = x_1^L - 1$$
$$\partial F_3/\partial N_2 = x_1^L$$
$$\partial F_3/\partial x_1^I = c_t^L k^L \Xi^L$$
$$\partial F_4/\partial y_1^I = x_1^I \partial K_1/\partial y_1^I - 1$$
$$\partial F_4/\partial y_2^I = x_1^I \partial K_1/\partial y_2^I$$
$$\partial F_4/\partial x_1^I = x_1^I \partial K_1/\partial x_1^I + K_1$$
$$\partial F_4/\partial x_2^I = x_1^I \partial K_1/\partial x_2^I$$
$$\partial F_4/\partial T^I = x_1^I \partial K_1/\partial T^I$$
$$\partial F_5/\partial y_1^I = x_2^I \partial K_2/\partial y_1^I$$
$$\partial F_5/\partial y_2^I = x_2^I \partial K_2/\partial y_2^I - 1$$
$$\partial F_5/\partial x_1^I = x_2^I \partial K_2/\partial x_1^I$$
$$\partial F_5/\partial x_2^I = x_2^I \partial K_2/\partial x_2^I + K_2$$
$$\partial F_5/\partial T^I = x_2^I \partial K_2/\partial T^I$$
$$\partial F_6/\partial y_1^I = 1$$
$$\partial F_6/\partial y_2^I = 1$$
$$\partial F_7/\partial x_1^I = 1$$
$$\partial F_7/\partial x_2^I = 1$$

When taking these partial derivatives it must be remembered that, in general, the molar densities, the mass transfer coefficients, and thermodynamic properties are functions of temperature, pressure, and composition. In addition, Ξ is a function of the molar fluxes. We have ignored most of these dependencies in deriving the expressions given above. The important exception is the dependence of the K values on temperature and composition that cannot be ignored. The derivatives of the K values with respect to the vapor mole fractions are zero in this case since the model used to evaluate the K values is independent of the vapor composition.

SOLUTION We are now in a position to be able to compute the fluxes, N_1 and N_2.

Step 1: Estimate all of the independent variables. For the first iteration we use the following values.

$$N_1 = -0.1 \text{ mol/m}^2\text{s} \qquad N_2 = 0.09 \text{ mol/m}^2\text{s}$$
$$y_1^I = y_2^I = 0.5$$
$$x_1^I = x_2^I = 0.5$$
$$T^I = 75°C$$

The fluxes were estimated simply by giving two small, nonzero numbers. A negative sign was put in front of the estimate of the flux of methanol because methanol is more volatile than water and will transfer from the liquid to the vapor phases. The mole fractions at the interface are given values of $1/n$, where n is the number of components. We will see later that it would have been more logical to estimate the interfacial mole fractions at their corresponding bulk values. The temperature is somewhere between the boiling point of methanol and that of water.

Step 2: Evaluation of the discrepancy functions. The molar density of the gas phase is computed from the ideal gas law using the temperature of the interface (which, by dropping the Δq term in F_2, is assumed equal to the bulk vapor and liquid temperatures)

$$c_t^V = 35.0 \text{ mol/m}^3$$

Next, the mass transfer rate factor for the vapor phase is

$$\Phi^V = (N_1 + N_2)/c_t^V k^V = -0.00357$$

and the vapor-phase high flux correction factor follows as:

$$\Xi^V = \Phi^V/(\exp(\Phi^V) - 1) = 1.002$$

which is close enough to unity to suggest that it may not be worth including it in the computations.

We now have everything we need to evaluate F_1. By simply substituting the current values for every quantity appearing in the expression for F_1 we get

$$\begin{aligned}
F_1 &= c_t^V k^V \Xi^V (y_1^V - y_1^I) + y_1^V(N_1 + N_2) - N_1 \\
&= 35.0 \times 0.08 \times 1.002 \times (0.5776 - 0.5) \\
&\quad + 0.5776 \times (-0.1 + 0.09) - (-0.1) \\
&= 0.3119
\end{aligned}$$

The energy balance discreprancy is quickly evaluated as

$$\begin{aligned}
F_1 &= N_1 \Delta H_{vap1} + N_2 \Delta H_{vap2} \\
&= -0.1 \times 36{,}000 + 0.09 \times 43{,}000 \\
&= 270
\end{aligned}$$

The mass transfer rate factor for the liquid phase is computed as

$$\begin{aligned}
\Phi^L &= (N_1 + N_2)/c_t^L k^L \\
&= (-0.1 + 0.09)/(36.58 \times 10^3 \times 4 \times 10^{-4}) \\
&= -0.00137
\end{aligned}$$

and the liquid-phase high flux correction factor follows as:

$$\begin{aligned}
\Xi^L &= \Phi^L \exp(\Phi^L)/(\exp(\Phi^L) - 1) \\
&= (-0.00137) \times \exp(-0.00137)/(\exp(-0.00137) - 1) \\
&= 0.9993
\end{aligned}$$

The liquid-phase mass transfer discrepancy is

$$F_3 = c_t^L k^L \Xi^L \left(x_1^I - x_1^L \right) + x_1^L (N_1 + N_2) - N_1$$
$$= 36.58 \times 10^3 \times 4 \times 10^{-4} \times 0.9993 \times (0.5 - 0.3105)$$
$$+ 0.3105 \times (-0.1 + 0.09) - (-0.1)$$
$$= 1.4833$$

Evaluating the equilibrium relations requires us to first compute the vapor pressures and activity coefficients. Note that these thermodynamic properties are computed using the temperature and composition at the interface. Expressions for the activity coefficients are given in Table D.2. Substituting the numerical values of x^I into those equations gives the following results:

$$\gamma_1 = 1.1232 \qquad \gamma_2 = 1.2373$$

The vapor pressures are

$$P_1^s = 155{,}274 \text{ Pa} \qquad P_2^S = 38{,}551 \text{ Pa}$$

The K values may now be computed as

$$K_1 = 1.7217 \qquad K_2 = 0.4708$$

We may now evaluate the departures from equilibrium.

$$F_4 = K_1 x_1^I - y_1^I$$
$$= 1.7217 \times 0.5 - 0.5$$
$$= 0.3606$$
$$F_5 \equiv K_2 x_2^I - y_2^I$$
$$= 0.4708 \times 0.5 - 0.5$$
$$= -0.2646$$

Finally, our initial guess of the interfacial mole fractions was such that the mole fractions summation equations are already satisfied

$$F_6 = 0 \qquad F_7 = 0$$

It is clear simply by inspection of the values of F_1–F_7 that their values are not sufficiently small so we must generate a new estimate of the independent variables.

Step 3: Evaluation of the Jacobian matrix [J]. The elements of [J] are obtained quite straightforwardly using the expressions derived above. The only derivatives that need further discussion are those of the equilibrium equations, F_4 and F_5.

In this particular problem the partial derivatives of the K values are given by

$$\partial K_1 / \partial \chi = K_1 (\partial \ln \gamma_1 / \partial \chi + \partial \ln P_1^s / \partial \chi)$$

with a similar expression for K_2. The parameter χ represents any mole fraction or temperature.

The composition derivatives of the activity coefficients are given by expressions in Table D.2, the numerical values are

$$\partial \ln \gamma_1 / \partial x_1^I = -0.4259 \qquad \partial \ln \gamma_1 / \partial x_2^I = 0.2324$$
$$\partial \ln \gamma_2 / \partial x_1^I = 0.4259 \qquad \partial \ln \gamma_2 / \partial x_2^I = -0.2324$$

The partial derivatives $\partial \ln \gamma_1 / \partial x_1^I$ and $-\partial \ln \gamma_2 / \partial x_1^I$ are equal only when the mixture is an equimolar one (as it is currently estimated to be).

The derivatives of vapor pressure with respect to temperature are

$$\partial \ln P_1^s / \partial T^I = 0.0365 \qquad \partial \ln P_2^s / \partial T^I = 0.0418$$

The vapor pressures are independent of composition.

The partial derivatives of the K values with respect to the temperature and composition at the interface, therefore, are

$$\partial K_1 / \partial x_1^I = -0.7330 \qquad \partial K_1 / \partial x_2^I = 0.4000$$

$$\partial K_2 / \partial x_1^I = 0.2004 \qquad \partial K_2 / \partial x_2^I = -0.1094$$

$$\partial K_1 / \partial T^I = 0.0628 \qquad \partial K_2 / \partial T^I = 0.0197$$

Finally, since the vapor phase is assumed ideal, the K values do not depend on the composition of the vapor phase. Hence,

$$\partial F_4 / \partial y_1^I = -1 \qquad \partial F_5 / \partial y_1^I = 0$$

$$\partial F_4 / \partial y_2^I = 0 \qquad \partial F_5 / \partial y_2^I = -1$$

Step 5: Computation of a new set of independent variables.

Following the calculation of [J] we solve the linear system (Eq. C.2.5) to obtain a new estimate. The above procedure is then repeated three times with the following converged results.

$$N_1 = -0.2598 \text{ mol/m}^2\text{s} \qquad N_2 = 0.2175 \text{ mol/m}^2\text{s}$$

$$y_1^I = 0.6617 \qquad y_2^I = 0.3383$$

$$x_1^I = 0.2767 \qquad x_2^I = 0.7233$$

$$T^I = 78.09°\text{C}$$

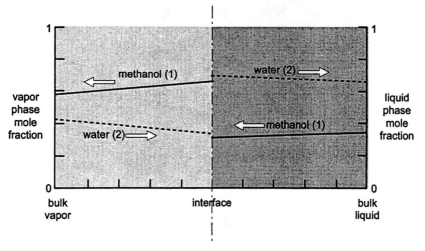

Figure 11.5. Composition profiles in the vapor and liquid phases during distillation of a methanol and water mixture. Arrows indicate the actual direction of mass transfer.

Composition profiles for both phases are shown in Figure 11.5. The profiles are not linear although that fact is hard to discern from Figure 11.5.

The final values of the high flux correction factors ($\Xi^V = 1.009$, $\Xi^L = 0.9999$) confirm our earlier remark that it was not worth including them in the calculations. This will often be true for distillation calculations but not for other processes, such as condensation and evaporation. ■

Example 11.5.3   Interphase Mass Transfer in the Presence of an Inert Gas

A series of experiments involving mass transfer in a wetted wall column were carried out by Modine (1963) with a view to investigating coupled diffusion effects. A schematic diagram of the wetted wall column is provided in Figure 11.6. The liquid mixture flowing down the inside of the tube contained acetone (1) and benzene (2). The downwards flowing gas–vapor mixture contained acetone, benzene, and an inert gas that was either nitrogen or helium. Modine's data for experiment No. 7, with helium (3) as the inert gas, are summarized below (from Krishna, 1981a).

Diameter of column, $d = 0.025019$ m.

Helium flow rate = 0.2438985 mol/s.

Vapor inlet temperature = 36°C.

Liquid flow rate = 5.5439×10^{-3} kg/s.

Liquid inlet temperature = 36.45°C.

Inlet pressure = 129.81 k Pa.

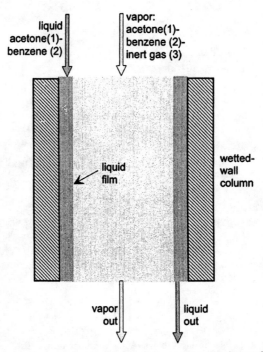

Figure 11.6. Schematic diagram of wetted wall column used in experiments by Modine (1963).

Inlet vapor composition (mole fraction)

$$y_1 = 0.052354 \qquad y_2 = 0.0 \qquad y_3 = 0.947646$$

Inlet liquid composition (mole fraction)

$$x_1 = 0.076583 \qquad x_2 = 0.923417$$

Estimate the mass fluxes at the top of the wetted wall column.

ANALYSIS We shall use the turbulent eddy diffusivity model to represent the transport processes in the vapor phase. The mass fluxes in the vapor phase will, therefore, be calculated from

$$n_1 = j_1 + \omega_1^I(n_1 + n_2)$$
$$n_2 = j_2 + \omega_2^I(n_1 + n_2)$$

The mass diffusion fluxes j_1 and j_2 are given by

$$j_1 = \rho_t^V k_{11}^{\bullet}(\omega_1^I - \omega_1^V) + \rho_t^V k_{12}^{\bullet}(\omega_2^I - \omega_2^V)$$
$$j_2 = \rho_t^V k_{21}^{\bullet}(\omega_1^I - \omega_1^V) + \rho_t^V k_{22}^{\bullet}(\omega_2^I - \omega_2^V)$$

where we have adopted the convention that transfers from the liquid to the vapor phase are positive. The flux of helium is zero so there is no contribution from n_3 in the convective contributions to n_1 and n_2.

The molar flux of acetone in the liquid phase is given by

$$N_1 = c_t^L k^L \Xi^L(x_1^L - x_1^I) + x_1^I(N_1 + N_2)$$

A similar equation for benzene is not needed since the liquid phase is a binary mixture.

The energy balance at the interface may be written in mass units as

$$h^L(T^L - T^I) + n_1 \overline{H}_1^L + n_2 \overline{H}_2^L = h^V \Xi_H(T^I - T^V) + n_1 \overline{H}_1^V + n_2 \overline{H}_2^V$$

where the $\overline{H}$ values are the partial specific enthalpies of the subscripted component in the superscripted phase. Helium does not contribute to the interphase energy-transfer process.

The model is completed by assuming equilibrium at the phase interface. This condition is represented by Eqs. (11.5.11, 11.5.25, and 11.5.26). Since helium is present only in the vapor phase, its mole fraction in the liquid phase is zero and, hence, its K value is infinite. Therefore, Eq. 11.5.11 for helium cannot be used in the calculations.

We shall, as before, use Newton's method to solve all the independent equations simultaneously. The independent variables that are to be determined by iteration are the fluxes and the interface compositions and temperature. However, the use of the turbulent eddy diffusion model for the vapor-phase mass transport means that the mass fluxes n_1 and n_2, and the molar fluxes N_1 and N_2, appear in the set of model equations. These fluxes are related by

$$n_i = N_i M_i$$

Furthermore, the interface composition is needed in both mass fraction and mole fraction units. The selection of mass or molar quantities as independent variables is not obvious; indeed, there are good reasons for using either set. We feel that it is preferable to retain the interface vapor mole fractions as independent variables, rather than the mass fractions that appear in the rate equations, because the interface equilibrium equations are simpler when expressed in molar quantities. We also prefer to use the mass fluxes n_1 and n_2, as independent variables in place of the molar fluxes.

The eight independent equations, written in the form $F(\chi) = 0$ in terms of the mass fluxes and the interface mole fractions, are listed below.

$$F_1 \equiv j_1 + \omega_1^I(n_1 + n_2) - n_1 = 0$$

$$F_2 \equiv j_2 + \omega_2^I(n_1 + n_2) - n_2 = 0$$

$$F_3 \equiv h^L(T^L - T^I) - h^V \Xi_H(T^I - T^V)$$
$$- n_1(\bar{H}_1^V - \bar{H}_1^L) - n_2(\bar{H}_2^V - \bar{H}_2^L) = 0$$

$$F_4 \equiv c_t^L k^L \Xi^L(x_1^L - x_1^I) + x_1^I(n_1/M_1 + n_2/M_2) - n_1/M_1 = 0$$

$$F_5 \equiv K_1 x_1^I - y_1^I = 0$$

$$F_6 \equiv K_2 x_2^I - y_2^I = 0$$

$$F_7 \equiv y_1^I + y_2^I + y_3^I - 1 = 0$$

$$F_8 \equiv x_1^I + x_2^I - 1 = 0$$

Note that only two equilibrium equations F_5 and F_6 are used, and that the mole fraction summation for the vapor phase is made over all three components, whereas that for the liquid is made over only the two species actually present in the liquid. The eight independent variables computed by solving these equations are

$$n_1, n_2, x_1^I, x_2^I, y_1^I, y_2^I, y_3^I, \text{ and } T^I$$

Step 4 of Algorithm 11.1 calls for the evaluation of the Jacobian matrix [J]. The elements of this matrix are obtained by differentiating the above equations with respect to the independent variables. Many of these partial derivatives are zero (or can be approximated as zero). The nonzero derivatives of F_1 and F_2 are as follows:

$$\partial F_1/\partial n_1 = \omega_1^I - 1$$

$$\partial F_1/\partial n_2 = \omega_1^I$$

$$\partial F_1/\partial y_1^I = (\rho_t^V k_{11}^\bullet + n_t)(\partial \omega_1^I/\partial y_1^I) + \rho_t^V k_{12}^\bullet(\partial \omega_2^I/\partial y_1^I)$$

$$\partial F_1/\partial y_2^I = (\rho_t^V k_{11}^\bullet + n_t)(\partial \omega_1^I/\partial y_2^I) + \rho_t^V k_{12}^\bullet(\partial \omega_2^I/\partial y_2^I)$$

$$\partial F_2/\partial n_1 = \omega_2^I$$

$$\partial F_2/\partial n_2 = \omega_2^I - 1$$

$$\partial F_2/\partial y_1^I = \rho_t^V k_{21}^\bullet(\partial \omega_1^I/\partial y_1^I) + (\rho_t^V k_{22}^\bullet + n_t)(\partial \omega_2^I/\partial y_1^I)$$

$$\partial F_2/\partial y_2^I = \rho_t^V k_{21}^\bullet(\partial \omega_1^I/\partial y_2^I) + (\rho_t^V k_{22}^\bullet + n_t)(\partial \omega_2^I/\partial y_2^I)$$

The partial derivatives of the mass fractions with respect to the mole fractions are given by

$$\partial \omega_i/\partial y_j = \delta_{ij} M_i/\bar{M} - y_i M_i M_j/\bar{M}^2$$

The above derivatives must be evaluated at the interface composition before use in computing the Jacobian elements. This additional complexity in evaluating the derivatives of the vapor-phase mass transfer rate equations arises because we have used mass fluxes and mole fractions as independent variables. If we had used mass fractions in place of mole factions the derivatives of the rate equations would be simpler, but the derivatives of the equilibrium equations would be more complicated. For simplicity, we have ignored the dependence of the mass transfer coefficients themselves on the mixture composition and on the fluxes.

The partial derivatives of the energy balance are given below

$$\partial F_3/\partial n_1 = \overline{H}_1^V - \overline{H}_1^L$$

$$\partial F_3/\partial n_2 = \overline{H}_2^V - \overline{H}_2^L$$

$$\partial F_3/\partial T^I = -h^L - h^V \Xi_H + n_1\left(C_{p1}^V - C_{p1}^L\right) + n_2\left(C_{p2}^V - C_{p2}^L\right)$$

where we have assumed the heat transfer coefficients and the flux correction factor to be constant. The C_p terms in the third equation are due to the temperature dependence of the vapor- and liquid-phase enthalpies.

The partial derivatives of the liquid-phase mass transfer rate equation are (assuming the liquid-phase density and mass transfer coefficient may be regarded as constant)

$$\partial F_4/\partial n_1 = \left(x_1^I - 1\right)/M_1$$

$$\partial F_4/\partial n_2 = x_1^I/M_2$$

$$\partial F_4/\partial x_1^I = -c_t^L k^L \Xi^L + n_1/M_1 + n_2/M_2$$

Expressions for the partial derivatives of the equilibrium equations with respect to the interface temperature and compositions are the same as those given in Example 11.5.2. The partial derivatives of the mole fraction summation equations are either unity or zero (cf. Example 11.5.2).

DATA   A good deal of physical property data is needed for interphase mass and energy-transfer calculations. The data and methods used for this example are summarized below.

Maxwell–Stefan diffusion coefficients in the vapor phase

$$\mathcal{D}_{12} = 2.93 \times 10^{-6} \text{ m}^2/\text{s}$$

$$\mathcal{D}_{13} = 31.8 \times 10^{-6} \text{ m}^2/\text{s}$$

$$\mathcal{D}_{23} = 29.0 \times 10^{-6} \text{ m}^2/\text{s}$$

Molar masses

$$M_1 = 0.05808 \qquad M_2 = 0.0781 \qquad M_3 = 0.004 \text{ kg/mol}$$

Vapor viscosity and thermal conductivity

$$\mu^V = 1.485 \times 10^{-5} \text{ Pa s}$$

$$\lambda^V = 0.025 \text{ W/mK}$$

The liquid-phase mass transfer coefficient is taken to be

$$k^L = 2.26 \times 10^{-4} \text{ m/s}$$

This is the value used by Krishna (1981a) in his simulations of Modine's experiments. This value is consistent with a penetration model of mass transfer in the liquid phase with a contact time of 0.065 s or a surface renewal frequency of 0.25 s^{-1}. The molar density of the liquid phase has been estimated as

$$c_t^L = 11.34 \text{ kmol/m}^3$$

The vapor-phase molar density may be calculated from the ideal gas law.

The liquid-phase heat transfer coefficient may be estimated using a correlation provided by Modine et al. (1963). The numerical value used in this example is

$$h^L = 2063 \text{ W/m}^2\text{K}$$

The K values of acetone and benzene may be calculated from

$$K_1 = \gamma_1 P_1^s / P \qquad K_2 = \gamma_2 P_2^s / P$$

The pure component vapor pressures may be estimated from the Antonie equation as follows:

$$\ln P_1^s = 21.625 - 2975.9/(T - 34.523)$$

$$\ln P_2^s = 21.068 - 2948.8/(T - 44.563)$$

where P_1^s and P_2^s are in pascals and T is in kelvin.

The liquid phase is a binary mixture so we may use the Margules equation (Table D.2) to estimate the activity coefficients. The interaction parameters are taken from Gmehling and Onken (1977ff, Vol. I/3 + 4, p. 197)

$$A_{12} = 0.4608 \qquad A_{21} = 0.3159$$

For this example we shall assume that the partial molar enthalpies are equal to the pure component enthalpies. The liquid enthalpy may be calculated from

$$\overline{H}_i^L(T) = C_{pi}^L(T - T_{\text{ref}})$$

and the enthalpy of the vapor species will be

$$\overline{H}_i^V(T) = C_{pi}^L(T_{b,i} - T_{\text{ref}}) + \Delta H_{\text{vap},i} + C_{pi}^V(T - T_{b,i})$$

where $T_{b,i}$ is the normal boiling point of species i. The reference temperature used in these calculations is 300 K. The pure component vapor-phase heat capacities have been estimated at the bulk vapor temperature as

$$C_{p1}^V = 1300 \qquad C_{p2}^V = 1100 \qquad C_{p3}^V = 5200 \text{ J/kg K}$$

and the pure component liquid heat capacities at the bulk liquid temperature are

$$C_{p1}^L = 2241 \qquad C_{p2}^L = 1773 \text{ J/kg K}$$

The latent heats of vaporization at the normal boiling points have the values

$$\Delta H_{\text{vap1}} = 501.7 \text{ kJ/kg} \qquad \Delta H_{\text{vap2}} = 394.1 \text{ kJ/kg}$$

and the normal boiling points are

$$T_{b1} = 329.43 \text{ K} \qquad T_{b2} = 353.25 \text{ K}$$

Viscosities, thermal conductivities, liquid density, and pure component heat capacities have been estimated at the temperature and composition in the respective bulk phases. These properties are assumed constant for this example. In view of the small temperature changes that are encountered here, this is a reasonable assumption.

SOLUTION We will illustrate the evaluation of the functions F_1–F_8 using the following "estimates" of the independent variables.

$$n_1 = 1.1467 \times 10^{-3} \text{ kg/m}^2\text{s} \qquad n_2 = 18.809 \times 10^{-3} \text{ kg/m}^2\text{s}$$

$$y_1^I = 0.03603 \qquad y_2^I = 0.12480 \qquad y_3^I = 0.83917$$

$$x_1^I = 0.07667 \qquad x_2^I = 0.92333$$

$$T^I = 32.39°\text{C}$$

The elements of the matrix of low flux mass transfer coefficients may be computed using Eqs. 10.4.25 and 10.4.30. This requires the matrix of Fick diffusion coefficients in the mass average velocity reference frame. This matrix can be computed with the help of Eqs. 4.2.2, from which $[D]$ is obtained in the molar average velocity reference frame, and Eqs. 3.2.11, which allows transformation to the mass average reference velocity frame. Thus, we need the Maxwell–Stefan diffusivities of the three binary pairs in the vapor phase and the molar masses of the three components.

The first step is to convert the mole fractions given above, as well as those in the bulk vapor, to mass fractions. The required formulas are given in Table 1.1 and the results are

$$\omega_1^V = 0.44512 \qquad \omega_2^V = 0.0 \qquad \omega_3^V = 0.55488$$

$$\omega_1^I = 0.13770 \qquad \omega_2^I = 0.64140 \qquad \omega_3^I = 0.22089$$

The arithmetic average mole fractions will be needed in the calculation of the matrix of Fick diffusivities $[D]$.

$$y_{av1} = 0.04419 \qquad y_{av2} = 0.06240 \qquad y_{av3} = 0.89341$$

Converting the average mole fractions to mass fractions yields

$$\omega_{av1} = 0.23304 \qquad \omega_{av2} = 0.44249 \qquad \omega_{av3} = 0.32447$$

The next step is to calculate $[D]$ in the molar average reference velocity frame directly from Eqs. 4.2.7 as

$$[D] = \begin{bmatrix} 22.063 & 6.288 \\ 8.793 & 23.322 \end{bmatrix} \times 10^{-6} \text{ m}^2/\text{s}$$

The elements of the reference frame transformation matrix $[B^{uo}]$ are given by Eqs. 1.2.25; the numerical values are

$$[B^{uo}] = \begin{bmatrix} 1.597 & 0.609 \\ 1.134 & 2.156 \end{bmatrix}$$

The inverse matrix is

$$[B^{ou}] = \begin{bmatrix} 0.783 & -0.221 \\ -0.412 & 0.580 \end{bmatrix}$$

The average mass and mole fractions given above were used in the computation of $[B^{uo}]$ and $[B^{uo}]$.

The matrix $[D]$ in the mass average reference velocity frame may now be calculated directly from Eqs. 3.2.11 as

$$[D] = \begin{bmatrix} 21.73 & 5.703 \\ 9.600 & 23.659 \end{bmatrix} \times 10^{-6} \text{ m}^2/\text{s}$$

The eigenvalues of this matrix are

$$\hat{D}_1 = 30.16 \times 10^{-6} \qquad \hat{D}_2 = 15.23 \times 10^{-6} \text{ m}^2/\text{s}$$

We require the density of the vapor mixture in order to calculate the low flux mass transfer coefficients. The molar density of the vapor may be estimated using the ideal gas law and, since the system is almost isothermal, may safely be assumed to be nearly constant. The mass density, however, is likely to vary considerably between the bulk and interface, since the molar masses of the three components in the vapor phase cover such a wide range. The mass density should, therefore, be evaluated with the average molar mass

$$\rho_t^V = \overline{M}P/RT^V$$

$$= 0.5562 \text{ kg/m}^3$$

where $\overline{M}$ is the molar mass of the vapor evaluated at the average molar composition.

The total vapor flow rate is the flow rate of inert helium divided by its bulk phase mole fraction

$$V = v_3/y_3^V = 0.2574 \text{ mol/s}$$

where v_3 is the molar flow rate of helium. The velocity of the vapor mixture u may be calculated from

$$V = c_t^V u A$$

where A is the cross-sectional area of the column (4.92×10^{-4} m^2). The velocity is found to be

$$u = 10.37 \text{ m/s}$$

The Reynolds number follows as:

$$\text{Re} = \rho_t^V u d/\mu^V$$

$$= 9714$$

The friction factor may be calculated from a correlation obtained by Modine (1963)

$$f = 2 \times 0.926(0.0007 + 0.0625/\text{Re}^{0.32})$$
$$= 0.00743$$

The eigenvalues of [Sc] are calculated from

$$\widehat{Sc_1} = \mu^V/\rho_t^V \hat{D}_1 = 0.8853$$
$$\widehat{Sc_1} = \mu^V/\rho_t^V \hat{D}_2 = 1.753$$

The eigenvalues of [St] may now be calculated from Eq. 10.3.27

$$\widehat{St_1} = 0.00397 \qquad \widehat{St_2} = 0.00270$$

[St] follows from Eq. 10.4.31 as

$$[St] = \begin{bmatrix} 3.252 & 0.489 \\ 0.824 & 3.418 \end{bmatrix} \times 10^{-3}$$

The matrix of low flux mass transfer coefficients is obtained by multiplying [St] by the velocity u

$$[k] = \begin{bmatrix} 3.369 & 0.502 \\ 0.859 & 3.545 \end{bmatrix} \times 10^{-2} \text{ m}^2/\text{s}$$

which has the following eigenvalues

$$\hat{k}_1 = 4.120 \times 10^{-2} \text{ m/s} \qquad \hat{k}_2 = 2.794 \times 10^{-2} \text{ m/s}$$

The next step is to compute the multicomponent mass transfer coefficients modified for the effect of a nonzero total flux. The total mass flux is needed for the evaluation of the high flux correction factor and is

$$n_t = n_1 + n_2 = 19.95 \times 10^{-3} \text{ kg/m}^2\text{s}$$

The eigenvalues of the rate factor matrix are given by Eq. 10.4.34 with numerical values

$$\hat{\Phi}_1 = 0.8708 \qquad \hat{\Phi}_2 = 1.2839$$

The eigenvalues of the correction factor matrix are obtained from the film theory expression (Eq. 10.4.35), and the eigenvalues of the high flux mass transfer coefficient matrix follow from Eq. 10.4.32

8.4.2S

$$\hat{\Xi}_1 = 0.6270 \qquad \qquad \hat{\Xi}_2 = 0.4917$$
$$\hat{k}_1^{\bullet} = 2.583 \times 10^{-2} \text{ m/s} \qquad \hat{k}_2^{\bullet} = 1.374 \times 10^{-2} \text{ m/s}$$

The matrix of high flux mass transfer coefficients $[k^{\bullet}]$ may now be computed from Eqs.

8.4.31 as

$$[k^\bullet] = \begin{bmatrix} 1.900 & 0.462 \\ 0.787 & 2.057 \end{bmatrix} \times 10^{-2} \text{ m/s}$$

The mass diffusion fluxes may be calculated from Eq. 10.4.19 as

$$j_1 = \rho_t^V\{k_{11}^\bullet(\omega_1^I - \omega_1^V) + k_{12}^\bullet(\omega_2^I - \omega_2^V)\}$$

$$= -1.6104 \times 10^{-3} \text{ kg/m}^2\text{s}$$

$$j_2 = \rho_t^V\{k_{21}^\bullet(\omega_1^I - \omega_1^V) + k_{22}^\bullet(\omega_2^I - \omega_2^V)\}$$

$$= 6.0087 \times 10^{-2} \text{ kg/m}^2\text{s}$$

$$j_3 = -j_1 - j_2$$

$$= -4.407 \times 10^{-2} \text{ kg/m}^2\text{s}$$

The first two discrepancy functions F_1 and F_2 may now be evaluated as

$$F_1 = j_1 + \omega_1^I(n_1 + n_2) - n_1$$

$$= -1.6104 \times 10^{-3} + 0.1377 \times (1.1467 + 18.809) \times 10^{-3}$$

$$- 1.1466 \times 10^{-3}$$

$$\approx 0$$

$$F_2 = j_2 + \omega_2^I(n_1 + n_2) - n_2$$

$$= 6.0087 \times 10^{-2} + 0.6440 \times (1.1467 + 18.809) \times 10^{-3}$$

$$- 18.809 \times 10^{-3}$$

$$\approx 0$$

To evaluate the energy balance we need the heat transfer coefficients and partial molar enthalpies in both phases. The vapor-phase heat transfer coefficient may be estimated from Eq. 11.4.30 as shown below.

The average heat capacity of the gas mixture is calculated at the mean composition between the interface and bulk gas phases

$$\bar{C}_p = \omega_{av1}C_{p1}^V + \omega_{av2}C_{p2}^V + \omega_{av3}C_{p3}^V$$

$$= 2476.1 \text{ J/kg K}$$

The Prandtl number of the vapor mixture is

$$\text{Pr} = \bar{C}_p\mu/\lambda$$

$$= 2476.1 \times 14.85 \times 10^{-6}/0.025$$

$$= 1.471$$

The Reynolds number and friction factor were calculated above. The Stanton number for heat transfer may now be evaluated from Eq. 11.4.29 as

$$\text{St}_H = 0.002984$$

and so the low flux heat transfer coefficient is

$$\overset{\vee}{h} = u\rho_t \overline{C}_p \mathrm{St}_H$$
$$= 10.37 \times 0.5562 \times 2476.1 \times 0.002984$$
$$= 42.62 \ \mathrm{W/m^2 K}$$

The heat transfer rate factor is

$$\Phi_H = (n_1 C_{p1} + n_2 C_{p2})/h^V$$
$$= 0.5204$$

and the high flux heat transfer coefficient is

$$h^\bullet = h^V \Phi_H/(\exp \Phi_H - 1)$$
$$= 32.49 \ \mathrm{W/m^2 K}$$

Evaluation of the enthalpies at the interface temperature using the formulas presented above yields the following numerical values

$$\overline{H}_1^V(T^I) = C_{p1}^L(T_{b1} - T_{\mathrm{ref}}) + \Delta H_{\mathrm{vap1}} + C_{p1}^V(T^I - T_{b1})$$
$$= 2.241 \times (329.43 - 300.0) + 501.7 + 1.300 \times (305.54 - 329.43)$$
$$= 536.6 \ \mathrm{kJ/kg}$$

$$\overline{H}_2^V(T^I) = C_{p2}^L(T_{b2} - T_{\mathrm{ref}}) + \Delta H_{\mathrm{vap2}} + C_{p2}^V(T^I - T_{b2})$$
$$= 1.773 \times (353.25 - 300.0) + 394.1 + 1.100 \times (305.54 - 353.25)$$
$$= 436.0 \ \mathrm{kJ/kg}$$

$$\overline{H}_1^L(T^I) = C_{p1}^L(T^I - T_{\mathrm{ref}})$$
$$= 2.241 \times (305.54 - 300.0)$$
$$= 12.43 \ \mathrm{kJ/kg}$$

$$\overline{H}_2^L(T^I) = C_{p1}^L(T^I - T_{\mathrm{ref}})$$
$$= 1.773 \times (305.54 - 300.0)$$
$$= 9.83 \ \mathrm{kJ/kg}$$

The use of constant physical properties is acceptable in this case since the composition and temperature changes over the length of the column are not substantial. The system is at a pressure slightly above atmospheric but that has been ignored in making the enthalpy calculations.

The energy balance function may now be evaluated as

$$F_3 \equiv h^L(T^L - T^I) - h^V \Xi_H(T^I - T^V)$$
$$\qquad - n_1(\overline{H}_1^V - \overline{H}_1^L) - n_2(\overline{H}_2^V - \overline{H}_2^L)$$
$$= 2096 \times (36.45 - 32.39) - 32.49 \times (32.39 - 36.0)$$
$$\qquad - 1.1467 \times 10^{-3} \times (536.6 - 12.43) \times 10^3$$
$$\qquad - 18.809 \times 10^{-3} \times (436.0 - 9.83) \times 10^3$$
$$\approx 0$$

We now turn our attention to the liquid-phase mass transfer rate equation. The high flux correction to the liquid-phase mass transfer coefficient is given by the second part of Eq. 8.2.13. However, its value is sufficiently close to unity to have no impact on the results. Accordingly, we shall set the liquid-phase mass transfer flux correction factor to unity.

We may now complete the evaluation of the liquid-phase mass transfer equation.

$$F_4 \equiv c_t^L k^L \Xi^L \left(x_1^L - x_1^I \right) + x_1^I (n_1/M_1 + n_2/M_2) - n_1/M_1$$

$$= 11.34 \times 10^3 \times 2.26 \times 10^{-4} \times 1 \times (0.07658 - 0.07667)$$

$$+ 0.07667 \times (1.14667/0.0581 + 18.809/0.0781) \times 10^{-3}$$

$$- 1.160 \times 10^{-3}/0.0581$$

$$\approx 0$$

To evaluate the equilibrium relations requires the computation of the K values. The pure component vapor pressures may be estimated from the Antoine equation. At the interface temperature the vapor pressures are

$$P_1^s = 41,961 \text{ Pa} \qquad P_2^s = 17,482 \text{ Pa}$$

Substituting the interface mole fractions x^I into the Margules equation for activity coefficients (Table D.2) gives the following results:

$$\gamma_1 = 1.4534 \qquad \gamma_2 = 1.0034$$

The K values may now be computed as

$$K_1 = 0.4699 \qquad K_2 = 0.1352$$

We may now evaluate the departures from equilibrium as

$$F_5 = K_1 x_1^I - y_1^I$$

$$= 0.4699 \times 0.07667 - 0.03603$$

$$\approx 0$$

$$F_6 = K_2 x_2^I - y_2^I$$

$$= 0.1352 \times 0.92333 - 0.12480$$

$$\approx 0$$

The interfacial mole fractions satisfy the mole fraction summation equations

$$F_7 = 0 \qquad F_8 = 0$$

It is apparent that the "initial" estimate with which we began these computations was, in fact, the correct solution. When Algorithm 11.1 was used to obtain these results, convergence was obtained in about 15 iterations from what was more or less the starting guess described in that algorithm. The interface mole fraction of benzene was given an initial value of 0.05 since a value of 0.0 (the bulk value) gave numerical problems during the computations. In some cases, negative mole fractions were obtained after the first iteration, but these caused no problems even when left uncorrected before continuing with the calculations. That might not always be the case; in general, it is a good idea to reset negative mole fractions to small nonzero values before continuing.

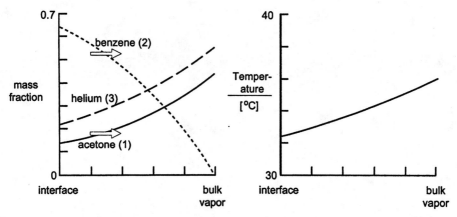

Figure 11.7. Composition and temperature profiles in the vapor phase at the top of the wetted wall column. Bulk composition corresponds to the conditions in an experiment carried out by Modine (1963). Arrows indicate the actual direction of mass transfer.

Composition and temperature profiles in the vapor phase at the top of the column are shown in Figure 11.7. Note the arrows show that acetone is evaporating even though the concentration of acetone is significantly higher in the bulk vapor than it is at the interface. Diffusional interactions are responsible only for part of the "reverse" mass flux of acetone; there is a convective contribution that provides the balance of the flux.   ■

PART III
Design

12 Multicomponent Distillation: Mass Transfer Models

> Distillation trays are so simple... Sieve tray decks are, after all, hardly more than sheets of metal with a few holes punched in them. This of course is part of the fascination—that the behavior of something so simple can be so difficult to predict with regard to its hydrodynamic and mass transfer performance.
>
> —M. J. Lockett (1986)

Distillation retains its position of supremacy among chemical engineering unit operations despite the emergence in recent years of many new separation techniques (e.g., membranes). In fact, when choosing a separation scheme the first question that is usually asked is Why not distillation? (King, 1980).

It is beyond the scope of this book to describe distillation equipment at any length; in depth treatments are available in, for example, Smith (1964), Billet (1979), King (1980), Fair (1984) Lockett (1986) and Kister (1992). However, some comments are needed to place the material in Chapters 13 and 14 in their proper context.

Distillation is most frequently carried out in multitray columns, although packed columns have long been the preferred alternative when pressure drop is an important consideration. In recent years, the development of highly efficient structured packings has led to increased use of packed columns in distillation.

The design of both types of distillation columns is a fascinating subject to which a great many books and papers have been devoted (some were cited above). The modeling of mass transfer on distillation trays and the use of these mass transfer models in the simulation of multicomponent distillation and absorption columns are the aspects of the process design function that we shall consider in this book.

Two different approaches have evolved for the simulation and design of multicomponent distillation columns. The conventional approach is through the use of an *equilibrium stage model* together with methods for estimating the *tray efficiency*. This approach is discussed in Chapter 13. An alternative approach based on direct use of matrix models of multicomponent mass transfer is developed in Chapter 14. This *nonequilibrium stage model* is also applicable, with only minor modification, to gas absorption and liquid–liquid extraction and to operations in trayed or packed columns.

In this chapter we set the scene for these later chapters by developing models that represent the overall mass transfer performance in distillation columns.

12.1 BINARY DISTILLATION IN TRAY COLUMNS

A schematic diagram of the two-phase dispersion on a distillation tray is shown in Figure 12.1, which serves to introduce some of the symbols we shall use in this and subsequent chapters. The composition of the vapor below the tray is y_{iE} and y_{iL} is the composition of

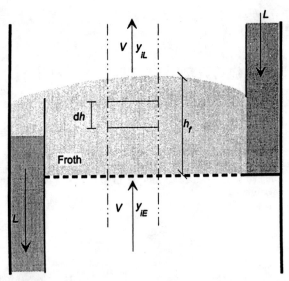

Figure 12.1. Schematic illustration of the froth on a distillation tray. The entering vapor-phase mole fractions are y_{iE}; y_{iL} represents the average exiting vapor-phase mole fractions.

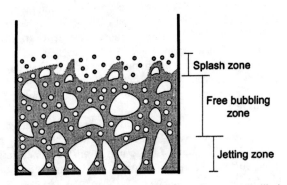

Figure 12.2. Idealized view of the free bubbling regime on a distillation tray.

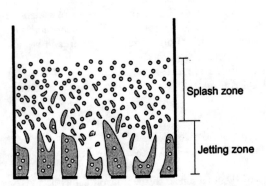

Figure 12.3. Idealized view of the spray regime on a distillation tray.

the vapor above the dispersion in the narrow slice of froth shown in the figure. The parameter h_f is the froth height.

Two fundamentally different regimes of operation on distillation trays have been distinguished (Lockett, 1986): the *free bubbling regime*, pictured in Figure 12.2, and the *spray regime*, depicted in Figure 12.3. Detailed modeling of the hydrodynamics and mass transfer coefficients for these regimes is postponed until Section 12.1.7. Below we present the material balance relations that are the starting point for all subsequent analyses.

12.1.1   Material Balance Relations

Let us concentrate our attention on the narrow, vertical slice of froth in Figure 12.1. The liquid phase in the vertical slice will be assumed to be well mixed and have a bulk phase composition x_i. The vapor phase is assumed to rise through the liquid in plug flow. The composition of the vapor phase depends on the distance above the tray.

If v_i represents the molar flow rate of component i in the vapor phase, $V = \Sigma\, v_i$ the total vapor flow rate, a the interfacial area per unit volume of froth, h_f the froth height, and A_b the active bubbling area, then the component material balance for the vapor phase may be written as

$$dv_i/dh = -N_i a A_b \qquad i = 1, 2, \ldots, n \tag{12.1.1}$$

where N_i is the molar flux of species i across the vapor–liquid interface. For the mixture we may sum Eqs. 12.1.1 to give

$$dV/dh = -N_t a A_b \tag{12.1.2}$$

Substituting $N_i = J_i^V + y_i N_t$ in Eq. 12.1.1 and writing $v_i \equiv y_i V$ we obtain

$$V\frac{dy_i}{dh} + y_i\frac{dV}{dh} = -J_i^V a A_b - y_i N_t a A_b \tag{12.1.3}$$

In view of Eq. 12.1.2, the right members on both sides of Eq. 12.1.3 cancel each other and we get

$$V(dy_i/dh) = -J_i^V a A_b \qquad i = 1, 2, \ldots, n \tag{12.1.4}$$

This result is valid even when $N_t \neq 0$, that is, when we have nonequimolar transfer.

Equations 12.1.4 must be integrated over the froth height to yield the composition profiles. The boundary conditions are

$$h = 0; \qquad y_i = y_{iE}$$
$$h = h_f; \qquad y_i = y_{iL} \tag{12.1.5}$$

The material balance relations presented above are valid for any number of components. We shall discuss solutions to this system of equations for binary mixtures in the remainder of this section of Chapter 12 before moving on to obtain generalized results for multicomponent systems in Section 12.2. In the analyses that follow we shall ignore the effects of heat transfer between the vapor and liquid phases.

12.1.2   Composition Profiles

For a binary system the material balance relation (Eq. 12.1.6) may be written for component 1 as

$$V\frac{dy_1}{dh} = -J_1^V a A_b \tag{12.1.6}$$

A similar expression for component 2 is not needed since $y_1 + y_2 = 1$.

To compute the diffusion flux we use Eq. 7.3.9 (simplified here by ignoring the flux correction)

$$J_1^V = c_t^V K_{OV}(y_1 - y_1^*) \tag{7.3.9a}$$

where K_{OV} is the overall mass transfer coefficient given by Eq. 7.3.15.

$$\frac{1}{K_{OV}} = \frac{1}{k^V} + \frac{c_t^V}{c_t^L}\frac{M}{k^L} \tag{7.3.15}$$

When the binary rate relation (Eq. 7.3.9a) is combined with the material balance Eq. 12.1.6 we obtain

$$V(dy_1/dh) = c_t^V K_{OV}(y_1^* - y_1)a A_b \tag{12.1.7}$$

Equation 12.1.7 may be solved, subject to the boundary conditions (Eq. 12.1.5) to give

$$(y_1^* - y_{1L}) = \exp(-\mathsf{N}_{OV})(y_1^* - y_{1E}) \tag{12.1.8}$$

where N_{OV}, the overall number of transfer units for the vapor phase, is defined by

$$\mathsf{N}_{OV} \equiv \int_0^{h_f}(c_t^V K_{OV} a A_b/V)\,dh \tag{12.1.9}$$

If we assume the integrand in Eq. 12.1.9 to be independent of froth height we may complete the integration to give the overall number of transfer units N_{OV} as

$$\mathsf{N}_{OV} \equiv c_t^V K_{OV} a h_f A_b/V \tag{12.1.10}$$

and the vapor composition at any point in the dispersion y_1 may be obtained from

$$(y_1^* - y_1) = \exp(-\mathsf{N}_{OV}\zeta)(y_1^* - y_{1E}) \tag{12.1.11}$$

where ζ is a dimensionless froth height defined by $\zeta = h/h_f$.

12.1.3   Mass Transfer Rates

The rates of interphase transfer vary as the vapor rises through the froth and the compositions change. The average molar flux of component 1 ($\bar{N}_1$) in the vertical section of the froth in Figure 12.1 may be found from

$$\bar{N}_1 = c_t^V K_{OV}(\bar{y}_1 - y_1^*) \tag{12.1.12}$$

where $\bar{y}_1$ is the average mole fraction of component 1 and is defined by

$$\bar{y}_1 \equiv \int_0^1 y(\zeta)\,d\zeta \tag{12.1.13}$$

We may use Eq. 12.1.11 for $y(\zeta)$ with the result

$$(y_1^* - \bar{y}_1) = (y_1^* - y_{1E})/\Omega^V \qquad (12.1.14)$$

where Ω^V is defined by

$$\Omega^V = (-\mathbb{N}_{OV})/(\exp(-\mathbb{N}_{OV}) - 1) \qquad (12.1.15)$$

Equation 12.1.15 may be combined with Eq. 12.1.12 to give

$$\bar{N}_1 = c_t^V K_{OV}(\Omega^V)^{-1}(y_{1E} - y_1^*) \qquad (12.1.16)$$

Equation 12.1.16 is based on the assumptions that K_{OV} may be considered constant over the height of the formation zone and that the total flux is zero. If the total flux is not zero the right-hand side of Eq. 12.1.16 must be multiplied by the bootstrap coefficient β^V and the calculation of the overall mass transfer coefficient modified (see Example 12.1.2).

12.1.4 Numbers of Transfer Units

Equation 12.1.8 relates the entering and exiting vapor mole fractions through the overall number of transfer units. To predict tray performance, therefore, we need to estimate this quantity. A working relationship for $\mathbb{N}_{OV}$ may be obtained by combining Eq. 12.1.10 with Eq. 7.3.15 for K_{OV} to give

$$\frac{1}{\mathbb{N}_{OV}} = \frac{1}{\mathbb{N}_V} + \frac{M(V/L)}{\mathbb{N}'_L} \qquad (12.1.17)$$

where $\mathbb{N}_V$ and $\mathbb{N}'_L$ are the numbers of transfer units for the vapor and liquid phases defined by

$$\mathbb{N}_V \equiv k^V a' t_V = k^V a h_f/u_s \qquad (12.1.18)$$

$$\mathbb{N}'_L \equiv k^L \bar{a} t_L = k^L a h_f Z/(Q_L/W) \qquad (12.1.19)$$

where h_f is the froth height, Z is the liquid flow path length, W is the weir length, Q_L is the volumetric liquid flow rate ($Q_L = L/c_t^L$) and u_s is the superficial vapor velocity based on the bubbling area of the tray

$$u_s = Q_V/A_b \qquad (12.1.20)$$

where Q_V is the volumetric vapor flow rate ($Q_V = V/c_t^V$), a' is the interfacial area per unit volume of vapor, and $\bar{a}$ is the interfacial area per unit volume of liquid. The area terms a' and $\bar{a}$ are related to a, the interfacial area per unit volume of froth, by

$$a' = a/(1 - \alpha)$$
$$\bar{a} = a/\alpha$$

where α is the relative froth density (h_L/h_f). The parameters t_V and t_L are the residence times of the vapor and liquid phases, respectively:

$$t_V \equiv (1 - \alpha)h_f/u_s \qquad (12.1.21)$$

$$t_L \equiv Z/u_L = h_L ZW/Q_L \qquad (12.1.22)$$

where u_L is the horizontal liquid velocity.

The prime on N'_L appears because an alternative definition of the number of transfer units for the liquid phase has been suggested by Lockett to deal with some fundamental deficiencies in the definition of N'_L. While Lockett is undoubtedly right that the conventional *derivations* of N_L are at fault, there is no ambiguity in the *calculation* of N'_L.

In practice, the numbers of transfer units for each phase may be obtained from either one of the following sources:

- Experimental data.
- Empirical correlations (of experimental data).
- Theoretical models.

12.1.5 Numbers of Transfer Units from Empirical Correlations

The *AIChE Bubble Tray Design Manual* (AIChE, 1958; Gerster et al., 1958) presented the first comprehensive estimation procedure for numbers of transfer units. For many years this remained the only such procedure available in the open literature; the work of organizations like Fractionation Research Incorporated (FRI) was available only to member companies. However, during the last 15 years or so there has been a revival of distillation research and other comprehensive estimation procedures have been published (e.g., Zuiderweg, 1982; Chan and Fair, 1984a). We summarize these methods below. The text by Lockett (1986) provides an excellent summary of what is available in the open literature on distillation tray design for those interested in further study.

1. *AIChE Method.* The *AIChE* correlation of the number of transfer units for the vapor phase on both bubble caps and sieve trays is

$$N_V = (0.776 + 4.57h_w - 0.238F_s + 104.8Q_L/W)\text{Sc}_V^{-0.5} \qquad (12.1.23)$$

where F_s is the superficial F factor defined by

$$F_s = u_s(\rho_t^V)^{0.5} \qquad (12.1.24)$$

where Sc_V is the vapor-phase Schmidt number ($\text{Sc}_V = \mu^V/\rho_t^V D^V$), D^V is the Fick diffusivity in the vapor phase in meters squared per second (m^2/s), h_w is the exit weir height in meters (m), W is the weir length in meters (m), and Q_L is the volumetric liquid flow rate in meters cubed per second (m^3/s).

The liquid-phase number of transfer units for sieve trays only is given by

$$N'_L = 19,700(D^L)^{0.5}(0.4F_s + 0.17)t_L \qquad (12.1.25)$$

where D^L is the Fick diffusivity in the liquid phase (in meters squared per second) and t_L is the liquid-phase residence time defined by Eq. 12.1.22.

The original *AIChE* method includes a correlation for the clear liquid height h_L; it is preferable, however, to use the more recent correlation of Bennett et al. (1983)

$$h_L = \alpha_e\{h_w + C(Q_L/(W\alpha_e))^{0.67}\} \qquad (12.1.26)$$

where

$$\alpha_e = \exp\left\{-12.55\left(u_s\{\rho_t^V/(\rho_t^L - \rho_t^V)\}^{0.5}\right)^{0.91}\right\} \qquad (12.1.27)$$

and

$$C = 0.50 + 0.438 \exp(-137.8h_w) \qquad (12.1.28)$$

2. *Chan and Fair Method.* Chan and Fair (1984a) retained the *AIChE* (1985) procedure for calculation of the $\mathbb{N}'_L$ but developed an alternative correlation for the number of transfer units in the vapor phase $\mathbb{N}_V$ as follows:

$$\mathbb{N}_V = (10{,}300 - 8670F_f)\, F_f (D^V)^{0.5}\, t_V/h_L^{0.5} \qquad (12.1.29)$$

where F_f is the fractional approach to flooding defined by

$$F_f = u_s/u_{sf} \qquad (12.1.30)$$

where u_{sf} is the superficial velocity under flooding conditions and t_V is given by

$$t_V = (1 - \alpha_e)h_L/(\alpha_e u_s) \qquad (12.1.31)$$

The parameters h_L and α_e are to be calculated using the method of Bennett et al. (1983) [see, also, Lockett (1986) for details].

3. *Zuiderweg Method.* Zuiderweg's (1982) method involves calculation of $\mathbb{N}_V$ and $\mathbb{N}'_L$ from separate correlations of k^V, k^L, and (ah_f). The correlation for the vapor-phase mass transfer coefficient is

$$k^V = 0.13/\rho_t^V - 0.065/(\rho_t^V)^2 \qquad (1 < \rho_t^V < 80 \text{ kg/m}^3) \qquad (12.1.32)$$

It is interesting to note that Zuiderweg's correlation for k^V is independent of the diffusion coefficient. The liquid-phase mass transfer coefficient is calculated from either

$$k^L = 2.6 \times 10^{-5}(\mu^L)^{-0.25} \qquad (12.1.33)$$

or

$$k^L = 0.024(D^L)^{0.25} \qquad (12.1.34)$$

The parameter (ah_f) is dependent on the regime of operation; for the spray regime

$$ah_f = \frac{40}{\phi^{0.3}}\left(\frac{u_s^2 \rho_t^V h_L \text{ FP}}{\sigma}\right)^{0.37} \qquad (12.1.35)$$

and for the mixed froth–emulsion flow regime

$$ah_f = \frac{43}{\phi^{0.3}}\left(\frac{u_s^2 \rho_t^V h_L \text{ FP}}{\sigma}\right)^{0.53} \qquad (12.1.36)$$

where ϕ is the fractional free area of the tray, $\phi = A_h/A_b$ where A_h is the total area of the holes, and A_b is the bubbling area of the tray. The flow parameter (FP) is defined by

$$\text{FP} = (M_L/M_V)(\rho_t^V/\rho_t^L)^{0.5} \qquad (12.1.37)$$

where M_L and M_V are the mass flow rates of liquid and vapor phases.

The transition from the spray regime to mixed froth–emulsion flow is described by

$$FP > 3.0bh_L$$

where b is the weir length per unit bubbling area

$$b = W/A_b \tag{12.1.38}$$

For estimation of the clear liquid height Zuiderweg (1982) gives the following correlation:

$$h_L = 0.6h_w^{0.5}(p\ FP/b)^{0.25} \tag{12.1.39}$$

where p is the hole pitch.

Example 12.1.1   Distillation of Toluene–Methylcyclohexane

Estimate the numbers of transfer units for the system toluene (1)–methylcyclohexane(2) for a distillation tray with the following specifications:

Column diameter	0.6 m
Weir length	0.457 m
Flow path length	0.374 m
Active tray bubbling area	0.2 m^2
Downcomer area	0.034 m^2
Total hole area	0.0185 m^2
Tray spacing	0.34 m
Hole diameter	4.8 mm
Hole pitch	12.7 mm
Exit weir height	50 mm

The molar flow rate of the vapor and liquid phases leaving the tray are

$$V = 4.54 \text{ mol/s} \qquad L = 4.80 \text{ mol/s}$$

The pressure is 101 kPa. The composition of the liquid leaving the tray is

$$x_1 = 0.52 \qquad x_2 = 0.48$$

The flooding velocity has been estimated to be

$$u_{sf} = 0.949 \text{ m/s}$$

DATA   Physical properties of the mixture at the bubble point temperature of the liquid have been estimated as follows:

Viscosity of vapor mixture: 3.373×10^{-5} Pa s.
Viscosity of liquid mixture: 2.203×10^{-4} Pa s.
Vapor density: 2.986 kg/m^3.
Liquid density: 726 kg/m^3.
Average molar mass of vapor: 0.0955 kg/mol.
Average molar mass of liquid: 0.0952 kg/mol.
The vapor-phase diffusivity: $D^V = 3.856 \times 10^{-6}$ m^2/s.
The liquid-phase diffusivity: $D^L = 7.082 \times 10^{-9}$ m^2/s.
Surface tension: $\sigma = 0.0169$ N/m.
The slope of the vapor–liquid equilibrium data: $M = 1.152$.

The properties are not particularly sensitive to the composition of the vapor and liquid mixtures.

SOLUTION We will begin by estimating the numbers of transfer units using the *AIChE* correlations.

The Schmidt number is calculated first as

$$Sc_V = \mu^V/\rho^V D^V$$
$$= 3.373 \times 10^{-5}/(2.986 \times 3.856 \times 10^{-6})$$
$$= 2.930$$

The gas velocity may be calculated from the vapor flow rate as

$$u_s = V/(A_b c_t^V)$$
$$= 4.54/\{0.2 \times (2.986/0.0955)\}$$
$$= 0.726 \text{ m/s}$$

The F factor is calculated next

$$F_s = u_s(\rho_t^V)^{0.5}$$
$$= 0.726 \times 2.986^{0.5}$$
$$= 1.255 \text{ (m/s)}(kg/m^3)^{0.5}$$

The volumetric liquid flow rate is determined as follows:

$$Q_L = L\overline{M}_L/\rho_t^L$$
$$= (4.80 \times 0.0952)/726$$
$$= 6.294 \times 10^{-4} \text{ m}^3/\text{s}$$

The number of transfer units N_V may now be computed using Eq. 12.1.23 as

$$N_V = (0.776 + 4.57 \times 0.05 - 0.238 \times 1.255 + 104.8 \times 6.294 \times 10^{-4}/0.457)/2.930^{0.5}$$
$$= 0.4968$$

The froth height is determined using the method of Bennett et al. (1983) as discussed above with the following results:

$$\alpha_e = 0.4623$$
$$h_f = 60.16 \text{ mm}$$
$$h_L = 27.81 \text{ mm}$$

The liquid-phase residence time may now be calculated from Eq. 12.1.22 as

$$t_L = h_L Z W/Q_L$$
$$= 27.81 \times 10^{-3} \times 0.374 \times 0.457/6.294 \times 10^{-4}$$
$$= 7.553 \text{ s}$$

The number of transfer units for the liquid phase may now be calculated as

$$N_L' = 1.97 \times 10^4 \times (7.082 \times 10^{-9})^{0.5}(0.4 \times 1.255 + 0.17) \times 7.553$$
$$= 8.412$$

The overall number of transfer units follows from Eq. 12.1.17

$$\frac{1}{N_{OV}} = \frac{1}{N_V} + \frac{M(V/L)}{N_L'}$$

The stripping factor $M(V/L)$ has the value 1.090. Thus N_{OV} is computed from

$$\frac{1}{N_{OV}} = \frac{1}{0.4968} + \frac{1.090}{8.412}$$

and so

$$N_{OV} = 0.4667$$

We will repeat the estimation of the numbers of transfer units using the Chan and Fair correlation for the number of transfer units for the vapor phase. The fractional approach to flooding is

$$F_f = u_s/u_{sf}$$
$$= 0.726/0.949$$
$$= 0.7650$$

The froth height and clear liquid height are as calculated above; thus the gas-phase residence time is

$$t_V = (1 - \alpha_e)h_f/u_s$$
$$= 0.0446 \text{ s}$$

The number of transfer units for the vapor phase may now be calculated as

$$N_V = (10,300 - 8670 \times 0.7650) \times 0.7650 \times (3.856 \times 10^{-6})^{0.5} \times 0.0446/(0.02781)^{0.5}$$
$$= 1.472$$

For the liquid-phase Chan and Fair make use of the *AIChE* correlation; thus, $N_L' = 8.412$, as calculated above. The overall number of transfer units follows from Eq. 12.1.17 as

$$N_{OV} = 1.236$$

In Zuiderweg's method we determine the mass transfer coefficients for the vapor phase from Eq. 12.1.32 as

$$k^V = 0.13/2.986 - 0.065/2.986^2$$
$$= 0.0362 \text{ m/s}$$

and for the liquid phase using Eq. 12.1.34

$$k^L = 0.024 \times (7.082 \times 10^{-9})^{0.25}$$
$$= 2.2017 \times 10^{-4} \text{ m/s}$$

The mass flow rate of the vapor and liquid phases are

$$M_V = 4.54 \times 0.0955 = 0.4336 \text{ kg/s}$$
$$M_L = 4.80 \times 0.0952 = 0.4570 \text{ kg/s}$$

The flow parameter is computed next

$$FP = (0.4570/0.4336) \times (2.986/726)^{0.5}$$
$$= 0.0676$$

The clear liquid height is calculated from Zuiderweg's correlation (Eq. 12.1.39), as

$$h_L = 0.6 \times (0.050)^{0.5} \times (0.0127 \times 0.0676/2.285)^{0.25}$$
$$= 0.0187 \text{ m}$$

The parameter $(3bh_L)$ has the value 0.128, which is greater than the value of the flow parameter FP. According to the criterion below Eq. 12.1.37, this means that the tray is operating in the spray regime. The interfacial area term is, therefore, calculated from Eq. 12.1.35 as

$$ah_f = 37.00$$

The number of transfer units for the vapor phase now follows from Eq. 12.1.18

$$\mathbb{N}_V = k^V ah_f/u_s$$
$$= 0.0362 \times 37.00/0.726$$
$$= 1.8474$$

and for the liquid phase from Eq. 12.1.19 as

$$\mathbb{N}'_L = k^L ah_f Z/(Q_L/W)$$
$$= 2.2017 \times 10^{-4} \times 37.00 \times 0.374/(6.294/0.457)$$
$$= 2.2122$$

The overall number of transfer units is

$$\mathbb{N}_{OV} = 0.9673$$

We shall continue these calculations in Example 13.2.1. ■

12.1.6 Numbers of Transfer Units—A Simplified Approach

An interesting simplification of the above procedure results when we use Eq. 7.3.7 for the linearized equilibria (for a binary system, $M = (K_1\Gamma)$ and Eqs. 8.8.16 for the mass transfer coefficients $(k = \kappa\Gamma)$ to get (the vapor phase is assumed ideal)

$$\frac{1}{K_{OV}} = \frac{1}{\kappa^V} + \frac{c_t^V}{c_t^L}\frac{K_1}{\kappa^L} \tag{8.8.21}$$

which is noteworthy for the absence of the thermodynamic factor Γ. If we combine Eq. 8.8.21 with Eq. 12.1.17 we may express the overall number of transfer units as

$$\frac{1}{\mathbb{N}_{OV}} = \frac{1}{\mathscr{N}_V} + \frac{K_1(V/L)}{\mathscr{N}'_L} \tag{12.1.40}$$

where $\mathcal{N}_V$ and $\mathcal{N}'_L$ are numbers of transfer units for the vapor and liquid phases defined as follows:

$$\mathcal{N}_V \equiv \kappa^V a' t_V \tag{12.1.41}$$

$$\mathcal{N}'_L \equiv \kappa^L \bar{a} t_L \tag{12.1.42}$$

Note the presence of the "ideal" mass transfer coefficients κ^V and κ^L in place of the conventional k^V and k^L.

We suggest that the numbers of transfer units $\mathcal{N}_V$ and $\mathcal{N}'_L$ be evaluated from the models presented above with the Maxwell–Stefan diffusivities $Ð^V$ and $Ð^L$ replacing the Fick diffusivities. The elimination of the thermodynamic factor will prove particularly useful when we come to adapt this method for estimating efficiencies of multicomponent systems.

12.1.7   A Fundamental Model of Tray Performance

As noted in Section 12.1.1, there are two quite different regimes of operation of distillation trays, the froth regime and the spray regime. In this section we develop detailed models of the hydrodynamic and mass transfer character of these regimes.

The froth regime on a distillation tray really consists of three zones.

1. Zone *I*   Jetting–bubble formation region.
2. Zone *II*   Free bubbling zone.
3. Zone *III*   The splash zone.

In Zone *I*, the jetting–bubbling formation zone, the vapor issues through the perforations in the tray in the form of jets, breaking into bubbles. These jets could be modeled as a set of parallel cylindrical vapor jets. The model parameters are

- The diameter of the jet—d_I.
- The velocity of the vapor in the jet—U_I.
- The height of the jetting zone—h_I.

In the free bubbling zone a distribution of bubble sizes is usually obtained. The model parameters for the bubbling zone are

- The bubble diameters—$d_{II,k}$.
- The bubble rise velocities—$U_{II,k}$.
- The height of the free bubbling zone—h_{II}.
- The fraction of vapor that is in each bubble population—$f_{II,k}$.

The fraction of vapor in each bubble population are given by

$$f_{II,k} = U_{II,k} \varepsilon_{II,k} / u_s \tag{12.1.43}$$

where u_s is the superficial velocity based on the bubbling area and $\varepsilon_{II,k}$ is the gas hold-up of the kth bubble population. The bubble population fractions sum to unity.

Available experimental data show that the assumption of a bimodal bubble size distribution is a good approximation (Lockett and Plaka, 1983; Lockett, 1986; Kaltenbacher, 1982; Hofer, 1983; Prado and Fair, 1990). Thus, we have fast-rising "large" bubbles and slow-rising "small" bubbles in the dispersion. This bimodal bubble size distribution is also a characteristic feature in gas–liquid bubble columns operating in the churn-turbulent regime (Vermeer and Krishna, 1981) and has hydrodynamic analogies with a gas–solid fluidised bed (Van Deemter, 1961). Typically, about 90% of the incoming vapor is transported by the large bubbles (this is a kind of vapor channelling) while the small bubbles despite their large interfacial area are not very effective in mass transfer contributing only about 10% of the total transfer (Krishna, 1985).

The large bubbles are of the order of 10–20 mm in diameter and have a rise velocity of about 1.5 m/s. The small bubbles, on the other hand, are 2–5 mm in diameter and have a rise velocity of about 0.25 m/s. The large bubbles rise through the froth virtually in plug flow. Prado and Fair (1990), in developing their tray hydrodynamics model, assumed the small bubble population to also rise in plug flow through the froth. If we draw an analogy between the hydrodynamics on a tray with a gas–liquid bubble column (Krishna, 1993b), we may consider the small bubbles to be entrained in the froth and, as a first approximation, to have the backmixing characteristics of the froth on the tray. Further experimental work needs to be carried out to settle this issue. For purposes of calculation later on we shall follow Prado and Fair (1990) and assume plug flow of the small bubble population.

The splash zone above the free bubbling zone consists of entrained droplets. We may model this zone as being made up of droplets of uniform size rising in plug flow through the splash zone. The model parameters are

- The diameter of the drop—d_{III}.
- The velocity of the entrained droplets—U_{III}.
- The height of the splash zone—h_{III}.

The contribution to the total mass transfer of the splash zone is generally negligibly small and is neglected in the working model for mass transfer calculations pictured in Figure 12.4.

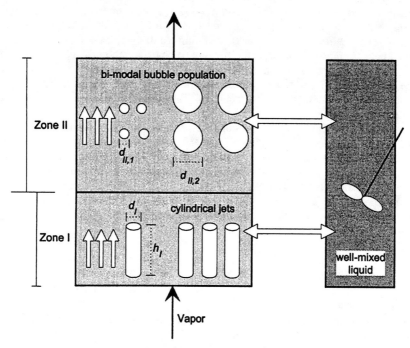

Figure 12.4. Model of the free bubbling regime on a distillation tray.

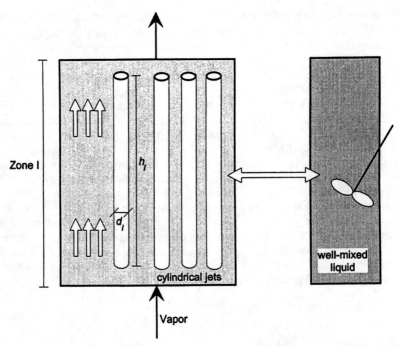

Figure 12.5. Model of the spray regime on a distillation tray.

It is interesting to note a further analogy between tray hydrodynamics and gas–solid fluidized beds: the splash zone is akin to the freeboard zone in the latter case.

At higher vapor velocities we attain the *spray regime* of operation, pictured in Figure 12.3. In the spray regime the vapor jets penetrate through the whole dispersion. Between these vapor jets we have liquid "pockets" containing entrained small vapor bubbles. Almost all the entering vapor is transported by the vapor jets, the small entrained bubbles contributing almost nothing to vapor through-flow and mass transfer.

The splash zone at the top of the dispersion has the same characteristics as the one described above for the free bubbling zone; its contribution to mass transfer can usually be neglected.

The model of the spray regime for mass transfer calculations is pictured in Figure 12.5.

The key to developing an expression for estimating the overall mass transfer performance of a single tray using the zone-model is to recognize that in each region of the froth (bubble formation zone and each bubble population) the composition change in the vapor is given by Eq. 12.1.8, where y_E and y_L refer to the mole fractions entering and leaving the respective region. The overall numbers of transfer units must, therefore, be found for each region; we shall use the simplified procedure described in Section 12.1.4 for estimating these numbers.

We define the following quantities:

y_{iE} The mole fraction of component i in the vapor entering the tray.

$y_{i,I}$ The mole fraction of component i in the vapor at the top of the bubble formation zone.

$y_{i,II,k}$ The mole fraction of component i in the vapor in the kth bubble population at the top of the bubble rise zone.

y_{iL} The average mole fraction of the vapor above the froth.

The bulk liquid is assumed completely mixed and so y_i^*, the mole fractions in a vapor in equilibrium with the bulk liquid, will be the same for all regions of the froth.

Bubble Formation Zone   The change in the vapor composition in the bubble formation zone is given by (cf., Eq. 12.1.8).

$$(y_1^* - y_{1I}) = Q_I(y_1^* - y_{1E}) \tag{12.1.44}$$

where

$$Q_I = \exp(-\mathbb{N}_{OV,I}) \tag{12.1.45}$$

with $\mathbb{N}_{OV,I}$ the overall number of transfer units for the bubble formation zone. The parameter $\mathbb{N}_{OV,I}$ is defined by

$$\mathbb{N}_{OV,I} \equiv K_{OV,I} a_I' t_I \tag{12.1.46}$$

where a_I' is the interfacial area per unit volume of vapor

$$a_I' = 4/d_I \tag{12.1.47}$$

and t_I is the residence time for the vapor

$$t_I = h_I/U_I \tag{12.1.48}$$

where d_I is the diameter of the jet, U_I is the velocity of the vapor in the jet, and h_I is the height of the jetting zone.

The overall mass transfer coefficient for the jetting zone is given by (cf. Eq. 8.8.21)

$$\frac{1}{K_{OV,I}} = \frac{1}{\kappa_I^V} + \frac{c_t^V}{c_t^L} \frac{K_1}{\kappa_I^L} \tag{12.1.49}$$

with the vapor-phase mass transfer coefficient for the bubble formation zone obtained from Eqs. 9.4.14 and 9.4.12

$$\kappa_I^V = -\ln(1 - F_I)/(a_I' t_I) \tag{12.1.50}$$

where

$$F_I = 1 - 4 \sum_{n=1}^{\infty} \frac{\exp\{-\lambda_n^2 r_I^2 \text{Fo}_I\}}{\lambda_n^2 r_I^2} \tag{12.1.51}$$

where r_I is the radius of the jet and with Fo_I the Fourier number for the jetting zone.

$$\text{Fo}_I = 4\text{Ð}^V t_I/d_I^2 \tag{12.1.52}$$

The penetration model is used for the mass transfer coefficient in the liquid phase

$$\kappa_I^L = 2(\text{Ð}^L/\pi t_I)^{1/2} \tag{12.1.53}$$

The average molar flux of component 1, $\bar{N}_{1I}$, in the bubble formation zone are given by (cf. Eq. 12.1.16)

$$\bar{N}_{1I} = c_t^V K_{OV,I}(\Omega_I^V)^{-1}(y_{1E} - y_1^*) \tag{12.1.54}$$

where

$$\Omega_I^V = (-\mathsf{N}_{OV,I})/(\exp(-\mathsf{N}_{OV,I}) - 1) \tag{12.1.55}$$

Equation 12.1.54 is subject to the same assumptions and modifications that apply to Eq. 12.1.16.

The kth Bubble Population   The change in composition in the kth bubble population as the bubbles rise from the top of the formation zone to the top of the froth is given by

$$(y_1^* - y_{1II,k}) = Q_{II,k}(y_1^* - y_{1I}) \tag{12.1.56}$$

where

$$Q_{II,k} = \exp(-\mathsf{N}_{OV,II,k}) \tag{12.1.57}$$

with $\mathsf{N}_{OV,II,k}$ the overall number of transfer units for the kth bubble population. The parameter $\mathsf{N}_{OV,II,k}$ is defined by

$$\mathsf{N}_{OV,II,k} \equiv K_{OV,II,k} a'_{II,k} t_{II,k} \tag{12.1.58}$$

where $a'_{II,k}$ is the interfacial area per unit volume of vapor in the kth bubble population

$$a'_{II,k} = 6/d_{II,k} \tag{12.1.59}$$

and $t_{II,k}$ is the residence time for the vapor in the kth bubble population

$$t_{II,k} = h_{II}/U_{II,k} \tag{12.1.60}$$

where $d_{II,k}$ is the diameter of the kth bubble population, $U_{II,k}$ is the rise velocity of the kth bubble population, and h_{II} is the height of the bubble rise zone.

$$h_{II} = h_f - h_I \tag{12.1.61}$$

The overall mass transfer coefficient for each bubble population is given by (cf. Eq. 8.8.21)

$$\frac{1}{K_{OV,II,k}} = \frac{1}{\kappa_{II,k}^V} + \frac{c_t^V}{c_t^L} \frac{K_1}{\kappa_{II,k}^L} \tag{12.1.62}$$

with the vapor-phase mass transfer coefficient for a rigid spherical bubble obtained from Eq. 9.4.5 as

$$\kappa_{II,k}^V = -\ln(1 - F_{II,k})/(a'_{II,k} t_{II,k}) \tag{12.1.63}$$

where $F_{II,k}$ is the fractional approach to equilibrium in the kth bubble population and is given by Eq. 9.4.6

$$F_{II,k} = 1 - \frac{6}{\pi^2} \sum_{n=1}^{\infty} \frac{1}{n^2} \exp\{-n^2\pi^2 \mathrm{Fo}_{II,k}\} \tag{12.1.64}$$

with $Fo_{II,k}$ the Fourier number for the kth bubble population

$$Fo_{II,k} = \mathcal{D}^V t_{II,k}/r_{II,k}^2 \tag{12.1.65}$$

where $r_{II,k}$ is the radius of the kth bubble population.

The penetration model is used for the mass transfer coefficient in the liquid phase; the contact time being the time required for the bubble to rise one diameter

$$\kappa_{II,k}^L = 2\left[\mathcal{D}^L/\pi(d_{II,k}/U_{II,k})\right]^{1/2} \tag{12.1.66}$$

The average molar flux of component 1, $\bar{N}_{1II,k}$, in the kth bubble population may be found from (cf. Eq. 12.1.16)

$$\bar{N}_{1II,k} = c_t^V K_{OV,II,k}\left(\Omega_{II,k}^V\right)^{-1}(y_{1I} - y_1^*) \tag{12.1.67}$$

where

$$\Omega_{II,k}^V = (-\mathbb{N}_{OV,II,k})/(\exp(-\mathbb{N}_{OV,II,k}) - 1) \tag{12.1.68}$$

The vapor composition at the top of the formation zone may be eliminated using Eq. 12.1.44 to give

$$\bar{N}_{1II,k} = c_t^V K_{OV,II,k}\left(\Omega_{II,k}^V\right)^{-1} Q_I(y_{1E} - y_1^*) \tag{12.1.69}$$

Overall Performance   The average composition of the vapor above the froth is given by

$$y_{i,L} = \sum_{k=1}^{p} f_{II,k} y_{i,II,k} \tag{12.1.70}$$

where p is the number of discrete bubble populations in the model and $f_{II,k}$ is the fraction of the vapor in the kth bubble population (both parameters of the model). When we combine Eq. 12.1.56 with Eq. 12.1.70 we find

$$(y_1^* - y_{1L}) = Q_{II}(y_1^* - y_{1I}) \tag{12.1.71}$$

where Q_{II} is defined by

$$Q_{II} = \left\{\sum_{k=1}^{p} f_{II,k} Q_{II,k}\right\} \tag{12.1.72}$$

The mole fraction difference at the top of the formation zone may be eliminated with Eq. 12.1.44 to give

$$(y_1^* - y_{1L}) = Q_{II} Q_I(y_1^* - y_{1E}) \tag{12.1.73}$$

The total mass transferred between phases may be determined by summing the contributions from the bubble formation zone and each bubble population. The total number of moles transferred in the kth bubble population is the product of the average molar flux given by Eq. 12.1.69 and the total interfacial area in that population $(a_{II,k}h_{II}A_b\varepsilon_{II,k})$. The total number of moles transferred in the bubble formation zone is the product of the average molar flux given by Eq. 12.1.54 and the total interfacial area in the formation zone $(a_I h_I A_b \varepsilon_I)$, where ε_I is the gas hold-up in Zone I.

Example 12.1.2   Regeneration of Triethylene Glycol

Estimate the composition of the vapor above the froth on a tray in a column for the regeneration of triethylene glycol(2) (TEG) from a mixture with water(1) under the following conditions:

Composition in the bulk vapor below tray

$$y_{1E} = 0.7216 \qquad y_{2E} = 0.2784$$

Composition in the liquid on the tray

$$x_1 = 0.258 \qquad x_2 = 0.742$$

Equilibrium vapor composition

$$y_1^* = 0.969 \qquad y_1^* = 0.031$$

The froth may be modeled with a jet–bubble formation zone and a bimodal bubble population with the following characteristics:

The height of the jetting zone: $h_I = 2$ mm.
The diameter of the jet: $d_I = 3$ mm.
The velocity of the vapor in the jet: $U_I = 5$ m/s.
The height of the free bubbling zone: $h_{II} = 80$ mm.
The small bubble diameter: $d_{II,1} = 3$ mm.
The small bubble rise velocity: $U_{II,1} = 0.23$ m/s.
The fraction of vapor in the small bubbles: $f_{II,1} = 0.05$.
The large bubble diameter: $d_{II,2} = 13$ mm.
The large bubble rise velocity: $U_{II,2} = 1.5$ m/s.
The fraction of vapor in the large bubbles: $f_{II,2} = 0.95$.

DATA   Physical properties have been estimated as follows:

Liquid molar density: 8.621 kmol/m³.
Vapor molar density: 32.25 mol/m³.
The vapor-phase Maxwell–Stefan diffusivity

$$Ð^V = 1.4 \times 10^{-5} \text{ m}^2/\text{s}$$

The liquid-phase Maxwell–Stefan diffusivity

$$Ð^L = 2.0 \times 10^{-9} \text{ m}^2/\text{s}$$

The K value: $K_1 = 3.756$.
Heat of vaporization of water: $\Delta H_{\text{vap}_1} = 40.5$ kJ/mol.
Heat of vaporization of TEG: $\Delta H_{\text{vap}_2} = 70$ kJ/mol.

SOLUTION   The heats of vaporization of TEG and water are sufficiently different as to make it necessary to consider departures from equimolar distillation in the evaluation of the overall mass transfer coefficients. The overall mass transfer coefficient in each region of the froth will, therefore, be evaluated from (cf. Eq. 8.8.21)

$$\frac{1}{K_{OV}} = \frac{1}{\kappa^V} + \frac{c_t^V}{c_t^L} \frac{\beta^V}{\beta^L} \frac{K_1}{\kappa^L}$$

where β^V and β^L are the bootstrap coefficients for the vapor and liquid phases given by (in this case) Eqs. 7.2.16, 7.2.14, and 7.2.18. The bootstrap coefficient for the liquid phase will be evaluated at the liquid composition as follows:

$$\beta^L = 1 - x_1 \Lambda_1^L$$

with

$$\Lambda_1^L = \frac{\Delta H_{vap1} - \Delta H_{vap2}}{x_1 \Delta H_{vap1} + x_2 \Delta H_{vap2}}$$

Hence,

$$\Lambda_1 = \frac{40.5 - 70}{0.258 \times 40.5 + 0.742 \times 70}$$

$$= -0.473$$

$$\beta^L = 1 - 0.258 \times (-0.473)$$

$$= 1.122$$

The bootstrap coefficient for the vapor phase will be evaluated at the average of the entering vapor and equilibrium vapor composition

$$\beta^V = 1 - y_{1av} \Lambda_1^V$$

with

$$\Lambda_1^V = \frac{\Delta H_{vap1} - \Delta H_{vap2}}{y_{1av} \Delta H_{vap1} + y_{2av} \Delta H_{vap2}}$$

The average vapor composition is

$$y_{1av} = 0.8455 \qquad y_{2av} = 0.1545$$

Hence,

$$\Lambda_1 = \frac{40.5 - 70}{0.8455 \times 40.5 + 0.1545 \times 70}$$

$$= -0.655$$

$$\beta^V = 1 - 0.8455 \times (-0.655)$$

$$= 1.553$$

The values of β^L and β^V computed above will be used in the determination of the overall mass transfer coefficients in all three regions of the dispersion: the bubble formation–jetting zone and both small and large bubble populations.

Bubble Formation–Jetting Zone   The interfacial area per unit volume of vapor in the jetting zone is

$$a'_I = 4/d_I$$
$$= 4/0.003$$
$$= 1333 \text{ m}^2/\text{m}^3$$

The residence time for the vapor in a jet is

$$t_I = h_I/U_I$$
$$= 0.002/5$$
$$= 0.0004 \text{ s}$$

The Fourier number for a jet is

$$\text{Fo}_I = 4 \, Đ^V t_I/(d_I)^2$$
$$= 4 \times 1.4 \times 10^{-5} \times 0.0004/0.003^2$$
$$= 2.489 \times 10^{-3}$$

The fractional approach to equilibrium in the jetting zone is given by Eq. 12.1.51 and is computed to be

$$F_I = 0.1101$$

The vapor-phase mass transfer coefficient may now be calculated from Eq. 12.1.50 as

$$\kappa_I^V = -\ln(1 - F_I)/(a'_I t_I)$$
$$= -\ln(1 - 0.1101)/(1333 \times 0.0004)$$
$$= 0.2187 \text{ m/s}$$

Turning to the liquid phase, we may evaluate the mass transfer coefficient with Eq. 12.1.53 as

$$\kappa_I^L = 2(Đ^L/\pi t_I)^{1/2}$$
$$= 2 \times (2 \times 10^{-9}/(\pi \times 0.0004))^{1/2}$$
$$= 2.523 \times 10^{-3} \text{ m/s}$$

We may now compute the overall mass transfer coefficient for the jetting zone as

$$\frac{1}{K_{OV,I}} = \frac{1}{0.2187} + \frac{32.23}{8621} \times \frac{1.553}{1.122} \times \frac{3.756}{2.523 \times 10^{-3}}$$

Hence,

$$K_{OV,I} = 0.08142 \text{ m/s}$$

The overall number of transfer units is

$$\mathbb{N}_{OV,I} = K_{OV,I} a'_I t_I$$
$$= 0.08142 \times 1333 \times 0.0004$$
$$= 0.04342$$

The departure from equilibrium in the jetting zone follows as:

$$Q_I = \exp(-\mathbb{N}_{OV,I})$$
$$= \exp(-0.04342)$$
$$= 0.9575$$

The molar fluxes in the formation zone are computed as follows (cf. Eq. 12.1.54)

$$\overline{N}_{1I} = c_t^V \beta_I^V K_{OV,I} (\Omega_I^V)^{-1} (y_{1E} - y_1^*)$$

with

$$\Omega_I^V = (-\mathbb{N}_{OV,I})/(\exp(-\mathbb{N}_{OV,I}) - 1)$$
$$= (-0.04342)/(\exp(-0.04342) - 1)$$
$$= 1.0219$$

Hence

$$\overline{N}_{1I} = 32.25 \times 1.553 \times 0.08142 \times (1.0219)^{-1}(0.7216 - 0.969)$$
$$= -0.9875 \text{ mol/s}$$

Small Bubble Population We now turn our attention to the determination of the mass transfer characteristics of the bubble rise zone, beginning with the small bubble population.

The interfacial area per unit volume of vapor in the small bubble population is

$$a'_{II,1} = 6/d_{II,1}$$
$$= 6/0.003$$
$$= 2000 \text{ m}^2/\text{m}^3$$

The residence time for the vapor in the small bubbles is

$$t_{II,1} = h_{II}/U_{II,1}$$
$$= 0.08/0.23$$
$$= 0.3478 \text{ s}$$

The Fourier number for the small bubbles is

$$\text{Fo}_{II,1} = 4 \, \mathcal{D}^V t_{II,1}/(d_{II,1})^2$$
$$= 4 \times 1.4 \times 10^{-5} \times 0.3478/0.003^2$$
$$= 2.1643$$

The fractional approach to equilibrium in the small bubbles is given by Eq. 12.1.64 and is computed as

$$1 - F_{II,1} = 3.215 \times 10^{-10}$$

The vapor-phase mass transfer coefficient is calculated from Eq. 12.1.63 as

$$\kappa_{II,1}^V = -\ln(1 - F_{II,1})/(a'_{II,1} t_{II,1})$$
$$= -\ln(3.215 \times 10^{-10})/(2000 \times 0.3478)$$
$$= 0.03142 \text{ m/s}$$

The mass transfer coefficient for the liquid phase is computed from Eq. 12.1.66 as

$$\kappa_{II,1}^{L} = 2\left(Đ^{L}/(\pi d_{II,1}/U_{II,1})\right)^{1/2}$$
$$= 2 \times \left(2 \times 10^{-9}/(\pi \times 0.003/0.23)\right)^{1/2}$$
$$= 4.419 \times 10^{-4} \text{ m/s}$$

The overall mass transfer coefficient for the small bubbles is calculated next

$$\frac{1}{K_{OV,II,1}} = \frac{1}{0.03142} + \frac{32.23}{8621} \times \frac{1.553}{1.122} \times \frac{3.756}{4.419 \times 10^{-4}}$$

Hence

$$K_{OV,II,1} = 0.01318 \text{ m/s}$$

The overall number of transfer units is

$$N_{OV,II,1} = K_{OV,II,1}a'_{II,1}t_{II,1}$$
$$= 0.01318 \times 2000 \times 0.3478$$
$$= 9.172$$

The departure from equilibrium in the small bubbles follows as:

$$Q_{II,1} = \exp(-N_{OV,II,1})$$
$$= \exp(-9.172)$$
$$= 1.0394 \times 10^{-4}$$

In other words, the small bubbles are close to equilibrium when they reach the top of the froth.

The molar fluxes in the small bubble population are computed next. The factor $\Omega_{II,1}^{V}$ is computed first as

$$\Omega_{II,1}^{V} = (-N_{OV,II,1})/(\exp(-N_{OV,II,1}-1)$$
$$= (-9.172)/(\exp(-9.172)-1)$$
$$= 9.173$$

The high value of $\Omega_{II,1}^{V}$ is another indication that the small bubble population is close to equilibrium. The average molar flux is

$$\bar{N}_{1II,1} = c_{t}^{V}\beta_{II,1}^{V}K_{OV,II,1}\left(\Omega_{II,1}^{V}\right)^{-1}Q_{I}(y_{1E}-y_{1}^{*})$$
$$= 32.25 \times 1.553 \times 0.01318 \times (9.173)^{-1} \times 0.9575 \times (0.7216 - 0.969)$$
$$= -0.01706 \text{ mol/s}$$

Large Bubble Population   The final part of the problem is the calculation of the mass transfer performance of the large bubble population.

The interfacial area per unit volume of vapor in the large bubble population is

$$a'_{II,2} = 6/d_{II,2}$$
$$= 6/0.013$$
$$= 461.5 \text{ m}^{2}/\text{m}^{3}$$

The residence time for the vapor in the large bubbles is

$$t_{II,2} = h_{II}/U_{II,2}$$
$$= 0.08/1.5$$
$$= 0.0533 \text{ s}$$

The Fourier number for the large bubbles is

$$\text{Fo}_{II,2} = 4 \, Ð^V t_{II,2}/(d_{II,2})^2$$
$$= 4 \times 1.4 \times 10^{-5} \times 0.0533/0.013^2$$
$$= 0.01767$$

The fractional approach to equilibrium in the large bubbles is computed as for the small bubble population with Eq. 12.1.64.

$$F_{II,2} = 0.3970$$

The vapor-phase mass transfer coefficient may now be calculated from Eq. 12.1.63 as

$$\kappa^V_{II,2} = -\ln(1 - F_{II,2})/(a'_{II,2} t_{II,2})$$
$$= -\ln(1 - 0.3970)/(461.5 \times 0.0533)$$
$$= 0.02055 \text{ m/s}$$

The mass transfer coefficient for the liquid phase is computed from Eq. 12.1.66 as for the small bubble population.

$$\kappa^L_{II,2} = 2\left(Ð^L/(\pi d_{II,2}/U_{II,2})\right)^{1/2}$$
$$= 2 \times \left(2 \times 10^{-9}/(\pi \times 0.013/1.5)\right)^{1/2}$$
$$= 5.421 \times 10^{-4} \text{ m/s}$$

We may now calculate the overall mass transfer coefficient for the large bubbles as

$$\frac{1}{K_{OV,II,2}} = \frac{1}{0.02055} + \frac{32.23}{8621} \times \frac{1.553}{1.122} \times \frac{3.756}{5.421 \times 10^{-4}}$$

Hence,

$$K_{OV,II,2} = 0.01183 \text{ m/s}$$

The overall number of transfer units for the large bubble population is

$$\mathbb{N}_{OV,II,2} = K_{OV,II,2} a'_{II,2} t_{II,2}$$
$$= 0.0193 \times 461.5 \times 0.0533$$
$$= 0.2912$$

The departure from equilibrium in the large bubbles follows as:

$$Q_{II,2} = \exp(-\mathbb{N}_{OV,II,2})$$
$$= \exp(-0.2912)$$
$$= 0.7474$$

The molar fluxes in the large bubble population may now be calculated. The factor $\Omega_{II,2}^V$ is computed

$$\Omega_{II,2}^V = (-\mathbb{N}_{OV,II,2})/(\exp(-\mathbb{N}_{OV,II,2}) - 1)$$

$$= (-0.2912)/(\exp(-0.2912) - 1)$$

$$= 1.1526$$

Hence,

$$\bar{N}_{1II,2} = c_t^V \beta_{II,2}^V K_{OV,II,2}(\Omega_{II,2}^V)^{-1} Q_I(y_{1E} - y_1^*)$$

$$= 32.25 \times 1.553 \times 0.01183 \times (1.1526)^{-1} \times 0.9575 \times (0.7216 - 0.969)$$

$$= -0.12178 \text{ mol/s}$$

Overall Performance The overall departure from equilibrium for the bubble rise zone is given by (cf. Eq. 12.1.72)

$$Q_{II} = f_1 Q_{II,1} + f_2 Q_{II,2}$$

However, the contribution of the small bubbles ($f_1 Q_{II,1}$), is essentially nothing and Q_{II} is well approximated by

$$Q_{II} \approx f_2 Q_{II,2}$$

$$= 0.95 \times 0.7474$$

$$= 0.7100$$

Armed with the value of Q_{II} we may calculate the composition of the vapor above the tray as

$$y_{1L} = y_1^* - Q_{II} Q_I(y_1^* - y_{1I})$$

$$= 0.969 - 0.7100 \times 0.9575 \times (0.969 - 0.7216)$$

$$= 0.8008$$

12.2 MULTICOMPONENT DISTILLATION IN TRAY COLUMNS

12.2.1 Composition Profiles

Let us now try to extend the model of binary distillation developed in Section 12.1.1 to multicomponent systems. The extension is based on the work of Toor (1964b) and the starting point is the material balance Eqs. 12.1.4, which must now be combined in $n - 1$ dimensional matrix form as

$$V\frac{d(y)}{dh} = -(J^V)aA_b \tag{12.2.1}$$

The vapor-phase diffusion fluxes at any point are expressed using Eq. 7.3.11 (once again ignoring the high flux correction)

$$(J^V) = c_t^V[K_{OV}](y - y^*) \tag{7.3.11a}$$

where $[K_{OV}]$ is the matrix of overall mass transfer coefficients and accounts for resistances to mass transfer in both vapor and liquid phases. Introducing the rate relations (Eq. 7.3.11a) into Eq. 12.2.1 gives

$$V\frac{d(y)}{dh} = c_t^V[K_{OV}](y^* - y)aA_b \tag{12.2.2}$$

Equation 12.2.2 may be integrated over the dispersion height to give

$$(\Delta y_L) = [Q](\Delta y_E) \tag{12.2.3}$$

where $(\Delta y_L) = (y^* - y_L)$, $(\Delta y_E) = (y^* - y_E)$, and

$$[Q] \equiv \exp[-[\mathbb{N}_{OV}]] \tag{12.2.4}$$

and where $[\mathbb{N}_{OV}]$, the overall number of transfer units for the vapor phase, is defined by

$$[\mathbb{N}_{OV}] \equiv \int_0^{h_f}\{c_t^V[K_{OV}]aA_b/V\}\,dh \tag{12.2.5}$$

As in the binary case, we assume the integrand in Eq. 12.2.5 to be independent of froth height in order to complete the integration to give the overall number of transfer units, $[\mathbb{N}_{OV}]$, as

$$[\mathbb{N}_{OV}] \equiv c_t^V[K_{OV}]ah_fA_b/V$$
$$\equiv c_t^V[K_{OV}]a't_V \tag{12.2.6}$$

and the composition of the froth at any point is obtained from the matrix generalization of Eq. 12.1.11 as

$$(y^* - y(\zeta)) = \exp[-[\mathbb{N}_{OV}]\zeta](y^* - y_E) \tag{12.2.7}$$

where $\zeta = h/h_f$.

If the matrix of overall numbers of transfer units $[\mathbb{N}_{OV}]$ is known, (a topic we address below) $[Q]$ can be evaluated using the truncated power series, Eq. A.6.3, by diagonalization (Eq. A.5.32) or by Sylvester's expansion formula (Eq. A.5.21). For a ternary system the latter two methods may be expressed as (assuming the two eigenvalues are distinct)

$$[Q] = \frac{\hat{Q}_1[[\mathbb{N}_{OV}] - \hat{\mathbb{N}}_{OV2}[I]]}{\hat{\mathbb{N}}_{OV1} - \hat{\mathbb{N}}_{OV2}} + \frac{\hat{Q}_2[[\mathbb{N}_{OV}] - \hat{\mathbb{N}}_{OV1}[I]]}{\hat{\mathbb{N}}_{OV2} - \hat{\mathbb{N}}_{OV1}} \tag{12.2.8}$$

where $\hat{\mathbb{N}}_{OV_1}$ and $\hat{\mathbb{N}}_{OV_2}$ are the eigenvalues of $[\mathbb{N}_{OV}]$

$$\hat{\mathbb{N}}_{OV1} = \tfrac{1}{2}\{\mathrm{tr}[\mathbb{N}_{OV}] + \sqrt{\mathrm{disc}[\mathbb{N}_{OV}]}\}$$
$$\hat{\mathbb{N}}_{OV2} = \tfrac{1}{2}\{\mathrm{tr}[\mathbb{N}_{OV}] - \sqrt{\mathrm{disc}[\mathbb{N}_{OV}]}\} \tag{12.2.9}$$

where

$$\text{tr}[\mathsf{N}_{OV}] = \mathsf{N}_{OV11} + \mathsf{N}_{OV22}$$

$$\text{disc}[\mathsf{N}_{OV}] = (\text{tr}[\mathsf{N}_{OV}])^2 - 4|\mathsf{N}_{OV}|$$

$$|\mathsf{N}_{OV}| = \mathsf{N}_{OV11}\mathsf{N}_{OV22} - \mathsf{N}_{OV12}\mathsf{N}_{OV21}$$

and where $\hat{Q}_i$ are the eigenvalues of $[Q]$.

$$\hat{Q}_i = \exp(-\hat{\mathsf{N}}_{OVi}) \tag{12.2.10}$$

Equation 12.2.8 can be expanded as follows:

$$Q_{11} = \frac{\hat{Q}_1(\mathsf{N}_{OV11} - \hat{\mathsf{N}}_{OV2})}{\hat{\mathsf{N}}_{OV1} - \hat{\mathsf{N}}_{OV2}} + \frac{\hat{Q}_2(\mathsf{N}_{OV11} - \hat{\mathsf{N}}_{OV1})}{\hat{\mathsf{N}}_{OV2} - \hat{\mathsf{N}}_{OV1}} \tag{12.2.8a}$$

$$Q_{12} = \frac{(\hat{Q}_1 - \hat{Q}_2)}{(\hat{\mathsf{N}}_{OV1} - \hat{\mathsf{N}}_{OV2})}\mathsf{N}_{OV12} \tag{12.2.8b}$$

$$Q_{21} = \frac{(\hat{Q}_1 - \hat{Q}_2)}{(\hat{\mathsf{N}}_{OV1} - \hat{\mathsf{N}}_{OV2})}\mathsf{N}_{OV21} \tag{12.2.8c}$$

$$Q_{22} = \frac{\hat{Q}_1(\mathsf{N}_{OV22} - \hat{\mathsf{N}}_{OV2})}{\hat{\mathsf{N}}_{OV1} - \hat{\mathsf{N}}_{OV2}} + \frac{\hat{Q}_2(\mathsf{N}_{OV22} - \hat{\mathsf{N}}_{OV1})}{\hat{\mathsf{N}}_{OV2} - \hat{\mathsf{N}}_{OV1}} \tag{12.2.8d}$$

12.2.2 Mass Transfer Rates

The rates of interphase transfer vary as the vapor rises through the froth and the bulk vapor composition changes. The average molar fluxes $\overline{N}_i$, are given by

$$(\overline{N}) = c_t^V[K_{OV}](\bar{y} - y^*) \tag{12.2.11}$$

where $(\bar{y})$ is the average vapor composition and is defined by

$$(\bar{y}) \equiv \int_0^1 (y(\zeta))\, d\zeta \tag{12.2.12}$$

We may use Eq. 12.2.7 for $(y(\zeta))$ with the result

$$(y^* - \bar{y}) = [\Omega^V]^{-1}(y^* - y_E) \tag{12.2.13}$$

where $[\Omega^V]$ is defined by

$$[\Omega^V] = [-[\mathsf{N}_{OV}]][\exp[-[\mathsf{N}_{OV}]] - [I]]^{-1}$$

$$= [-[\mathsf{N}_{OV}]][[Q] - [I]]^{-1} \tag{12.2.14}$$

Equation 12.2.14 may be combined with Eq. 12.2.11 to give

$$(\overline{N}) = c_t^V[K_{OV}][\Omega^V]^{-1}(y_E - y^*) \tag{12.2.15}$$

Equation 12.2.15 is based on the assumptions that the matrix of overall mass transfer coefficients $[K_{OV}]$ may be considered constant over the height of the two-phase dispersion and that the total flux is zero. If the total flux is not zero the right-hand side of Eq. 12.2.11 must be multiplied by the bootstrap matrix $[\beta^V]$ and the calculation of the overall mass transfer coefficient modified.

The total mass transferred is the product of the average flux and the total interfacial area $(ah_f A_b)$. These expressions for the mass transfer rates in distillation are useful in the prediction of the performance of distillation columns (Chapter 14).

12.2.3 Numbers of Transfer Units for Multicomponent Systems

Let us now consider the prediction of the matrix $[\mathbb{N}_{OV}]$ for multicomponent systems. The starting point for the development that follows is Eq. 8.8.20 for $[K_{OV}]$ and Eq. 8.8.19 for $[R^{OV}]$

$$[K_{OV}]^{-1} = [R^V] + (c_t^V/c_t^L)[K][R^L] \tag{12.2.16}$$

The matrices $[R^V]$ and $[R^L]$ have elements defined by (cf. Eqs. 8.3.25)

$$R_{ii} = \frac{z_i}{\kappa_{in}} + \sum_{\substack{k=1 \\ k \neq i}}^{n} \frac{z_k}{\kappa_{ik}} \tag{12.2.17}$$

$$R_{ij} = -z_i \left(\frac{1}{\kappa_{ij}} - \frac{1}{\kappa_{in}} \right) \tag{12.2.18}$$

where the z_i are the mole fractions of the appropriate phase. The κ_{ij} are the "ideal" mass transfer coefficients of the binary $i \neq j$ pair for the same phase.

If we combine Eq. 12.2.16 for $[K_{OV}]$ with Eq. 12.2.6 for $[\mathbb{N}_{OV}]$ we have

$$[\mathbb{N}_{OV}]^{-1} = [\mathbb{N}_V]^{-1} + (V/L)[K][\mathbb{N}_L]^{-1} \tag{12.2.19}$$

where $[K]$ is a diagonal matrix of the first $n-1$ equilibrium K values. The matrices $[\mathbb{N}_V]$ and $[\mathbb{N}_L]$ are matrices of numbers of transfer units for the vapor and liquid phases, respectively. The inverse matrices are defined by

$$[\mathbb{N}_V]^{-1} \equiv [R^V]/a't_V \tag{12.2.20}$$

$$[\mathbb{N}_L]^{-1} \equiv [R^L]/\bar{a}t_L \tag{12.2.21}$$

When we carry out the multiplications required by Eqs. 12.2.18–12.2.21, we obtain explicit expressions for the elements of the inverse matrices $[\mathbb{N}_V]^{-1}$ and $[\mathbb{N}_L]^{-1}$ in terms of the numbers of mass transfer units of each binary pair as follows:

$$\mathbb{N}_{ii}^{-1} = \frac{z_i}{\mathscr{N}_{in}} + \sum_{\substack{k=1 \\ k \neq i}}^{n} \frac{z_k}{\mathscr{N}_{ik}} \tag{12.2.22}$$

$$\mathbb{N}_{ij}^{-1} = -z_i \left(\frac{1}{\mathscr{N}_{ij}} - \frac{1}{\mathscr{N}_{in}} \right) \tag{12.2.23}$$

where $\mathscr{N}_{ij}$ are *binary* numbers of transfer units for the phase in question defined by Eqs. 12.1.41 or 12.1.42. The superscript -1 on the elements $\mathbb{N}_{ij}$ indicates that these quantities are the elements of the inverse matrices $[\mathbb{N}]^{-1}$. Thus, to calculate $[\mathbb{N}_V]^{-1}$ and $[\mathbb{N}_L]^{-1}$ requires nothing more complicated than the determination of the binary numbers of transfer units $\mathscr{N}_{ij}$ from an appropriate correlation, theoretical model or experimental data, and the use of Eqs. 12.2.22 and 12.2.23 directly.

For ternary systems $[N]^{-1}$ may be inverted explicitly with the following results (cf. the inversion of $[R]$ in Section 8.3.1)

$$N_{11} = \mathcal{N}_{13}(z_1\mathcal{N}_{23} + (1 - z_1)\mathcal{N}_{12})/S$$
$$N_{12} = z_1\mathcal{N}_{23}(\mathcal{N}_{13} - \mathcal{N}_{12})/S$$
$$N_{21} = z_2\mathcal{N}_{13}(\mathcal{N}_{23} - \mathcal{N}_{12})/S \qquad (12.2.24)$$
$$N_{22} = \mathcal{N}_{23}(z_2\mathcal{N}_{13} + (1 - z_2)\mathcal{N}_{12})/S$$

where

$$S = z_1\mathcal{N}_{23} + z_2\mathcal{N}_{13} + z_3\mathcal{N}_{12} \qquad (12.2.25)$$

The numbers of transfer units for each binary pair may be obtained as described in Section 12.1.5 or from experimental data and these binary numbers of transfer units used directly in the estimation of the matrices of numbers of transfer units for multicomponent systems as Example 12.2.3 demonstrates.

Example 12.2.1   Numbers of Transfer Units for the Methanol–1–Propanol–Water System

Biddulph and Kalbassi (1988) investigated the distillation of the ternary system methanol(1)–1-propanol(2)–water(3). In separate experiments they determined the numbers of transfer units for each binary pair that makes up the ternary system. Estimate the number of transfer units for the ternary system at total reflux if the composition of the liquid leaving the tray is

$$x_{1L} = 0.1533 \qquad x_{2L} = 0.5231 \qquad x_{3L} = 0.3236$$

The experiments were carried out at 101.3 kPa.

DATA   The numbers of transfer units for each phase reported by Biddulph and Kalbassi are as follows:

System	$\mathcal{N}_V$	$\mathcal{N}_L$
MeOH–1-PrOH	1.61	5.83
MeOH–H$_2$O	2.56	12.5
1-PrOH–H$_2$O	1.88	7.05

The equilibrium vapor composition y^* has been estimated from a bubble point calculation on x_L to be

$$y_1^* = 0.35434 \qquad y_2^* = 0.33373 \qquad y_3^* = 0.31193$$

and the equilibrium ratios are

$$K_1 = 2.3114 \qquad K_2 = 0.63799 \qquad K_3 = 0.96394$$

The vapor–liquid equilibria of this system was represented by the Wilson model for the activity coefficients and the Antoine equation for the vapor pressures. The binary Wilson model parameters are (quoted by Biddulph and Kalbassi).

MeOH–1-PrOH	421.821	245.905	cal/mol
MeOH–H$_2$O	216.851	468.601	cal/mol
1-PrOH–H$_2$O	906.526	1396.6398	cal/mol

The molar volumes used in the Wilson model are (from Gmehling and Onken, 1977)

$$V_1 = 40.73 \qquad V_2 = 75.14 \qquad V_3 = 18.07 \text{ cm}^3/\text{mol}$$

SOLUTION At total reflux the average composition of the vapor just below the tray is equal to the composition of the liquid leaving the tray. Thus, $y_E = x_L$

$$y_{1E} = 0.1533 \qquad y_{2E} = 0.5231 \qquad y_{3E} = 0.3236$$

We shall compute the overall number of transfer units matrix from the simplified Eq. 12.2.19. We shall use the average of the entering and equilibrium vapor mole fractions as representative of the average vapor composition in the dispersion

$$y_1 = 0.2538 \qquad y_2 = 0.4284 \qquad y_3 = 0.3178$$

The elements of $[\mathsf{N}_V]^{-1}$ may be calculated directly from Eqs. 12.2.22 and 12.2.23 using the measured binary numbers of transfer units for the vapor phase. Thus,

$$\mathsf{N}_{V11}^{-1} = \frac{y_1}{\mathscr{N}_{V13}} + \frac{y_2}{\mathscr{N}_{V12}} + \frac{y_3}{\mathscr{N}_{V13}}$$

$$= \frac{0.2538}{2.56} + \frac{0.4284}{1.61} + \frac{0.3178}{2.56}$$

$$= 0.4894$$

$$\mathsf{N}_{V12}^{-1} = -y_1 \left(\frac{1}{\mathscr{N}_{V12}} - \frac{1}{\mathscr{N}_{V13}} \right)$$

$$= -0.2358 \times \left(\frac{1}{1.61} - \frac{1}{2.56} \right)$$

$$= -0.0585$$

$$\mathsf{N}_{V21}^{-1} = -y_2 \left(\frac{1}{\mathscr{N}_{V12}} - \frac{1}{\mathscr{N}_{V23}} \right)$$

$$= -0.4284 \times \left(\frac{1}{1.61} - \frac{1}{1.88} \right)$$

$$= -0.0382$$

$$\mathsf{N}_{V22}^{-1} = \frac{y_1}{\mathscr{N}_{V12}} + \frac{y_2}{\mathscr{N}_{V23}} + \frac{y_3}{\mathscr{N}_{V23}}$$

$$= \frac{0.2538}{1.61} + \frac{0.4284}{1.88} + \frac{0.3178}{1.88}$$

$$= 0.5546$$

We may collect these elements together in matrix notation as

$$[\mathsf{N}_V]^{-1} = \begin{bmatrix} 0.4894 & -0.0585 \\ -0.0382 & 0.5546 \end{bmatrix}$$

The elements of $[N_L]^{-1}$ may be calculated in a similar way using the liquid composition and the measured binary numbers of transfer units for the liquid phase

$$N_{L11}^{-1} = \frac{x_1}{\mathcal{N}_{L13}} + \frac{x_2}{\mathcal{N}_{L12}} + \frac{x_3}{\mathcal{N}_{L13}}$$

$$= \frac{0.1533}{12.5} + \frac{0.5231}{5.83} + \frac{0.3236}{12.5}$$

$$= 0.1279$$

$$N_{L12}^{-1} = -x_1\left(\frac{1}{\mathcal{N}_{L12}} - \frac{1}{\mathcal{N}_{L13}}\right)$$

$$= -0.1533 \times \left(\frac{1}{5.83} - \frac{1}{12.5}\right)$$

$$= -0.014$$

$$N_{L21}^{-1} = -x_2\left(\frac{1}{\mathcal{N}_{L12}} - \frac{1}{\mathcal{N}_{L23}}\right)$$

$$= -0.5231 \times \left(\frac{1}{5.83} - \frac{1}{7.05}\right)$$

$$= -0.0155$$

$$N_{L22}^{-1} = \frac{x_1}{\mathcal{N}_{L12}} + \frac{x_2}{\mathcal{N}_{L23}} + \frac{x_3}{\mathcal{N}_{L23}}$$

$$= \frac{0.1533}{5.83} + \frac{0.5231}{7.05} + \frac{0.3236}{7.05}$$

$$= 0.1464$$

or, in matrix notation

$$[N_L]^{-1} = \begin{bmatrix} 0.1279 & -0.014 \\ -0.0155 & 0.1464 \end{bmatrix}$$

The matrix $[N_{OV}]^{-1}$ is now readily computed from Eq. 12.2.19 as

$$[N_{OV}]^{-1} = \begin{bmatrix} 0.7849 & -0.0909 \\ -0.0481 & 0.6480 \end{bmatrix}$$

and the matrix of overall numbers of transfer units is obtained on inverting as

$$[N_{OV}] = \begin{bmatrix} 1.2850 & 0.1803 \\ 0.0954 & 1.5567 \end{bmatrix}$$

The composition of the vapor above the froth may be estimated using Eq. 12.2.3 following the computation of $[Q]$ using Eqs. 12.2.8. We leave this as an exercise for our readers to complete. ∎

12.2.4   A Fundamental Model of Mass Transfer in Multicomponent Distillation

For multicomponent systems we may extend the hydrodynamic model described in Section 12.1.7 to obtain the following matrix generalization of Eq. 12.1.73

$$(y^* - y_L) = [Q_{II}][Q_I](y^* - y_E) \tag{12.2.26}$$

where $[Q_{II}]$ and $[Q_I]$ are the matrices representing the departure from equilibrium in the free bubbling Zone *II* and the bubble formation–jetting Zone *I*, respectively. It should be noted the order of multiplication $[Q_{II}][Q_I]$ is important. The matrices $[Q_I]$ and $[Q_{II}]$ are obtained from

$$[Q_I] = \exp[-[\mathbb{N}_{OV, I}]] \tag{12.2.27}$$

for the jetting zone and

$$[Q_{II}] = \left[\sum_{k=1}^{p} f_{II, k}[Q_{II, k}]\right] \tag{12.2.28}$$

for the bubbling zone with

$$[Q_{II, k}] = \exp[-[\mathbb{N}_{OV, II, k}]] \tag{12.2.29}$$

for the kth bubble population.

The matrices of numbers of transfer units for each region of the froth may be obtained from their defining equation

$$[\mathbb{N}_{OV}] \equiv [K_{OV}]a't \tag{12.2.30}$$

where a' is the interfacial area per unit volume of vapor and t is the residence time for the vapor in the particular region of the froth. Equation 12.2.30 applies to the bubble formation zone and all bubble populations.

The matrices of multicomponent mass transfer coefficients in all three zones may be estimated using the procedure developed in Section 8.8. The overall mass transfer coefficient matrix for each zone may be obtained from Eq. 8.8.20

$$[K_{OV}] = [R^{OV}]^{-1} \tag{8.8.20}$$

where

$$[R^{OV}] = [R^V] + \frac{c_t^V}{c_t^L}[K][R^L] \tag{8.8.19}$$

where $[K]$ is a diagonal matrix of the first $n-1$ K values. Both $[R]$ matrices are calculated from Eqs. 8.3.25

$$R_{ii} = \frac{z_i}{\kappa_{in}} + \sum_{\substack{k=1 \\ k \neq i}}^{n} \frac{z_k}{\kappa_{ik}}$$

$$R_{ij} = -z_i\left(\frac{1}{\kappa_{ij}} - \frac{1}{\kappa_{in}}\right) \tag{8.3.25}$$

using the mole fractions z_i, and binary Maxwell–Stefan mass transfer coefficients κ_{ij} of the appropriate phase.

Binary mass transfer coefficients for the vapor in the jetting–bubble formation zone may be computed from

$$\kappa^V_{I_{ij}} = -\ln(1 - F_{I_{ij}})/(a'_I t_I) \tag{12.2.31}$$

where

$$F_{I_{ij}} = 1 - \sum_{m=1}^{\infty} \frac{4}{j_m^2} \exp\left\{-j_m^2 \, \text{Fo}_I \frac{Ð_{ij}^V}{Ð_{\text{ref}}}\right\} \qquad (12.2.32)$$

where the j_m are the roots of the Bessel function $J_0(j_m) = 0$ and Fo_I is the Fourier number for the jetting zone defined in terms of a reference diffusivity $Ð_{\text{ref}}$

$$\text{Fo}_I = 4Ð_{\text{ref}} t_I / d_I^2 \qquad (12.2.33)$$

The binary mass transfer coefficients for the liquid phase may be evaluated with a penetration model

$$\kappa_{I_{ij}}^L = 2\left(Ð_{ij}^L / \pi t_I\right)^{1/2} \qquad (12.2.34)$$

The residence time of the vapor is given by

$$t_I = h_I / U_I \qquad (12.1.48)$$

and the interfacial area per unit volume of vapor is

$$a_I' = 4/d_I \qquad (12.1.47)$$

The average molar fluxes in the bubble formation zone are obtained from (cf. Eq. 12.2.11)

$$(\bar{N}_I) = c_t^V [K_{OV,I}](\bar{y}_I - y^*) \qquad (12.2.35)$$

where $(\bar{y}_I)$ is the average composition in the bubble formation zone and is determined from (cf. Eq. 12.2.13)

$$(y^* - \bar{y}_I) = \left[\Omega_I^V\right]^{-1}(y^* - y_E) \qquad (12.2.36)$$

$[\Omega_I^V]$ is given by

$$\left[\Omega_I^V\right] = \left[-[N_{OV,I}]\right]\left[\exp[-[N_{OV,I}]] - [I]\right]^{-1} \qquad (12.2.37)$$

Equations 12.2.35–12.2.37 may be combined to give

$$(\bar{N}_I) = c_t^V [K_{OV,I}][\Omega_I^V]^{-1}(y_E - y^*) \qquad (12.2.38)$$

The total mass transferred in the bubble formation zone is the product of the average flux and the total interfacial area in the bubble formation zone.

Binary mass transfer coefficients for the vapor in the kth bubble population may be computed from

$$\kappa_{II,k_{ij}}^V = -\ln\left(1 - F_{II,k_{ij}}\right)/\left(a_{II,k}' t_{II,k}\right) \qquad (12.2.39)$$

where

$$F_{II,k_{ij}} = 1 - \frac{6}{\pi^2} \sum_{m=1}^{\infty} \frac{1}{m^2} \exp\left\{ -m^2\pi^2 \, \text{Fo}_{II,k} \frac{Ð_{ij}^V}{Ð_{\text{ref}}} \right\} \qquad (12.2.40)$$

with $\text{Fo}_{II,k}$ the Fourier number for the kth bubble population defined using the reference diffusivity

$$\text{Fo}_{II,k} = 4Ð_{\text{ref}}t_{II,k}/d_{II,k}^2 \qquad (12.2.41)$$

with the residence time for the vapor inside each bubble population given by Eq. 12.1.60

$$t_{II,k} = h_{II,k}/U_{II,k} \qquad (12.1.60)$$

and the interfacial areas per unit volume of vapor from Eq. 12.1.59

$$a'_{II,k} = 6/d_{II,k} \qquad (12.1.59)$$

The penetration model is used for the mass transfer coefficients in the liquid phase; the contact time being the time required for the bubble to rise one diameter

$$\kappa_{II,k_{ij}}^L = 2\left(Ð_{ij}^L/\pi(d_{II,k}/U_{II,k}) \right)^{1/2} \qquad (12.2.42)$$

The average molar fluxes in each bubble population are obtained from (cf. Eq. 12.2.11)

$$(\bar{N}_{II,k}) = c_t^V[K_{OV,II,k}](\bar{y}_{II,k} - y^*) \qquad (12.2.43)$$

where $(\bar{y}_{II,k})$ is the average composition in the kth bubble population and is determined from (cf. Eq. 12.2.13)

$$(y^* - \bar{y}_{II,k}) = [\Omega_{II,k}^V]^{-1}(y^* - y_I) \qquad (12.2.44)$$

(y_I) is the composition of the vapor at the top of the bubble formation zone and $[\Omega_{II,k}^V]$ is defined by

$$[\Omega_{II,k}^V] = [-[N_{OV,II,k}]][\exp[-[N_{OV,II,k}]] - [I]]^{-1} \qquad (12.2.45)$$

Equations 12.2.43–12.2.45 may be combined to give

$$(\bar{N}_{II,k}) = c_t^V[K_{OV,II,k}][\Omega_{II,k}^V]^{-1}(y_I - y^*) \qquad (12.2.46)$$

The mole fractions at the top of the formation zone may be eliminated to yield (cf. Eq. 12.1.69)

$$(\bar{N}_{II,k}) = c_t^V[K_{OV,II,k}][\Omega_{II,k}^V]^{-1}[Q_I](y_E - y^*) \qquad (12.2.47)$$

Example 12.2.2   Distillation of Ethanol–tert-Butyl Alcohol*–Water in a Sieve Tray Column

A mixture of ethanol(1)–t-butyl alcohol(2)–water(3) is being distilled in a sieve tray column operating in the froth regime. Estimate the mass transfer coefficients in a dispersion

---

*tert-Butyl alcohol is the common name for 2-methyl-2-propanol.

with the following characteristics:

The height of the jetting zone: h_I = 10 mm.
The diameter of the jet: d_I = 3 mm.
The velocity of the vapor in the jet: U_I = 6.4 m/s.
The height of the free bubbling zone: h_{II} = 65 mm.
The small bubble diameter: $d_{II,1}$ = 5 mm.
The small bubble rise velocity: $U_{II,1}$ = 0.3 m/s.
The fraction of vapor in the small bubbles: $f_{II,1}$ = 0.10.
The large bubble diameter: $d_{II,2}$ = 12.5 mm.
The large bubble rise velocity: $U_{II,2}$ = 1.5 m/s.
The fraction of vapor in the large bubbles: $f_{II,2}$ = 0.90.

These values are consistent with a superficial velocity of 1 m/s and a total gas hold-up ε of 0.8 with $\varepsilon = \varepsilon_{II,1} + \varepsilon_{II,2}$ and $\varepsilon_{II,1} = \frac{1}{6}$.

The composition of the bulk vapor below the tray is

$$y_1 = 0.5558 \qquad y_1 = 0.1353 \qquad y_3 = 0.3089$$

At total reflux, the composition of the liquid leaving the tray is the same as the composition of the vapor entering the tray

$$x_1 = 0.5558 \qquad x_2 = 0.1353 \qquad x_3 = 0.3089$$

The liquid is considered to be well mixed. The equilibrium vapor composition is determined from a bubble point calculation to be

$$y_1^* = 0.6040 \qquad y_1^* = 0.1335 \qquad y_3^* = 0.2625$$

and the equilibrium ratios are

$$K_1 = 1.0867 \qquad K_2 = 0.98669 \qquad K_3 = 0.84979$$

DATA   Other physical property data is summarized below.

Liquid molar density: c_t^V = 19.665 kmol/m^3.
Vapor molar density: c_t^L = 34.07 mol/m^3.
The Maxwell–Stefan diffusivities in the vapor phase [units are 10^{-5} m^2/s]

$$\DJ_{12}^V = 0.799$$

$$\DJ_{13}^V = 2.14$$

$$\DJ_{23}^V = 1.65$$

The infinite dilution diffusivities in the liquid phase [units are 10^{-9} m^2/s]

$$\DJ_{12}^0 = 4.08 \qquad \DJ_{21}^0 = 2.60$$

$$\DJ_{13}^0 = 3.711 \qquad \DJ_{31}^0 = 4.064$$

$$\DJ_{23}^0 = 2.652 \qquad \DJ_{32}^0 = 3.143$$

The molar latent heats of vaporization may be taken to be equal.

SOLUTION In preparation for the estimation of the mass transfer coefficients in the liquid phase we must first compute the Maxwell–Stefan diffusion coefficients. Equation 4.2.18 is used for this task as illustrated below.

$$\mathcal{D}_{12}^{L} = (\mathcal{D}_{12}^{\circ})^{(1+x_2-x_1)/2}(\mathcal{D}_{21}^{\circ})^{(1+x_1-x_2)/2}$$

$$= (4.08 \times 10^{-9})^{0.28975} \times (2.60 \times 10^{-9})^{0.71025}$$

$$= 2.9626 \times 10^{-9} \text{ m}^2/\text{s}$$

Similarly,

$$\mathcal{D}_{13}^{L} = 3.9273 \times 10^{-9} \text{ m}^2/\text{s}$$

$$\mathcal{D}_{23}^{L} = 2.84483 \times 10^{-9} \text{ m}^2/\text{s}$$

Bubble Formation–Jetting Zone The interfacial area per unit volume of vapor in the jetting zone is

$$a_I' = 4/d_I$$

$$= 4/0.003$$

$$= 1333 \text{ m}^2/\text{m}^3$$

and the contact time is

$$t_I = h_I/U_I$$

$$= 0.01/6.4$$

$$= 0.0015625 \text{ s}$$

For the estimation of the vapor-phase mass transfer coefficients we use Eqs. 12.2.31–12.2.33 and use $\mathcal{D}_{12}^{V}$ as the reference diffusivity. The Fourier number for the jetting zone is calculated as

$$\text{Fo}_I = 4\mathcal{D}_{\text{ref}}t_I/d_I^2$$

$$= 4 \times 0.799 \times 10^{-5} \times 0.0015625/0.003^2$$

$$= 5.549 \times 10^{-3}$$

The binary mass transfer coefficients for the vapor phase are found to be

$$\kappa_{I_{12}}^{V} = 0.0851 \text{ m/s}$$

$$\kappa_{I_{13}}^{V} = 0.1445 \text{ m/s}$$

$$\kappa_{I_{23}}^{V} = 0.1254 \text{ m/s}$$

The matrix of inverted mass transfer coefficients for the vapor phase $[R_I^V]$ is computed below:

$$R_{I_{11}}^V = \frac{y_1}{\kappa_{I_{13}}^V} + \frac{y_2}{\kappa_{I_{12}}^V} + \frac{y_3}{\kappa_{I_{13}}^V}$$

$$= \frac{0.5558}{0.1445} + \frac{0.1353}{0.0851} + \frac{0.3089}{0.1445}$$

$$= 7.576 \ (m/s)^{-1}$$

$$R_{I_{12}}^V = -y_1 \left(\frac{1}{\kappa_{I_{12}}^V} - \frac{1}{\kappa_{I_{13}}^V} \right)$$

$$= -0.5558 \times \left(\frac{1}{0.0851} - \frac{1}{0.1445} \right)$$

$$= -2.683 \ (m/s)^{-1}$$

$$R_{I_{21}}^V = -y_2 \left(\frac{1}{\kappa_{I_{12}}^V} - \frac{1}{\kappa_{I_{23}}^V} \right)$$

$$= -0.1353 \times \left(\frac{1}{0.0851} - \frac{1}{0.1254} \right)$$

$$= -0.511 \ (m/s)^{-1}$$

$$R_{I_{22}}^V = \frac{y_1}{\kappa_{I_{12}}^V} + \frac{y_2}{\kappa_{I_{23}}^V} + \frac{y_3}{\kappa_{I_{23}}^V}$$

$$= \frac{0.5558}{0.0851} + \frac{0.1353}{0.1254} + \frac{0.3089}{0.1254}$$

$$= 10.07 \ (m/s)^{-1}$$

We may collect these elements together in matrix notation as

$$[R_I^V] = \begin{bmatrix} 7.576 & -2.683 \\ -0.511 & 10.07 \end{bmatrix} (m/s)^{-1}$$

The binary mass transfer coefficients for the liquid phase are evaluated using Eq. 12.2.34

$$\kappa_{I_{12}}^L = 2 \left(Ð_{12}^L / (\pi t_I) \right)^{1/2}$$

$$= 2 \times \left(2.9626 \times 10^{-9} / (\pi \times 0.0015625) \right)^{1/2}$$

$$= 1.554 \times 10^{-3} \ m/s$$

The remaining binary mass transfer coefficients are computed in a similar way to give

$$\kappa_{I_{13}}^L = 1.789 \times 10^{-3} \ m/s$$

$$\kappa_{I_{23}}^L = 1.523 \times 10^{-3} \ m/s$$

The matrix of inverse mass transfer coefficients for the liquid phase in the jetting zone may now be computed just as the matrix $[R^V]$ was computed:

$$R^L_{I_{11}} = \frac{x_1}{\kappa^L_{I_{13}}} + \frac{x_2}{\kappa^L_{I_{12}}} + \frac{x_3}{\kappa^L_{I_{13}}}$$

$$= \frac{0.5558}{1.789 \times 10^{-3}} + \frac{0.1353}{1.554 \times 10^{-3}} + \frac{0.3089}{1.789 \times 10^{-3}}$$

$$= 570.4 \; (\text{m/s})^{-1}$$

The complete matrix $[R^L_I]$ is

$$[R^L_I] = \begin{bmatrix} 570.4 & -47.01 \\ 1.784 & 649.5 \end{bmatrix} (\text{m/s})^{-1}$$

We may now compute $[R^{OV}]$ from Eq. 8.8.19

$$R^{OV}_{11} = 7.576 + (34.07/19665) \times 1.0867 \times 570.4$$

$$= 8.650 \; (\text{m/s})^{-1}$$

The complete matrix is

$$[R^{OV}_I] = \begin{bmatrix} 8.650 & -2.772 \\ -0.508 & 11.184 \end{bmatrix} (\text{m/s})^{-1}$$

The matrix of overall mass transfer coefficients for the jetting region is, therefore

$$[K_{OV,I}] = [R^{OV}_I]^{-1}$$

$$= \begin{bmatrix} 0.1173 & 0.0291 \\ 0.00532 & 0.0907 \end{bmatrix} \text{m/s}$$

The matrix of overall number of transfer units for the bubble formation zone may now be computed from

$$[\mathbb{N}_{OV,I}] = [K_{OV,I}]a'_I t_I$$

$$= \begin{bmatrix} 0.2444 & 0.0606 \\ 0.0111 & 0.1890 \end{bmatrix}$$

The departure from equilibrium in the jetting zone follows as:

$$[Q_I] = \exp[-[\mathbb{N}_{OV,I}]]$$

$$= \begin{bmatrix} 0.7834 & -0.0488 \\ -0.0089 & 0.8280 \end{bmatrix}$$

The composition of the vapor as it leaves the jetting zone (y_I) is given by

$$(y^* - y_I) = [Q_I](y^* - y_E)$$

So

$$y_{1I} = 0.56615 \qquad y_{2I} = 0.13542 \qquad y_{3I} = 0.29843$$

To compute the average molar fluxes in the formation zone we first compute $[\Omega_I^V]$ using Eq. 12.2.37.

$$[\Omega_I^V] = [-[N_{OV,I}]][[Q_I] - [I]]^{-1}$$

$$= \begin{bmatrix} 1.1272 & 0.0325 \\ 0.0059 & 1.0975 \end{bmatrix}$$

The average molar fluxes follow from Eq. 12.2.38 as

$$(\bar{N}_I) = c_t^V[K_{OV,I}][\Omega_I^V]^{-1}(y_E - y^*)$$

Thus

$$\bar{N}_{I1} = 0.1692 \text{ mol/m}^2\text{s} \qquad \bar{N}_{I2} = 0.00197 \text{ mol/m}^2\text{s}$$

The molar flux of component 3 is obtained from the assumption of equimolar countertransport.

Small Bubble Population We now turn our attention to the determination of the mass transfer characteristics of the bubble rise zone, beginning with the small bubble population. The interfacial area per unit volume of vapor in the small bubble population is

$$a'_{II,1} = 6/d_{II,1}$$
$$= 6/0.005$$
$$= 1200 \text{ m}^2/\text{m}^3$$

The residence time for the vapor in the small bubbles is

$$t_{II,1} = h_{II}/U_{II,1}$$
$$= 0.065/0.3$$
$$= 0.21667 \text{ s}$$

The Fourier number for the small bubbles is

$$\text{Fo}_{II,1} = 4D_{\text{ref}}t_{II,1}/(d_{II,1})^2$$
$$= 4 \times 0.799 \times 10^{-5} \times 0.21667/0.005^2$$
$$= 0.27699$$

The binary mass transfer coefficients for the vapor phase are calculated using Eqs. 12.2.39 as

$$\kappa_{II,1_{12}}^V = 0.01243 \text{ m/s}$$
$$\kappa_{II,1_{13}}^V = 0.03008 \text{ m/s}$$
$$\kappa_{II,1_{23}}^V = 0.02363 \text{ m/s}$$

The matrix of inverted mass transfer coefficients for the vapor phase $[R_{II,1}^V]$ is computed in exactly the same way that $[R_I^V]$ was found earlier.

$$[R_{II,1}^V] = \begin{bmatrix} 39.64 & -26.73 \\ -5.16 & 63.92 \end{bmatrix} (\text{m/s})^{-1}$$

The binary mass transfer coefficients for the liquid phase are evaluated with the help of Eq. 12.2.42.

$$\kappa^L_{II,1_{12}} = 2\left(\mathcal{D}^L_{12}/(\pi d_{II,1}/U_{II,1})\right)^{1/2}$$

$$= 2 \times \left(2.9626 \times 10^{-9}/(\pi \times 0.005/0.3)\right)^{1/2}$$

$$= 0.4757 \times 10^{-3} \text{ m/s}$$

The remaining binary mass transfer coefficients are computed in a similar way to give

$$\kappa^L_{II,1_{13}} = 0.5477 \times 10^{-3} \text{ m/s}$$

$$\kappa^L_{II,1_{23}} = 0.4662 \times 10^{-3} \text{ m/s}$$

The matrix of inverse liquid-phase mass transfer coefficients for the small bubbles may now be computed

$$\left[R^L_{II,1}\right] = \begin{bmatrix} 1863.1 & -153.58 \\ 5.827 & 2121.1 \end{bmatrix} (\text{m/s})^{-1}$$

$[R^{OV}_{II,1}]$ is calculated from Eq. 8.8.19.

$$\left[R^{OV}_{II,1}\right] = \begin{bmatrix} 43.145 & -26.529 \\ -5.150 & 67.146 \end{bmatrix} (\text{m/s})^{-1}$$

The matrix of overall mass transfer coefficients for the small bubbles is computed next

$$[K_{OV,II,1}] = \left[R^{OV}_{II,1}\right]^{-1}$$

$$= \begin{bmatrix} 0.02434 & 0.00974 \\ 0.00186 & 0.01555 \end{bmatrix} \text{m/s}$$

Thus, the overall number of transfer units for the small bubble population is

$$[\mathbb{N}_{OV,II,1}] = [K_{OV,II,1}]a'_{II,1}t_{II,1}$$

$$= \begin{bmatrix} 6.3277 & 2.5312 \\ 0.4829 & 4.0427 \end{bmatrix}$$

The departure from equilibrium in the small population follows as:

$$[Q_{II,1}] = \exp[-[\mathbb{N}_{OV,II,1}]]$$

$$= \begin{bmatrix} 0.0048 & -0.0209 \\ -0.0040 & 0.0238 \end{bmatrix}$$

indicating that the small bubble population is almost at equilibrium with the bulk liquid.

To compute the average molar fluxes in the small bubble population we compute $[\Omega^V_{II,1}]$ as follows:

$$[\Omega^V_{II,1}] = [-[\mathbb{N}_{OV,II,1}]][[Q_{II,1}] - [I]]^{-1}$$

$$= \begin{bmatrix} 6.3487 & 2.4566 \\ 0.4687 & 4.1310 \end{bmatrix}$$

The average molar fluxes follow from Eq. 12.2.47

$$(\bar{N}_{II,1}) = c_t^V [K_{OV,II,1}][\Omega_{II,1}^V]^{-1}[Q_I](y_E - y^*)$$

Thus

$$\bar{N}_{II,1_1} = -0.004929 \text{ mol/m}^2\text{s} \qquad \bar{N}_{II,1_2} = 0.0002262 \text{ mol/m}^2\text{s}$$

Large Bubble Population The interfacial area per unit volume of vapor in the large bubble population is

$$
\begin{aligned}
a'_{II,2} &= 6/d_{II,2} \\
&= 6/0.0125 \\
&= 480 \text{ m}^2/\text{m}^3
\end{aligned}
$$

The residence time for the vapor in the large bubbles is

$$
\begin{aligned}
t_{II,2} &= h_{II}/U_{II,2} \\
&= 0.065/1.5 \\
&= 0.04333 \text{ s}
\end{aligned}
$$

Note the very low residence time of the large bubbles.
The Fourier number for the large bubbles is

$$
\begin{aligned}
\text{Fo}_{II,2} &= 4\mathcal{D}_{\text{ref}}t_{II,2}/(d_{II,2})^2 \\
&= 4 \times 0.799 \times 10^{-5} \times 0.04333/0.0125^2 \\
&= 0.00886
\end{aligned}
$$

As for the small bubble population, the binary mass transfer coefficients for the vapor phase are calculated using Eqs. 12.2.39

$$\kappa_{II,2_{12}}^V = 0.01661 \text{ m/s}$$

$$\kappa_{II,2_{13}}^V = 0.02877 \text{ m/s}$$

$$\kappa_{II,2_{23}}^V = 0.02481 \text{ m/s}$$

The matrix of inverted mass transfer coefficients for the vapor phase $[R_{II,2}^V]$ is computed to be

$$[R_{II,2}^V] = \begin{bmatrix} 38.20 & -14.41 \\ -2.69 & 51.58 \end{bmatrix} (\text{m/s})^{-1}$$

The binary mass transfer coefficients for the liquid phase are evaluated using Eq. 12.2.42

$$
\begin{aligned}
\kappa_{II,2_{12}}^L &= 2\big(\mathcal{D}_{12}^L/(\pi d_{II,2}/U_{II,2})\big)^{1/2} \\
&= 2 \times \big(2.9626 \times 10^{-9}/(\pi \times 0.0125/1.5)\big)^{1/2} \\
&= 0.6728 \times 10^{-3} \text{ m/s}
\end{aligned}
$$

and

$$\kappa^L_{II,2_{13}} = 0.7746 \times 10^{-3} \text{ m/s}$$

$$\kappa^L_{II,2_{23}} = 0.6593 \times 10^{-3} \text{ m/s}$$

The matrix of inverse liquid-phase mass transfer coefficients for the large bubbles may now be computed

$$[R^L_{II,2}] = \begin{bmatrix} 1317.4 & -108.60 \\ 4.120 & 1499.9 \end{bmatrix} (\text{m/s})^{-1}$$

$[R^{OV}_{II,2}]$ is calculated from Eq. 8.8.19

$$[R^{OV}_{II,2}] = \begin{bmatrix} 40.682 & -14.615 \\ -2.687 & 54.141 \end{bmatrix} (\text{m/s})^{-1}$$

The matrix of overall mass transfer coefficients for the large bubbles is computed next.

$$[K_{OV,II,2}] = [R^{OV}_{II,2}]^{-1}$$

$$= \begin{bmatrix} 0.02503 & 0.00676 \\ 0.00124 & 0.01881 \end{bmatrix} \text{m/s}$$

The overall number of transfer units for the large bubble population is

$$[\mathsf{N}_{OV,II,2}] = [K_{OV,II,2}]a'_{II,2}t_{II,2}$$

$$= \begin{bmatrix} 0.5206 & 0.1405 \\ 0.0258 & 0.3912 \end{bmatrix}$$

The departure from equilibrium in the large population follows as:

$$[Q_{II,2}] = \exp[-[\mathsf{N}_{OV,II,2}]]$$

$$= \begin{bmatrix} 0.5953 & -0.0892 \\ -0.0164 & 0.6775 \end{bmatrix}$$

The molar fluxes in the large bubble population are computed following the procedure used above for the small bubble population.

$$[\Omega^V_{II,2}] = [-[\mathsf{N}_{OV,II,2}]][[Q_{II,2}] - [I]]^{-1}$$

$$= \begin{bmatrix} 1.2830 & 0.0809 \\ 0.0149 & 1.2086 \end{bmatrix}$$

The average molar fluxes follow from Eq. 12.2.47

$$(\overline{N}_{II,2}) = c^V_t[K_{OV,II,2}][\Omega^V_{II,2}]^{-1}[Q_I](y_E - y^*)$$

with numerical results

$$\overline{N}_{II,2_1} = -0.0248 \text{ mol/m}^2\text{s}$$

$$\overline{N}_{II,2_2} = -1.2173 \times 10^{-6} \text{ mol/m}^2\text{s}$$

Overall Performance   The overall departure from equilibrium for the bubble rise zone is given by (cf. Eq. 12.2.28)

$$[Q_{II}] = f_{II,1}[Q_{II,1}] + f_{II,2}[Q_{II,2}]$$

$$= \begin{bmatrix} 0.5363 & -0.0824 \\ -0.0152 & 0.6121 \end{bmatrix}$$

The combined matrix is

$$[Q] = [Q_{II}][Q_I]$$

$$= \begin{bmatrix} 0.4212 & -0.0944 \\ -0.0174 & 0.5079 \end{bmatrix}$$

which allows the calculation of the composition of the vapor above the froth.

It is interesting to note that although the transfer process is predominantly gas-phase mass transfer controlled, there is a finite contribution from the liquid-phase transfer resistance. In multicomponent distillation, it is our experience that it is not "safe" to ignore the liquid-phase resistance even when for similar operating conditions for a *binary* system, the liquid-phase resistance is negligible.

It should be noted that $[R^V]$ was computed using the composition of the vapor below the tray. It would have been better to use an average of the mole fractions in the appropriate zones in the computations of the three vapor phase $[R]$ matrices. However, this would have required an iterative procedure which, in view of the length of the calculations, we wished to avoid. The above calculations would serve as a first iteration if it is desired to repeat the calculations. ■

12.3 DISTILLATION IN PACKED COLUMNS

The purpose of packing in a column is to provide a large surface area for interphase mass transfer. With that objective in mind, almost any material that will survive the environment in the column while providing adequate surface area at a moderate to low pressure drop could be used. Indeed, Fair (1987) reports that in the 1930s chain and carpet tacks were tested as column packing material.

Today, there are two main types of packing commonly employed in distillation and absorption columns:

- Random or dumped packings.
- Structured packing.

Random packings, illustrated in Figure 12.6, may be of metal, plastic, or ceramic construction. Typical packing element sizes are from 1 to 5 cm with the larger sizes favored in commercial operations. The older designs of the simple ring type have largely been superseded by the more open rings and saddles that allow fluid to flow through as well as around the packing elements. Structured packings are depicted in Figure 12.7. Structured packings have not been used in large scale packed columns for as long as dumped packings. Although they are more expensive than dumped packings, they have better pressure drop and mass transfer characteristics than dumped packings and are, therefore, popular for fitting in new columns and for revamping older columns to improve performance (including refitting columns equipped with trays).

Figure 12.6(a). Example of dumped packing. This illustration shows Hypak$^{(R)}$ packing. Photograph courtesy of Norton Chemical Process Products Corporation, Stow, OH.

Figure 12.6(b). Example of dumped packing. This illustration shows ceramic Intalox$^{(R)}$ saddles. Photograph courtesy of Norton Chemical Process Products Corporation, Stow, OH.

Figure 12.6(c). Example of dumped packing. This illustration shows Intalox[R] Snowflake[R] packing. Photograph courtesy of Norton Chemical Process Products Corporation, Stow, OH.

Figure 12.7. Structured packing. This illustration shows Intalox[R] structured packing. Photograph courtesy of Norton Chemical Process Products Corporation, Stow, OH.

Vapor and liquid flows in a packed column truly flow in opposite directions. Contrast this with the flow in a tray column where the vapor–liquid contact is between the vapor as it rises up through a liquid that is flowing laterally across the column prior to flowing down to the tray below.

12.3.1 Material and Energy Balance Relations

For continuous contact equipment the material and energy balances are written around a section of column of differential height as shown in Figure 12.8. For the vapor phase the component material balance reads

$$v_i|_{z+\Delta z} - v_i|_z - N_i^V a' A_c \, \Delta z = 0 \tag{12.3.1}$$

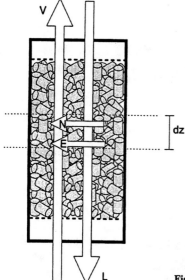

Figure 12.8. Differential section of a packed column.

where a' is the interfacial area per unit volume and A_c is the cross-sectional area of the column. Thus, $A_c \, \Delta z$ is the volume of the small section of packing we are considering. Dividing by Δz and taking the limit as Δz goes to zero gives

$$\frac{dv_i}{dz} = N_i^V a' A_c \qquad i = 1, 2, \ldots, n \tag{12.3.2}$$

The total material balance is obtained simply by summing Eqs. 12.3.2 over all n species

$$\frac{dV}{dz} = N_t^V \tag{12.3.3}$$

The differential material balances for the liquid phase are obtained in a similar way: for each component we have

$$\ell_i|_{z+\Delta z} - \ell_i|_z - N_i^L a' A_c \, \Delta z = 0 \tag{12.3.4}$$

Note that the liquid is flowing countercurrent to the vapor phase. When we divide by z and take the limit as Δz goes to zero we obtain

$$\frac{d\ell_i}{dz} = N_i^L a' A_c \qquad i = 1, 2, \ldots, n \tag{12.3.5}$$

and the total material balance is

$$\frac{dL}{dz} = N_t^L a' A_c \tag{12.3.6}$$

where $L = \Sigma \ell_i$ is the total liquid flow down the column.

The terms on the right-hand sides of Eqs. 12.3.1 and 12.3.4 are the molar fluxes of species i in the vapor and liquid phases, respectively; we assume that transfers from the vapor to the liquid phase are positive.

The energy balance for the vapor phase can be derived as follows. First, we write down the energy flows into and out of the differential element of packing

$$VH^V|_{z+\Delta z} - VH^V|_z + E^V a' A_c \, \Delta z = 0 \tag{12.3.7}$$

Once again, we divide by Δz and take the limit as Δz goes to zero to give

$$\frac{d(VH^V)}{dz} = E^V a' A_c \tag{12.3.8}$$

In a similar way we can derive expressions for the energy balance for the liquid phase as

$$\frac{d(LH^L)}{dz} = E^L a' A_c \tag{12.3.9}$$

where E^L is the energy flux into the liquid.

At any position in the packed column we have continuity of the molar fluxes across the vapor–liquid interface

$$N_i^L = N_i = N_i^V \tag{11.5.1}$$

and

$$N_t^L = N_t = N_t^V \tag{11.5.2}$$

These fluxes are made up of diffusive and convective contributions as

$$N_i^L = J_i^L + x_i N_t^L = N_i = J_i^V + y_i N_t^V = N_i^V \tag{11.5.3}$$

The diffusion fluxes are given in terms of the mass transfer coefficients as

$$(J^L) = c_t^L [k_L^\bullet](x^I - x^L) \tag{11.5.4}$$

$$(J^V) = c_t^V [k_V^\bullet](y^V - y^I) \tag{11.5.5}$$

with the multicomponent mass transfer coefficients $[k^\bullet]$ determined from one or other of the models described in Chapters 8–10.

We also have continuity of the energy flux across the vapor–liquid interface.

$$E^V = E^L = E \tag{11.5.6}$$

which may be expressed as

$$q^V + \sum_{i=1}^n N_i^V \bar{H}_i^V(T^V) = q^L + \sum_{i=1}^n N_i^L \bar{H}_i^L(T^L) \tag{11.5.7}$$

with the heat fluxes in the two phases given by

$$q^V = h_V^\bullet(T^V - T^I) \tag{11.5.8}$$

$$q^L = h_L^\bullet(T^I - T^L) \tag{11.5.9}$$

where $h_V^\bullet$ and $h_L^\bullet$ are the finite flux heat transfer coefficients for the vapor and liquid phases, respectively (see Chapter 11 for further discussion).

The set of differential and algebraic equations given above for modeling multicomponent distillation in a packed column must be integrated numerically in general. The complexity and nonlinearity of the above equations precludes analytical solution in most cases of practical importance. Moreover, because the vapor and liquid streams flow in opposite directions means that, in all but one circumstance—total reflux—several integrations may be required in order to properly solve the equations. An alternative method of solving approximate forms of these equations is discussed in Chapter 14.

Let us proceed by writing the differential material balances (Eqs. 12.2.2) in the following equivalent form (cf. the derivation in Section 12.2.1):

$$V\frac{dy_i}{dz} = -J_i^V aA_c \qquad i = 1, 2, \ldots, n \tag{12.3.10}$$

which is valid when $N_t \neq 0$, that is, when we have nonequimolar transfer. These equations may be combined in $n - 1$ dimensional matrix form as

$$V\frac{d(y)}{dh} = -(J^V)aA_c \tag{12.3.11}$$

Introducing the rate relations (Eq. 7.3.11a)

$$(J^V) = c_t^V[K_{OV}](y - y^*) \tag{7.3.11a}$$

into Eq. 12.3.11 gives

$$V\frac{d(y)}{dz} = c_t^V[K_{OV}](y^* - y)aA_c \tag{12.3.12}$$

12.3.2 Transfer Units for Binary Systems

It is traditional for chemical engineers to model packed columns through the concept of transfer units (in much the same way as we used transfer units in the treatment of transfer in the froth on a tray in Sections 12.1.2 and 12.2.1). For two component systems Eq. 12.3.12 simplifies to (cf Eq. 12.1.7)

$$V\frac{dy_1}{dh} = c_t^V K_{OV}(y_1^* - y_1)a'A_c \tag{12.3.13}$$

which may be written in dimensionless form as

$$\frac{dy_1}{d\zeta} = \mathbb{N}_{OV}(y_1^* - y_1) \tag{12.3.14}$$

where $\zeta = z/H$ is a dimensionless coordinate with H the total height of packing. The parameter $\mathbb{N}_{OV}$, the overall number of transfer units for the vapor phase, may be defined by

$$\mathbb{N}_{OV} \equiv c_t^V K_{OV} a' HA_c/V \tag{12.3.15}$$

Equation 12.3.13 is similar to the corresponding expressions for binary mass transfer in the froth on a distillation tray, Eqs. 12.1.7. In the model of mass transfer in the froth on a tray, the liquid is assumed to be well mixed vertically. Hence, y^* may be assumed constant. A similar assumption cannot be justified for a packed column where both liquid and vapor truly flow in opposite directions and the equilibrium vapor composition y^* changes with elevation as the liquid composition changes. Analytical solutions of Eqs. 12.3.14 for binary systems are available in the literature for certain special cases (equimolar countertransfer or Stefan diffusion where the equilibrium data can be approximated by the equation of a straight line, as in the two component—scalar—form of Eq. 7.3.8, for the entire height of the packing). In general, however, numerical or graphical techniques are used to solve Eqs. 12.3.14 (see, e.g., Sherwood, et al., 1975).

To evaluate the overall number of transfer units we may proceed to combine Eq. 7.3.15 for K_{OV} with Eq. 12.3.15 for N_{OV} to give (cf. Eq. 12.1.17)

$$\frac{1}{N_{OV}} = \frac{1}{N_V} + \frac{M(V/L)}{N'_L} \tag{12.3.16}$$

where N_V and N'_L are the numbers of transfer units for the vapor and liquid phases defined by

$$N_V \equiv k^V a' H / u_V \tag{12.3.17}$$

$$N'_L \equiv k^L a' H / u_L \tag{12.3.18}$$

where $u_V = V/(A_c c_t^V)$ and $u_L = L/(A_c c_t^L)$ are the superficial vapor and liquid velocities, respectively.

Mass transfer coefficients for packed columns sometimes are expressed as the height of a transfer unit (HTU). Thus, for the vapor and liquid phases, respectively,

$$H_V = H/N_V = u_V/(k^V a') \tag{12.3.19}$$

$$H_L = H/N'_L = u_L/(k^L a') \tag{12.3.20}$$

The overall height of a transfer unit for a two-component system is

$$H_{OV} = H/N_{OV} = u_V/(K_{OV} a') \tag{12.3.21}$$

If we use Eq. 7.3.15 for K_{OV}, we obtain the following relationship between H_{OV}, H_V, and H_L

$$H_{OV} = H_V + \Lambda H_L \tag{12.3.22}$$

where Λ is the stripping factor

$$\Lambda = MV/L \tag{12.3.23}$$

where M is the slope of the equilibrium line, V is the molar flow rate of the vapor, and L is the molar flow rate of the liquid. If, as alternative to Eq. 7.3.15, we use the simplified Eq. 8.8.21 for K_{OV}, we recover Eq. 12.3.22 for H_{OV} with $\Lambda = K_1 V/L$ and K_1 the K value.

The performance of a packed column is often expressed in terms of the height equivalent to a theoretical plate (HETP). The HETP is related to the height of packing by

$$\text{HETP} = H/N_{eqm} \tag{12.3.24}$$

where N_{eqm} is the number of equilibrium stages needed to accomplish the same separation possible in a real packed column of height H. The equilibrium stage model of distillation and absorption is reviewed briefly in Chapter 13 and in more detail in a number of textbooks (see, e.g., King, 1981; Henley and Seader, 1981). If the equilibrium line may be assumed straight, the HETP may be related to the HTU values by

$$HETP = \mathbb{H}_{OV}\frac{\ln \Lambda}{\Lambda - 1} \tag{12.3.25}$$

The evaluation of the HTU and HETP values is illustrated in Examples 12.3.1 and 12.3.2.

12.3.3 Mass Transfer Coefficients for Packed Columns

Three methods of estimating binary mass transfer coefficients in packed columns are presented below.

- Correlation for random packings by Onda et al. (1968).
- Correlation for random packings by Bravo and Fair (1982).
- Correlation for structured packings by Bravo et al. (1985).

Many other correlations for packed columns have been presented in the literature; they are reviewed by Ponter and Au-Yeung (1986). Mass transfer coefficients for two component systems in packed columns sometimes are reported as a correlation of the height of a transfer unit (see, e.g., Fair, 1984).

The methods of Onda et al. (1968) and of Bravo and Fair (1982) are illustrated in Example 12.3.1; mass transfer coefficients in structured packing are calculated in Example 12.3.2.

Onda's Correlations for Randomly Packed Columns Onda et al. (1968) developed correlations of mass transfer coefficients for gas absorption, desorption, and vaporization in randomly packed columns. The vapor-phase mass transfer coefficient is obtained from

$$k^V/\left(a_p D^V\right) = A \, \mathrm{Re}_V^{0.7} \mathrm{Sc}_V^{0.333}\left(a_p d_p\right)^{-2} \tag{12.3.26}$$

where d_p is the nominal packing size, and a_p is the specific surface area of the packing (m^2/m^3 of packing). The parameter A is a constant that takes the numerical value 2.0 if the nominal packing size, d_p, is less than 0.012 m and has the value 5.23 if the nominal packing size is greater than (or equal to) 0.012 m. The parameter Re_V is the vapor-phase Reynolds number

$$\mathrm{Re}_V = \rho_t^V u_V/\left(\mu^V a_p\right) \tag{12.3.27}$$

where u_V is the superficial velocity of the vapor. The vapor-phase Schmidt number is

$$\mathrm{Sc}_V = \mu^V/\left(\rho_t^V D^V\right) \tag{12.3.28}$$

The liquid-phase mass transfer coefficient is obtained from the following equation:

$$k^L\left(\rho_t^L/\mu_L g\right)^{0.333} = 0.0051(\mathrm{Re}_L')^{0.667}\mathrm{Sc}_L^{-0.5}\left(a_p d_p\right)^{0.4} \tag{12.3.29}$$

where Sc_L is the Schmidt number for the liquid phase

$$\mathrm{Sc}_L = \mu^L/\left(\rho_t^L D^L\right) \tag{12.3.30}$$

and Re'_L is the liquid-phase Reynolds number based on the interfacial area

$$Re'_L = \rho_t^L u_L / (\mu^L a') \tag{12.3.31}$$

The parameter a' is the interfacial area density (m^2/m^3 packing) and is obtained from the third part of Onda's correlation

$$a' = a_p \Big(1 - \exp\big\{ -1.45 (\sigma_c/\sigma)^{0.75} Re_L^{0.1} Fr_L^{-0.05} We_L^{0.2} \big\} \Big) \tag{12.3.32}$$

where σ is the surface tension of the liquid and σ_c is the critical surface tension of the packing. Values of the critical surface tension are tabulated by Onda et al. (1968). The liquid-phase Reynolds (Re_L) number based on the specific surface area (cf. Eq. 12.3.27), is

$$Re_L = \rho_t^L u_L / (\mu^L a_p) \tag{12.3.33}$$

The liquid-phase Froude (Fr_L) number is

$$Fr_L = a_p u_L^2 / g \tag{12.3.34}$$

and the Weber (We_L) number is

$$We_L = \rho_t^L u_L^2 / (a_p \sigma) \tag{12.3.35}$$

Bravo and Fair's Correlation for Randomly Packed Columns   Bravo and Fair (1982) developed a method of estimating the mass transfer characteristics for distillation in randomly packed columns. Their method is based on Onda's Eqs. 12.3.26 and 12.3.29 for the vapor- and liquid-phase mass transfer coefficients, but uses an alternative correlation for the interfacial area density

$$a' = 19.78 a_p (Ca_L Re_V)^{0.392} \sigma^{0.5} / H^{0.4} \tag{12.3.36}$$

where the vapor-phase Reynolds number is as defined in Eq. 12.3.27, H is the height of the packed section, and Ca_L is the capillary number

$$Ca_L = u_L^2 \mu^L / \sigma \tag{12.3.37}$$

Although the method of Bravo and Fair makes use of Onda's equations for k^V and k^L, it should be noted that the liquid-phase mass transfer coefficients estimated from the two procedures will be different since Eq. 12.3.29 for k^L depends on the interfacial area density (through the liquid-phase Reynolds number).

Correlation of Bravo, Rocha, and Fair for Structured Packings   Bravo et al. (1985, 1992) developed correlations for the prediction of mass transfer coefficients for structured packing with the geometry shown in Figure 12.9. Bravo et al. (1985) use the correlation of Johnstone and Pigford (1942) to estimate the Sherwood (Sh_V) number for the vapor phase

$$Sh_V = 0.0338 Re_V^{0.8} Sc_V^{0.333} \tag{12.3.38}$$

where the Sherwood number is defined by

$$Sh_V = k^V d_{eq} / D^V \tag{12.3.39}$$

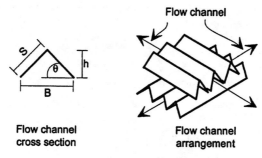

Figure 12.9. Geometry of structured packing.

Here d_{eq} is the equivalent diameter of a channel and is given by

$$d_{eq} = Bh[1/(B + 2S) + 1/2S]$$ (12.3.40)

where B is the base of the triangle (channel base), S is the corrugation spacing (channel side), and h is the height of the triangle (crimp height) (consult Fig. 12.9).

The vapor-phase Reynolds Re_V number is defined by

$$Re_V = d_{eq}\rho_t^V(u_{Ve} + u_{Le})/\mu^V$$ (12.3.41)

The effective velocity of vapor through the channel u_{Ve} is defined by

$$u_{Ve} = u_V/(\varepsilon \sin \theta)$$ (12.3.42)

The superficial vapor velocity is u_V, ε is the void fraction of the packing, and θ is the angle of the channel with respect to the horizontal. The effective liquid velocity u_{Le} is based on the relationship for laminar flow in a falling film

$$u_{Le} = \frac{3\Gamma}{2\rho_t^L}\left(\frac{(\rho_t^L)^2 g}{3\mu^L\Gamma}\right)^{0.333}$$ (12.3.43)

where Γ is the liquid flow rate per unit length of perimeter

$$\Gamma = \rho_t^L u_L/(PA_t)$$ (12.3.44)

where A_t is the cross-sectional area of the column, P is the perimeter determined by

$$P = (4S + B)/Bh$$ (12.3.45)

The penetration model (Eq. 9.2.11) is used to predict the liquid-phase mass transfer coefficients with the exposure time assumed to be the time required for the liquid to flow between corrugations (a distance equal to the channel side)

$$k^L = 2(D^L u_{Le}/\pi S)^{1/2}$$ (12.3.46)

The method of Bravo et al. (1985) is based on the assumption that the surface is completely wetted. That is the interfacial area density is equal to the specific packing surface: $a' = a_p$.

The method presented above has been revised by Bravo et al. (1992). In the revised correlations the equivalent diameter is taken to be equal to the channel side S. The velocities are adjusted to allow for liquid holdup and the assumption that the packing is completely wetted has been dispensed with.

Example 12.3.1   Distillation of Acetone and Water in a Packed Column

Estimate the mass transfer coefficients, interfacial area, and the heights of transfer units for the system acetone (1)–water (2) in a packed column under the following conditions prevailing near the top of the column:

Column diameter	$d_t = 2$ m
Column cross-sectional area	$A_t = 3.142$ m^2
Height of packing	$H = 2$ m
Packing type	50-mm steel Pall rings
Specific packing surface	$a_p = 105$ m^2/m^3
Critical surface tension	$\sigma_c = 0.075$ N/m

The molar flow rates of the vapor and liquid phases are

$$V = 100 \text{ mol/s} \qquad L = 25 \text{ mol/s}$$

The pressure and temperature are 101 kPa and 330 K, respectively. The composition of the liquid and vapor at the point in the packed section under consideration

$$x_1 = 0.84189 \qquad x_2 = 0.15811$$
$$y_1 = 0.91000 \qquad y_2 = 0.09000$$

DATA   Physical properties of the mixture at the bubble point temperature of the liquid have been estimated as follows:

Viscosity of vapor mixture: $\mu^V = 1.0 \times 10^{-5}$ Pa s.
Viscosity of liquid mixture: $\mu^L = 2.8 \times 10^{-4}$ Pa s.
Vapor density: $\rho_t^V = 2.0$ kg/m^3.
Liquid density: $\rho_t^L = 780$ kg/m^3.
Molar mass of acetone: $M_1 = 0.05808$ kg/mol.
Molar mass of water: $M_2 = 0.0181$ kg/mol.
The vapor-phase diffusivity: $D^V = 1.67 \times 10^{-5}$ m^2/s.
The liquid-phase diffusivity: $D^L = 4.3 \times 10^{-9}$ m^2/s.
Surface tension: $\sigma = 0.035$ N/m.
The K value: $K_1 = 1.081$.

SOLUTION   We will begin by estimating the mass transfer coefficients and interfacial area using the correlations of Onda et al., Eqs. 12.3.26–12.3.35. To use these correlations we first compute a number of dimensionless groups. These calculations are summarized below.

The molar mass of the vapor is calculated first.

$$M^V = y_1 M_1 + y_2 M_2$$
$$= 0.91 \times 0.05808 + 0.09 \times 0.0181$$
$$= 0.05448$$

The molar mass of the liquid is calculated in the same way with the result

$$M^L = x_1 M_1 + x_2 M_2$$
$$= 0.05176$$

The gas velocity may be calculated from the vapor flow rate as

$$u_V = V M^V / (A_t \rho_t^V)$$
$$= 100 \times 0.05448 / (3.142 \times 2)$$
$$= 0.8671 \text{ m/s}$$

The liquid velocity is determined in a similar way.

$$u_L = L M^L / (A_t \rho_t^L)$$
$$= 25 \times 0.05176 / (3.142 \times 780)$$
$$= 5.279 \times 10^{-4} \text{ m/s}$$

The vapor-phase Reynolds number is

$$\text{Re}_V = \rho_t^V u_V / (\mu^V a_p)$$
$$= 2 \times 0.8671 / (1 \times 10^{-5} \times 105)$$
$$= 1651$$

The liquid-phase Reynolds number is

$$\text{Re}_L = \rho_t^L u_L / (\mu^L a_p)$$
$$= 780 \times 5.279 \times 10^{-4} / (2.8 \times 10^{-4} \times 105)$$
$$= 14.01$$

The liquid-phase Froude number is

$$\text{Fr}_L = a_p u_L^2 / g$$
$$= 105 \times (5.279 \times 10^{-4})^2 / 9.81$$
$$= 2.985 \times 10^{-5}$$

and the Weber number is

$$\text{We}_L = \rho_t^L u_L^2 / (a_p \sigma)$$
$$= 780 \times (5.279 \times 10^{-4})^2 / (105 \times 0.035)$$
$$= 5.918 \times 10^{-5}$$

The Schmidt number for the vapor is found to be

$$\text{Sc}_V = \mu^V / (\rho_t^V D^V)$$
$$= 1.0 \times 10^{-5} / (2.0 \times 1.67 \times 10^{-5})$$
$$= 0.299$$

and the Schmidt number for the liquid is

$$Sc_L = \mu^L/(\rho_t^L D^L)$$
$$= 2.8 \times 10^{-4}/(780 \times 4.3 \times 10^{-9})$$
$$= 83.48$$

The interfacial area density may now be obtained from Eq. 12.3.32 as

$$a' = a_p\left(1 - \exp\left\{-1.45(\sigma_c/\sigma)^{0.75} \, Re_L^{0.1} \, Fr_L^{-0.05} \, We_L^{0.2}\right\}\right)$$
$$= 62.37 \, m^2/m^3$$

The liquid-phase Reynolds number based on the interfacial area may now be calculated as

$$Re'_L = \rho_t^L u_L/(\mu^L a')$$
$$= 780 \times 5.279 \times 10^{-4}/(2.8 \times 10^{-4} \times 62.37)$$
$$= 23.58$$

The mass transfer coefficient for the vapor phase is computed from Eq. 12.3.26 with the coefficient A equal to 5.23, since the packing size d_p is larger than 0.012 m,

$$k^V = 5.23 \, Re_V^{0.7} \, Sc_V^{0.333} (a_p d_p)^{-2} (a_p D^V)$$
$$= 0.03982 \, m/s$$

The mass transfer coefficient for the liquid phase is computed from Eq. 12.3.29

$$k^L = 0.0051(Re'_L)^{0.667} \, Sc_L^{-0.5} (a_p d_p)^{0.4} (\mu_L g/\rho_t^L)^{0.333}$$
$$= 1.363 \times 10^{-4} \, m/s$$

The heights of the transfer units follow as:

$$H_V = u_V/(k^V a')$$
$$= 0.8671/(0.03982 \times 62.368)$$
$$= 0.349 \, m$$
$$H_L = u_L/(k^L a')$$
$$= 5.279 \times 10^{-5}/(1.363 \times 10^{-4} \times 62.368)$$
$$= 0.062 \, m$$

The stripping factor Λ is 4.324 and the overall height of a transfer unit is

$$H_{OV} = H_V + \Lambda H_L$$
$$= 0.349 + 4.324 \times 0.062$$
$$= 0.618$$

Finally, the HETP is calculated from

$$HETP = H_{OV} \ln \Lambda/(\Lambda - 1)$$
$$= 0.618 \ln(4.324)/(4.324 - 1)$$
$$= 0.272$$

We now repeat the calculation using the Bravo and Fair correlations. The vapor-phase mass transfer coefficient is exactly the same as determined above. We need only compute the interfacial area density and the liquid-phase mass transfer coefficient. We begin by computing the capillary number

$$Ca_L = u_L \mu^L / \sigma$$
$$= 5.279 \times 10^{-4} \times 2.8 \times 10^{-4} / 0.035$$
$$= 4.224 \times 10^{-6}$$

The interfacial area density follows from Eq. 12.3.36

$$a' = 19.78 a_p (Ca_L \, Re_V)^{0.392} \sigma^{0.5} / H^{0.4}$$
$$= 19.78 \times 105 \times (4.224 \times 10^{-6} \times 1651)^{0.392} 0.035^{0.5} / 2^{0.4}$$
$$= 42.045 \; m^2/m^3$$

The liquid-phase Reynolds number based on the interfacial area may now be recalculated as

$$Re'_L = \rho_t^L u_L / (\mu^L a')$$
$$= 780 \times 5.279 \times 10^{-4} / (2.8 \times 10^{-4} \times 42.045)$$
$$= 34.98$$

The mass transfer coefficient for the liquid phase is computed from Eq. 12.3.29

$$k^L = 0.0051 (Re'_L)^{0.667} \, Sc_L^{-0.5} (a_p d_p)^{0.4} (\mu_L g / \rho_t^L)^{0.333}$$
$$= 1.773 \times 10^{-4} \; m/s$$

The heights of the transfer units follow as:

$$\mathbb{H}_V = u_V / (k^V a')$$
$$= 0.8671 / (0.03982 \times 42.045)$$
$$= 0.518 \; m$$
$$\mathbb{H}_L = u_L / (k^L a')$$
$$= 5.279 \times 10^{-5} / (1.363 \times 10^{-4} \times 42.045)$$
$$= 0.0709 \; m$$

The stripping factor Λ is 4.324 and the overall height of a transfer unit is

$$\mathbb{H}_{OV} = \mathbb{H}_V + \Lambda \mathbb{H}_L$$
$$= 0.518 + 4.324 \times 0.0709$$
$$= 0.824$$

The HETP is calculated as

$$HETP = \mathbb{H}_{OV} \ln \Lambda / (\Lambda - 1)$$
$$= 0.824 \ln(4.324) / (4.324 - 1)$$
$$= 0.363$$

∎

Example 12.3.2   Mass Transfer Coefficients in a Column with Structured Packing

Estimate the mass transfer coefficients, interfacial area, and the heights of transfer units for the system in Example 12.3.1, namely, acetone (1)–water (2), in a column equipped with Sulzer BX structured packing. The characteristics of this packing are (Bravo et al., 1985).

Specific packing surface	a_p	$= 492 \text{ m}^2/\text{m}^3$
Crimp height	h	$= 6.4 \text{ mm}$
Channel base	B	$= 12.7 \text{ mm}$
Channel side	S	$= 8.9 \text{ mm}$
Void fraction	ε	$= 0.90$
Channel flow angle	θ	$= 60°$

All other details of the operation are as in Example 12.3.1.

SOLUTION   We begin by computing the equivalent diameter from Eq. 12.3.40

$$d_{eq} = Bh(1/(B + 2S) + 1/2S)$$

$$= 12.7 \times 6.4 \times (1/(12.7 + 2 \times 6.4) + 1/(2 \times 6.4))$$

$$= 7.231 \text{ mm}$$

The wetted perimeter is computed next

$$P = (4S + B)/Bh$$

$$= (4 \times 6.4 + 12.7)/(12.7 \times 6.4)$$

$$= 0.59424 \text{ mm/mm}^2$$

$$= 594.24 \text{ m/m}^2$$

The superficial vapor velocity was determined in Example 12.3.1 to be

$$u_V = VM^V/(A_t\rho_t^V)$$

$$= 0.867 \text{ m/s}$$

The effective vapor velocity follows from Eq. 12.3.42

$$u_{V_e} = u_V/(\varepsilon \sin \theta)$$

$$= 0.867/(0.9 \times \sin(\pi/3))$$

$$= 1.1125 \text{ m/s}$$

The superficial liquid velocity is (from Example 12.3.1)

$$u_L = LM^L/(A_t\rho_t^L)$$

$$= 5.279 \times 10^{-4} \text{ m/s}$$

Thus, the liquid flow rate per unit length of perimeter is

$$\Gamma = \rho_t^L u_L / (PA_t)$$

$$= 780 \times 5.279 \times 10^{-4} / (594.24 \times 3.142)$$

$$= 2.206 \times 10^{-4}$$

and the effective liquid velocity, computed from Eq. 12.3.43, is

$$u_{Le} = \frac{3\Gamma}{2\rho_t^L} \left(\frac{(\rho_t^L)^2 g}{3\mu^L \Gamma} \right)^{0.333}$$

$$= \frac{3 \times 2.206 \times 10^{-4}}{2 \times 780} \left(\frac{780^2 \times 9.81}{3 \times 2.8 \times 10^{-4} \times 2.206 \times 10^{-4}} \right)^{0.333}$$

$$= 0.01336 \text{ m/s}$$

The vapor-phase Reynolds number is

$$\text{Re}_V = d_{eq} \rho_t^V (u_{Ve} + u_{Le}) / \mu^V$$

$$= 7.231 \times 10^{-3} \times 2.0 \times (1.1125 + 0.01336) / (1.0 \times 10^{-5})$$

$$= 1628$$

The Schmidt number for the vapor is (from Example 12.3.1)

$$\text{Sc}_V = \mu^V / (\rho_t^V D^V)$$

$$= 0.299$$

The Sherwood number may now be determined as

$$\text{Sh}_V = 0.0338 \, \text{Re}_V^{0.8} \, \text{Sc}_V^{0.333}$$

$$= 0.0338 \times 1628^{0.8} \times 0.299^{0.333}$$

$$= 8.392$$

The mass transfer coefficient for the vapor phase is computed from the definition of the Sherwood number (Eq. 12.3.39), as

$$k^V = \text{Sh}^V D^V / d_{eq}$$

$$= 8.392 \times 1.67 \times 10^{-5} / 7.231 \times 10^{-3}$$

$$= 0.01938 \text{ m/s}$$

The mass transfer coefficient for the liquid phase is computed from Eq. 12.3.46

$$k^L = 2 (D^L u_{Le} / \pi S)^{1/2}$$

$$= 9.065 \times 10^{-5} \text{ m/s}$$

The heights of the transfer units follow as:

$$\mathbb{H}_V = u_V/\left(k^V a_p\right)$$

$$= 0.867/(0.01938 \times 492)$$

$$= 0.091 \text{ m}$$

$$\mathbb{H}_L = u_L/\left(k^L a_p\right)$$

$$= 5.279 \times 10^{-5}/(9.065 \times 10^{-5} \times 492)$$

$$= 0.012 \text{ m}$$

$$\mathbb{H}_{OV} = H_V + \Lambda H_L$$

$$= 0.091 + 4.324 \times 0.012$$

$$= 0.142$$

Notice that we use the specific area of the packing in calculating $\mathbb{H}_V$ and $\mathbb{H}_L$ as required by the method.

Finally, the HETP is calculated from

$$\text{HETP} = \mathbb{H}_{OV} \ln \Lambda/(\Lambda - 1)$$

$$= 0.142 \ln(4.324)/(4.324 - 1)$$

$$= 0.063$$

The height of the transfer units is much lower for the Sulzer BX packing than it was for the 50-mm Pall rings considered in Example 12.3.1. The practical implication of this calculation is that a much shorter column would be needed to separate acetone–water mixtures if the column were fitted with this kind of structured packing instead of Pall rings. The reduction in the height of the transfer units is brought about not by increasing the mass transfer coefficients (in fact, k^V and k^L are lower in this example than they were in Example 12.3.1) but by an enormous increase in the interfacial area. ∎

12.3.4   Transfer Units for Multicomponent Systems

Equation 12.3.12 may be written in dimensionless form as

$$\frac{d(y)}{d\zeta} = [\mathbb{N}_{OV}](y^* - y) \tag{12.3.47}$$

where $\zeta = z/H$. The overall number of transfer units $[\mathbb{N}_{OV}]$ is for the vapor phase and may be defined by

$$[\mathbb{N}_{OV}] \equiv c_t^V [K_{OV}] a' H A_c/V \equiv [K_{OV}] a' H/u_V \tag{12.3.48}$$

The composition of the vapor along the length may be determined by integrating (numerically) Eq. 12.3.47. Each step of the integration requires the estimation of the matrix of overall number of transfer units.

To evaluate the overall number of transfer units we may proceed to combine Eq. 12.2.16 for $[K_{OV}]$ with Eq. 12.3.48 for $[\mathbb{N}_{OV}]$ to give (cf. Eq. 12.2.19)

$$[\mathbb{N}_{OV}]^{-1} = [\mathbb{N}_V]^{-1} + (V/L)[K][\mathbb{N}_L]^{-1} \tag{12.3.49}$$

where $[K]$ is a diagonal matrix of the first $n - 1$ equilibrium K values. The matrices $[\mathbb{N}_V]$ and $[\mathbb{N}_L]$ represent the numbers of transfer units for the vapor and liquid phases, respectively. The matrices of numbers of transfer units for each phase may be expressed in terms of the mass transfer coefficients for each phase as

$$[\mathbb{N}_V] \equiv [k^V]a'H/u_V \tag{12.3.50}$$

$$[\mathbb{N}_L] \equiv [k^L]a'H/u_L \tag{12.3.51}$$

If we use the simplified models discussed in Section 8.8 for the matrices of mass transfer coefficients we may relate the inverse matrices of numbers of transfer units to the matrices of inverse binary mass transfer coefficients as (cf. Eqs. 12.2.20 and 12.2.21)

$$[\mathbb{N}_V]^{-1} \equiv [R^V]u_V/(a'H) \tag{12.3.52}$$

$$[\mathbb{N}_L]^{-1} \equiv [R^L]u_L/(a'H) \tag{12.3.53}$$

For multicomponent systems it is possible to define matrices of HTU values. Thus, for the vapor phase

$$[\mathbb{H}_V] = (u_V/a')[k^V]^{-1} \tag{12.3.54}$$

and for the liquid phase

$$[\mathbb{H}_L] = (u_L/a')[k^L]^{-1} \tag{12.3.55}$$

The multicomponent generalization of Eq. 12.3.22 is

$$[\mathbb{H}_{OV}] = [\mathbb{H}_V] + (V/L)[K][\mathbb{H}_L] \tag{12.3.56}$$

To evaluate the matrices of heights or numbers of transfer units we may use the empirical methods of Section 12.3.3 to estimate the binary (Maxwell–Stefan) mass transfer coefficients as functions of the Maxwell–Stefan diffusion coefficients. The elements of the $[R]$ matrices may then be computed with the aid of Eqs. 12.2.17 and 12.2.18 as illustrated in Example 12.3.3.

Experimental studies carried out with a view to testing these models have been reported by Dribicka and Sandall (1979), by Gorak and co-workers (1983, 1985, 1987, 1988, 1990, 1991), by Krishna et al. (1981), and by Arwickar (1981). All of these groups assumed equimolar overflow in the analysis of their data. There are additional differences between the equations integrated by these groups that the interested reader can discern for themselves.

Example 12.3.3 Distillation of a Quaternary System in a Sulzer Packed Column

Gorak (1991) conducted a number of distillation experiments with the system acetone(1)–methanol(2)–2-propanol(3)–water(4) in a column of 0.1-m internal diameter and height 0.8 m fitted with Sulzer BX structured packing. The characteristics of Sulzer BX packing are (Bravo et al., 1985).

Specific packing surface	$a_p = 450 \text{ m}^2/\text{m}^3$
Crimp height	$h = 6.4 \text{ mm}$
Channel base	$B = 12.7 \text{ mm}$
Channel side	$S = 8.9 \text{ mm}$
Void fraction	$\varepsilon = 0.90$
Channel flow angle	$\theta = 60°$

Estimate the mass transfer coefficients and the numbers of transfer units for this system at the following conditions that existed at the bottom of the packing in one of Gorak's experiments.

Flow rate of the vapor   28.3305 mol/m^{2}s
Pressure   101 kPa.

Composition of the vapor below the packing:

$$y_1 = 0.003157 \quad y_2 = 0.164701 \quad y_3 = 0.394518 \quad y_4 = 0.437623$$

The experiments were carried out at total reflux.

DATA   Physical properties of the vapor and liquid phases at the composition reported above and at the bubble point temperature of the liquid have been estimated as follows:

Viscosity of vapor mixture: $\mu^V = 1.113 \times 10^{-5}$ Pa s.
Viscosity of liquid mixture: $\mu^L = 3.814 \times 10^{-4}$ Pa s.
Vapor density: $\rho_t^V = 1.253$ kg/m^3.
Liquid density: $\rho_t^L = 823.7$ kg/m^3.
Average molar mass of vapor: $M^V = 0.03705$ kg/mol.
Surface tension: $\sigma = 0.048$ N m.

The Maxwell–Stefan diffusivities in the vapor phase [units are 10^{-5} m^2/s]

$$\mathcal{D}_{12}^V = 1.258$$
$$\mathcal{D}_{13}^V = 0.8084$$
$$\mathcal{D}_{14}^V = 1.944$$
$$\mathcal{D}_{23}^V = 1.220$$
$$\mathcal{D}_{24}^V = 2.896$$
$$\mathcal{D}_{34}^V = 1.883$$

The Maxwell–Stefan diffusivities in the liquid phase [units are 10^{-9} m^2/s]

$$\mathcal{D}_{12}^L = 3.423$$
$$\mathcal{D}_{13}^L = 2.925$$
$$\mathcal{D}_{14}^L = 3.742$$
$$\mathcal{D}_{23}^L = 3.637$$
$$\mathcal{D}_{24}^L = 5.029$$
$$\mathcal{D}_{34}^L = 4.358$$

Equimolar countertransfer may be assumed to prevail.
    The K values determined at the bubble point of the liquid with a composition equal to that of the vapor are

$$K_1 = 3.5416 \quad K_2 = 1.5581 \quad K_3 = 1.0865 \quad K_4 = 0.6936$$

K values were estimated using Eq. 7.3.6 with activity coefficients calculated with the NRTL model and vapor pressures determined with the Antoine equation.

SOLUTION We begin by performing a few geometric calculations. The cross-sectional area of the column is

$$A_t = \pi d_c^2/4$$

$$= 7.854 \times 10^{-3} \text{ m}^2$$

The equivalent diameter is found using Eq. 12.3.40

$$d_{eq} = Bh(1/(B + 2S) + 1/2S)$$

$$= 12.7 \times 6.4 \times (1/(12.7 + 2 \times 6.4) + 1/(2 \times 6.4))$$

$$= 7.231 \text{ mm}$$

and the wetted perimeter is

$$P = (4S + B)/Bh$$

$$= (4 \times 6.4 + 12.7)/(12.7 \times 6.4)$$

$$= 0.59424 \text{ mm/mm}^2$$

$$= 594.24 \text{ m/m}^2$$

The molar flow rate of the vapor below the packing is

$$V = 0.0283305 \times A_t$$

$$= 0.22251 \text{ mol/s}$$

At total reflux the flow and composition of the liquid leaving the bottom of the packing is the same as the flow and composition of the vapor entering the column. Thus,

$$L = 0.22251 \text{ mol/s}$$

and

$$x_1 = 0.003157 \qquad x_2 = 0.164701 \qquad x_3 = 0.394518 \qquad x_4 = 0.437623$$

The molar mass of the liquid will equal the molar mass of the vapor; $M^V = M^L$.
   The superficial vapor velocity is

$$u_V = VM^V/(A_t\rho_t^V)$$

$$= 0.8379 \text{ m/s}$$

The effective vapor velocity follows from Eq. 12.3.42

$$u_{Ve} = u_V/(\varepsilon \sin \theta)$$

$$= 0.8379/(0.9 \times \sin(\pi/3))$$

$$= 1.085 \text{ m/s}$$

The superficial liquid velocity is

$$u_L = LM^L/(A_t \rho_t^L)$$
$$= 1.2745 \times 10^{-3} \text{ m/s}$$

Thus, the liquid flow rate per unit length of perimeter is

$$\Gamma = \rho_t^L u_L/(PA_t)$$
$$= 823.7 \times 1.2745 \times 10^{-3}/(594.24 \times 7.854 \times 10^{-3})$$
$$= 0.227$$

and the effective liquid velocity, computed from Eq. 12.3.43 is

$$u_{Le} = 3\Gamma/2\rho_t^L \left((\rho_t^L)^2 g/3\mu^L \Gamma\right)^{0.333}$$
$$= \frac{3 \times 0.227}{2 \times 823.7} \left(\frac{823.7^2 \times 9.81}{3 \times 3.814 \times 10^{-4} \times 0.227}\right)^{0.333}$$
$$= 1.2017 \text{ m/s}$$

The vapor-phase Reynolds number is

$$\text{Re}_V = d_{eq}\rho_t^V(u_{Ve} + u_{Le})/\mu^V$$
$$= 7.231 \times 10^{-3} \times 1.253 \times (1.0750 + 1.2017)/(1.113 \times 10^{-5})$$
$$= 1853.4$$

For each binary pair in the mixture we must determine the Schmidt number and the Sherwood number before computing the mass transfer coefficient. For the 1–2 pair the Schmidt number for the vapor is (from Example 12.3.1)

$$\text{Sc}_{V12} = \mu^V/(\rho_t^V \mathcal{D}_{12}^V)$$
$$= 0.7061$$

The Sherwood number for the 1–2 pair is

$$\text{Sh}_{V12} = 0.0338 \, \text{Re}_V^{0.8} \, \text{Sc}_{V12}^{0.333}$$
$$= 0.0338 \times 1853.4^{0.8} \times 0.7061^{0.333}$$
$$= 12.387$$

The vapor-phase mass transfer coefficient for the 1–2 pair is computed from the definition of the Sherwood number (Eq. 12.3.39) as

$$\kappa_{12}^V = \text{Sh}_{V12}\mathcal{D}_{12}^V/d_{eq}$$
$$= 12.387 \times 1.258 \times 10^{-5}/7.231 \times 10^{-3}$$
$$= 0.0216 \text{ m/s}$$

In the same way we compute the mass transfer coefficients for the other binary pairs

$$\kappa_{13}^V = 0.0160 \text{ m/s}$$

$$\kappa_{14}^V = 0.0288 \text{ m/s}$$

$$\kappa_{23}^V = 0.0211 \text{ m/s}$$

$$\kappa_{24}^V = 0.0376 \text{ m/s}$$

$$\kappa_{34}^V = 0.0282 \text{ m/s}$$

The matrix of inverted mass transfer coefficients for the vapor phase $[R^V]$ is computed using Eqs. 12.2.17 and 12.2.18. We illustrate by calculating just two elements

$$R_{11}^V = \frac{y_1}{\kappa_{14}^V} + \frac{y_2}{\kappa_{12}^V} + \frac{y_3}{\kappa_{13}^V} + \frac{y_4}{\kappa_{14}^V}$$

$$= \frac{0.003157}{0.0288} + \frac{0.1647}{0.0216} + \frac{0.3945}{0.0160} + \frac{0.4376}{0.0288}$$

$$= 47.5323 \text{ (m/s)}^{-1}$$

$$R_{12}^V = -y_1\left(\frac{1}{\kappa_{12}^V} - \frac{1}{\kappa_{14}^V}\right)$$

$$= -0.003157 \times \left(\frac{1}{0.0216} - \frac{1}{0.0288}\right)$$

$$= -0.0367 \text{ (m/s)}^{-1}$$

The remaining elements are computed with the following result:

$$[R^V] = \begin{bmatrix} 47.5323 & -0.0367 & -0.0872 \\ -3.2604 & 34.8600 & -3.4184 \\ -10.599 & -4.6981 & 37.505 \end{bmatrix} \text{(m/s)}^{-1}$$

Turning, now, to the liquid phase we have, for the 1–2 binary

$$\kappa_{12}^L = 2\left(Ð_{12}^L u_{Le}/\pi S\right)^{1/2}$$

$$= 2\left(3.423 \times 10^{-9} \times 1.2017/(\pi \times 8.9 \times 10^{-3})\right)^{1/2}$$

$$= 0.7671 \times 10^{-3} \text{ m/s}$$

The liquid-phase mass transfer coefficients are computed in the same way to give

$$\kappa_{13}^L = 0.7091 \times 10^{-3} \text{ m/s}$$

$$\kappa_{14}^L = 0.8021 \times 10^{-3} \text{ m/s}$$

$$\kappa_{23}^L = 0.7907 \times 10^{-3} \text{ m/s}$$

$$\kappa_{24}^L = 0.9298 \times 10^{-3} \text{ m/s}$$

$$\kappa_{34}^L = 0.8656 \times 10^{-3} \text{ m/s}$$

The matrix of inverted mass transfer coefficients for the liquid phase $[R^L]$ is computed from Eqs. 12.2.17 and 12.2.18 using $z_i = x_i$ and the liquid-phase mass transfer coefficients computed above. The result is

$$[R^L] = \begin{bmatrix} 13.206 & -0.0018 & -0.0052 \\ -0.3757 & 11.508 & -0.3116 \\ -1.006 & -0.4314 & 11.741 \end{bmatrix} \times 10^2 \ (m/s)^{-1}$$

We may now compute $[R^{OV}]$ from Eq. 8.8.19. The complete matrix is

$$[R^{OV}] = \begin{bmatrix} 54.6470 & -0.0379 & -0.0900 \\ -3.3494 & 37.5876 & -3.4922 \\ -10.7655 & -4.7681 & 39.4452 \end{bmatrix} (m/s)^{-1}$$

The matrix of overall mass transfer coefficients is, therefore

$$[K_{OV}] = [R^{OV}]^{-1}$$

$$= \begin{bmatrix} 18.3094 & 0.0240 & 0.0439 \\ 2.1196 & 26.9095 & 2.3872 \\ 5.2533 & 3.2593 & 25.6521 \end{bmatrix} \times 10^{-3} \ m/s$$

Finally, the matrix of overall numbers of transfer units may now be computed from

$$[\mathbb{N}_{OV}] \equiv c_t^V [K_{OV}] a' H A_c / V$$

$$= \begin{bmatrix} 7.8669 & 0.0103 & 0.0188 \\ 0.9107 & 11.5621 & 1.0257 \\ 2.2572 & 1.4004 & 11.0218 \end{bmatrix}$$

This matrix can be used in the determination of the composition of the vapor a short distance above the bottom of the packing by integrating (numerically) Eq. 12.3.47.

The molar fluxes at the bottom of the packing may be evaluated using Eq. 7.3.11a (noting that $N_t = 0$ and, therefore, $N_i = J_i$)

$$(N) = c_t^V [K_{OV}](y - y^*)$$

which yields the following values for the molar fluxes

$$N_1 = -5.093 \times 10^{-3} \ mol/m^2 s$$

$$N_2 = -86.97 \times 10^{-3} \ mol/m^2 s$$

$$N_3 = -41.16 \times 10^{-3} \ mol/m^2 s$$

$$N_4 = 0.1332 \times 10^{-3} \ mol/m^2 s$$

showing, as might be expected, that acetone, methanol, 2-propanol (components 1–3) are transferring from the liquid to the vapor phase. The equilibrium vapor mole fractions y^* needed for this calculation were determined from $y_i^* = K_i x_i$ with the K values provided in the problem statement. ∎

13 Multicomponent Distillation: Efficiency Models

The equilibrium stage concept makes possible the adequate design of separation processes despite our inability to deal adequately with the complex heat and mass transfer operations that occur in an actual contact stage. A hypothetical process whose contact stages are all true equilibrium stages is created on paper to accomplish the separation desired in the actual plant process. The number of equilibrium stages required in the hypothetical process is related to the required number of actual contact stages by proportionality factors (stage efficiencies) which describe the extent to which the performance of an actual contact stage duplicates the performance of an equilibrium stage.

—B. D. Smith (1964)

13.1 INTRODUCTION

Simulation and design of multicomponent distillation and other multistage separation operations usually is carried out using the *equilibrium stage model*. The key assumption of this model is that the vapor and liquid streams leaving a stage are in equilibrium with each other. A complete distillation or absorption column may be modeled as a sequence of these stages. The first methods for solving the equilibrium stage model equations were graphical in nature. Modern methods for solving equilibrium stage separation process problems are computer based. Indeed, since the late 1950s, hardly a year has gone by without the publication of at least one computer-based method for solving the equilibrium stage model equations. Developments to about 1980 have been described in a number of textbooks (see, e.g., Holland, 1975, 1981; King, 1980; Henley and Seader, 1981). Seader (1985) has written as interesting history of equilibrium stage simulation. Computer programs implementing one or more simulation methods are commonly used in industry.

The trays in a real distillation column are not equilibrium stages. In other words, the vapor above the froth is not in equilibrium with the liquid leaving the tray. The degree of separation on a real tray is, of course, determined as much by mass and energy transfer between the phases being contacted on a tray as it is by thermodynamic equilibrium considerations. This is where an *efficiency* comes into the picture.

13.1.1 Definitions of Efficiency

An efficiency is a measure of how close to the equilibrium separation the real column or tray comes. The simplest and still a widely used approach is to use an overall column or section efficiency defined by

$$E_o = N_{eqm}/N_{act} \tag{13.1.1}$$

where N_{eqm} and N_{act} are, respectively, the number of ideal or equilibrium stages and the number of actual trays in the section of the column under consideration.

Overall column or section efficiencies are complicated functions of tray design, fluid properties, and operating conditions. Some empirical correlations of overall column efficiency have been developed but are capable only of rough estimates of efficiency that may at best be useful in preliminary design studies (Lockett, 1986).

Some column simulation programs require the user to provide an efficiency for each tray. There are many different definitions of *tray efficiency*: Murphree (1925), Hausen (1953), vaporization (Holland, 1975) and generalized Hausen (Standart, 1965). There is by no means a consensus on which is the most useful. Arguments for and against the various definitions are presented by, among others, Standart (1965, 1971), Holland and McMahon (1970) and by Medina et al. (1978, 1979). King (1980), Lockett (1986) and Seader (1989) conveniently summarized many of these arguments. Possibly the most soundly based efficiency, the generalized Hausen efficiency of Standart (1965), is the most difficult to use (see, however, Fletcher, 1987). The Murphree tray efficiency is the one most widely used in separation process calculations.

The *Murphree vapor-phase tray efficiency* is defined by (refer to Fig. 12.1, which pictures vapor–liquid contacting on a distillation tray)

$$E_i^{MV} = (\bar{y}_{iL} - y_{iE})/(y_{iL}^* - y_{iE}) \tag{13.1.2}$$

where the composition of vapor below the tray is y_{iE}, $\bar{y}_{iL}$ is the average composition of the vapor above the froth and y_{iL}^* is the mole fraction of component i in a vapor in equilibrium with the liquid that is leaving the tray. The composition y_{iL}^* is determined from a bubble-point calculation for a liquid of composition x_{iL}.

In practice the tray efficiency is estimated from a model that combines information about the tray hydrodynamics, with a model for the *point efficiency* that is defined analogously to the tray efficiency but applied to a narrow vertical slice of the froth (see Fig. 12.1). The *Murphree vapor-phase point efficiency*, for example, is defined by

$$E_{OV,i} = \frac{y_{iL} - y_{iE}}{y_i^* - y_{iE}} = 1 - \frac{\Delta y_{iL}}{\Delta y_{iE}} \tag{13.1.3}$$

where the composition of vapor below the tray is y_{iE}, y_{iL} is the composition of the vapor above the vertical slice of froth under consideration, and y_i^* is the composition of a vapor that would be in equilibrium with the liquid at the point on the tray in question. The composition y_i^* is determined from a bubble-point calculation for a liquid of composition x_i and Δy_i is the difference between the equilibrium composition and the actual vapor composition $\Delta y_i = y_i^* - y_i$.

There are basically two situations that might require a process engineer to estimate the distillation tray efficiency. First, when he/she is involved in the *design* of the column, the tray efficiency is required for estimation of the actual number of trays to be installed. The second situation arises when the operating column does not meet the desired purity requirements and one would like to know the tray efficiency for the given set of operating conditions in order to determine the precise cause for the less than desired purity; for example, to establish whether the column is operating with excessive entrainment or weeping or whether there are "hardware" problems in the column (blown valves, plugged holes, etc.). The estimation of the tray efficiency in the second situation (trouble shooting) requires a much higher level of understanding of the factors affecting the tray efficiency than the design situation. The reason is that at the design stage it is possible to select the operating conditions; one does not need to bother about, for example, weeping and entrainment, one simply tries to minimize their negative influence.

Most published procedures deal primarily with binary systems. For a binary system the fact that the mole fractions sum to unity means that the Murphree efficiencies of both components are the same on any given tray. These values may, however, vary from tray to tray. Multicomponent systems are, however, far more common in industrial processes. For a multicomponent system there are $n - 1$ independent component efficiencies for each tray and there is no requirement that the efficiencies of all components be equal. In fact, it is possible to have negative efficiencies for one or more components in a multicomponent mixture. Conventional practice is to use the same value for all component efficiencies. This approach to separation process design and simulation may be expected to lead to over- or underdesign when, for whatever reason, the efficiencies of different components are quite distinct (as they may be in mixtures of species with widely differing physical properties), when efficiencies are low or vary widely from stage to stage (as in hydrocarbon absorbers and in high purity separations, such as azeotropic and extractive distillation), or when heat effects are important (as they are in several gas absorption processes).

In fact, through use of matrix models of mass transfer in multicomponent systems (as opposed to effective diffusivity methods) it is possible to develop methods for estimating point and tray efficiencies in multicomponent systems that, when combined with an equilibrium stage model, overcome some of the limitations of conventional design methods. The purpose of this chapter is to develop these methods. We look briefly at ways of solving the set of equations that model an entire distillation column and close with a review of experimental and simulation studies that have been carried out with a view to testing multicomponent efficiency models.

13.2   EFFICIENCIES OF BINARY SYSTEMS

13.2.1   Point Efficiency for Binary Systems

The composition of the vapor above any point of the dispersion is given by Eq. 12.1.8

$$(y_1^* - y_{1L}) = \exp(-\mathbb{N}_{OV})(y_1^* - y_{1E}) \tag{12.1.8}$$

which allows us to express the point efficiency defined by Eq. 13.1.3 as

$$E_{OV1} = \frac{y_{1L} - y_{1E}}{y_{1L}^* - y_{1E}} = 1 - \frac{\Delta y_{1L}}{\Delta y_{1E}} = 1 - \exp(-\mathbb{N}_{OV}) \tag{13.2.1}$$

E_{OV1} has a value ranging between 0 and 1 because $\mathbb{N}_{OV}$ is a real and positive quantity. Also, for a two component system there is only one independent mole fraction y_1 and only one independent point efficiency E_{OV1}, which equals the component efficiency of Component 2: $E_{OV1} = E_{OV2}$. The prediction of the number of transfer units was discussed at length in Section 12.1.5.

If the fundamental model described in Section 12.1.7 is used to characterize the mass transfer properties of the dispersion, the overall composition change in the vapor phase is given by Eq. 12.1.73.

$$(y_1^* - y_{1L}) = Q_{II}Q_I(y_1^* - y_{1E}) \tag{12.1.73}$$

where Q_I and Q_{II}, for the bubble formation and free bubbling zones, respectively, are obtained in terms of the numbers of transfer units for each region of the froth as discussed in Section 12.1.7. The point efficiency is, therefore,

$$E_{OV1} = 1 - Q_{II}Q_I \tag{13.2.2}$$

Example 13.2.1   Point Efficiency of Toluene–Methylcyclohexane

Estimate the Murphree point efficiency for the system toluene (1)–methylcyclohexane (2) considered in Example 12.1.1.

The number of overall transfer units was estimated for this system using three different empirical methods. The results were as follows:

AIChE method $N_{OV} = 0.4667$
Chan and Fair method $N_{OV} = 1.236$
Zuiderweg method $N_{OV} = 0.9673$

SOLUTION   Using the results of the AIChE correlation, the point efficiency follows from Eq. 13.2.1

$$E_{OV} = 1 - \exp(-N_{OV})$$
$$= 1 - \exp(-0.4667)$$
$$= 0.3729$$

We will repeat the estimation of the point efficiency using the Chan and Fair correlation for the number of transfer units for the vapor phase.

The point efficiency again follows from Eq. 13.2.1

$$E_{OV} = 1 - \exp(-N_{OV})$$
$$= 1 - \exp(-1.236)$$
$$= 0.7095$$

Using Zuiderweg's method we determine the point efficiency as

$$E_{OV} = 1 - \exp(-N_{OV})$$
$$= 1 - \exp(-0.9673)$$
$$= 0.6199$$

The AIChE and Chan and Fair methods have given quite different estimates of the point efficiency for this system; Zuiderweg's method falls in between. In fact, this example is based on experiments carried out by Plaka et al. (1989) and the column dimensions are for their column. The point efficiency determined by Plaka et al. in a number of experiments is around 75–80%. In this case we see that the method of Chan and Fair has given an acceptable estimate of the experimentally determined value, whereas the AIChE method severely underpredicts the point efficiency. The latter result is in accord with the findings of Plaka et al. ∎

Example 13.2.2   Point Efficiency for the Regeneration of Triethylene Glycol

Let us continue with Example 12.1.2 and estimate the point efficiency for the regeneration of triethylene glycol(2) (TEG) from a mixture with water(1).

Composition in the bulk vapor below tray.

$$y_{1E} = 0.7216 \qquad y_{2E} = 0.2784$$

Composition in the liquid on the tray.

$$x_1 = 0.258 \qquad x_2 = 0.742$$

Equilibrium vapor composition.

$$y_1^* = 0.969 \qquad y_2^* = 0.031$$

SOLUTION The values of both Q_I and Q_{II} were determined in Example 12.1.2. The values were

$$Q_I = 0.9575$$

$$Q_{II} = 0.7100$$

The point efficiency may be calculated from Eq. 13.2.2 as

$$E_{OV} = 1 - Q_{II}Q_I$$
$$= 1 - 0.7100 \times 0.9576$$
$$= 0.3201$$

Armed with the point efficiency we may calculate the composition of the vapor above the tray as

$$y_{1L} = y_{1E} + E_{OV}(y_1^* - y_{1E})$$
$$= 0.7216 + 0.3201 \times (0.969 - 0.7216)$$
$$= 0.801 \qquad \blacksquare$$

13.2.2 Tray Efficiency

A horizontal concentration gradient will develop in the liquid due to mass transfer into and from the liquid as the liquid flows across the tray. Thus, the composition of the vapor above the froth will change as we traverse the tray even if the composition of the vapor just below the tray is uniform. The point efficiency defined in the preceding section models the mass transfer processes at a particular point on the tray but does not take into account the fact that the liquid may have a significant concentration change as it crosses the tray. Thus, the point efficiency must be related to the tray efficiency before it can be used in column design calculations.

There are many different models for liquid flow across a distillation tray. We consider the two simplest here.

If the liquid is completely mixed in the horizontal direction (a reasonable approximation for small diameter columns) then the tray efficiency and the point efficiency are one and the same (all points being equal as it were) and $E_i^{MV} = E_{OV,i}$.

Lewis (1936) was the first to provide a model for relating the point efficiency to the tray efficiency for the special case in which the liquid flows across the tray in plug flow (e.g., with no horizontal mixing). In the so-called Lewis Case *I*, the vapor entering the tray is assumed to be well mixed and we obtain (see, e.g., Lockett, 1986):

$$E^{MV}/E_{OV} = (\exp(E_{OV}\Lambda) - 1)/E_{OV}\Lambda \qquad (13.2.3)$$

where

$$\Lambda = MV/L \qquad (13.2.4)$$

is the stripping factor.

13.3 EFFICIENCIES OF MULTICOMPONENT SYSTEMS

13.3.1 Point Efficiency of Multicomponent Systems

For multicomponent systems the composition above the froth is given by Eq. 12.2.3 as

$$(\Delta y_L) = [Q](\Delta y_E) \qquad (12.2.3)$$

where $(\Delta y_L) = (y^* - y_L), (\Delta y_E) = (y^* - y_E)$ and where $[Q]$ is defined by Eq. 12.2.4

$$[Q] \equiv \exp\{-[\mathsf{N}_{OV}]\} \tag{12.2.4}$$

The Murphree point efficiencies conventionally defined by Eq. 13.1.3 may be expressed in terms of the elements of the matrices $[Q]$. For a ternary mixture there are two independent compositions and two independent efficiencies. Using Eq. 12.2.3 we can show that the component efficiencies E_{OV1} and E_{OV2} are given by

$$E_{OV1} = \frac{y_{1L} - y_{1E}}{y_1^* - y_{1E}} = 1 - \frac{\Delta y_{1L}}{\Delta y_{1E}} = 1 - Q_{11} - Q_{12}/\alpha \tag{13.3.1}$$

$$E_{OV2} = \frac{y_{2L} - y_{2E}}{y_2^* - y_{2E}} = 1 - \frac{\Delta y_{2L}}{\Delta y_{2E}} = 1 - Q_{22} - Q_{21}\alpha \tag{13.3.2}$$

where α is defined by

$$\alpha = \Delta y_{1E}/\Delta y_{2E} \tag{13.3.3}$$

The efficiency of component 3 can be expressed in terms of the efficiency of the other two components as

$$E_{OV3} = \frac{y_{3L} - y_{3E}}{y_3^* - y_{3E}} = 1 - \frac{\Delta y_{1L} + \Delta y_{2L}}{\Delta y_{1E} + \Delta y_{2E}} = \frac{\alpha E_{OV1} + E_{OV2}}{\alpha + 1} \tag{13.3.4}$$

Only when $[Q]$ is a diagonal matrix with all elements on the main diagonal equal to one another (i.e., $[Q]$ reduces to the form $Q[I]$) will the three component efficiencies E_{OV1}, E_{OV2}, and E_{OV3} have the same value. This will be the case in mixtures made up of components of a similar nature (e.g., close boiling hydrocarbons or mixtures of isomers). For mixtures made up of chemically dissimilar species, that is, mixtures with large differences between the binary pair diffusivities, we must except $[\mathsf{N}_{OV}]$ to have significant nondiagonal elements (through $[K_{OV}]$). When this is the case, $[Q]$ will have significant cross-coefficients and the point efficiencies will not be equal to each other.

Examination of Eqs. 13.3.1–13.3.4 shows that the ratio of driving forces $\alpha = \Delta y_{1E}/\Delta y_{2E}$ plays a key role in determining the relative magnitudes of the $E_{OV,i}$. Now, in a multicomponent system, α may take any value in the range $-\infty$ to $+\infty$ (contrast this with a binary system for which $\Delta y_1/\Delta y_2 = -1$). Thus, the component $E_{OV,i}$ are unbounded and could exhibit values ranging anywhere from $-\infty$ to $+\infty$!

Example 13.3.1   Point Efficiency in the Distillation of the Methanol–1-Propanol–Water System

Estimate the point efficiency for the ternary system methanol(1)–1-propanol(2)–water(3) studied in Example 12.2.1.

DATA   The average composition of the vapor just below the tray is (at total reflux) equal to the composition of the liquid leaving the tray. Thus, $y_E = x_L$ with

$$y_{1E} = 0.1533 \qquad y_{2E} = 0.5231 \qquad y_{3E} = 0.3236$$

The matrix of overall numbers of transfer units was obtained in Example 12.2.1 as

$$[\mathsf{N}_{OV}] = \begin{bmatrix} 1.2850 & 0.1803 \\ 0.0954 & 1.5567 \end{bmatrix}$$

The equilibrium vapor composition y^* has been estimated from a bubble point calculation on x_L to be

$$y_1^* = 0.35434 \qquad y_2^* = 0.33373 \qquad y_3^* = 0.31193$$

SOLUTION   The matrix $[Q]$ may be computed from Eqs. 12.2.8 with the following results:

$$[Q] = \begin{bmatrix} 0.2788 & -0.0438 \\ -0.0232 & 0.2128 \end{bmatrix}$$

The ratio of driving forces α is found to be

$$\begin{aligned} \alpha &= \Delta y_{1E}/\Delta y_{2E} \\ &= (y_1^* - y_{1E})/(y_2^* - y_{2E}) \\ &= (0.35434 - 0.1533)/(0.33373 - 0.5231) \\ &= -1.0616 \end{aligned}$$

The point efficiencies of components 1 and 2 may be computed directly from Eqs. 13.3.1 and 13.3.2)

$$\begin{aligned} E_{OV1} &= 1 - Q_{11} - Q_{12}/\alpha \\ &= 1 - 0.2788 - (-0.0438)/(-1.0616) \\ &= 0.6799 \\ E_{OV2} &= 1 - Q_{22} - Q_{21}\alpha \\ &= 1 - 0.2129 - (-0.0232) \times (-1.0616) \\ &= 0.7626 \end{aligned}$$

The point efficiency of component 3 follows from Eq. 13.3.4

$$\begin{aligned} E_{OV3} &= \frac{\alpha E_{OV1} + E_{OV2}}{\alpha + 1} \\ &= \frac{(-1.0616) \times 0.6799 + 0.7626}{(-1.0616) + 1} \\ &= -0.6614 \end{aligned}$$

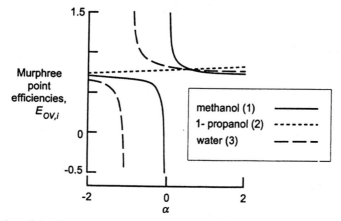

Figure 13.1. Point efficiencies for the system methanol–1-propanol–water as a function of the ratio of driving forces.

It can be seen from Eq. 13.3.4 that since $\alpha \approx -1$, E_{OV3} is particularly sensitive (in magnitude and sign) to α. In this case E_{OV3} is negative in sign. It should also be noted that the point efficiencies may be rather sensitive to the composition of the vapor and liquid phases (even though $[Q]$ is not so sensitive to composition variations). The component efficiencies are shown as function of the ratio of driving forces α in Figure 13.1. The efficiencies of methanol and water are particularly sensitive to the value of α. ∎

Example 13.3.2   Point Efficiencies of Ethanol–tert-Butyl Alcohol–Water System*

The distillation of a mixture of ethanol(1)–t-butyl alcohol(2)–water(3) in a sieve tray column was considered in Example 12.2.2. Here, we continue that problem by determining the point efficiencies for this system under the conditions specified in the prior example. For the record the composition of the vapor entering the tray is

$$y_{1E} = 0.5558 \qquad y_{2E} = 0.1353 \qquad y_{3E} = 0.3089$$

The distillation is at total reflux; thus, the composition of the liquid leaving the tray is the same as that of the entering vapor.

$$x_1 = 0.5558 \qquad x_2 = 0.1353 \qquad x_3 = 0.3089$$

The liquid is considered to be well mixed. The equilibrium vapor composition is

$$y_1^* = 0.6040 \qquad y_2^* = 0.1335 \qquad y_3^* = 0.2625$$

SOLUTION   First, we recall some of the important results from Example 12.2.2. The departure from equilibrium in the jetting zone is

$$[Q_I] = \exp(-[\mathbb{N}_{OV,I}])$$

$$= \begin{bmatrix} 0.7834 & -0.0488 \\ -0.0089 & 0.8280 \end{bmatrix}$$

The departure from equilibrium in the small population is

$$[Q_{II,1}] = \exp(-[\mathbb{N}_{OV,II,1}])$$

$$= \begin{bmatrix} 0.0048 & -0.0209 \\ -0.0040 & 0.0238 \end{bmatrix}$$

The departure from equilibrium in the large population follows as:

$$[Q_{II,2}] = \exp(-[\mathbb{N}_{OV,II,2}])$$

$$= \begin{bmatrix} 0.5953 & -0.0892 \\ -0.0164 & 0.6775 \end{bmatrix}$$

The overall departure from equilibrium for the bubble rise zone is given by (cf. Eq. 12.2.27)

$$[Q_{II}] = f_{II,1}[Q_{II,1}] + f_{II,2}[Q_{II,2}]$$

$$= \begin{bmatrix} 0.5363 & -0.0824 \\ -0.0152 & 0.6121 \end{bmatrix}$$

---

*tert-Butyl alcohol is the common name for 2-methyl-2-propanol.

The combined inefficiency matrix is

$$[Q] = [Q_{II}][Q_I]$$
$$= \begin{bmatrix} 0.4212 & -0.0944 \\ -0.0174 & 0.5079 \end{bmatrix}$$

The composition of the vapor above the froth may now be calculated from

$$(y^* - y_L) = [Q](y^* - y_E)$$

Thus,

$$y_{1L} = 0.58352 \qquad y_{2L} = 0.13525 \qquad y_{3L} = 0.28123$$

The ratio of driving forces α is found to be

$$\alpha = \Delta y_{1E}/\Delta y_{2E}$$
$$= (y_1^* - y_{1E})/(y_2^* - y_{1E})$$
$$= (0.60399 - 0.5558)/(0.1335 - 0.1353)$$
$$= -26.759$$

Finally, the point efficiencies for each component may be calculated as

$$E_{OV1} = 1 - Q_{11} - Q_{12}/\alpha$$
$$= 1 - 0.4212 - (-0.0944)/(-26.759)$$
$$= 0.5753$$
$$E_{OV2} = 1 - Q_{22} - Q_{21}\alpha$$
$$= 1 - 0.5079 - (-0.0174) \times (-26.759)$$
$$= -0.0277$$

The point efficiency of component 3 follows from Eq. 12.3.4

$$E_{OV3} = \frac{\alpha E_{OV1} + E_{OV2}}{\alpha + 1}$$
$$= \frac{(-26.759) \times 0.5753 + 0.0277}{(-26.759) + 1}$$
$$= 0.5966$$

The interesting consequences for the point efficiencies that result when the model parameters are varied are left as exercises for our readers. ∎

13.3.3 Tray Efficiency for Multicomponent Systems

In this section we show how the matrix generalization of the binary tray efficiency equation (Eq. 13.2.3) may be obtained.

Consider the distillation tray illustrated in Figure 13.2. The parameter w is the fractional coordinate direction across the tray starting from the outlet weir. Note that the liquid is flowing in the negative w direction. The composition of the vapor at any point above the froth is given by Eq. 12.2.3, which we rearrange to give

$$[E](y^* - y_E) = (y_L - y_E) \tag{13.3.5}$$

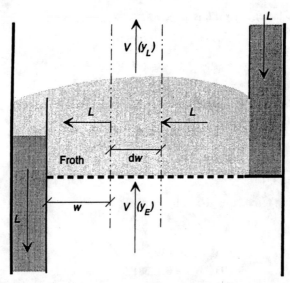

Figure 13.2. Schematic diagram of the froth on a distillation tray showing the control volume used for the material balances. The parameter w is the coordinate direction starting from the weir.

where

$$[E] = [I] - [Q] \tag{13.3.6}$$

is the matrix generalization of Eq. 13.2.1.

We differentiate Eq. 13.3.5 with respect to w (assuming the matrix $[E]$ is constant) to get

$$[E]\frac{d(y^*)}{dw} = \frac{d(y_L)}{dw} + [[E] - [I]]\frac{d(y_E)}{dw} \tag{13.3.7}$$

In order to proceed further we must say something about how the composition of the vapor below the tray varies with coordinate w. In the so-called Lewis Case I, the vapor entering the tray is assumed to be well mixed: $d(y_E)/dw = (0)$ in which case Eq. 13.3.7 simplifies to

$$[E]\frac{d(y^*)}{dw} = \frac{d(y_L)}{dw} \tag{13.3.8}$$

To eliminate the equilibrium vapor composition (y^*) we use the linear equilibrium relationship

$$(y^*) = [M](x) + (b) \tag{7.3.8a}$$

which we differentiate with respect to w (now assuming $[M]$ and (b) to be constant)

$$\frac{d(y^*)}{dw} = [M]\frac{d(x)}{dw} \tag{13.3.9}$$

We have replaced the problem of not knowing the gradient of (y^*) with the more easily solved problem of not knowing how the liquid-phase composition changes as it flows across the tray. In fact, we do know (or have assumed we know) exactly how the liquid flows across the tray (in plug flow). This assumption allows us to draw up the material balances around the differential section of tray in Figure 13.2.

$$V(y)_E \, dw + L(x)|_{w+dw} = V(y)_L \, dw + L(x)|_w \qquad (13.3.10)$$

Dividing by dw and taking the limit as dw goes to zero gives

$$L \frac{d(x)}{dw} = V(y_L - y_E) \qquad (13.3.11)$$

which may be combined with Eq. 13.3.9 to give

$$\frac{d(y^*)}{dw} = [\Lambda](y_L - y_E) \qquad (13.3.12)$$

$[\Lambda]$ is a matrix of stripping factors defined by

$$[\Lambda] = (V/L)[M] \qquad (13.3.13)$$

Substitution of Eq. 13.3.12 into Eq. 13.3.8 gives

$$\frac{d(y_L)}{dw} = [E][\Lambda](y_L - y_E) \qquad (13.3.14)$$

Equation 13.3.14 is a first-order matrix differential equation with constant coefficients (we have already assumed that $[E]$ and $[M]$ are constant matrices). The solution is

$$(y_L - y_E) = \exp[[E][\Lambda]w](y_{L0} - y_E) \qquad (13.3.15)$$

where y_{L0} is the composition of the vapor above the liquid at the tray exit, $w = 0$.
    The average composition of the vapor above the liquid is defined by

$$(\bar{y}_L) = \int_0^1 (y_L) \, dw \qquad (13.3.16)$$

Insertion of Eq. 13.3.15 into Eq. 13.3.16 and carrying out the resulting integration gives

$$(\bar{y}_L - y_E) = [\exp[[E][\Lambda]] - [I]][\Lambda]^{-1}[E]^{-1}(y_{L0} - y_E) \qquad (13.3.17)$$

Equation 13.3.5 is used to express (y_{L0}) in terms of (y_E) and (y_0^*)

$$(y_{L0} - y_E) = [E](y_0^* - y_E) \qquad (13.3.18)$$

where (y_0^*) is the composition of the vapor in equilibrium with the liquid leaving the tray. Combination of this result with Eq. 13.3.18 gives

$$(\bar{y}_L - y_E) = [\exp([E][\Lambda]) - [I]][\Lambda]^{-1}(y_0^* - y_E) \qquad (13.3.19)$$

which we rewrite as

$$(\bar{y}_L - y_E) = [E^{MV}](y_0^* - y_E) \qquad (13.3.20)$$

where $[E^{MV}]$ is a square matrix of dimension $n - 1 \times n - 1$ of multicomponent Murphree tray efficiencies defined by

$$[E^{MV}] = [\exp[[E][\Lambda]] - [I]][\Lambda]^{-1} \qquad (13.3.21)$$

Equation 13.3.21 may also be written as

$$[E^{MV}][E]^{-1} = [\exp[[E][\Lambda]] - [I]][[E][\Lambda]]^{-1} \qquad (13.3.22)$$

which is the generalization of the well-known result for binary systems Eq. 13.2.3.

The Murphree tray efficiency for component i as defined by Eq. 13.1.2 can be expressed in terms of the elements of the matrix $[E^{MV}]$ if desired.

This extension to multicomponent systems of Lewis's tray efficiency model is due to Toor (1964b). Many more complicated models of liquid flow across a tray have been developed, and used for the determination of the tray efficiency for binary distillation. We recommend the book by Lockett (1986) for those interested in reading further. Few of these more advanced models have been extended to multicomponent systems.

Example 13.3.3   *Tray Efficiency in the Distillation of the Methanol–1-Propanol–Water System*

Estimate the tray efficiency in the distillation of the ternary system methanol(1)–1-propanol(2)–water(3) considered in Examples 12.2.1 and 13.3.1.

DATA   The composition of the vapor below the tray is

$$y_{1E} = 0.1533 \qquad y_{2E} = 0.5231 \qquad y_{3E} = 0.3236$$

The equilibrium vapor composition y^* is

$$y_1^* = 0.35434 \qquad y_2^* = 0.33373 \qquad y_3^* = 0.31193$$

and the equilibrium ratios are

$$K_1 = 2.3114 \qquad K_2 = 0.63799 \qquad K_3 = 0.96394$$

The matrix of thermodynamic factors computed from the Wilson model with the parameters in Example 12.2.1 and at the liquid bubble point temperature of 80.57°C is

$$[\Gamma] = \begin{bmatrix} 0.7875 & -0.0875 \\ 0.0748 & 1.0433 \end{bmatrix}$$

SOLUTION   We will use the Lewis Case *I* to compute the tray efficiency. Specifically, Eq. 13.3.20 is used to estimate the average composition of the vapor above the froth and Eq. 13.3.21 is used to obtain $[E^{MV}]$.

The matrix $[Q]$ was obtained in Example 12.3.1 as

$$[Q] = \begin{bmatrix} 0.2788 & -0.0438 \\ -0.0232 & 0.2128 \end{bmatrix}$$

The matrix $[E] = [I] - [Q]$ is found next as

$$[E] = \begin{bmatrix} 0.7212 & 0.0438 \\ 0.0232 & 0.7872 \end{bmatrix}$$

The matrix $[M]$ is given by Eq. 7.3.7

$$[M] = [K][\Gamma]$$

where $[K]$ is a diagonal matrix of the first $n - 1$ K values. The computed results are

$$[M] = \begin{bmatrix} 1.8202 & -0.2022 \\ 0.0477 & 0.6656 \end{bmatrix}$$

At total reflux the ratio of vapor and liquid flows is unity, thus $[\Lambda] = [M]$.

To compute $[E^{MV}]$ we postmultiply $[E]$ by $[\Lambda]$, then evaluate its exponential (again, we used the power series method), subtract an identity matrix and, finally, postmultiply the result of the foregoing by the inverse of the stripping factor matrix. The result is

$$[E^{MV}] = \begin{bmatrix} 1.489 & 0.0029 \\ 0.0853 & 1.0331 \end{bmatrix}$$

The average composition of the vapor above the froth may now be computed directly from Eq. 13.3.20

$$(\bar{y}_L - y_E) = [E^{MV}](y_0^* - y_E)$$

as

$$\bar{y}_{1L} = y_{1E} = E_{11}^{MV}(y_1^* - y_{1E}) + E_{12}^{MV}(y_2^* - y_{2E})$$
$$= 0.1533 + 1.489 \times (0.35434 - 0.1533) + 0.0029 \times (0.33373 - 0.5231)$$
$$= 0.4521$$
$$\bar{y}_{2L} = y_{2E} + E_{21}^{MV}(y_1^* - y_{1E}) + E_{22}^{MV}(y_2^* - y_{2E})$$
$$= 0.5231 + 0.0853 \times (0.35434 - 0.1533) + 1.0331 \times (0.33373 - 0.5231)$$
$$= 0.3446$$
$$\bar{y}_{3L} = 1 - y_{1L} - y_{2L}$$
$$= 0.2033$$

The tray efficiencies follow from the definition of the Murphree tray efficiency (Eq. 13.1.2)

$$E_i^{MV} = \frac{\bar{y}_{iL} - y_{iE}}{y_i^* - y_{iE}}$$

with numerical results

$$E_1^{MV} = 1.4863$$
$$E_2^{MV} = 0.9426$$
$$E_3^{MV} = 10.31$$

As in Example 13.3.1 we find that E_3^{MV} is particularly sensitive to the magnitude of the ratio of driving forces α defined in Eq. 13.3.3. Note that though we calculate $E_3^{MV} = 1031\%$, the estimation of the exiting vapor composition y_{3L}, is insensitive to the sign and magnitude of the E_3^{MV}.

It is worth pointing out that the composition of the vapor above the froth can be calculated without computing the component efficiencies. As noted later, the use of component efficiencies may well be the cause of convergence difficulties that have been encountered when using these multicomponent tray efficiency models in column simulations. ∎

13.4 COLUMN SIMULATION

The next task is to set up the equations that model a complete distillation column. As noted in the introduction to this chapter, simulation of multicomponent distillation operations usually is carried out using the equilibrium stage model introduced below.

13.4.1 The Equilibrium Stage Model

A schematic diagram of a single equilibrium stage is shown in Figure 13.3, which also serves to introduce the notation we shall use here. A complete distillation column is taken to be a sequence of these stages (Fig. 13.4). The equations that model equilibrium stages have been termed the MESH equations, MESH being an acronym referring to the different types of equations that form the mathematical model. The M equations are the Material balance

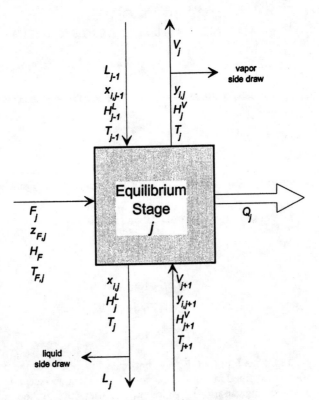

Figure 13.3. Equilibrium stage model.

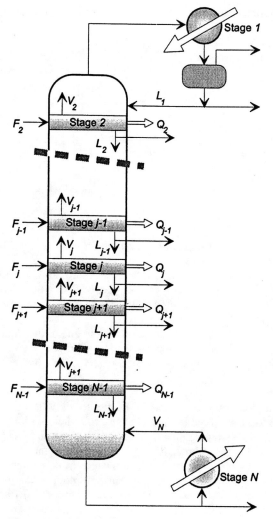

Figure 13.4. Schematic diagram of a distillation column.

equations, of which there are two types: the total material balance for stage j

$$M_j^T \equiv V_j + L_j - V_{j+1} - L_{j-1} - F_j = 0 \qquad (13.4.1)$$

and the component material balances

$$M_{ij} \equiv V_j y_{ij} + L_j x_{ij} - V_{j+1} y_{i,j+1}$$
$$-L_{j-1} x_{i,j-1} - F_j z_{ij} = 0 \qquad (13.4.2)$$

The E equations are the Equilibrium relations, here modified to include the Murphree efficiencies defined by Eq. 13.1.5

$$E_{ij} \equiv E_{ij}^{MV} K_{ij} x_{ij} - y_{ij} - \left(1 - E_{ij}^{MV}\right) y_{i,j+1} = 0 \qquad (13.4.3)$$

The S or Summation equations

$$S_j^V \equiv \sum_{i=1}^{n} y_{ij} - 1 = 0 \tag{13.4.4}$$

$$S_j^L \equiv \sum_{i=1}^{n} x_{ij} - 1 = 0 \tag{13.4.5}$$

are required to force the mole fractions to sum to unity.

The H equations are the ent$H$alpy balance equations

$$H_j \equiv V_j H_j^V + L_j H_j^L - V_{j+1} H_{j+1}^V$$

$$-L_{j-1} H_{j-1}^L - F_j H_j^F + Q_j = 0 \tag{13.4.6}$$

where H_j^V and H_j^L are the enthalpies of the vapor and liquid streams leaving the jth stage; H_j^F is the enthalpy of the feed stream.

If we count the equations listed, we will find that there are $2n + 4$ equations per stage. However, only $2n + 3$ of these equations are independent. These independent equations are generally taken to be the n component mass balance equations, the n equilibrium relations, the enthalpy balance, and two more equations. These two equations can be the two summation equations or the total mass balance and one of the summation equations (or an equivalent form). The $2n + 3$ unknown variables determined by the equations are the n vapor mole fractions y_{ij}; the n liquid mole fractions, x_{ij}; the stage temperature T_j and the vapor and liquid flow rates V_j and L_j. Thus, for a column of s stages, we must solve $s(2n + 3)$ equations.

The MESH equations can be applied as written to all of the trays in the column. In addition to these stages, the reboiler and condenser (if they are included) for the column must be considered. These stages differ from the other stages in the column in that they are a heat source or sink for the column. The MESH equations may be used more or less as written above to model these stages exactly as you wold any other stage in the column. For a total condenser at the top of a distillation column, for example, the distillate is U_1 and the reflux ratio is $1/r_1^L$. For a partial condenser, the vapor product is V_1 and the "reflux ratio" is L_1/V_1. For a partial reboiler at the base of a column, the bottoms flow rate is L_s. The efficiency of condensers and reboilers would normally be unity. In other words, the liquid condensate is usually in equilibrium with the vapor entering the condenser. Similarly, the vapor leaving the reboiler is in equilibrium with the liquid entering it. Equations 13.4.3, therefore, reduce to

$$E_{ij} \equiv K_{ij} x_{ij} - y_{ij} = 0 \tag{13.4.7}$$

For these special stages it is common to use some specification equation instead of the enthalpy balance. Common specifications include

1. The flow rate of the distillate–bottoms product stream.
2. The mole fraction of a given component in either the distillate or bottoms product stream.
3. A component flow rate in either the distillate or bottoms product stream.
4. A reflux/reboil ratio or rate.
5. A heat duty to the condenser or reboiler.

In the case of a total condenser, the vapor-phase compositions used in the calculation of the equilibrium relations and the summation equations are those that would be in equilibrium with the liquid stream that actually exists. That is, for a total condenser, the vapor composition used in the equilibrium relations is the vapor composition determined during a bubble point calculation based on the actual pressure and liquid compositions found in the condenser. These compositions are not used in the component mass balances since there is no vapor stream from a total condenser.

13.4.2 Solving the Model Equations

The class of simultaneous solution methods in which all of the model equations are solved simultaneously using Newton's method (or a modification thereof) is one class of methods for solving the MESH equations that allow the user to incorporate efficiencies that differ from unity. Simultaneous solution methods have long been used for solving equilibrium stage simulation problems (see, e.g., Whitehouse, 1964; Stainthorp and Whitehouse, 1967; Naphtali, 1965; Goldstein and Stanfield, 1970; Naphtali and Sandholm, 1971). Simultaneous solution methods are discussed at length in the textbook by Henley and Seader (1981) and by Seader (1986).

Newton's method for solving systems of equations is summarized in Algorithm C.2 and has been used elsewhere in this book for solving mass and energy transport problems. For solving the modified MESH equations we may identify the vector of variables (x) as

$$(x)^T \equiv \left((x_1)^T \quad (x_2)^T \cdots (x_N)^T \right)$$

where N is the number of stages. The vector (x_j) is a vector of unknown variables for stage j.

$$(x_j)^T \equiv \left(V_j, y_{1j}, y_{2j} \cdots y_{nj}, T_j, x_{1j}, x_{2j} \cdots x_{nj}, L_j \right)$$

The vector of discrepancy functions for the column as a whole is given by

$$(F)^T \equiv \left((F_1)^T \quad (F_2)^T \cdots (F_N)^T \right)$$

The vector of functions for the jth stage (F_j) corresponding to (x_j) is

$$(F_j)^T \equiv \left(M_j^T, M_{1j}, M_{2j} \cdots M_{nj}, H_j, E_{1j}, E_{2j} \cdots E_{nj}, S_j^{L-V} \right)$$

where S_j^{L-V} is

$$S_j^{L-V} \equiv \sum_{i=1}^{n} (x_{ij} - y_{ij}) = 0 \qquad (13.4.8)$$

To the best of our knowledge, the only attempt to combine the full matrix efficiency prediction models with a general purpose column simulation code has been made by Aittamaa (1981). Aittamaa integrated a program for computing efficiencies from the methods described above with an equilibrium stage simulation program that solved the process model equations simultaneously using Newton's method. Aittamaa's procedure was first to solve the column model equations with an efficiency of 75% for all components on all stages. On convergence, the tray efficiencies were computed from models similar to those described in Sections 13.2 and 13.3. The simulation was then repeated with these new values for the component efficiencies. This cycle of efficiency estimation—column simulation is continued until complete convergence is obtained. The procedure is summarized in Algorithm 13.1. Some results of Aittamaa's calculations are presented in Section 13.5.

Aittamaa (1981) reports that the complete calculation required about three times the computer time required by a simulation in which the column efficiencies were kept constant even though the efficiency estimation method accounted for only about 15% of the total computation time. The reason appeared to be that many more Newton iterations were required each time the column model equations were solved for the case where tray efficiencies vary from component to component. Convergence also seemed to depend on the model used to predict the tray efficiency from the point efficiency. In one case involving a six component mixture, convergence was easily obtained when the liquid was assumed well mixed but the program failed to converge when the plug flow model was used. We suspect that the convergence difficulties experienced by Aittamaa may be due to the way in which the efficiency models are used. It may be that using Eq. 13.3.20 directly, rather than the more conventional Eqs. 13.1.2 would be a better approach; an observation also made by Toor (1964).

Algorithm 13.1   Procedure for Solving Equilibrium Model Equations with Multicomponent Tray Efficiency Model Based on Method of Aittamaa (1981)

Step 1:	Specify: Feed conditions.
	Column configuration and design (as needed).
	Column pressures and heat duties.
Step 2:	Generate Initial Estimates of
	Flow rates
	Temperatures
	Mole fractions of vapor and liquid phases
	Set efficiency of all components to 0.75.
Step 3:	Compute, for each Stage j.
	Thermodynamic properties (K values, enthalpies).
	Vector of stage functions (F_j).
	Jacobian matrix of partial derivatives.
Step 4:	Check for convergence, if not obtained continue with Step 5.
Step 5:	Solve linear system (Eq. C.2.5) for new (X) vector.
	Return to Step 3.
Step 6:	If this is the first time through continue with Step 7.
	Otherwise compare composition and temperature profiles from last converged column calculation.
	If not converged continue with Step 7.
Step 7:	Compute stage efficiencies from multicomponent model.
	Return to Step 3.

13.5   SIMULATION AND EXPERIMENTAL RESULTS

There is ample experimental evidence to show that the efficiencies of different components in a multicomponent system are not all equal. The first clear statement of this fact can be found in a paper by Walter and Sherwood (1941) who, on the basis of an extensive experimental study of Murphree vapor and liquid efficiencies for absorption, desorption, and rectification operations, concluded "The results indicate that different efficiencies should be used for each component in the design of absorbers for natural gasoline and refinery gases." Since the publication of their paper many others have provided additional data to confirm this view [see Krishna and Standart (1979) for a list of references]. We review some of these data below.

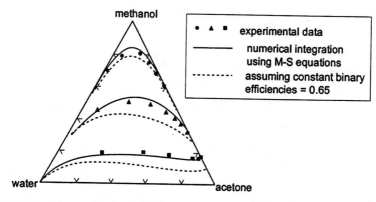

Figure 13.5. Composition profiles in distillation of acetone–methanol–water system. Calculations and data from Vogelpohl (1979).

13.5.1 Two Nonideal Systems at Total Reflux

Vogelpohl (1979) presented some results for the distillation of two ternary systems: acetone, methanol, water and methanol, 2-propanol, water. The experiments were carried out in a 38 bubble cap tray column of 0.3-m diameter with 0.2 m between the trays. Due to the ease of separating these particular mixtures only up to 13 trays were active for the experiments for which composition profiles and flow rates are reported. A portion of Vogelpohl's experimental data for the system acetone–methanol–water is shown in Figure 13.5, which also shows the composition profiles computed using a theoretical model based on the Maxwell–Stefan equations and those obtained using a model based on equal efficiencies for all components. Figure 13.5 shows that the assumption of equal component efficiencies gives rise to large differences between the predicted and measured composition profiles.

Component Murphree efficiencies for two of Vogelpohl's experiments are shown in Figures 13.6 and 13.7, where it can be seen that the efficiencies vary from tray to tray and are not the same for each component. The behavior anticipated in Figure 13.1 (in which the component efficiencies are shown to go through a singular point depending on the driving forces) is in evidence here.

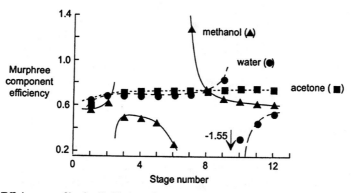

Figure 13.6. Efficiency profiles in distillation of acetone–methanol–water system. Data from Vogelpohl (1979).

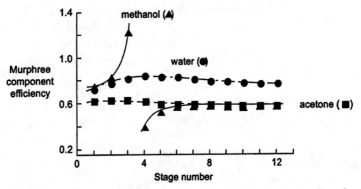

Figure 13.7. Efficiency profiles in distillation of acetone–methanol–water system. Data from Vogelpohl (1979).

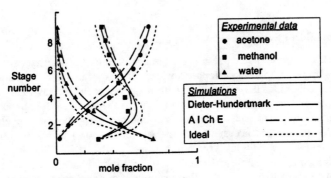

Figure 13.8. Composition profiles in distillation of acetone–methanol–water system. Calculations by Aittamaa (1981), data from Vogelpohl (1979).

Vogelpohl's data have been used by a number of other investigators to test three quite different column simulation models (Aittamaa, 1981; Burghardt and Warmuzinski, 1983, 1984; Burghardt et al., 1983; Krishnamurthy and Taylor, 1985b); we present Aittamaa's results below.

Aittamaa used a number of different methods to calculate the number of transfer units. Profiles calculated using the AIChE method of Section 12.1.5 and a method due to Dieter and Hundertmark (1963) are shown in Figure 13.8 along with the profiles obtained with an equilibrium stage (ideal) model. The Dieter and Hundertmark correlation was found to give the best agreement with the measured profiles for the acetone–methanol–water system. Composition profiles obtained using the Dieter and Hundertmark (1963) correlation for the system methanol–2-propanol–water are shown in Figure 13.9. It can be seen that there is good agreement between predicted and measure profiles.

13.5.2 Industrial Scale Columns

Few investigators have used large industrial scale equipment to test multicomponent efficiency models; an exception is the work reported by Ognisty and Sakata (1987). Tests with a mixture of propane, isobutane,* and n-butane, were carried out in a column of 2.4-m

---

*Isobutane is the common name for 2-methylpropane.

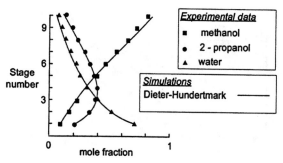

Figure 13.9. Composition profiles in distillation of methanol–2-propanol–water system. Calculations by Aittamaa (1981), data from Vogelpohl (1979).

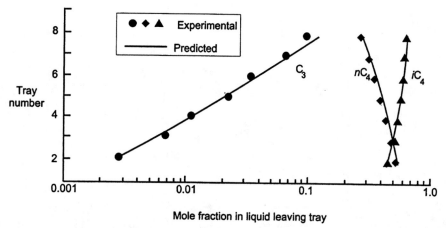

Figure 13.10. Composition and efficiency profiles in distillation of propane–isobutane–n-butane. Calculations and data from Ognisty and Sakata (1987).

diameter with 8 active trays and in a column with two-pass bubble cap trays of 3-m diameter with 36 trays. Composition profiles for one of these tests is shown in Figure 13.10. Interaction effects were minimal in this system, as is only to be expected for this thermodynamically ideal system made up of components with not too dissimilar molar masses. For this reason Ognisty and Sakata also investigated the acetone–methanol–water system. Composition profiles for this system are shown in Figure 13.11. The full matrix method of predicting the efficiencies was not found by these authors to be significantly better than the simpler effective diffusivity type of method.

13.5.3 Simulations of Aittamaa

Aittamaa (1981) simulated a number of experiments with the systems ethanol–benzene–n-heptane, and chloroform–benzene–n-heptane (data were obtained at Hoffmann-La Roche in a fair size pilot scale column) and 1-butanol–ethanol–water in a 12 sieve-tray column. The Hoffman-La Roche data were taken in a column having 24 sieve trays and 30-cm inside diameter. Unlike many studies of distillation efficiency, these experiments were not carried out at total reflux. The measured flow rates and compositions of the feed, distillate, and

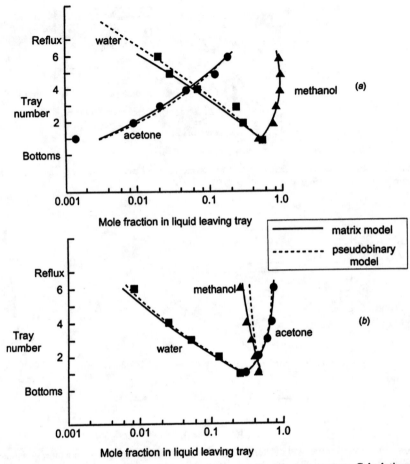

Figure 13.11. Composition profiles in distillation of acetone–methanol–water system. Calculations and data from Ognisty and Sakata (1987).

TABLE 13.1 Specifications and Results for Distillation of Ethanol–Benzene–n-Heptane

Column has 26 Stages (includes reboiler and condenser)
Tray type: Sieve
Average column width = 0.3 m
Total condenser, equilibrium reboiler
Reflux ratio = 4.6
Bottoms flow rate = 55.2 kg/h
Feed to stage 15 from top

Feed Composition (Weight Percent)		
	Measured	Adjusted
Ethanol	50.86	50.88
Benzene	23.12	23.10
n-Heptane	25.96	26.01

Product Compositions (Weight Percent)		
Component	Overhead	Bottoms
Ethanol	36.75	80.70
Benzene	33.79	0.56
n-Heptane	24.96	18.74

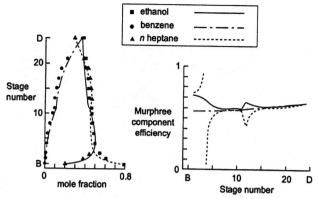

Figure 13.12. Composition and efficiency profiles in distillation of ethanol–benzene–n-heptane system. Calculations by Aittamaa (1981).

bottoms are given in Table 13.1 for one of these experiments. Composition and efficiency profiles for the ethanol–benzene–n-heptane system are shown in Figure 13.12, for the chloroform–benzene–n-heptane system in Figure 13.13, and for the system ethanol–1-butanol–water in Figure 13.14. The correlation of Hughmark was used to calculate the numbers of transfer units. The liquid phase was assumed completely mixed.

For the ethanol–benzene–n-heptane system, the calculated product flows agree well with the measured values. There are some deviations in the concentrations on some of the trays but most of the deviations are less than the accuracy of the measurements. The deviations between simulation and experiment are a little larger for benzene on some of the intermediate trays for the ethanol–benzene–n-heptane system but the calculated product flows agree very well with the measured values.

It can be seen that there are wild fluctuations between values of the efficiency for different components on each tray and from tray to tray for each component. The variation is

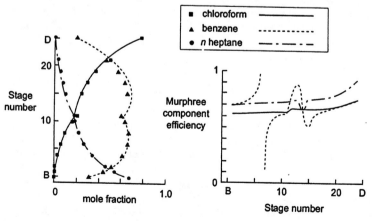

Figure 13.13. Composition and efficiency profiles in distillation of chloroform–benzene–n-heptane system. Calculations by Aittamaa (1981).

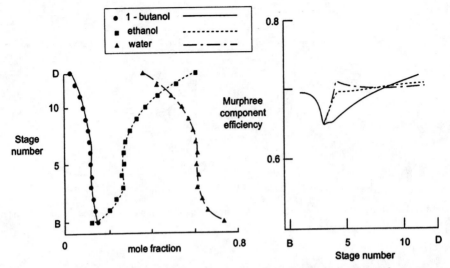

Figure 13.14. Composition and efficiency profiles in distillation of ethanol–1-butanol–water system. Calculations by Aittamaa (1981).

particularly strong around the feed tray or when the mole fraction of a component passes through a maximum somewhere in the column.

13.5.4 Other Studies

Other studies carried out with a view to testing these methods of predicting multicomponent efficiency have been reported by Diener and Gerster (1968), Krishna et al. (1977), Medina et al. (1979), Aittamaa (1981), Chan and Fair (1984b), Bunke et al. (1987), and Biddulph and Kalbassi (1988). Diener and Gerster investigated the system acetone–methanol–water in a small split flow sieve tray. Krishna et al. used the system ethanol–*t*-butyl alcohol–water in a small sieve tray column and found widely varying efficiencies that they attributed to diffusional interactions. Aittamaa investigated the same system. Chan and Fair worked with the systems methyl acetate–chloroform–benzene and methanol–methyl acetate–chloroform at total reflux in an Oldershaw column [see the thesis by Chan (1983) for complete details]. Bunke et al. (1987) used the system methanol–acetone–methyl acetate at total reflux in a small laboratory scale column.

13.5.5 Design Calculations

As far back as 1960, Toor and Burchard performed a theoretical calculation to show that the neglect of interaction effects in design calculations could lead to severe underdesign. For systems under complete vapor-phase control, equations were developed based on a generalized driving force pseudobinary approach (Toor, 1957), for the effects of diffusional interactions among the components on their respective plate efficiency. A design calculation was made for the separation of methanol from 2-propanol and water. For the hypothetical case in which all binary efficiencies were assumed to be 40%, consideration of interactions gave a column requiring 117 plates compared to 84 plates for the case where diffusional interactions were neglected.

Chan and Fair (1984b) carried out a case study involving a six component de-ethanizer. The full matrix method presented above was compared to pseudobinary methods of estimating the component efficiencies. They found insignificant differences between the matrix and pseudobinary efficiency methods. However, considering the nature of their system, this should not come as a great surprise.

Using an entirely different approach to the modeling of multicomponent mass transfer in distillation (an approach that we describe in Chapter 14), Krishnamurthy and Taylor (1985c) found significant differences in design calculations involving nonideal systems. For an almost ideal system (a hydrocarbon mixture), pseudobinary methods were found to be essentially equivalent to a more rigorous model that accounted for diffusional interaction effects.

13.6   CONCLUDING REMARKS

The results of the many investigations to indeed confirm that individual component efficiencies are likely to be different. It is less clear that the matrix methods provide uniformly better predictions of component efficiencies than simpler effective diffusivity type approaches; as shown in, for example, the data of Chan and Fair (1984b) and Ognisty and Sakata (1987).

Aittamaa (1981) notes that column composition profiles are not very sensitive to wildly fluctuating component efficiencies simply because large negative or positive efficiencies of a component occur only when next to no transport of that component is taking place and, therefore, the mole fraction of that component does not change greatly from stage to stage.

It should be remembered that unequal component tray efficiencies are not due solely to diffusional interaction effects via off-diagonal elements of the matrices of multicomponent mass transfer coefficients. Unequal component efficiencies may also be due to differing volatilities of the various components. If the resistance in either phase is totally unimportant, unequal component point efficiencies could be said definitely to be due to diffusional interactions. Negative or large positive efficiencies arise only for those components whose driving forces Δy are close to zero. Whenever this happens, for example, when the mole fraction of a component of intermediate volatility passes through a maximum somewhere in the column, the efficiency is exceptionally sensitive to the measured composition (see, e.g., Biddulph and Kalbassi, 1988). Furthermore, the component efficiencies are also very sensitive to the computed equilibrium composition y^*. It is quite possible to compute efficiencies using one particular method for computing y^* (this involves a bubble point calculation) that show the gross fluctuations expected of strongly interacting systems and obtain completely different efficiencies with a different method of computing y^*. This can also happen when the same thermodynamic model is used with a different set of interaction parameters even when both models or parameter sets give equally good representations of the vapor–liquid equilibrium behavior of the system. Few of the published studies that have attempted to test these matrix efficiency models have noted this sensitivity of the component efficiencies to the vapor–liquid equilibrium data (see, however, Ognisty and Sakata, 1987).

In any event, we hope it is now well understood that mass transfer in multicomponent systems is described better by the full set of Maxwell–Stefan or generalized Fick's law equations than by a pseudobinary method. A pseudobinary method cannot be capable of superior predictions of efficiency. For a simpler method to provide consistently better predictions of efficiency than a more rigorous method could mean that an inappropriate model of point or tray efficiency is being employed. In addition, uncertainties in the estimation of the necessary transport and thermodynamic properties could easily mask more subtle diffusional interaction effects in the estimation of multicomponent tray efficiencies. It should also be borne in mind that a pseudobinary approach to the prediction of efficiency requires the a priori selection of the pair of components that are representative of the system. The best choice of key components may change in different parts of the column (Chan and Fair, 1984b) and it may not always be obvious which pair of components should be selected. The full matrix methods require no a priori designation of the key components.

In view of the large influence of interaction effects found by Toor and Burchard (1960) it is a little surprising that there have been so few design calculations reported in the literature. More experience with these models is required before definitive conclusions can be made regarding the use of complicated efficiency models in sophisticated distillation codes. The whole issue of multicomponent mass transfer models in distillation column simulation is taken up again in Chapter 14.

14 Multicomponent Distillation: A Nonequilibrium Stage Model

"...the concept of plate efficiency of individual components in a multicomponent mixture is of doubtful validity and is retained only on account of its simplicity."

—H. Sawistowksi (1980)

In Chapter 13 we presented the equilibrium stage–stage efficiency approach to the modeling of multicomponent separation processes. However, the matrix models of efficiency are rarely employed in industrial design practice. If an efficiency model is used at all, it is common to assume the efficiency is the same for all components on any given tray (and, often, constant through the column as well). The limitations of conventional equilibrium stage–stage efficiency calculations have been recognized for a very long time. As far back as 1925 E. V. Murphree wrote:

For three-component mixtures, the approach to equilibrium would not in general be equal for the two volatile components, and hence the theoretical plate cannot be used as a basis of calculation.

In this chapter we discuss some work that has been done towards the development of what we shall refer to as *nonequilibrium models* of multicomponent distillation (and absorption and extraction) operations. The building blocks of any *nonequilibrium model* will include

- Material balances.
- Energy balances.
- Equilibrium relations.
- Mass and energy transfer models.

The first three of the above items are also used in building equilibrium stage models; however, there is a crucial difference in the way in which the conservation and equilibrium equations are used in the two models. In equilibrium stage models the balance equations are written around the stage as a whole and the composition of the leaving streams related through an assumption that they are in equilibrium or by use of an efficiency equation. In a nonequilibrium model, separate balance equations are written for each phase. The conservation equations for each phase are linked by material balances around the interface; whatever material is lost by the vapor phase is gained by the liquid phase. The energy balance for the stage as a whole is treated in a similar way, split into two parts—one for each phase, each part containing a term for the rate of energy transfer across the phase interface.

In equilibrium stage calculations the equilibrium equations are used to relate the composition of the streams leaving the stage. K values are evaluated at the composition of the two exiting streams and the stage temperature (usually assumed to be the same in both phases). In a nonequilibrium model the equilibrium relations are used to relate the compositions on either side of the phase interface; the K values being evaluated at the interface compositions and temperature. The interface composition and temperature must, therefore, be determined (by calculation) as part of a nonequilibrium column simulation.

A number of nonequilibrium models fall into the general framework described above. The differences between models are due primarily to the models of flow and mass transfer on a tray (or within a section of packed column). Young and Stewart (1990), for example, use collocation techniques to solve a boundary layer model of cross-flow on a tray. An alternative approach that builds on the models of mass and energy transfer described in Chapters 11 and 12 has been developed in a series of papers by Taylor and co-workers (Krishnamurthy and Taylor, 1985a–c, 1986; Taylor et al., 1992). The latter model and some illustrations of its use are presented in this chapter.

14.1 A NONEQUILIBRIUM MODEL

A schematic illustration of a nonequilibrium stage is provided in Figure 14.1. Vapor from the stage below is brought into contact with liquid from the stage above and allowed to exchange mass and energy across their common interface represented in the diagram by the wavy line. Provision is made for vapor and liquid feed streams. Sidestream drawoffs of vapor

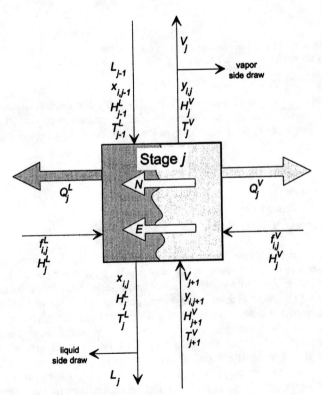

Figure 14.1. Schematic diagram of a nonequilibrium stage. This stage represents a tray in a trayed column or a section of packing in a packed column.

and/or liquid may also be accounted for if necessary. The entire column is taken to consist of a sequence of such stages. Stages are numbered starting at the top with the condenser, if required, being stage number 1.

The nonequilibrium stage in Figure 14.1 may represent either a single tray or a section of packing in a packed column. In the models described in this chapter the same equations are used to model both types of equipment and the only difference between these two simulation problems is that different expressions must be used for estimating the binary mass transfer coefficients and interfacial areas.

For packed columns each model stage represents a section of packing and the height of that section must be specified in advance. Shorter section heights mean more stages are needed to represent a specific total height, therefore, more calculations and longer computation times. Smaller stages mean more accurate results. It is not always easy to know in advance how many stages should be used in any given situation. To some extent, experience plays a role in choosing an appropriate height of packing for each model stage. The problem is akin to choosing the appropriate step size for the numerical integration of a set of differential equations.

In writing down the equations that model the behavior of this nonequilibrium stage, the flow rates of vapor and liquid phases leaving the jth stage are denoted by V_j and L_j, respectively. The mole fractions in these streams are y_{ij} and x_{ij}. The $\mathcal{N}_{ij}$ are the rates of mass transfer of species i on stage j. The temperature of the vapor and liquid phases are not assumed to be equal and we must allow for heat transfer as well as mass transfer across the interface. The symbol $\mathscr{E}_j$ represents the rate of energy transfer across the phase boundary.

The equations used to model the behavior of this stage are presented below.

14.1.1 The Conservation Equations

The component material balance equations for the vapor phase may be written as follows:

$$M_{ij}^V \equiv \left(1 + r_j^V\right)V_j y_{ij} - V_{j+1}y_{i,j+1} - f_{ij}^V + \mathcal{N}_{ij}^V$$
$$= 0 \qquad i = 1, 2, \ldots, c \qquad (14.1.1)$$

where f_{ij}^V and is the feed flow rate of component i to stage j in the vapor phase.

The component material balance for the liquid phase is

$$M_{ij}^L \equiv \left(1 + r_j^L\right)L_j x_{ij} - L_{j-1}x_{i,j-1} - f_{ij}^L - \mathcal{N}_{ij}^L$$
$$= 0 \qquad i = 1, 2, \ldots, c \qquad (14.1.2)$$

where f_{ij}^L and is the feed flow rate of component i to stage j in the liquid phase.

The last term in the left-hand side of Eqs. 14.1.1 and 14.1.2 represents the net loss or gain of component i on stage j due to interphase transport. Without loss of generality we may define $\mathcal{N}_{ij}^V$ and $\mathcal{N}_{ij}^L$ as

$$\mathcal{N}_{ij}^V \equiv \int N_{ij}^V \, da_j \qquad (14.1.3)$$

and

$$\mathcal{N}_{ij}^L \equiv \int N_{ij}^L \, da_j \qquad (14.1.4)$$

where N_{ij}^V and N_{ij}^L are the molar fluxes of component i at a particular point in the two phase dispersion and da_j is the elemental interfacial area through which that flux passes. We have adopted the convention that transfers from the vapor to the liquid phases are positive in sign. With a mechanistic model, such as that developed in Sections 12.1.7 and 12.2.4, we can account for the separate contributions to mass transfer in the bubble formation zone, in the bulk froth zone on a tray and to entrained droplets in the spray. An alternative (and somewhat simpler) approach based on the use of the empirical methods in Section 12.1.5 for tray columns and Section 12.3.3 for packed columns is described below.

It follows from Eqs. 14.1.3–14.1.4 that

$$M_{ij}^I \equiv \mathcal{N}_{ij}^V - \mathcal{N}_{ij}^L = 0 \qquad i = 1, 2, \ldots, n \tag{14.1.5}$$

Equation 14.1.5 may also be derived by constructing a material balance around the entire interface.

The total material balances for the two phases are obtained by summing Eqs. 14.1.1 and 14.1.2 over the component index i.

$$M_{t_j}^V \equiv \left(1 + r_j^V\right)V_j - V_{j+1} - F_j^V + \mathcal{N}_{t_j}^V = 0 \tag{14.1.6}$$

$$M_{t_j}^L \equiv \left(1 + r_j^L\right)L_j - L_{j-1} - F_j^L - \mathcal{N}_{t_j}^L = 0 \tag{14.1.7}$$

where F_j denotes the total feed flow rate for stage j, $F_j = \sum_{i=1}^{c} f_{ij}$. The use of total material balances and the component balances in the solution procedure ensures the the bulk phase mole fractions sum to unity.

The energy balance for the vapor phase is

$$E_j^V \equiv \left(1 + r_j^V\right)V_j H_j^V - V_{j+1} H_{j+1}^V - F_j^V H_j^{VF} + Q_j^V + \mathcal{E}_j^V = 0 \tag{14.1.8}$$

where V_j is the total vapor flow leaving stage j. The energy balance for the liquid phase is

$$E_j^L \equiv \left(1 + r_j^L\right)L_j H_j^L - L_{j-1} H_{j-1}^L - F_j^L H_j^{LF} + Q_j^L - \mathcal{E}_j^L = 0 \tag{14.1.9}$$

The last terms on the left-hand side of Eqs. 14.1.8 and 14.1.9 represent the energy loss or gain due to interphase transfer. We may define $\mathcal{E}_j^V$ and $\mathcal{E}_j^L$ by the following equations:

$$\mathcal{E}_j^V \equiv \int E_j^V \, da_j \tag{14.1.10}$$

and

$$\mathcal{E}_j^L \equiv \int E_j^L \, da_j \tag{14.1.11}$$

where E_j is the energy flux at a particular point in the dispersion.

An energy balance around the interface yields

$$E_j^I \equiv \mathcal{E}_j^V - \mathcal{E}_j^L = 0 \tag{14.1.12}$$

14.1.2 The Rate Equations

The mass transfer *rates* can be evaluated from a model of mass and energy transfer in distillation such as those developed in Chapters 11 and 12. We review the necessary material here for convenience. The molar fluxes in each phase are given by

$$N_i^V = J_i^V + N_t^V y_i^V \tag{11.5.3a}$$

$$N_i^L = J_i^L + N_t^L x_i^L \tag{11.5.3b}$$

where y_i^V is the mole fraction of species i, is the bulk vapor, and x_i^L is the mole fraction of species i in the bulk liquid. The diffusion fluxes J_i^V and J_i^L are given by Eqs. 11.5.4 and 11.5.5.

$$(J^V) = c_t^V [k^V](y^V - y^I) \tag{11.5.5a}$$

$$(J^L) = c_t^L [k^L](x^I - x^L) \tag{11.5.4a}$$

Powers et al. (1988) found that the high flux correction factor is not important in distillation and it has been ignored in rewriting Eqs. 11.5.4 and 11.5.5.

In the nonequilibrium stage model of Krishnamurthy and Taylor (1985a) the total mass transfer *rates* are obtained by combining Eqs. 11.5.3a and 11.5.5a and multiplying by the interfacial area available for mass transfer

$$\left(\mathcal{N}_j^V\right) = c_{t_j}^V \left[k_j^V\right] a_j \left(y_j - y_j^I\right) + \mathcal{N}_{t_j}^V (y_j) \tag{14.1.13}$$

For the liquid phase we have the analogous relation

$$\left(\mathcal{N}_j^L\right) = c_{t_j}^L \left[k_j^L\right] a_j \left(x_j^I - x_j\right) + \mathcal{N}_{t_j}^L (x_j) \tag{14.1.14}$$

where $\mathcal{N}_{t_j}$ is the total mass transfer rate. Following Krishnamurthy and Taylor (1985a), we have used the composition of the phases leaving stage j as the bulk phase composition for the purposes of evaluating the mass transfer rates. This is equivalent to assuming the bulk phases are completely mixed.

The matrices of multicomponent low-flux mass transfer coefficients $[k^V]$ and $[k^L]$ may be calculated from Eqs. 8.8.17 and 8.8.18, respectively, using binary mass transfer coefficients obtained from the correlations in Sections 12.1.5 and 12.3.2. It is important to recognize that the binary mass transfer coefficients obtained from these correlations are functions of the tray design and layout, or of the packing type and size, as well as of the operating conditions. This means that equipment design parameters must be known so that the nonequilibrium model equations can be solved. Equipment design parameters may be specified in advance (along with other necessary specifications) or determined during the solution by carrying out column design calculations simultaneously with the solution of all the model equations (Taylor et al., 1992).

For tray columns the net interfacial area is $a = a' h_f A_b$, where a' is the interfacial area per unit volume of froth, h_f is the froth height, and A_b is the bubbling area. For tray columns the interfacial area is $a' h A_c$, where a' is the interfacial area per unit volume, h is the height of a section of packing, and A_c is the cross-sectional area of the column.

The energy flux is defined by Eq. 11.1.4, written for each phase as

$$E_j^V = h_j^V\left(T_j^V - T_j^I\right) + \sum_{i=1}^{c} N_{ij}\overline{H}_{ij}^V \tag{14.1.15}$$

and

$$E_j^L = h_j^L\left(T_j^I - T_j^L\right) + \sum_{i=1}^{c} N_{ij}\overline{H}_{ij}^L \tag{14.1.16}$$

where h_j is the heat transfer coefficient, and $\overline{H}_{ij}$ is the partial molar enthalpy of component i for stage j. Heat transfer coefficients may be estimated from correlations or analogies as discussed in Section 11.4.4.

The energy transfer rates in the vapor and liquid phases are obtained after multiplying the energy fluxes by the interfacial area a_j

$$\mathscr{E}_j^V = h_j^V a_j\left(T_j^V - T_j^I\right) + \sum_{i=1}^{c} \mathscr{N}_{ij}\overline{H}_{ij}^V \tag{14.1.17}$$

$$\mathscr{E}_j^L = h_j^L a_j\left(T_j^I - T_j^L\right) + \sum_{i=1}^{c} \mathscr{N}_{ij}\overline{H}_{ij}^L \tag{14.1.18}$$

14.1.3   The Interface Model

Phase equilibrium is assumed to exist only at the interface with the mole fractions in both phases related by

$$Q_{ij}^I \equiv K_{ij}x_{ij}^I - y_{ij}^I = 0 \qquad i = 1, 2, \ldots, c \tag{14.1.19}$$

where K_{ij} is the equilibrium ratio for component i on stage j. The K_{ij} must be evaluated at the temperature, pressure, and mole fractions at the interface. The mole fractions at the interface must sum to unity

$$S_j^{VI} \equiv \sum_{i=1}^{c} y_{ij}^I - 1 = 0 \tag{14.1.20}$$

$$S_j^{LI} \equiv \sum_{i=1}^{c} x_{ij}^I - 1 = 0 \tag{14.1.21}$$

14.1.4   The Hydraulic Equations

Column pressure drop is a function of tray (or packing) type and design and column operating conditions, information that is required for the estimation of the mass transfer coefficients. It is, therefore, possible to add a pressure drop equation to the set of independent equations for each stage and to make the pressure of each stage (tray or packed section) an unknown variable. The stage is assumed to be at mechanical equilibrium so $P_j^V = P_j^L = P_j$.

The pressure of the top tray (or top of the packing) must be specified along with the pressure of any condenser. The pressure of trays (or packed sections) below the topmost are calculated from the pressure of the stage above and the pressure drop on that tray (or over that packed section).

If the column has a condenser, which is numbered as stage 1 here, the hydraulic equations are expressed as follows:

$$P_1 \equiv P_c - P_1 = 0 \tag{14.1.22}$$

$$P_2 \equiv P_{spec} - P_2 = 0 \tag{14.1.23}$$

$$P_j \equiv P_j - P_{j-1} - (\Delta P_{j-1}) = 0 \qquad j = 3, 4, \ldots, n \tag{14.1.24}$$

where P_c is the condenser pressure, P_{spec} is the specified pressure of the tray or section of packing at the top of the column. The term ΔP_{j-1} is the pressure drop per tray or section of packing from section–stage $j - 1$ to section–stage j.

If the top stage is not a condenser, the hydraulic equations are expressed as follows:

$$P_1 \equiv P_{spec} - P_1 = 0 \tag{14.1.25}$$

$$P_j \equiv P_j - P_{j-1} - (\Delta P_{j-1}) = 0 \qquad j = 2, 3, \ldots, n \tag{14.1.26}$$

The pressure drop over sieve and valve trays may be estimated using correlations presented by Lockett (1986) and Kister (1992). For bubble cap trays the procedures described by Bolles (1963) can be adapted for computer based calculation. Kister (1992) reviews the methods available for estimating the pressure drop in dumped packed columns. The pressure drop in structured packed columns may be estimated using the method of Bravo et al. (1986).

14.1.5 Specifications for Nonequilibrium Simulation

Most of the quantities normally specified in an equilibrium stage simulation (number of stages, feed stage location, feed flow rates and composition, reflux ratio, distillate or bottoms rate, and so on) must be specified for a comparable nonequilibrium simulation. By including the hydraulic or pressure drop equations in the model it is not necessary to specify the pressure of each stage for a nonequilibrium simulation; only the pressure of the top stage and of the condenser need be specified.

As noted above, a number of equipment parameters must normally be specified so that the mass transfer coefficients can be estimated correctly. For example, the diameter of all columns must be known. For trayed columns, the tray type, weir height, liquid flow path length, and bubbling area must be known; for packed columns, the packing type, size, and material must be known. It may also be necessary to allow for different diameters, tray or packing type, or other tray or packing parameters in different parts of the same column.

The actual process flow rates are important in nonequilibrium model simulations, whereas in most equilibrium stage simulations, a simulation with a feed flow rate of 1 unit is as meaningful as a simulation with a feed flow of 10, 100, or 573 units. In real columns the flow rates influence the mass transfer coefficients as well as the tray hydraulics. An inappropriate flow specification may mean the column will flood or, just as likely, dump all the liquid through the holes in the tray. Thus, it is important to ensure that the specified (or calculated) flows and tray or packing characteristics are consistent with the satisfactory operation of the column.

14.2 SOLVING THE MODEL EQUATIONS

Newton's method has often been the method of choice for solving equilibrium stage separation process problems (see, e.g., Henley and Seader, 1981). We have also made extensive use of Newton's method throughout this book. Thus, it will come as no surprise to our readers that we advocate the use of Newton's method for solving the nonlinear algebraic equations of the nonequilibrium stage model.

14.2.1 Variables and Equations for a Nonequilibrium Stage

There are $6c + 8$ unknown variables for each nonequilibrium stage.

- Vapor and liquid flow rates $(V_j, L_j, 2)$.
- Vapor and liquid-phase compositions $(y_{ij}, x_{ij}; 2c)$.
- Vapor and liquid temperatures $(T_j^V, T_j^L; 2)$.
- Vapor and liquid interface compositions $(y_{ij}^I, x_{ij}^I; 2c)$.
- Interface temperature $(T_j^I; 1)$.
- Mass transfer rates $(\mathcal{N}_{ij}^V, \mathcal{N}_{ij}^L; 2c)$.
- Energy transfer rates $(\mathcal{E}_j^V, \mathcal{E}_j^L; 2)$.
- Stage pressure $(P_j; 1)$.

The $6c + 8$ equations for the jth stage, collectively referred to as the *MERSHQ* equations (after the letters used to identify the different equations), are

- *M*: *M*aterial balances for the vapor $(M_{ij}^V, M_{t_j}^V; c + 1)$.
- *M*: *M*aterial balances for the liquid $(M_{ij}^L, M_{t_j}^L; c + 1)$.
- *M*: *M*aterial balances for the interface $(M_{ij}^I; c)$.
- *E*: *E*nergy balance equations $(E_j^V, E_j^L, E_j^I; 3)$.
- *R*: transfer *R*ate equations $(2c - 2)$.
- *R*: energy transfer *R*ate equations (2).
- *S*: *S*ummation equations $(S_j^{VI}, S_j^{LI}; 2)$.
- *H*: *H*ydraulic equations $(P_j; 1)$.
- *Q*: interface e*Q*uilibrium equations $(Q_{ij}^I; c)$.

A modest reduction in the rather large number of variables and equations can be obtained by recognizing that there is really only one set of independent mass transfer rates, $\mathcal{N}_{ij} = \mathcal{N}_{ij}^V = \mathcal{N}_{ij}^L$, and we can eliminate either the vapor- or the liquid-phase mass transfer rates from Eqs. 14.1.1 and 14.1.2 and by combining the mass transfer rate equations (Eqs. 14.1.13 and 14.1.14) and the interface material balances (Eqs. 14.1.5) to give

$$R_{ij}^V \equiv \mathcal{N}_{ij} - \mathcal{N}_{ij}^V = 0 \qquad i = 1, 2, \ldots, n - 1 \qquad (14.2.1)$$

and

$$R_{ij}^L \equiv \mathcal{N}_{ij} - \mathcal{N}_{ij}^L = 0 \qquad i = 1, 2, \ldots, n - 1 \qquad (14.2.2)$$

where

$$\mathcal{N}_{i,j}^V = N_{i,j}^V\left(k_{ik,j}^V, a_j, y_{k,j}^I, y_{k,j}, \mathcal{N}_{kj}^V, \qquad k = 1, 2, \ldots, n\right) \qquad (14.2.3)$$

$$\mathcal{N}_{i,j}^L = \mathcal{N}_{i,j}^L\left(k_{ik,j}^L, a_j, x_{k,j}^I, x_{k,j}, \mathcal{N}_{k,j}^L, \qquad k = 1, 2, \ldots, n\right) \qquad (14.2.4)$$

are the mass transfer rates written as a function of mole fractions, mass transfer coefficients, and the mass transfer rates themselves.

The final working form of the interface energy balance E_j^I is obtained on substituting Eqs. 14.1.17 and 14.1.18 for the energy-transfer rates into Eq. 14.1.12 to give

$$E_j^I \equiv h_j^V a_j \left(T_j^V - T_j^I \right) - h_j^L a_j \left(T_j^I - T_j^L \right) + \sum_{i=1}^{c} \mathscr{N}_{ij} \left(\bar{H}_{ij}^V - \bar{H}_{ij}^L \right) = 0 \quad (14.2.5)$$

Note that the energy-transfer rates $\mathscr{E}_j^V$ and $\mathscr{E}_j^L$ are eliminated once this substitution has been made.

There remain only $5c + 6$ independent variables and they are ordered into a vector (x_j) as follows:

$$(x_j)^T \equiv \left(V_j, y_{1j}, y_{2j}, \ldots, y_{cj}, T_j^V, y_{1j}^I, y_{2j}^I, \ldots, y_{cj}^I, x_{1j}^I, x_{2j}^I, \ldots, x_{cj}^I, T_j^I, L_j, \right.$$

$$\left. x_{1j}, x_{2j}, \ldots, x_{cj}, T_j^L, \mathscr{N}_{ij}, \mathscr{N}_{2j}, \ldots, \mathscr{N}_{cj}, P_j \right) \quad (14.2.6)$$

The corresponding $5c + 6$ equations per stage are ordered into a vector (F_j) as follows:

$$(F_j)^T \equiv \left(M_{t_j}^V, M_{1j}^V, M_{2j}^V, \ldots, M_{cj}^V, E_j^V, R_{1j}^V, R_{2j}^V, \ldots, R_{c-1,j}^V, S_j^{VI}, \right.$$

$$Q_{1j}^I, Q_{2j}^I, \ldots, Q_{cj}^I, E_j^I, M_{t_j}^L, M_{1j}^L, M_{2j}^L, \ldots, M_{cj}^L, E_j^L,$$

$$\left. R_{1j}^L, R_{2j}^L, \ldots, R_{c-1,j}^L, S_j^{LI}, P_j \right) \quad (14.2.7)$$

14.2.2 Condensers and Reboilers

Condensers and reboilers may be modeled as equilibrium stages. The independent variables for such a stage are the mole fractions $(x_{ij}, y_{ij} : 2c)$, the temperature $(T_j : 1)$, the flow rates $(V_j, L_j : 2)$, and the stage pressure $(P_j : 1)$. The corresponding $2c + 4$ independent equations are the MESH equations (of Section 13.4.1) plus a pressure drop equation.

The $2c + 4$ variables for an equilibrium stage are ordered into a vector (x_j) as follows:

$$(x_j)^T \equiv \left(V_j, y_{1j}, y_{2j}, \ldots, y_{cj}, T_j, x_{1j}, x_{2j}, \ldots, x_{cj}, L_j, P_j \right) \quad (14.2.8)$$

The corresponding $2c + 4$ discrepancy equations for an equilibrium stage are ordered into a vector (F_j) as follows:

$$(F_j)^T \equiv \left(M_{t_j}, M_{1j}, M_{2j}, \ldots, M_{cj}, H_j, E_{1j}, E_{2j}, \ldots, E_{cj}, S_j, P_j \right) \quad (14.2.9)$$

where E_{ij} are the equilibrium relationships (Eqs. 13.4.7), and S_j is the summation equation for an equilibrium stage and is expressed as

$$S_j \equiv \sum_{i=1}^{c} \left(x_{ij} - y_{ij} \right) = 0 \quad (13.4.8)$$

14.2.3 Equations and Variables for a Multistage Column

The nonequilibrium model equations for an entire column can be expressed in the general functional form

$$(F(x)) = (0) \quad (14.2.10)$$

where (F)

$$(F)^T = \left((F_1)^T, (F_2)^T, \ldots, (F_n)^T \right) \quad (14.2.11)$$

is the vector of equations to be solved and where (x) is the vector of unknown variables

$$(x)^T = ((x_1)^T, (x_2)^T, \ldots, (x_n)^T) \tag{14.2.12}$$

(F_j) is the vector of model equations for stage j and (x_j) is a vector of variables for stage j.

14.2.4 Solution of the Model Equations

As noted above, Newton's method can be used to solve the entire set of nonlinear equations simultaneously. To use Newton's method, we repeatedly solve Eq. 14.2.10 linearized about a current guess (x_k) of the vector (x).

$$[J_k]\Delta(x_k) = -(F(x_k)) \tag{14.2.13}$$

$[J_k]$ is the Jacobian matrix at the kth iteration with elements

$$J_{ij} = \frac{\partial F_i}{\partial x_j} \tag{14.2.14}$$

and

$$\Delta(x_k) = (x_{k+1}) - (x_k) \tag{14.2.15}$$

The method may be assumed to have converged when either of the following two criteria are satisfied

$$\sqrt{\sum_{j=1}^{N} \sum_{i=1}^{e_j} F_{ij}^2} < \varepsilon \tag{14.2.16}$$

$$\sum_{j=1}^{n} \sum_{i=1}^{e_j} |\Delta x_{ij}| / x_{ij} < \varepsilon \tag{14.2.17}$$

where N is the number of stages, e_j is the number of equations for the jth stage, and ε is a small number (10^{-3} in the examples below).

Powers et al. (1988) developed some guidelines to provide a set of initial estimates for nonequilibrium models. With reasonable initial estimates x_0 Newton's method will usually converge quickly. However, it is a good idea to limit the variable changes computed from Eq. (14.2.15). Temperature changes should be restricted to 10 K per stage per iteration, flow changes to 50% of the flow itself, and changes in composition that would take a mole fraction outside the range from 0 to 1 to $\frac{1}{2}$ of the step that would take it to the boundary. Limiting variable corrections is particularly important when the column specifications do not constrain the internal flows. Nonequilibrium model simulations are about as difficult to converge as a corresponding equilibrium stage simulation provided that column design specifications (diameter, tray type, weir height, length, etc.) are consistent with the satisfactory operation of a real column.

Newton's method requires the evaluation of the partial derivatives of all equations with respect to all variables. The partial derivatives of thermodynamic properties with respect to temperature, pressure, and composition are most awkward to obtain (and the ones that have the most influence on the rate of convergence). Since pressure is an unknown variable in this model, the derivatives of K values and enthalpies with respect to pressure must be evaluated. Neglect of these derivatives (even though they are often small) can lead to convergence difficulties.

For single, simple columns, the Jacobian matrix has the block tridiagonal structure shown below.

$$[J] = \begin{bmatrix} [B_1] & [C_1] & & & & \\ [A_2] & [B_2] & [C_2] & & & \\ \cdots & \cdots & \cdots & \cdots & & \\ & [A_j] & [B_j] & [C_j] & & \\ & \cdots & \cdots & \cdots & \cdots & \\ & & [A_{n-1}] & [B_{n-1}] & [C_{n-1}] \\ & & & [A_n] & [B_n] \end{bmatrix} \qquad (14.2.18)$$

The submatrices $[A_j]$, $[B_j]$, and $[C_j]$ are defined by

$$[A_j] = \frac{\partial(F_j)}{\partial(x_{j-1})} \qquad j = 2, 3, \ldots, n \qquad (14.2.19)$$

$$[B_j] = \frac{\partial(F_j)}{\partial(x_j)} \qquad j = 1, 2, \ldots, n \qquad (14.2.20)$$

$$[C_j] = \frac{\partial(F_j)}{\partial(x_{j+1})} \qquad j = 1, 2, \ldots, n - 1 \qquad (14.2.21)$$

The structure of the submatrices $[A_j]$, $[B_j]$, and $[C_j]$ is given by Hung (1991).

Linear systems with a block tridiagonal coefficient matrix can be solved quite efficiently using a matrix generalization of the well-known Thomas algorithm. Henley and Seader (1981) give the steps of the algorithm. The number of equations per nonequilibrium stage ($5c + 6$) is not the same as the number of equations for equilibrium stages ($2c + 4$) and this must be allowed for when implementing the Block Thomas algorithm. In fact, even the diagonal blocks are rather empty and it is better to devise a special elimination procedure that exploits the structure of the nonequilibrium model or use a general sparse equation solver.

14.3 DESIGN STUDIES

In this section we present a number of examples designed to illustrate the use of a nonequilibrium model as a design tool. In view of the large number of equations that must be solved it is impossible to present illustrative examples of the application of the nonequilibrium model that are as detailed as the examples in prior chapters. In the examples that follow we confine ourselves to a brief summary of the problem specifications and the results obtained from a computer solution of the model equations. In most cases several different column configurations were simulated before the results presented below were obtained.

Example 14.3.1   Depropanizer Column Design

A depropanizer is a distillation operation encountered in almost all oil refineries. Our task here is to design a column to separate 1000 mol/s of a four component mixture containing 300 mol/s n-propane and 500 mol/s n-butane so that there is no more than 3.5 mol/s of n-propane present in the bottom product and no more than 3.5 mol/s of n-butane is

TABLE 14.1 Specifications and Stream Table for Depropanizer

Operation:
  Simple distillation
  Partial (vapor product) condenser
  Partial (liquid product) reboiler
  35 Stages
  Feeds to stages 16

Properties:
  EOS K model
  Peng–Robinson Cubic EOS
  Excess enthalpy from EOS

Specifications:
  Column pressure
    Condenser pressure 15.00 (atm)
    Top pressure 15.00 (atm)
  Condenser
    Reflux ratio = 2.500 (−)
  Reboiler
    Bottom product flow rate = 600.0 (mol/s)

Stream	Feed 1	Top	Bottom
Stage	16	1	35
Pressure (atm)	15.00	15.00	15.12
Vapor fraction (−)	0.0000	1.000	0.0000
Temperature (°C)	25.00	34.92	105.7
Mole flows (mol/s)			
Ethane	100.0	100.0	0.0001485
Propane	300.0	296.7	3.329
n-Butane	500.0	3.328	496.7
n-Pentane	100.0	0.0009178	100.00
Total molar flow	1000	400.0	600.0

TABLE 14.2 Depropanizer Column Design

Number of sections	2	
System factor (−)	1.000	
Default flooding factor (−)	0.7000	

Section	1	2
First stage	2	16
Last stage	15	34
Column internals	Sieve tray	Sieve tray
Column diameter (m)	4.820	6.170
Total tray area (m²)	18.25	29.90
Number of flow passes	5	5
Tray spacing (m)	0.5000	0.5000
Liquid flow path length (m)	0.7945	1.037
Active area (m²)	14.96	24.51
Total hole area (m²)	0.8813	1.355
Downcomer area (m²)	1.645	2.696
Hole diameter (m)	0.004763	0.004219
Hole pitch (m)	0.01807	0.01685
Weir length (m)	17.60	22.97
Weir height (m)	0.05080	0.03734
Downcomer clearance (m)	0.03810	0.03810
Deck thickness (m)	0.002540	0.002540

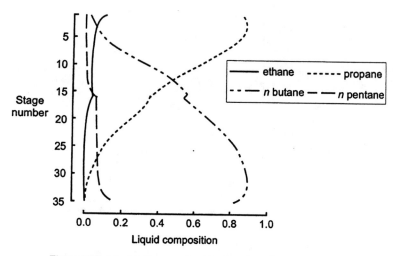

Figure 14.2. Liquid-phase composition profiles in a depropanizer.

present in the distillate. The operation takes place at the moderately elevated pressure of 15 atm.

A number of simulations were carried out for different column configurations (number of stages and feed stage location) and for differing operation specifications (reflux ratio and bottom product rate). Only the specifications and results of our final simulation are reported here.

The problem specification and computed product flows are provided in Table 14.1. The column is divided into two sections, the section above the feed has a diameter of 4.82 m and the section below the feed has a diameter of 6.17 m. Additional column design parameters are provided in Table 14.2.

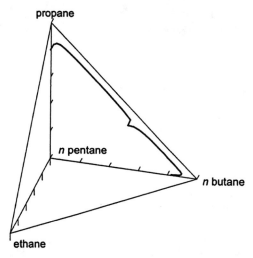

Figure 14.3. Liquid-phase composition profiles in a depropanizer shown as a tetrahedron. The four corners represent the pure components.

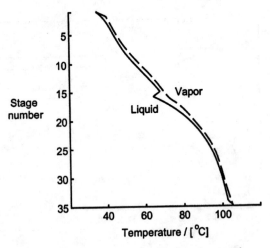

Figure 14.4. Temperature profiles in a depropanizer.

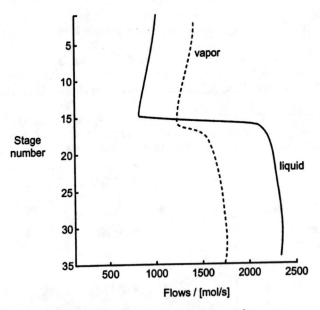

Figure 14.5. Flow profiles in a depropanizer.

Composition profiles are shown in Figures 14.2 and 14.3. Figure 14.4 shows the vapor- and liquid-phase temperatures. In this particular example these two temperatures are not too different. Flow profiles are shown in Figure 14.5.

A pseudo McCabe–Thiele diagram is shown in Figure 14.6. Note that the steps that represent the trays in the column do not reach the equilibrium line. This kind of diagram is invaluable in the design process as an aide to determining the optimum feed stage location. Murphree efficiencies of each component on each tray, computed from Eq. 13.1.2, are shown in Figure 14.7. The discontinuities in the efficiency profiles arise at or around feed stages or where the mole fractions of one of the components passes through a maximum. It

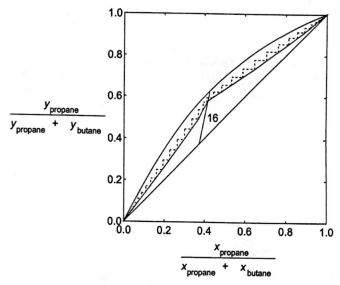

Figure 14.6. McCabe–Thiele diagram for propane and *n*-butane in a depropanizer.

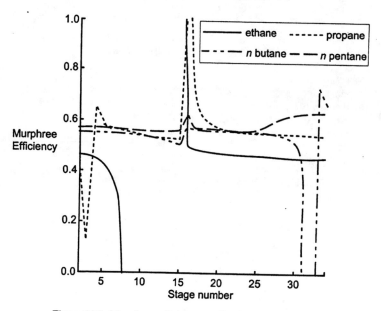

Figure 14.7. Murphree efficiency profiles in a depropanizer.

is important to emphasize that these efficiencies are calculated following a nonequilibrium simulation but are not used during the calculations. ∎

Example 14.3.2   *Extractive Distillation Column Design*

In Section 14.4.2 we shall show that the nonequilibrium model provides excellent predictions of the product compositions in the extractive distillation of an acetone (1)–methanol

TABLE 14.3 Specifications and Stream Summary for Extractive Distillation

Operation:
  Simple distillation
  Total (liquid product) condenser
  Total (liquid product) reboiler
  80 Stages
  Feeds to stages 13, 68

Properties:
  DECHEMA K model
  Original UNIQUAC model
  Antoine vapor pressure
  Excess enthalpy from UNIQUAC model

Specifications:
  Column pressure
    Condenser pressure 0.1013 (MPa)
    Top pressure 0.1013 (MPa)
  Condenser
    Reflux ratio = 10.00 (−)
  Reboiler
    Bottom product flow rate = 88.89 (mol/s)

Stream	Feed 1	Feed 2	Top	Bottom
Stage	13	68	1	80
Pressure (MPa)	0.1013	0.1013	0.1013	0.1560
Vapor fraction (−)	0.0000	0.0000	0.0000	1.000
Temperature (K)	373.2	333.5	328.5	360.9
Mole flows (mol/s)				
Acetone	0.0000	22.22	19.88	2.344
Methanol	0.0000	66.67	2.321	64.35
Water	22.22	0.0000	0.01960	22.20
Total molar flow	22.22	88.89	22.22	88.89

TABLE 14.4 Extractive Distillation Column Design

Number of sections	1
System factor (−)	1.000
Default flooding factor (−)	0.7000
Section	1
First stage	2
Last stage	79
Column internals	Sieve tray
Column diameter (m)	2.320
Total tray area (m²)	4.227
Number of flow passes	2
Tray spacing (m)	0.5000
Liquid flow path length (m)	0.7930
Active area (m²)	3.550
Total hole area (m²)	0.2366
Downcomer area (m²)	0.2842
Hole diameter (m)	0.004667
Hole pitch (m)	0.01406
Weir length (m)	2.984
Weir height (m)	0.05029

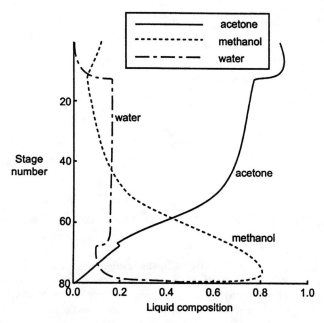

Figure 14.8. Liquid-phase composition profiles in extractive distillation column.

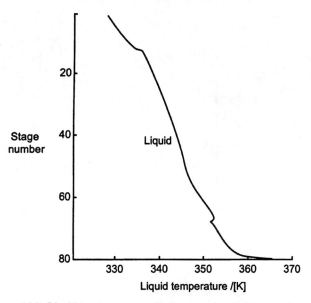

Figure 14.9. Liquid temperature profile in extractive distillation column.

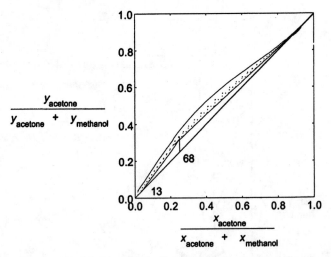

Figure 14.10. McCabe–Thiele diagram for acetone and methanol.

(2) mixture using water (3) as the solvent. In this example, we are asked to design a large scale column that can recover about 90 mol% of the acetone in the distillate product and more than 95 mol% of the methanol in the bottom product. The mole fraction of acetone in the acetone–methanol feed is 0.25 and the feed flow is 88.89 mol/s. The column is to operate at a pressure of 101.3 kPa.

A great many simulations are required to solve a problem of this kind. Since there are two separate feeds to the column, there are many more possible column configurations that need to be investigated. In addition, the separation is strongly influenced by the flow rate of

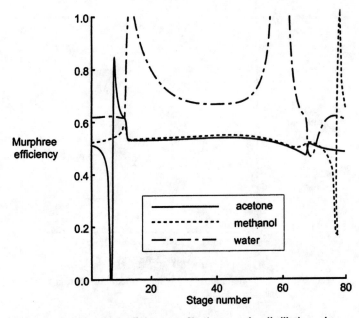

Figure 14.11. Murphree efficiency profiles in extractive distillation column.

the solvent, water. A column configuration and operation specifications that meets the design objective is provided in Table 14.3. There are, no doubt, many other configurations and operating conditions that would be as good if not better. The column was simulated as a single section with a uniform diameter; the equipment design parameters are given in Table 14.4.

The liquid-phase composition profiles are shown in Figure 14.8, the temperature profiles in Figure 14.9. In this example, the liquid and vapor temperatures are almost equal. The pressure profile is not shown since it was more or less a straight line between the specified top tray pressure and the computed bottom pressure reported in Table 14.3. The pseudo McCabe–Thiele diagram and efficiency profiles are shown in Figures 14.10 and 14.11. In this case we see that the efficiencies of acetone and methanol are essentially equal over most of the column with the efficiency of water somewhat higher. Interestingly, the efficiencies decrease in the lower portion of the column below the acetone–methanol feed.

∎

Example 14.3.3 Ethylbenzene–Styrene Distillation Column Design

In this example we are asked to design a column to separate 60 mol/s of a mixture of 45% styrene and 55% ethylbenzene. The column should produce an overhead product that is

TABLE 14.5 Specifications and Stream Table for Packed Column

Operation:
  Simple distillation
  Total (liquid product) condenser
  Partial (liquid product) reboiler
  152 Stages
  Feeds to stages 43

Properties:
  Raoult's law K model
  Lee Kesler vapor pressure
  Ideal enthalpy

Specifications:
  Column pressure
    Condenser pressure 50.00 (torr)
    Top pressure 50.00 (torr)
  Condenser
    Reflux ratio = 12.00 (−)
  Reboiler
    Bottom product flow rate = 25.25 (mol/s)

Stream	Feed 1	Top	Bottom
Stage	43	1	152
Pressure (torr)	65.00	50.00	104.9
Vapor fraction (−)	0.0000	0.0000	0.0000
Temperature (°F)	151.9	137.4	179.5
Mole flows (mol/s)			
Ethylbenzene	33.00	32.99	0.009403
Styrene	27.00	1.759	25.24
Total molar flow	60.00	34.75	25.25
Mole fractions (−)			
Ethylbenzene	0.5500	0.9494	0.0003724
Styrene	0.4500	0.05063	0.9996

TABLE 14.6   Packed Column Design

Number of sections	2
System factor ($-$)	1.000
Default flooding factor ($-$)	0.2800
Column internals	Koch Flexipac II
Packed height above feed (m)	14.000
Packed height below feed (m)	36.000
Column diameter above feed (m)	7.270
Column diameter below feed (m)	6.610
Specific packing surface (1/m)	223.0
Void fraction ($-$)	0.9500
Channel base (m)	0.02590
Crimp height (m)	0.01240
Channel side (m)	0.01800
Equivalent diameter (m)	0.01411
Channel flow angle (rad)	0.7854
Packing factor ($-$)	72.00

approximately 5% styrene and a bottoms product that is less than 400 ppm ethylbenzene (Strigle, 1987). This operation must be carried out at low pressure to reduce the polymerization of styrene that takes place at higher temperatures. A vacuum system at 50 mmHg is available for this operation. The column will use Koch Flexipac II structured packing.

A number of simulations were needed to design this column. The most important parameters were the overall height of packing, the feed stage location, and the reflux ratio. The purity specifications were not used in any simulation since nonstandard specifications of that kind can make the simulations harder to converge.

The configuration and operation specifications for a column that we found would produce the desired products are shown in Table 14.5. The total height of packing is 50 m

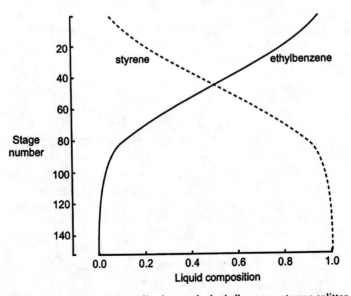

Figure 14.12. Composition profiles in a packed ethylbenzene–styrene splitter.

TABLE 14.7   Specifications and Results for Hydrocarbon Absorber

Operation
  Simple absorber/stripper
  12 Stages
  Feeds to stages 1, 12

Properties:
  EOS K model
  Peng–Robinson cubic EOS
  Excess enthalpy from EOS

Specifications:
  Column pressure
    Top pressure 4.000 (atm)

Stream	Feed 1	Feed 2	Top	Bottom
Stage	1	12	1	12
Pressure (atm)	4.000	4.000	4.000	4.103
Vapor fraction (−)	0.0000	0.9547	1.000	0.0000
Temperature (K)	305.2	300.0	309.0	321.7
Mole flows (kmol/s)				
Methane	0.0000	0.01430	0.01371	0.0005901
Ethane	0.0000	0.007850	0.006382	0.001468
Propane	0.0000	0.01200	0.005216	0.006784
n-Butane	0.001104	0.008450	0.0006009	0.008953
n-Pentane	0.002760	0.007400	0.0003280	0.009832
n-Tridecane	0.05134	0.0000	1.365E-06	0.05133
Total molar flow	0.05520	0.05000	0.02624	0.07896

TABLE 14.8   Absorber Column Design

Number of sections	1
System factor (−)	1.000
Default flooding factor (−)	0.7000
Section	´ 1
First stage	1
Last stage	12
Column internals	Sieve tray
Column diameter (m)	1.140
Total tray area (m^2)	1.021
Number of flow passes	1
Tray spacing (m)	0.5000
Liquid flow path length (m)	0.7412
Active area (m^2)	0.7886
Total hole area (m^2)	0.03803
Downcomer area (m)	0.1161
Hole diameter (m)	0.003737
Hole pitch (m)	0.01592
Weir length (m)	1.026
Weir height (m)	0.02745
Downcomer clearance (m)	0.03810
Deck Thickness (m)	0.002540

with 14 m above the feed and 36 m below. The reflux ratio is 12. The column diameter was 7.27 m above the feed and 6.61 m below the feed (Table 14.6). Larger column diameters above the feed are typical of these operations (King, 1980; Strigle, 1987). The complete column design is summarized in Table 14.6. A total of 150 nonequilibrium stages were used to model the 50 m of packing.

The predicted bottom pressure (Table 14.5) is 105 mmHg at a temperature of 180°F. This compares to 104 mmHg and 182°F as suggested by Strigle (1987). The column was designed with a flooding factor of 0.28 in order to meet the pressure drop limit mentioned by Strigle (1987).

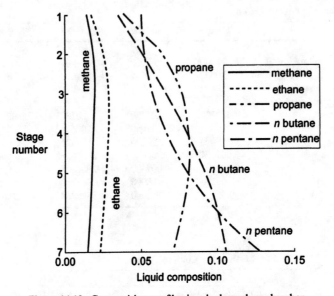

Figure 14.13. Composition profiles in a hydrocarbon absorber.

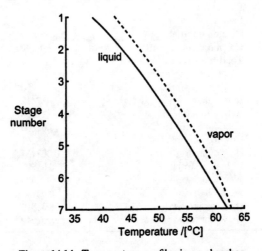

Figure 14.14. Temperature profiles in an absorber.

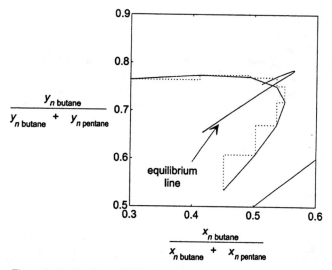

Figure 14.15. McCabe–Thiele diagram for n-butane and n-pentane.

Composition profiles are shown in Figure 14.12. Note the hour glass shape that asymptotes towards the vertical axes at the bottom of the column to meet the 400 ppm specification on the ethylbenzene purity. ∎

Example 14.3.4   Gas Absorption

The nonequilibrium model is not limited to simulating distillation operations; with no fundamental change, it can be applied to absorption operations as well. A design case study is presented here by way of illustrating the model. This problem is adapted from an application discussed by Krishnamurthy and Taylor (1986).

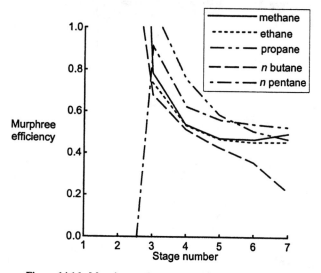

Figure 14.16. Murphree efficiency profiles for an absorber.

A hydrocarbon gas mixture is to have its propane content reduced by 45% by absorption in a heavy oil at a pressure of 4 atm. Tridecane is used to represent the oil in this illustration. Feed flows are summarized in Table 14.7.

The important design variables in this problem are the number of trays and the oil flow rate. Seven (or more) sieve trays were needed to reach the desired reduction in the propane content of the gas mixture. A complete set of configuration and operation specifications is given in Table 14.7. Calculated product stream flows and compositions are also given. The column and tray design is summarized in Table 14.8.

Composition and temperature profiles are shown in Figures 14.13 and 14.14. In this example we see that the vapor and liquid temperatures are rather different. This result is quite typical of absorption columns; indeed, it is possible for temperature differences to be over the order of 20 K. The McCabe–Thiele diagram is shown in Figure 14.15 and the component efficiencies in Figure 14.16. The efficiencies tend to be lower than in the distillation operations considered above.

It is possible to achieve more or less the same separation achieved in the seven tray column in two equilibrium stages. Thus, the overall efficiency is about 28%. Industrial absorbers typically operate at overall efficiencies between 15 and 40%. ■

14.4 EXPERIMENTAL STUDIES

In this section we present several comparisons between the predictions of a nonequilibrium model and actual experimental data. Simulations of a variety of operations are described including small and industrial scale and trayed and packed towers.

14.4.1 Multicomponent Distillation

In Section 13.5.1 we presented the data of Vogelpohl for the distillation of two ternary systems: acetone–methanol–water and methanol–2-propanol–water in a bubble cap column. Krishnamurthy and Taylor (1985b) simulated these experiments using a nonequilibrium stage model similar to the one described above. The *AIChE* correlations were used to calculate the mass transfer coefficients. Thermodynamic properties were calculated with the models described by Prausnitz et al. (1980).

Predicted and measured composition profiles for the acetone–methanol–water system are compared in Figure 14.17. The average discrepancy between predicted and measured

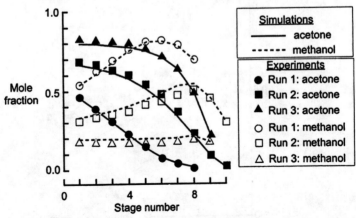

Figure 14.17. Composition profiles in the distillation of acetone–methanol–water system in a bubble cap column at total reflux. Data of Vogelpohl (1979). Calculations by Krishnamurthy and Taylor (1985b).

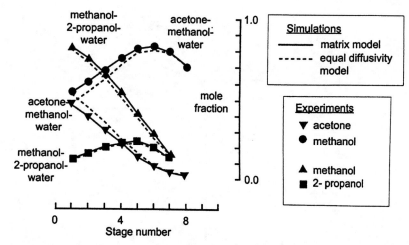

Figure 14.18. Composition profiles in the distillation of acetone–methanol–water and methanol–2-propanol–water systems in a bubble cap column at total reflux. Profiles obtained using matrix models and an effective (equal) diffusivity model of mass transfer. Data of Vogelpohl (1979). Calculations by Krishnamurthy and Taylor (1985b).

mole fractions is 1.122 mol% for the acetone–methanol–water system and 2.211 mol% for the methanol–2-propanol–water system.

Let us consider in more detail whether there is any support for a matrix model of mass transfer in these data. Krishnamurthy and Taylor (1985b) found that all multicomponent (i.e., matrix) models gave identical results. However, a model in which a single average value of the diffusion coefficient was used for all binary pairs does not do a particularly good job of predicting the composition profiles in Vogelpohl's experiments as shown in Figure 14.18. To give a feel for the magnitude of this discrepancy we note that for run 1 of the acetone–methanol–water system, the measured mole fractions of acetone and methanol on the top stage are 0.46 and 0.54, respectively. The mole fractions predicted using a matrix model of mass transfer are 0.4613 and 0.5334, respectively, whereas the mole fractions predicted using the equal diffusivity model are 0.5215 and 0.4716. The multicomponent matrix models are clearly superior in this case.

14.4.2 Extractive Distillation

Acetone and methanol are impossible to separate by simple distillation due to the presence of an azeotrope. However, the addition of water near the top of a column allows these two components to be separated. Five sets of steady-state operating data for the extractive distillation of an acetone–methanol azeotrope in a laboratory scale column have been provided by Kumar et al. (1984). A schematic diagram of the column is provided in Figure 14.19. The column had a diameter of 15 cm and was fitted with 13 bubble cap trays, a total condenser and a thermosiphon (equilibrium) reboiler. Unlike many experimental distillation studies, these experiments were not carried out at total reflux; the acetone–methanol feed entered the column on the eleventh stage from the top (the condenser counts as the first stage) and the water was introduced on stage six. The column was operated at atmospheric pressure for all five runs. Additional details of the column, operational specifications, and computed product compositions for one of these experiments can be found in Table 14.9.

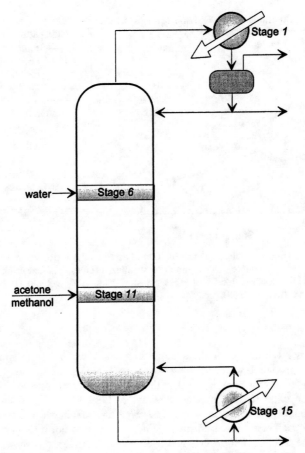

Figure 14.19. Schematic diagram of extractive distillation operation.

Predicted composition profiles are shown in Figure 14.20. Flow and temperature profiles are shown in Figures 14.21 and 14.22 respectively. The location of the feeds is easily seen in these profiles. Murphree efficiencies, computed from Eq. 13.2.1 using the results of a simulation, are shown in Figure 14.23. As has been the case in many of our comparisons, the Murphree efficiencies vary widely from component to component and from tray to tray as well.

A comparison between predicted and measured top and bottom product mole fractions for all five experiments is provided in Figure 14.24 (Taylor et al., 1987). In all five cases, the agreement between predicted and measured mole fractions is excellent. Kumar et al. simulated the behavior of the column using an equilibrium stage model. They fitted (but only to two significant figures) efficiency values for each component on each stage; a total of 39 quantities. The mole fractions predicted by the nonequilibrium stage model are not significantly different from theirs!

14.4.3 Alcohol Wash Columns

Simulations of two commercial scale alcohol wash columns have been described by Taylor et al. (1992). A schematic flowsheet of the wash columns is shown in Figure 14.25. Column 1

TABLE 14.9   Specifications and Stream Table for Extractive Distillation

Operation:
  Simple distillation
  Total condenser
  Total reboiler
  15 Stages
  Feeds to stages 6 and 11

Properties:
  DECHEMA K model
  Original UNIQUAC model
  Antoine vapor pressure
  Excess enthalpy

UNIQUAC Parameters (cal/mol)					
i	j	A_{ij}	A_{ji}	Component	Component
1	2	1236.206	−456.272	Acetone	Methanol
1	3	329.056	154.059	Acetone	Water
2	3	−284.892	194.508	Methanol	Water

Specifications:
  Column pressure
    Condenser pressure 0.1013 (MPa)
    Top pressure 0.1013 (MPa)
  Column design:
  Number of sections   1
  System factor ($-$)   1.000
  Default flooding factor ($-$)   0.7000
    Section   1
    First stage   2
    Last stage   .   14
    Column internals   Bubble cap tray
    Mass transfer coefficient   AIChE
    Column diameter (m)   0.1524
    Total tray area (m^2)   0.01824
    Tray spacing (m)   0.3048
    Liquid flow path length (m)   0.1050
    Active area (m^2)   0.01420
    Weir length (m)   0.1275
    Weir height (m)   0.02250
  Condenser
    Reflux ratio = 1.600 ($-$)
  Reboiler
    Bottom product flow rate = 0.2008 (kmol/h)

Stream	Feed 1	Feed 2	Top	Bottom
Stage	6	11	1	15
Pressure (MPa)	0.1013	0.1013	0.1013	0.1025
Vapor fraction ($-$)	0.0000	0.0000	0.0000	1.000
Temperature (K)	333.4	313.5	329.6	370.1
Total molar flow	0.1776	0.09876	0.07556	0.2008
Mole fractions ($-$)				
Acetone	0.0000	0.6286	0.8029	0.007039
Methanol	0.0000	0.3714	0.1296	0.1339
Water	1.000	0.0000	0.06751	0.8591

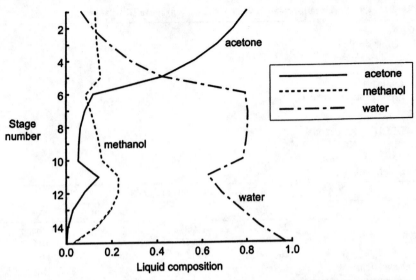

Figure 14.20. Composition profiles predicted by nonequilibrium model. Calculations by Taylor et al. (1987).

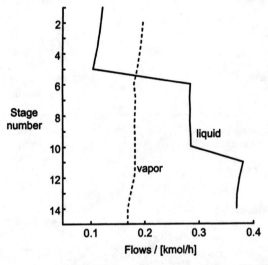

Figure 14.21. Calculated flow profiles in experimental extractive distillation column.

has 75 trays plus a reboiler. The alcohol feed is near the top of the column and there is a vapor side stream withdrawn near the bottom of the column. The column is swaged with a maximum diameter of 3 m. The main feed to column 1 is a four-component mixture containing ethanol (~ 20 mass%) and water (~ 80 mass%) along with small quantities of acetaldehyde and 2-butanol. There is an additional feed of water to the top tray in column 1. Column 2 processes the overhead from column 1 and has 60 trays plus a condenser and reboiler. The column has a diameter of around 1.5 m. There is a water feed near the top of the column and the stream from column 1 enters near the middle of the column.

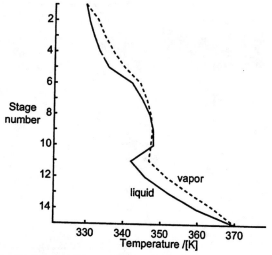

Figure 14.22. Calculated temperature profiles in experimental extractive distillation column.

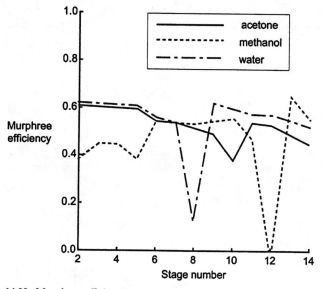

Figure 14.23. Murphree efficiencies for experimental extractive distillation column.

The nonequilibrium model of column 1 consists of 75 trays plus an equilibrium reboiler with feeds to and the side stream taken from the trays where they exist in the actual plant. The nonequilibrium model of column 2 consists of 60 trays plus an equilibrium reboiler and a total condenser. The column model takes into account the different diameters in different parts of the column. Representative average values of the weir height and length, bubbling area, liquid flow path length, and other equipment design parameters were used for each main section of the columns. No attempt was made to incorporate all of the very many design variations in each section into the simulation (even though this was possible). Minor variations in, for example, weir length do not have a major impact on the results of a simulation.

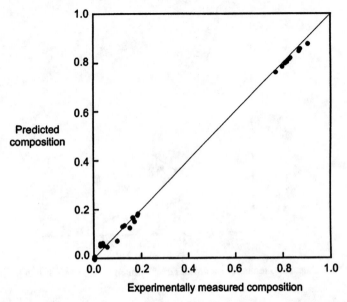

Figure 14.24. Predicted versus experimental top and bottom product compositions in distillation of acetone–methanol–water system. Calculations by Taylor et al. (1987). Data from Kumar et al. (1984).

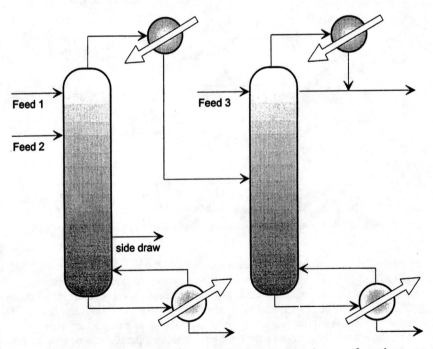

Figure 14.25. Schematic diagram of industrial alcohol wash column configuration.

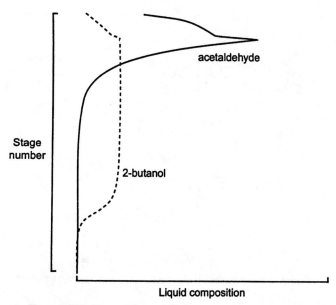

Figure 14.26. Composition profiles in alcohol wash column 1.

The bottom product rates were specified in molar units. For column 1 the side stream flow rate also was specified in molar units. Specification of the molar flow rates makes the simulations converge more easily. Nonstandard specifications can be harder for a nonequilibrium model to converge than for a corresponding equilibrium model because nonstandard specifications are more likely to lead to large variations in vapor and liquid flows from iteration to iteration. Since mass transfer coefficient and pressure drop calculations may

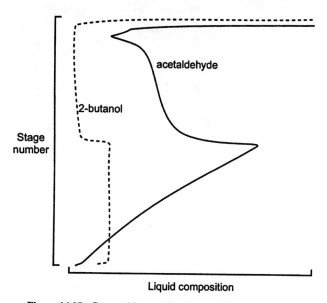

Figure 14.27. Composition profiles in alcohol wash column 2.

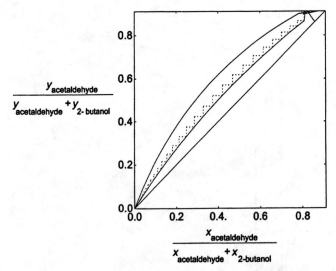

Figure 14.28. McCabe–Thiele diagram for alcohol wash column 1.

make use of empirical correlations for the estimation of tray–packing performance, it is possible that large changes in the flows could lead to the use of design correlations in regions for which they are invalid or to the prediction of flow regimes that are undesirable, unlikely, or simply impossible. For the same reason, it is essential to check at all stages of the calculations that the internal flows are consistent with the satisfactory performance of the column. K values were estimated using the NRTL model.

Predicted composition profiles for the two columns are shown in Figures 14.26 and 14.27. In column 1 the concentration of 2-butanol falls off rapidly below the side stream. Acetaldehyde, however, appears to be essentially absent from the column below about tray

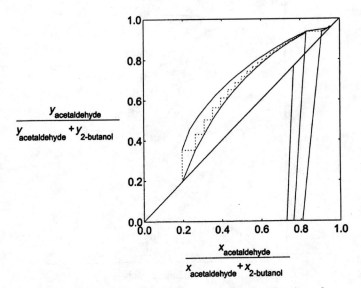

Figure 14.29. McCabe–Thiele diagram for alcohol wash column 2.

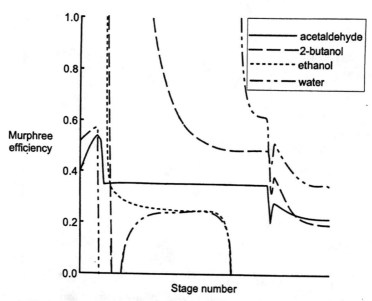

Figure 14.30. Murphree efficiency profiles for alcohol wash column 1.

40. In column 2 the acetaldehyde concentration falls off below the feed, whereas the 2-butanol concentration does not change a great deal.

Modified McCabe–Thiele diagrams for acetaldehyde and 2-butanol are shown in Figure 14.28 and 14.29. Note that the staircase construction that represents the trays in the real column does not touch the equilibrium line. The length of each horizontal step compared to the distance from the rather curved operating line to the equilibrium line is a measure of

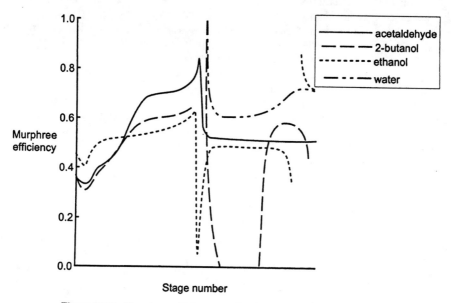

Figure 14.31. Murphree efficiency profiles for alcohol wash column 2.

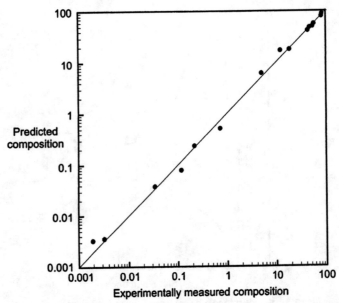

Figure 14.32. Comparison between predicted and observed product stream compositions for two industrial alcohol wash columns.

the component efficiency. However, it should be emphasized that this diagram is constructed from the simulation results directly and does not employ the efficiency concept sometimes used in the construction of pseudoequilibrium lines on McCabe–Thiele diagrams.

The Murphree efficiencies for each component have, in fact, been back calculated from the results of the simulation and are shown in Figures 14.30 and 14.31. It is to be noted that the efficiencies are not the same for each component and vary significantly from tray to tray as well. Consequently, it is difficult to know in advance of a simulation just how many equilibrium stages should be used as well as where the side streams and feeds are to be located (in terms of the number of equilibrium stages).

Figure 14.32 provides an indication of how well the model matches the plant data. The predicted product compositions appear to be quite good in general. The one major discrepancy between the plant data and the simulation results is the concentration of acetaldehyde in the side stream from column 1. This data point is not shown in Figure 14.32 since the model predicted no acetaldehyde in the sidestream (in contrast to the data) and could not be plotted in Figure 14.32 because it employs a log scale. In any event, there is no precise experimental measurement of the side stream composition; the mole fractions of the side stream are, indeed, inferred from measurements of other streams.

14.4.4   A Packed C4 Splitter

Simulations of an industrial scale column with structured packing have been reported by Taylor et al. (1992). They modeled a packed C4 splitter that had an internal diameter of about 2.5 m and five beds of structured packing with a total height of approximately 37 m as shown in Figure 14.33. The feed, which contains predominantly isobutane* and *n*-butane

---

*Isobutane is the common name for 2-methylbutane.

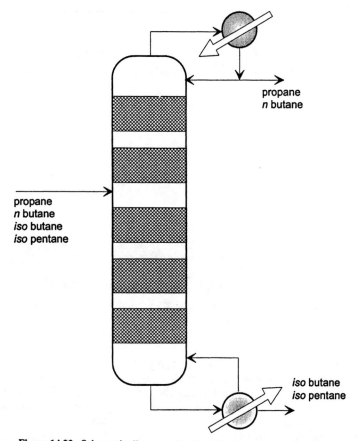

Figure 14.33. Schematic diagram of an industrial packed C4 splitter.

with small amounts of propane and isopentane* is introduced between the second and third packed sections. Feed and product flows, compositions and temperatures were available for a small number of plant experiments and these data were used as the basis for the simulations.

For the purposes of simulating a packed column the packing was divided into a number of sections each of which is modeled as a nonequilibrium stage as discussed above. The C4 splitter was modeled using 150 nonequilibrium sections, an equilibrium reboiler and a total condenser. The bottom product rate in molar units and the reflux ratio were fixed at the values observed in the plant tests. Additional specifications included the component feed flows and the column pressures.

Mass transfer coefficients and interfacial areas were computed using the model of Bravo et al. (1985) described in Section 12.3.3. K values and enthalpies were estimated using the Soave–Redlich–Kwong equations of state (see, e.g., Walas, 1985).

A parity plot comparing predicted overhead and bottoms compositions is provided in Figure 14.34. All four product compositions at both ends of the column are included in this illustration. It can be seen that the product compositions are predicted quite well. Product temperatures were predicted to within one-half a degree Celsius.

A sensitivity study was carried out by varying some of the important parameters in the mass transfer model and the column specifications. It was discovered that changing the

---

*Isopentane is the common name for 2-methylbutane.

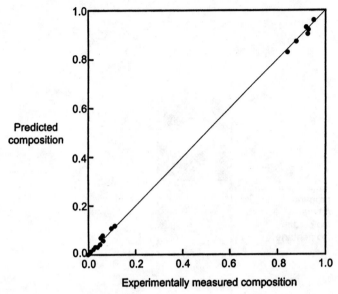

Figure 14.34. Comparison between predicted and measured distillate and bottoms compositions for a large scale packed C4 splitter. Calculations by Taylor et al. (1992).

bottoms flow rate, the reflux ratio, or the degree of feed vaporization by modest amounts had very little effect on the predicted product compositions. The column pressure has a more substantial effect. The parameter with the greatest influence, however, is the interfacial area which, in the model of Bravo et al. (1985), is assumed to be equal to the specific surface area of the packing. There is some evidence suggesting that the assumption of equal interfacial and specific surface areas is not valid for sheet metal packings (see Bravo et al., 1992). A modest change in the ratio of the interfacial area to specific surface area of the packing would allow the product compositions reported for each plant test to be matched exactly.

Equilibrium stage simulations of the C4 splitter were carried out in parallel with the nonequilibrium calculations. Several simulations of each data set were performed in which both the number of equilibrium stages and the location of the feed stage were adjusted until the product compositions could be matched at both ends of the column.

To achieve the separation reported for Test 1 35 equilibrium stages (plus an equilibrium reboiler and a condenser) were required (with the feed to stage 15). The corresponding HETP is, therefore, more than 1 m (assuming constant average HETP). To attain the separation reported for Test 2 required 45 equilibrium stages (plus condenser and reboiler) with the feed to stage 18. This suggests that the HETP for this test is about 80 cm. To achieve the separation reported for Test 3 30 equilibrium stages (plus condenser and reboiler) were needed. The corresponding HETP, therefore, is about 1.2 m. It is interesting to note that the number of equilibrium stages needed (and, hence, the HETP) is not the same for each test. As in the preceding examples, one level of guesswork is removed from the task of simulating an existing column by using a nonequilibrium model.

14.4.5 Other Applications

A number of other papers providing comparisons between plant, pilot, or laboratory scale operations have been published in the literature. A brief review is provided here.

A process that assumed enormous importance during World War II involved the separation of C4 hydrocarbons for the manufacture of synthetic rubber that was used to make tires. Furfural is used as an extractive distillation agent for three separations involved in this process: isobutane* from 1-butene, *n*-butane from 2-butenes, and 1-butene from butadiene. These separations were carried out in very large columns (10–14-ft diameter two-piece columns each of 50 stages) using a mixture of furfural and water as the extractive solvent (Happel et al., 1946, Gerster et al., 1955, Buell and Boatright, 1947). The stage efficiency of these columns was of the order of 25–45%. Taylor et al. (1987) simulated a number of pilot plant tests in which the separation of isobutane from 1-butene is made possible by the addition of furfural (McCartney, 1948). In this process, relatively pure isobutane leaves from the top of the column while most of the 1-butene and virtually all of the furfural leave as bottoms product. The apparatus consisted of a 13-in. diameter column with 10 bubble cap trays, a total condenser, and a once through reboiler with a separator (again modeled as an equilibrium reboiler). The column was operated as the stripping section of an extractive distillation column. This operation was accomplished by operating the column with no reflux and feeding the mixed overhead and bottoms products to the top of the column (stage two). The column was operated at a pressure of about 5.5 bar. The agreement between experimental measurement and simulation prediction was very good indeed.

Gorak and Vogelpohl (1985) provided experimental data for the distillation of methanol–2-propanol–water mixtures in a column filled with Sulzer CY wire mesh packing. The experiments were carried out at total reflux. Gorak and Vogelpohl simulated their own experiments using a nonequilibrium model devised for simulating distillation at total reflux in packed columns. In their simulations, binary mass transfer coefficients were obtained from a correlation due to Zogg (1972). A number of different models were used by Gorak and Vogelpohl to obtain the activity coefficients. The distillate compositions predicted with these models are significantly different from the measured values. Taylor et al. (1987) simulated these experiments using the nonequilibrium stage model. Using the correlation of Bravo et al. (1985) to estimate the binary mass transfer coefficients. The results obtained by Taylor et al. were in reasonable agreement with the data.

Arwickar (1981) reported some results for distillation under total reflux conditions of the system acetone–methyl acetate–methanol. The experiments were carried out in a laboratory scale column of 7.62 cm diameter packed with 0.635 cm Raschig rings. The simulation of total reflux operations using the nonequilibrium model is discussed by Krishnamurthy and Taylor (1985a). In simulations of Arwickar's experiments Taylor et al. used the correlations of Onda et al. (1968) to estimate the mass transfer coefficients in each phase and the effective interfacial area. The average absolute discrepancy between predicted and measured mole fractions was less than 2 mol% for acetone and methyl acetate and less than 4 mol% for methanol.

Gorak et al. (1991) and Wozny et al. (1991) presented a brief description of their use of a nonequilibrium state model to simulate vacuum distillation of fatty alcohols in columns fitted with structured packing. They found that the compositions predicted by the nonequilibrium model were closer to the experimental data than were the results of an equilibrium stage–HETP calculation.

Krishnamurthy and Andrecovich (1989) used a nonequilibrium model to simulate cyrogenic distillation processes.

McNulty and Chatterjee (1992) discuss the use of nonequilibrium models to design packed bed pumparound zones of crude distillation towers.

As noted above, nonequilibrium models can be used for modeling absorption operations. Krishnamurthy and Taylor (1986) present results for the absorption of ammonia in water,

---

*Isobutane is the common name for 2-methylpropane.

the absorption of acid gases in cold methanol and the design of a tray column to reduce the amount of propane in a hydrocarbon gas stream. Krishnamurthy and Taylor (1985b) provide results for two industrial scale packed columns processing hydrocarbon mixtures, one of which is a large absorption column.

Nonequilibrium models have also been developed for simulating liquid–liquid extraction operations. Descriptions of the models and some comparisons of simulation predictions with experimental data have been reported by Lao et al. (1989) and by Zimmerman et al. (1992).

14.5   CONCLUDING REMARKS—NONEQUILIBRIUM OR EQUILIBRIUM MODEL?

It has been demonstrated that nonequilibrium models offer several advantages over conventional (equilibrium stage) approaches for simulating existing columns. There is no need to guess the number of equilibrium stages needed or the location of feed or side streams when simulating an existing column. A priori computations of stage efficiency (for tray columns) and HETP (for packed columns) are entirely avoided. Nonequilibrium models are especially useful for the following classes of problem:

- Packed columns (including structured packings).
- Strongly nonideal systems (where efficiencies are uncertain and vary widely from tray to tray and component to component).
- Systems involving trace components, which include the processing of environmentally sensitive mixtures.
- Columns with profiles that contain maxima or change rapidly over a small section of the column, for example, azeotropic distillation or nonisothermal gas absorption. Nonequilibrium models can locate such features with greater accuracy.
- Columns with a complicated configuration (multiple feeds and sidestreams).
- Any columns with unknown efficiencies (in order to estimate overall efficiencies for use in equilibrium stage simulations).

Since a nonequilibrium model requires the equipment design parameters (column diameter, tray or packing type and design, etc.) to be available, they can be used to diagnose operating and design problems. Nonequilibrium models may also be used to identify equipment design parameters that can be altered to improve column performance.

15 Condensation of Vapor Mixtures

It is much the same for condensers: the presence of air ensures that the temperature of the gas phase does not remain uniform throughout, for it must fall as the steam concentration (and therefore partial pressure also) diminishes. Often, however, designers will ignore such effects, or take account of them only approximately. Similar problems arise in condensers for multicomponent mixtures.

—D. B. Spalding (1983)

Condensation of vapor mixtures is an operation of great significance in the process industries. The term vapor mixture covers a wide range of situations. One limit of this range is one in which all components have boiling points above the maximum coolant temperature; in this case the mixture can be totally condensed. The other limit is a mixture in which at least one component in the initial vapor stream has a boiling point lower than the minimum coolant temperature and is negligibly soluble in the liquid condensate formed by the remaining components and, hence, cannot be condensed at all. An intermediate case of some importance is typified by a mixture of light hydrocarbons, in which the lightest members often cannot be condensed as pure components. In each of the three cases, the vapor mixture may form a partially or totally immiscible condensate. In practice, condensation takes place in a variety of equipment, most notably, however, in shell and tube heat exchangers. Modeling multicomponent condensation is the subject of this chapter. We begin with an analysis of mass and energy transfer in condensation in Section 15.1 before moving on to develop the design equations in Section 15.2. A number of design examples are discussed in Section 15.3 and a comparison of model predictions with experimental data can be found in Section 15.4.

15.1 MASS AND ENERGY TRANSFER IN CONDENSATION

15.1.1 Condensation Flow Patterns

Condensation occurs whenever a vapor, a vapor mixture, or a vapor containing a noncondensable gas is brought into contact with a surface below the dew point or saturation temperature of the vapor. The condensed liquid is most likely to form a continuous film covering the cooled surface. In some cases, however, dropwise condensation is possible if the fluid does not wet the surface. Filmwise condensation is encountered in most industrial applications and is the only mode of condensation we shall consider further.

The cooled surface may be of any orientation although vertical and horizontal arrangements are most common. Condensation operations often are carried out inside shell and tube heat exchangers and the condensing vapors may be fed either to the shell side or to the tube side depending on the nature of the fluids involved, their pressure, and their corrosion and fouling characteristics (Webb and McNaught, 1980).

In vertical devices with condensation on the inside or on the outside of the tubes the condensed liquid-film falls under the influence of gravity. The thickness of the condensed liquid film increases in the direction of flow due to the increased liquid load caused by

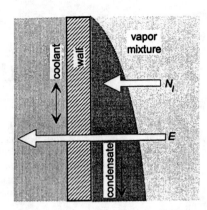

Figure 15.1. Condensation on vertical surfaces.

continuing condensation (Figure 15.1). At the top of the cooled surface, the film flow will be laminar but becomes turbulent if the liquid flow rate becomes high enough. The vapor may flow cocurrent with or countercurrent to the downwards flowing condensed liquid. The velocity and direction of the vapor flow may have a significant influence on the condensed liquid-film flow regime and, hence, on the condensate heat transfer coefficient (see Webb and McNaught, 1980).

On perfectly horizontal tubes a condensed film will increase its thickness towards the bottom of the tube (Fig. 15.2). Some of the condensed liquid will fall onto lower tubes, increasing the liquid load, and decreasing the heat transfer coefficient on those tubes. Even a slight inclination of the tube is sufficient to cause the condensate to drain in the direction of the slope. A horizontal shell and tube condenser is likely to be baffled so as to force the vapor to flow horizontally across the tubes. Other flow arrangements are, however, possible.

Condensation may also take place inside horizontal tubes, the flow regimes depend strongly on the velocity of the vapor. At low vapor flows, the condensate film tends to collect

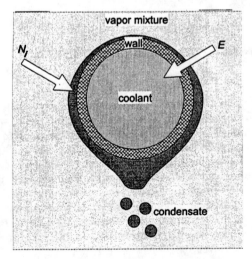

Figure 15.2. Condensation on horizontal tubes.

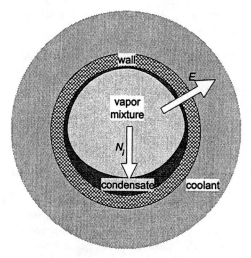

Figure 15.3. Condensation inside horizontal tubes.

at the bottom of the tube, whereas at very high vapor flows rates, a condensate film of more uniform thickness forms in the tube (Fig. 15.3).

15.1.2 Mass and Energy Transfer

Typical composition and temperature profiles in condensation operations are shown in Figure 15.4. The analysis of mass and energy transfer in condensation that follows is an extension of the general analysis of simultaneous mass and energy transfer presented in Section 11.5. The additional complication here is that we must account for energy transfer (but not mass transfer) across the tube wall into the coolant.

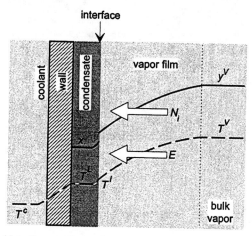

Figure 15.4. Composition and temperature profiles in condensation.

The molar fluxes on each side of the vapor–liquid interface must be equal, so Eqs. 11.5.3–11.5.5 apply here as well

$$N_i^L = J_i^L + x_i N_t^L = N_i = J_i^V + y_i N_t^V = N_i^V \tag{11.5.3}$$

with the diffusion fluxes in each phase are given by

$$(J^L) = c_t^L [k_L^\bullet](x^I - x^L) \tag{11.5.4}$$

$$(J^V) = c_t^V [k_V^\bullet](y^V - y^I) \tag{11.5.5}$$

We consider transfer from the "V" phase to the "L" phase as leading to a positive flux.

In most models of multicomponent condensation the mass transfer coefficients for the vapor phase are obtained using methods derived from film theory. The low flux coefficients commonly are estimated using the empirical procedures described in Section 8.8; the use of the Chilton–Colburn analogy is particularly common (see Webb and McNaught, 1980).

It is possible to circumvent the need to calculate liquid-phase mass transfer coefficients by assuming that either

1. The condensate is completely mixed with regard to composition (but not with regard to temperature) and so

$$x_i^I = x_i^L \tag{15.1.1}$$

2. The condensate is completely unmixed with the composition at the interface given by the relative rates of condensation

$$x_i^I = N_i/N_t \tag{15.1.2}$$

The former corresponds to liquid-phase mass transfer coefficients that are effectively infinite (no liquid-phase resistance), the latter to zero liquid-phase mass transfer coefficients (infinite liquid-phase resistance). Clearly, the truth lies somewhere between these two limiting cases. There is some evidence to show that if there is a noncondensing component in the vapor phase, then these two limiting cases give essentially identical results (Sections 15.3 and 15.4).

We also have continuity of the energy fluxes across the vapor–liquid interface, liquid–wall and wall–coolant interfaces. If we adopt a one-dimensional (film) model of the transport processes, then all of these energy fluxes are equal. Changes in the interfacial area due to curvature have been ignored; such differences may be accounted for with the corrections shown in Section 8.2.4.

$$E^V = E^I = E^L = E^W = E^C = E \tag{15.1.3}$$

with the energy flux defined by the one-dimensional form of Eq. 11.2.4

$$E \equiv q + \sum_{i=1}^{n} N_i \bar{H}_i \tag{15.1.4}$$

There is no convective (i.e., mass transfer) contribution to the energy flux for the coolant E^C

$$E^C = q^C \tag{15.1.5}$$

We may set the energy flux in the bulk vapor E^V equal to that in the liquid at the interface E^I to give

$$q^V + \sum_{i=1}^{n} N_i^V \overline{H}_i^V(T^V) = q^I + \sum_{i=1}^{n} N_i^L \overline{H}_i^L(T^I) \qquad (15.1.6)$$

or

$$q^I = q^V + \sum_{i=1}^{n} N_i \{ \overline{H}_i^V(T^V) - \overline{H}_i^L(T^I) \} \qquad (15.1.7)$$

The heat flux in the vapor phase is given by Eq. 11.5.8

$$q^V = h_V^\bullet (T^V - T^I) \qquad (11.5.8)$$

The finite flux heat transfer coefficient in the vapor phase $h_V^\bullet$ is related to the zero flux heat transfer coefficient h^V by

$$h_V^\bullet = h^V \frac{\Phi_H^V}{\exp \Phi_H^V - 1} \qquad (11.5.10)$$

where we have used the film model correction factor defined by Eq. 11.4.12 with Φ_H^V defined by Eq. 11.4.16

$$\Phi_H^V = \frac{\sum_{i=1}^{n} N_i C_{pi}^V}{h^V} \qquad (11.4.16a)$$

In most applications, the low flux heat transfer coefficient for the vapor phase h^V is estimated from the Chilton–Colburn analogy (Eq. 11.4.35).

If we neglect subcooling of the liquid condensate, the conductive heat flux in the condensed liquid film q^I, will also equal the heat flux through the wall and into the coolant. Thus, we may write

$$q^I = q^W = h_0(T^I - T^C) \qquad (15.1.8)$$

where h_0 is the heat transfer coefficient incorporating the resistances of the condensed liquid film, wall, coolant, and any fouling resistance. The heat transfer coefficient for the condensed liquid may be estimated from Nusselt's equation or modifications thereof as discussed by, for example, Webb and McNaught (1980). The heat transfer coefficient for the coolant may be estimated from standard correlations, such as the Dittus–Boelter correlation for flow in pipes.

To proceed further we break up the enthalpy difference in Eq. 15.1.7 into the sum of two contributions

$$\{ \overline{H}_i^V(T^V) - \overline{H}_i^L(T^I) \} = \{ \overline{H}_i^V(T^V) - \overline{H}_i^V(T^I) \} + \{ \overline{H}_i^V(T^I) - \overline{H}_i^L(T^I) \}$$

$$= \{ \overline{H}_i^V(T^V) - \overline{H}_i^V(T^I) \} + \Delta H_{\text{vap},i} \qquad (15.1.9)$$

where $\Delta H_{\text{vap},i}$ is the molar heat of vaporization of species i defined by

$$\Delta H_{\text{vap},i} \equiv \{ \overline{H}_i^V(T^I) - \overline{H}_i^L(T^I) \}.$$

We further assume that we are dealing with ideal vapor mixtures and that the ideal gas heat capacities can be considered constant over the temperature range of interest and so

$$\{\bar{H}_i^V(T^V) - \bar{H}_i^V(T^I)\} = C_{pi}^V(T^V - T^I) \tag{15.1.10}$$

Equation 15.1.7 may now be expressed as

$$
\begin{aligned}
q^W &= h_0(T^I - T^C) \\
&= h^V \frac{\Phi_H^V}{\exp \Phi_H^V - 1}(T^V - T^I) + \sum_{i=1}^{n} N_i C_{pi}^V(T^V - T^I) + \sum_{i=1}^{n} N_i \Delta H_{vap,i} \\
&= h^V \frac{\Phi_H^V \exp \Phi_H^V}{\exp \Phi_H^V - 1}(T^V - T^I) + \sum_{i=1}^{n} N_i \Delta H_{vap,i} \tag{15.1.11}
\end{aligned}
$$

The model is completed by equations that represent phase equilibrium at the vapor–liquid interface,

$$y_i^I = K_i x_i^I \qquad i = 1, 2, \ldots, n \tag{11.5.11}$$

and the two mole fraction summation equations (Eqs. 11.5.26 and 11.5.27)

$$S^V = \sum_{i=1}^{n} y_i^I - 1 = 0 \tag{11.5.26}$$

$$S^L = \sum_{i=1}^{n} x_i^I - 1 = 0 \tag{11.5.27}$$

A number of algorithms have been published for computing rates of condensation from the mathematical model developed above. Krishna et al. (1976) solved the mass and energy transfer rate equations (by repeated substitution of the N_i) and the vapor–liquid equilibrium equations (standard bubble point calculation) within an outer loop that, in effect, solved the liquid mixing equations using Newton's method. The outer loop variables were the $n - 1$ interfacial compositions x_i^I. Other algorithms, those of Price and Bell (1974) and of Webb and co-workers (Webb and McNaught, 1980; Webb and Taylor, 1982; Webb 1982) involve up to three levels of iteration loop. For the general multicomponent case we recommend solving all of the model equations simultaneously using Newton's method. This approach, as described in Section 15.1.3, represents an extension of Algorithm 11.5.1 for the general interphase mass transfer problem. Special algorithms for solving problems when the vapor phase contains two species are discussed in Sections 15.1.4 and 15.1.5.

15.1.3 Computation of the Fluxes in Multicomponent Systems

The number of independent equations that model the general multicomponent condensation problem is $3n + 1$. These equations are

- $n - 1$ mass transfer rate equations for the vapor phase.
- $n - 1$ equations representing the liquid phase.
- n equilibrium relations for the vapor–liquid interface.
- Two mole fraction summation equations (Eqs. 11.5.26 and 11.5.27).
- One interfacial energy balance (Eq. 15.1.11).

As noted above, we recommend solving all $3n + 1$ equations simultaneously using Newton's method. The $3n + 1$ variables that may be determined by solving this set of equations numerically are as follows:

- $2n$ mole fractions on each side of the interface y_i^I and x_i^I.
- n molar fluxes N_i.
- The vapor–liquid interface temperature T^I.

The problem formulation for condensation is very similar to the interphase mass transfer problem discussed at some length in Section 11.5. The set of independent equations, ordered into a vector of functions (F), is as follows:

$$(F)^T = \left(R_1^V, R_2^V, \ldots, R_{n-1}^V, E^I, X_1^L, X_2^L, \ldots, X_{n-1}^L, \right.$$
$$\left. Q_1^I, Q_2^I, \ldots, Q_{n-1}^I, Q_n^I, S^V, S^L \right)$$

where the R^V represent the mass transfer rate equations for the vapor phase given by

$$(R^V) \equiv c_t^V [k_V^{\bullet}] (y^V - y^I) + N_t (y^V) - (N) = (0) \qquad (11.5.28)$$

The interfacial energy balance (Eq. 11.5.7), is replaced by Eq. 15.1.11 rearranged as

$$E^I \equiv h_0(T^I - T^C) - h^V (\Xi_H + \Phi_H^V)(T^V - T^I) - \sum_{i=1}^{n} N_i \Delta H_{\text{vap},i} = 0 \quad (15.1.12)$$

The $n - 1$ equations for the liquid phase are either the $n - 1$ mass transfer rate equations (Eqs. 11.5.29), or $n - 1$ mixing equations for the liquid phase

$$X_i^L \equiv x_i^I N_t - N_i = 0 \qquad \text{no mixing} \qquad (15.1.13)$$

or

$$X_i^L \equiv x_i^I - x_i^L = 0 \qquad \text{complete mixing} \qquad (15.1.14)$$

The Q_i^I represent the equilibrium equations

$$Q_i^I \equiv K_i x_i^I - y_i^I = 0 \qquad i = 1, 2, \ldots, n \qquad (11.5.30)$$

The vector of independent variables remains as defined in Section 11.5.

$$(\chi)^T = \left(N_1, N_2, \ldots, N_{n-1}, N_n, x_1^I, x_2^I, \ldots, x_n^I, \right.$$
$$\left. y_1^I, y_2^I, \ldots, y_n^I, T^I \right)$$

Algorithm 11.1 may be adapted as Algorithm 15.1 to determine the set of temperatures, mole fractions, and fluxes that makes the discrepancy functions close to zero. An illustration of this algorithm is provided by Example 15.1.1. One of the advantages of the simultaneous solution procedure is that, as shown in Example 15.1.2, only minor modifications of it are needed if the vapor contains noncondensing gases.

Algorithm 15.1   Procedure for Determination of Mass and Energy Fluxes in Multicomponent Condensation

Given:	Bulk vapor conditions, coolant temperature.
	Mass and heat transfer coefficient models.
	Thermodynamic and physical property models.
Step 1:	Generate initial estimates of all unknown variables.
	Set vapor–liquid interface temperature to the average of vapor and coolant temperatures.
	Set interface mole fractions equal to bulk vapor composition.
	Set molar fluxes to small, nonzero, values.
Step 2:	Compute
	Transport properties.
	Thermodynamic properties (K values, enthalpies).
	Mass transfer coefficients.
	Heat transfer coefficients.
	Vector of discrepancy functions (F).
Step 3:	Compute elements of Jacobian matrix [J].
Step 4:	Check for Convergence, if not obtained continue with Step 5.
Step 5:	Compute new set of unknown variables by solving Eq. C.2.5. Return to Step 2.

Example 15.1.1   Condensation of a Methanol–Water Mixture

A mixture of methanol (1) and water (2) vapors at 90°C, containing 40 mol% methanol, is flowing downwards through a water cooled vertical tube as illustrated in Figure 15.5. Find the composition of the condensate formed at the top if the cooling water temperature at that position in the condenser is 39.1°C.

ANALYSIS   For two component mixtures the mathematical model developed above consists of seven equations.

$$F_1 \equiv c_t^V k^V \Xi^V \left(y_1^V - y_1^I \right) + y_1^V (N_1 + N_2) - N_1 = 0$$

$$F_2 \equiv h_0 (T^I - T^C) - h^V \Xi_H (T^V - T^I) - h^V \Phi_H (T^V - T^I) - N_1 \Delta H_{vap1} - N_2 \Delta H_{vap2} = 0$$

$$F_3 \equiv x_1^I (N_1 + N_2) - N_1 = 0$$

$$F_4 \equiv K_1 x_1^I - y_1^I = 0$$

$$F_5 \equiv K_2 x_2^I - y_2^I = 0$$

$$F_6 \equiv y_1^I + y_2^I - 1 = 0$$

$$F_7 \equiv x_1^I + x_2^I - 1 = 0$$

The function F_1 is the mass transfer rate equation for the vapor phase F_2 is the interfacial energy balance F_3 is valid if the liquid phase may be assumed unmixed with

$$k^L = 0 \qquad J_1 = 0 \qquad N_1 = x_1^I N_t$$

F_4–F_7 represent phase equilibrium at the interface.

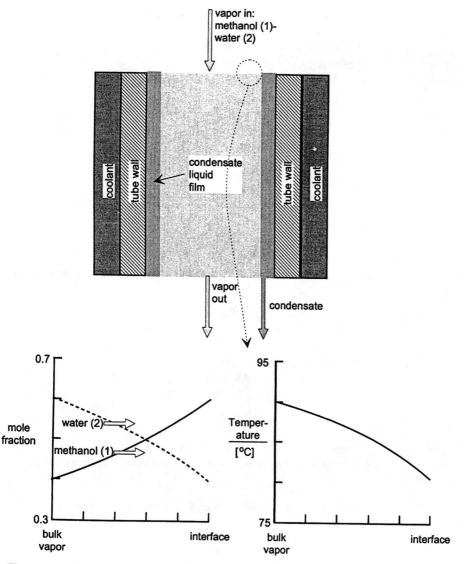

Figure 15.5. Schematic diagram of vertical condenser tube. Inset shows composition profiles in the vapor film at the top of the tube.

The seven variables to be computed are

$$N_1, N_2, x_1^I, x_2^I, y_1^I, y_2^I, T^I$$

All other quantities appearing in the above equations must be specified or calculated. Specified quantities include the coolant temperature, the bulk vapor composition, and the bulk vapor temperature. All physical and thermodynamic properties including the mass and heat transfer coefficients must be known or methods for estimating them in terms of specified or calculated variables must be available.

DATA

Molar heat capacity of methanol: $C_{p1} = 45$ J/mol K.

Molar heat capacity of water: $C_{p2} = 34$ J/mol K.

Latent heat of vaporization of methanol: $\Delta H_{vap1} = 36$ kJ/mol.

Latent heat of vaporization of water: $\Delta H_{vap2} = 43$ kJ/mol.

Vapor-phase heat transfer coefficient: $h^V = 60$ W/m²K.

Vapor-phase mass transfer coefficient: $k^V = 0.08$ m/s.

Heat transfer coefficient for

$$(\text{condensate} + \text{wall} + \text{coolant}): h_0 = 2000 \text{ W/m}^2\text{K}$$

The vapor–liquid equilibrium behavior of the methanol–water system may be represented by

$$K_1 = \gamma_1 P_1^s/P \qquad K_2 = \gamma_2 P_2^s/P$$

The pure component vapor pressures may be estimated from the Antoine equation as follows:

$$\ln P_1^s = 23.402 - 3593.4/(T - 34.29)$$

$$\ln P_2^s = 23.196 - 3816.4/(T - 46.13)$$

where P_1^s and P_2^s are in pascals and T is in kelvin.

The activity coefficients may be estimated from the Margules equation (Table D.2) with parameters from Gmehling and Onken (1977ff, Vol. 1/1a, p. 53):

$$A_{12} = 0.8517 \qquad A_{21} = 0.4648$$

SOLUTION This problem is very similar to Example 11.5.2 where a mixture of methanol and water was distilled. Essentially the same procedure is used here to determine the rates of condensation.

Step 1: Estimation of all of the independent variables. For the first iteration we use the following values:

$$N_1 = 0.4 \text{ mol/m}^2\text{s} \qquad N_2 = 0.6 \text{ mol/m}^2\text{s}$$

$$y_1^I = 0.4 \qquad y_2^I = 0.6$$

$$x_1^I = 0.4 \qquad x_2^I = 0.6$$

$$T^I = 65°\text{C}$$

The vapor-phase mole fractions at the interface are set equal to the bulk-phase mole fractions; the liquid-phase mole fractions are given the same values as the interface vapor mole fractions. The interface temperature is estimated midway between the vapor phase and coolant temperatures. Finally, the molar fluxes in moles per second (mol/s) were set equal to the liquid mole fractions. This set of initial values is not a particularly good set of initial values. In other words, it will take more iterations to converge to a final set of variables from this set of initial values than it might from some other, better set of values. The advantage of this procedure is that the rules outlined here are simple and easy to apply to any similar problem.

Step 2: Evaluation of the discrepancy functions. The evaluation of the functions F_1–F_7 follows much the same path that we took in Example 11.5.2.

The molar density of the gas phase is computed from the ideal gas law using the average of the bulk vapor and interface temperatures

$$c_t^V = 34.75 \text{ mol/m}^3$$

The mass transfer rate factor for the vapor phase may now be computed

$$\Phi^V = (N_1 + N_2)/c_t^V k^V$$
$$= (0.4 + 0.6)/(34.75 \times 0.08)$$
$$= 0.3597$$

The vapor-phase high flux correction factor follows as:

$$\Xi^V = \Phi^V/(\exp \Phi^V - 1)$$
$$= 0.3597/(\exp(0.3597) - 1)$$
$$= 0.8309$$

The function F_1 may now be evaluated as

$$F_1 \equiv c_t^V k^V \Xi^V (y_1^V - y_1^I) + y_1^V(N_1 + N_2) - N_1$$
$$= 34.75 \times 0.08 \times 0.8309 \times (0.4 - 0.4) + 0.4 \times (0.4 + 0.6) - 0.4$$
$$= 0$$

Note that our initial estimates were chosen in a way that makes F_1 equal to zero regardless of the values of c_t^V, Φ^V, and Ξ^V.

We now turn our attention to evaluation of the energy balance F_2. The heat transfer rate factor is computed first

$$\Phi_H^V = (N_1 C_{p1} + N_2 C_{p2})/h^V$$
$$= (0.4 \times 45 + 0.6 \times 34)/60$$
$$= 0.64$$

The heat transfer correction factor may now be calculated

$$\Xi_H = \Phi_H^V/(\exp \Phi_H^V - 1)$$
$$= 0.64/(\exp(0.64) - 1)$$
$$= 0.7139$$

The energy balance function F_2 may now be evaluated as

$$F_2 \equiv h_0(T^I - T^C) - h^V(\Xi_H + \Phi_H^V)(T^V - T^I)$$
$$\quad - N_1 \Delta H_{vap1} - N_2 \Delta H_{vap2}$$
$$= 2000 \times (65 - 39.1) - 60 \times (0.7139 + 0.64) \times (90 - 65)$$
$$\quad - 0.4 \times 36,000 - 0.6 \times 43,000$$
$$= 9569$$

The liquid-phase mixing equation is evaluated next

$$F_3 \equiv x_1^I(N_1 + N_2) - N_1$$
$$= 0.4 \times (0.4 + 0.6) - 0.4$$
$$= 0.0$$

As was the case with F_1, our initial estimates of the fluxes and liquid composition were chosen in a way that makes $F_3 = 0$.

The equilibrium relations F_4 and F_5 are evaluated in exactly the same way as in Example 11.5.2. Expressions for the activity coefficients are given in Table D.2. At the current estimates of the interface composition and temperature the activity coefficients have the values

$$\gamma_1 = 1.2155 \qquad \gamma_2 = 1.1603$$

The vapor pressures are

$$P_1^s = 106{,}527 \text{ Pa} \qquad P_2^s = 25{,}010 \text{ Pa}$$

and the K values are

$$K_1 = \gamma_1 P_1^s / P$$
$$= 1.2155 \times 106{,}527/101{,}325$$
$$= 1.278$$
$$K_2 = \gamma_2 P_2^s / P$$
$$= 0.2865$$

We may now evaluate the departures from equilibrium as

$$F_4 \equiv K_1 x_1^I - y_1^I$$
$$= 1.278 \times 0.4 - 0.4$$
$$= 0.0954$$
$$F_5 \equiv K_2 x_2^I - y_2^I = 0$$
$$= 0.2865 \times 0.6 - 0.6$$
$$= -0.4261$$

Finally, we evaluate the mole fraction summation functions.

$$F_6 \equiv y_1^I + y_2^I - 1$$
$$= 0.4 + 0.6 - 1$$
$$= 0$$
$$F_7 \equiv x_1^I + x_2^I - 1$$
$$= 0.4 + 0.6 - 1$$
$$= 0$$

Step 3: Evaluation of the Jacobian matrix [J]. The elements of [J] are obtained by differentiating the functions above with respect to each of the independent variables. The derivatives of F_1 and of F_4–F_7 are given in Example 11.5.2 and are not repeated here. The partial derivatives of F_2 and F_3 are given below

$$\partial F_2/\partial N_1 = \Delta H_{vap1} + C_{p1}(T^V - T^I)$$

$$\partial F_2/\partial N_2 = \Delta H_{vap2} + C_{p2}(T^V - T^I)$$

$$\partial F_2/\partial T^I = h_0 + h^V \Xi_H + h^V \Phi_H$$

$$\partial F_3/\partial N_1 = x_1^I - 1$$

$$\partial F_3/\partial N_2 = x_1^I$$

$$\partial F_3/\partial x_1^I = N_1 + N_2$$

We have assumed the high flux correction to the heat transfer coefficient can be considered a constant for the purpose of deriving the expressions given above.

We do not have the space to provide numerical values for all the elements of the Jacobian matrix here.

Step 5: Computation of a new set of independent variables.

Following the calculation of [J] we solve the linear system (Eq. C.2.5) and obtain the following new estimates of the independent variables.

$$N_1 = 0.4613 \text{ mol/m}^2\text{s} \qquad N_2 = 1.7565 \text{ mol/m}^2\text{s}$$

$$y_1^I = 0.5843 \qquad y_2^I = 0.4157$$

$$x_1^I = -0.0258 \qquad x_2^I = 1.0258$$

$$T^I = 359.01 \text{ K}$$

Although the new estimates of the liquid-phase mole fractions satisfy the mole fraction summation equation, the values given above clearly are physically impossible. Since negative mole fractions may cause problems computing physical properties it is advisable to reset them so that they are positive. We proceed with the values

$$x_1^I = 0.001 \qquad x_2^I = 0.999$$

A second computation of the functions F_1–F_7 yields

$$F_1 = 0.1049 \qquad F_2 = 937.3 \qquad F_3 = -0.5185$$

$$F_4 = -0.7294 \qquad F_5 = 0.1901$$

$$F_6 = 0 \qquad F_7 = 0$$

The energy balance is much closer to being satisfied but other functions show greater discrepancies than before. This is, in fact, no more or less than we should have expected. Since the equations are not dimensionless, we will find that the energy balance is always the last equation to be converged. The next set of values of the independent variables is found by solving Eq. C.2.5 again

$$N_1 = 0.2242 \text{ mol/m}^2\text{s} \qquad N_2 = 1.3578 \text{ mol/m}^2\text{s}$$

$$y_1^I = 0.6347 \qquad y_2^I = 0.3653$$

$$x_1^I = 0.0937 \qquad x_2^I = 0.9063$$

$$T^I = 346.34 \text{ K}$$

After about eight iterations we converge to the following values:

$$N_1 = 0.4134 \text{ mol}/m^2s \qquad N_2 = 1.5610 \text{ mol}/m^2s$$

$$y_1^I = 0.6033 \qquad y_2^I = 0.3467$$

$$x_1^I = 0.2094 \qquad x_2^I = 0.7906$$

$$T^I = 353.74 \text{ K}$$

The final values of the mass transfer rate factor and high flux correction factor are

$$\Phi^V = 0.726 \qquad \Xi^V = 0.6806$$

The corresponding heat transfer parameters are

$$\Phi_H^V = 1.195 \qquad \Xi_H = 0.5189$$

It is obvious that the flux correction factors cannot be ignored in this particular problem. This is likely to be the case in many condensation problems.

Composition and temperature profiles in the vapor phase are shown in Figure 15.5.   ∎

Example 15.1.2   Condensation of a Binary Vapor in the Presence of an Inert Gas

During the manufacture of methyl ethyl ketone (MEK) from 2-butanol it is necessary to condense a stream of MEK(1) and 2-butanol(2) in the presence of hydrogen (3). In this example, adapted from Austin and Jeffreys (1979), the gas–vapor mixture is fed to the shell side of a shell and tube heat exchanger as shown in Figure 15.6. The heat exchanger has the

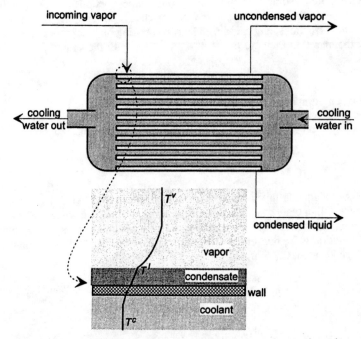

Figure 15.6. Schematic diagram of shell and tube condenser used for condensation of MEK and 2-butanol in the presence of hydrogen.

following dimensions:

Number of tubes	104
Length of tubes	2.55 m
Outside diameter of tubes	0.0191 m
Inside diameter of tubes	0.0158 m
Shell side cross-sectional area	0.0292 m^2
Shell side equivalent diameter (d_{eq})	0.0239 m
Heat transfer surface area	15.913 m^2

The gas–vapor mixture enters the condenser at the following conditions:

Vapor flow rate $M_V = 0.43293$ kg/s.
Vapor inlet temperature $T^V = 398.2$ K.
Inlet pressure = 100.0 kPa.
Inlet vapor composition (mole fraction)

$$y_1^V = 0.4737 \qquad y_2^V = 0.0579$$

The coolant temperature at the vapor inlet end of the condenser is 299 K. Estimate the rates of condensation at the inlet to the condenser.

ANALYSIS   The molar fluxes in the vapor phase will be calculated from

$$N_1 = J_1 + y_1^V(N_1 + N_2)$$

$$N_2 = J_2 + y_2^V(N_1 + N_2)$$

The molar diffusion fluxes J_1 and J_2 are given by

$$J_1 = c_t^V k_{11}^\bullet(y_1^V - y_1^I) + c_t^V k_{12}^\bullet(y_2^V - y_2^I)$$

$$J_2 = c_t^V k_{21}^\bullet(y_1^V - y_1^I) + c_t^V k_{22}^\bullet(y_2^V - y_2^I)$$

where we have adopted the convention that transfers from the vapor to the liquid phase are positive. Hydrogen does not condense so $N_3 = 0$ and there is no contribution from N_3 in the convective contributions to N_1 and N_2. For this example we shall use the Krishna–Standart (1976) method of estimating the multicomponent mass transfer coefficients. This method is described in detail in Section 8.3.3.

The energy balance relation (Eq. 15.1.11), simplifies to

$$q^W = h_0(T^I - T^C)$$

$$= h^V(\Xi_H + \Phi_H)(T^V - T^I) + N_1 \Delta H_{vap1} + N_2 \Delta H_{vap2}$$

The model is completed by assuming equilibrium at the phase interface. This condition is represented by Eqs. (11.5.11, 11.5.25, and 11.5.26).

We shall, as before, use Newton's method to solve all the independent equations simultaneously. The independent equations, written in the form $F(\chi) = 0$, are listed below

$$F_1 \equiv J_1 + y_1^V(N_1 + N_2) - N_1 = 0$$

$$F_2 \equiv J_2 + y_2^V(N_1 + N_2) - N_2 = 0$$

$$F_3 \equiv h_0(T^I - T^C) - h^V \Xi_H(T^V - T^I) - h^V \Phi_H^V(T^V - T^I) - N_1 \Delta H_{vap1} - N_2 \Delta H_{vap2} = 0$$

$$F_4 \equiv x_1^I(N_1 + N_2) - N_1$$

$$F_5 \equiv K_1 x_1^I - y_1^I = 0$$

$$F_6 \equiv K_2 x_2^I - y_2^I = 0$$

$$F_7 \equiv y_1^I + y_2^I + y_3^I - 1 = 0$$

$$F_8 \equiv x_1^I + x_2^I - 1 = 0$$

Note that only two equilibrium equations F_5 and F_6 are used. Since hydrogen is present only in the vapor phase, its mole fraction in the liquid phase is zero and, hence, its K value is infinite. Therefore, Eq. 11.5.11 for hydrogen cannot be used in the calculations.

The eight independent variables computed by solving these equations are

$$N_1, N_2, x_1^I, x_2^I, y_1^I, y_2^I, y_3^I, \text{ and } T^I$$

Step 4 of Algorithm 15.1.1 calls for the evaluation of the Jacobian matrix [J]. The elements of this matrix are obtained by differentiating the above equations with respect to the independent variables. Many of these partial derivatives are zero (or can be approximated as zero). The nonzero derivatives of F_1 and F_2 are as follows:

$$\partial F_1 / \partial N_1 = y_1^V - 1$$

$$\partial F_1 / \partial N_2 = y_1^V$$

$$\partial F_1 / \partial y_1^I = -c_t^V k_{11}^\bullet$$

$$\partial F_1 / \partial y_2^I = -c_t^V k_{12}^\bullet$$

$$\partial F_2 / \partial N_1 = y_2^V$$

$$\partial F_2 / \partial N_2 = y_2^V - 1$$

$$\partial F_2 / \partial y_1^I = c_t^V k_{21}^\bullet$$

$$\partial F_2 / \partial y_2^I = c_t^V k_{22}^\bullet$$

For simplicity, we have ignored the dependence of the mass transfer coefficients themselves on the mixture composition and on the fluxes.

The partial derivatives of the energy balance are given below

$$\partial F_3 / \partial N_1 = -\Delta H_{vap1} - C_{p1}(T^V - T^I)$$

$$\partial F_3 / \partial N_2 = -\Delta H_{vap2} - C_{p2}(T^V - T^I)$$

$$\partial F_3 / \partial T^I = h_0 + h^V \Xi_H + h^V \Phi_H^V$$

where we have assumed the heat transfer coefficients and the flux correction factor to be constant for the purpose of deriving expressions for the partial derivatives.

The partial derivatives of the liquid mixing equation are

$$\partial F_3/\partial N_1 = x_1^I - 1$$

$$\partial F_3/\partial N_2 = x_1^I$$

$$\partial F_3/\partial x_1^I = N_1 + N_2$$

The vapor phase is assumed ideal and so, the K values are independent of the composition of the vapor phase. The partial derivatives of the equilibrium equations with respect to the interface temperature and compositions are the same as those in Example 11.5.1. The partial derivatives of the mole fraction summation equations are either unity or zero (cf. Example 11.5.1).

DATA Maxwell–Stefan diffusion coefficients in the vapor phase.

$$Ð_{12} = 6.39 \times 10^{-6} \, \text{m}^2/\text{s}$$

$$Ð_{13} = 54.5 \times 10^{-6} \, \text{m}^2/\text{s}$$

$$Ð_{23} = 53.4 \times 10^{-6} \, \text{m}^2/\text{s}$$

Molar masses

$$M_1 = 0.072107 \quad M_2 = 0.07412 \quad M_3 = 0.002016 \, \text{kg/mol}$$

Vapor viscosity and thermal conductivity

$$\mu^V = 0.959 \times 10^{-5} \, \text{Pa s}$$

$$\lambda^V = 0.0203 \, \text{W/(m K)}$$

The vapor-phase molar density may be calculated from the ideal gas law.

The heat transfer coefficient for the vapor phase may be estimated from the following correlation:

$$h^V = 0.0035 \, \text{Re}_V \, \text{Pr}_V^{(1/3)} \left(\lambda^V/d_{eq} \right)$$

where the Reynolds number is defined by

$$\text{Re} = Gd_{eq}/\mu^V$$

where G is the mass velocity of the gas–vapor mixture and d_{eq} is the shell side equivalent diameter. The heat transfer coefficient that accounts for the resistances to heat transfer in the condensate, wall, and coolant is assumed constant with the value

$$h_0 = 1116 \, \text{W/m}^2\text{K}$$

Mass transfer coefficients for the gas–vapor phase may be estimated using the Chilton–Colburn analogy Eqs. 11.4.35 and 8.8.7

$$\text{St} \, \text{Sc}_V^{2/3} = \text{St}_H \, \text{Pr}_V^{2/3}$$

which can be arranged to give

$$\kappa = h/\left(c_t^V C_p^V \right)(\text{Pr}_V/\text{Sc}_V)^{2/3}$$

The K values of MEK and 2-butanol may be calculated from

$$K_1 = \gamma_1 P_1^s/P \qquad K_2 = \gamma_2 P_2^s/P$$

The pure component vapor pressures may be estimated from the Antoine equation as follows:

$$\log P_1^s = 7.06356 - 1261.340/(T - 51.181)$$

$$\log P_2^s = 7.47429 - 1314.188/(T - 86.65)$$

where P_1^s and P_2^s are in millimeters mercury and T is in kelvin.

The liquid phase is a binary mixture because hydrogen does not condense. We may, therefore, use the Margules equation (Table D.2) to estimate the activity coefficients for the liquid phase. The interaction parameters are

$$A_{12} = 0.3084 \qquad A_{21} = 0.3315$$

The pure component vapor-phase heat capacities have been estimated at the bulk vapor temperature to be

$$C_{p1}^V = 120 \qquad C_{p2}^V = 130.5 \qquad C_{p3}^V = 29.53 \text{ J/mol K}$$

The latent heats of vaporization at the normal boiling points have the values

$$\Delta H_{\text{vap1}} = 33.3 \text{ kJ/mol} \qquad \Delta H_{\text{vap2}} = 33.3 \text{ kJ/mol}$$

The vapor viscosity, thermal conductivity, and pure component capacities have been taken from the detailed calculations of Austin and Jeffreys (1979). For the purpose of this particular example these properties are assumed to be independent of temperature and composition. This may not be a particularly good assumption in this case since the temperature and composition changes are relatively large and the properties of the pure components also differ quite widely.

SOLUTION   We will illustrate the evaluation of the functions F_1–F_8 using the following "estimates" of the independent variables.

$$N_1 = 0.6594 \text{ mol/m}^2\text{s} \qquad N_2 = 0.1091 \text{ mol/m}^2\text{s}$$

$$y_1^I = 0.4227 \qquad y_2^I = 0.0308 \qquad y_3^I = 0.5465$$

$$x_1^I = 0.8581 \qquad x_2^I = 0.1419$$

$$T^I = 331.54 \text{ K}$$

The first step is the computation of a number of physical properties and dimensionless groups. These will be needed in the estimation of the heat transfer coefficient that must precede the calculation of any mass transfer coefficients.

The molar density of the vapor may be estimated using the ideal gas law at the average vapor temperature (364.87 K).

$$c_t^V = P/RT_{\text{av}}$$

$$= 32.964 \text{ mol/m}^3$$

The mass density is

$$\rho_t^V = c_t^V \overline{M}$$

$$= 1.2985 \text{ kg/m}^3$$

where $\overline{M} = 0.0394$ kg/mol was evaluated at the bulk gas composition.

The average heat capacity of the gas mixture is calculated using the bulk gas composition as

$$\overline{C}_p = y_1 C_{p1} + y_2 C_{p2} + y_3 C_{p3}$$

$$= 78.23 \text{ J/mol K}$$

The Prandtl number of the vapor mixture may now be evaluated as

$$\text{Pr}_V = \left(\overline{C}_p / \overline{M}\right) \mu / \lambda$$

$$= (78.23/0.0394) \times 0.959 \times 10^{-5} / 0.0203$$

$$= 0.9382$$

The mass velocity of the vapor is computed as follows:

$$G = M_V / A_s$$

$$= 0.43293 / 0.0292$$

$$= 14.826 \text{ kg/m}^2\text{s}$$

where A_s is the cross-sectional area for flow on the shell side. The Reynolds number follows as;

$$\text{Re} = G d_{eq} / \mu^V$$

$$= 14.826 \times 0.0239 / 0.959 \times 10^{-5}$$

$$= 36,950$$

The heat transfer coefficient in the vapor phase may now be calculated from

$$h^V = 0.0035 \, \text{Re}_V \, \text{Pr}_V^{1/3} \left(\lambda^V / d_{eq}\right)$$

$$= 0.0035 \times 36,950 \times 0.9382^{(1/3)} \times (0.0203/0.0239)$$

$$= 107.53 \text{ W/m}^2\text{K}$$

The next step is to compute the binary mass transfer coefficients. For this example we must make use of the Chilton–Colburn analogy as discussed above. The Schmidt numbers for the 1–2 binary pair is computed first

$$\text{Sc}_{V12} = \mu^V / \left(\rho_t^V \mathcal{D}_{12}\right)$$

$$= 0.959 \times 10^{-5} / (1.2985 \times 6.39 \times 10^{-6})$$

$$= 1.1558$$

The Schmidt numbers for the 1–3 and 2–3 binaries are found in the same way

$$\text{Sc}_{V13} = 0.1355$$

$$\text{Sc}_{V23} = 0.1383$$

The binary pair mass transfer coefficients may now be calculated

$$\kappa_{12} = h^V / (c_t C_p)(Pr_V / Sc_{V12})^{2/3}$$
$$= 107.53 / (32.964 \times 78.23)(0.9382/1.1558)^{2/3}$$
$$= 0.03629 \text{ m/s}$$

κ_{13} and κ_{23} are calculated in the same way

$$\kappa_{13} = 0.1515 \text{ m/s}$$
$$\kappa_{23} = 0.1494 \text{ m/s}$$

The multicomponent mass transfer coefficients may be evaluated with the help of Eqs. 8.3.30 and 8.3.31 using the binary mass transfer coefficients determined above and the bulk gas–vapor composition. The result of the calculation is the following matrix:

$$[k] = \begin{bmatrix} 0.1410 & 0.0844 \\ 0.0102 & 0.0665 \end{bmatrix} \text{m/s}$$

The elements of the matrix of mass transfer rate factors are computed next using Eqs. 8.3.32

$$[\Phi] = \begin{bmatrix} 0.2233 & -0.4192 \\ -0.0691 & 0.5734 \end{bmatrix} \text{m/s}$$

The eigenvalues of the rate factor matrix are given by Eqs. 8.3.48 and have the following numerical values:

$$\hat{\Phi}_1 = 0.6425 \qquad \hat{\Phi}_2 = 0.1542$$

The eigenvalues of the correction factor matrix are obtained from Eq. 8.3.45.

$$\hat{\Xi}_1 = 0.7129 \qquad \hat{\Xi}_2 = 0.9249$$

The matrix of correction factors follows from Eqs. 8.3.46.

$$[\Xi] = \begin{bmatrix} 0.8949 & 0.1820 \\ 0.0300 & 0.7429 \end{bmatrix} \text{m/s}$$

The matrix of high flux mass transfer coefficients is obtained by carrying out the matrix multiplication $[k][\Xi]$ to yield

$$[k^\bullet] = \begin{bmatrix} 0.1287 & 0.0884 \\ 0.0112 & 0.0513 \end{bmatrix} \text{m/s}$$

The diffusion fluxes may be calculated as

$$J_1 = c_t^V \{ k_{11}^\bullet (y_1^V - y_1^I) + k_{12}^\bullet (y_2^V - y_2^I) \}$$
$$= 0.2954 \text{ mol/m}^2\text{s}$$
$$J_2 = c_t^V \{ k_{21}^\bullet (y_1^V - y_1^I) + k_{22}^\bullet (y_2^V - y_2^I) \}$$
$$= 0.0646 \text{ mol/m}^2\text{s}$$
$$J_3 = -J_1 - J_2$$
$$= -0.3600 \text{ mol/m}^2\text{s}$$

The first two discrepancy functions F_1 and F_2 may now be evaluated as

$$F_1 = J_1 + y_1^V(N_1 + N_2) - N_1$$
$$= 0.2954 + 0.1377 \times (0.6954 + 0.1091) - 0.6594$$
$$\approx 0$$
$$F_2 = J_2 + y_2^V(N_1 + N_2) - N_2$$
$$= 0.0646 + 0.0597 \times (0.6954 + 0.1091) - 0.1091$$
$$\approx 0$$

Let us turn our attention to the evaluation of the energy balance. The heat transfer rate factor is

$$\Phi_H = (N_1 C_{p1} + N_2 C_{p2})/h^V$$
$$= 0.8682$$

and the Ackermann correction factor is

$$\Xi_H = \Phi_H^V/(\exp \Phi_H - 1)$$
$$= 0.6279$$

The energy balance function may now be evaluated as

$$F_3 \equiv h_0(T^I - T^C) - h^V(\Xi_H + \Phi_H^V)(T^V - T^I)$$
$$- N_1 \Delta H_{vap1} - N_2 \Delta H_{vap2}$$
$$= 1116 \times (331.45 - 299.0)$$
$$- 107.53 \times (0.6279 + 0.8682) \times (398.2 - 299.0)$$
$$- 0.6954 \times 33.3 \times 10^3$$
$$- 0.1091 \times 33.3 \times 10^3$$
$$\approx 0$$

Next, we complete the evaluation of the liquid-phase mixing equation

$$F_4 \equiv x_1^I(N_1 + N_2) - N_1$$
$$= 0.8581 \times (0.6594 + 0.1091) - 0.6594$$
$$\approx 0$$

To evaluate the equilibrium relations requires the computation of the K values. The pure component vapor pressures may be estimated from the Antoine equation. At the interface temperature the vapor pressures are

$$P_1^s = 48,917 \text{ Pa} \qquad P_2^s = 17,091 \text{ Pa}$$

Substituting the interface mole fractions x^I into the Margules equation for activity coefficients (Table D.2) gives the following results

$$y_1 = 1.0070 \qquad y_2 = 1.2703$$

The K values may now be computed as

$$K_1 = 0.4926 \qquad K_2 = 0.2171$$

We may now evaluate the departures from equilibrium as

$$\begin{aligned}
F_5 &= K_1 x_1^I - y_1^I \\
&= 0.4926 \times 0.8581 - 0.4227 \\
&\approx 0 \\
F_6 &= K_2 x_2^I - y_2^I \\
&= 0.2171 \times 0.1419 - 0.0308 \\
&\approx 0
\end{aligned}$$

The interfacial mole fractions satisfy the mole fraction summation equations.

$$F_7 = 0 \qquad F_8 = 0$$

It will now be clear that our initial "estimate" was, in fact, the final converged solution. Thus, all the calculations reported provide the final values of the appropriate quantities. Convergence was readily obtained from a set of initial values estimated as follows. The interface mole fractions for both vapor and liquid were set equal to the bulk vapor mole fractions. The interface temperature was estimated as the average of the coolant and bulk vapor temperatures. The molar fluxes in moles per second were estimated to be equal to the bulk-phase mole fractions.

Composition and temperature profiles are shown in Figure 15.7 where it is apparent that there is a large composition and temperature change over the vapor "film."

Diffusional interaction effects are quite important in this example. We leave it as an exercise for our readers to determine the molar rates of condensation using an effective diffusivity model. It is worth pointing out, however, that the rates are quite different from those calculated here.

The use of constant physical properties is somewhat questionable in this example since the composition and temperature changes over the film are substantial. Using the average

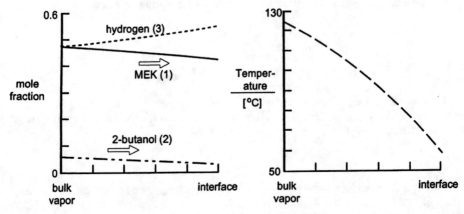

Figure 15.7. Composition and temperature profiles in the vapor film at the inlet to the MEK, 2-butanol condenser.

composition in the evaluation of the molar mass of the mixture and the average heat capacity does not have a significant effect on the final results. However, using constant average values for the diffusion coefficients and the mixture viscosity is a more serious offense. More exact calculation would have taken both temperature and composition dependence into consideration. ■

15.1.4 Condensation of a Binary Vapor Mixture

For two component mixtures we may, in fact, devise alternative procedures suitable for hand calculations.

A conventional bubble point calculation involves the specification of the liquid mole fractions and pressure the subsequent computation of the vapor-phase mole fractions and the system temperature. For a binary system (and only for a binary system) we may specify the temperature and pressure and compute the mole fractions of both phases. Thus, our first step is to estimate the interface temperature T^I. The second step is to solve the equilibrium equations for the mole fractions on either side of the interface. This step is, in fact, equivalent to reading the composition of both phases from a T–x–y equilibrium diagram.

If the liquid phase may be considered unmixed, the relative rates of condensation are given by

$$z_1 = x_1^I = N_1/(N_1 + N_2) \tag{15.1.15}$$

which permits the fluxes N_1 and N_2 to be calculated explicitly from Eq. 8.2.20.

$$\Phi = \ln\left\{\frac{1 - y_1^I/x_1^I}{1 - y_1^V/x_1^I}\right\} = \frac{N_t}{c_t^V k^V} = \frac{N_1/x_1^I}{c_t^V k^V} = \frac{N_2/x_2^I}{c_t^V k^V} \tag{15.1.16}$$

We may now solve the energy balance (Eq. 15.1.12) for the interface temperature that may be compared with the previous estimate. If the two values of T^I differ to any extent, we may use the newer value or some weighted average of the new and old values and go back to solving the interface equilibrium equations.

The steps of this computational procedure are summarized in Algorithm 15.2. It is left as an exercise for our readers to show that this simple procedure yields appropriate numerical results when applied to the system in Example 15.1.1.

Algorithm 15.2   Condensation of a Binary Vapor Mixture

Given:	Bulk vapor temperature and composition T^V, y^V.
	Coolant temperature T^C.
	Correlations for all necessary physical properties and transport coefficients.
Find:	Condensation rate, energy flux.
Step 1:	Estimate interface temperature T^I.
Step 2:	Calculate y_1^I, x_1^I by solving equilibrium equations.
Step 3:	Set the ratio of fluxes equal to the interface liquid mole fraction.

$$z_1 = x_1^I = N_1/(N_1 + N_2)$$

Step 4:	Calculate N_1 and N_2 from

$$\Phi = \ln\left\{\frac{1 - y_1^I/x_1^I}{1 - y_1^V/x_1^I}\right\} = \frac{N_t}{c_t^V k^V} = \frac{N_1/x_1^I}{c_t^V k^V} = \frac{N_2/x_2^I}{c_t^V k^V}$$

Step 5:	Check energy balance (Eq. 15.1.11).
	If not satisfied, return to Step 1.

15.1.5 Condensation of a Single Vapor in the Presence of an Inert Gas

The simplest case of practical interest to us involves the condensation of a single vapor in the presence of a noncondensing gas. Condensation of steam in the presence of air is an important practical application. The equations developed in the preceding section may also be used here to determine the rate of condensation of the vapor. However, some simplifications are possible since the liquid phase will be a pure component.

The vapor species (1) has to diffuse through the inert gaseous species (2) to the vapor–liquid interface before it can condense and release its heat to the coolant. Since the vapor is a binary mixture we may use Eqs. 8.2.14 and 8.2.16 to calculate the fluxes. However, since species 2 does not condense $N_2 = 0$, the rate of condensation of the vapor species 1 is given explicitly by Eq. 8.2.19.

$$N_1 = c_t^V k^V \ln\left\{\frac{1 - y_1^I}{1 - y_1^V}\right\} \tag{15.1.17}$$

The energy balance relation (Eq. 15.1.11) also simplifies to

$$E^W = q^W = h_0(T^I - T^C)$$
$$= h^V(\Xi_H + \Phi_H^V)(T^V - T^I) + N_1 \Delta H_{\mathrm{vap}1} \tag{15.1.18}$$

We can adopt a very simple strategy for calculating the rates of condensation. The calculation procedure is summarized in Algorithm 15.3 and illustrated in Example 15.1.2.

Algorithm 15.3 Condensation of Single Vapor in the Presence of Inert Gas

Given:	Bulk vapor temperature and composition T^V, y^V.
	Coolant temperature T^C.
	Correlations for all necessary physical properties and transport coefficients.
Find:	Condensation rate, energy flux.
Step 1:	Estimate interface temperature T^I.
Step 2:	Calculate interface vapor composition at T^I.

$$y_1^I = P_1^S/P \qquad y_2^I = 1 - y_1^I$$

Step 3: Calculate molar flux of condensing species.

$$N_1 = c_t^V k^V \ln\left\{\frac{1 - y_1^I}{1 - y_1^V}\right\}$$

Step 4: Calculate energy fluxes.

$$E^W = h_0(T^I - T^C)$$
$$E^V = h^V(\Xi_H + \Phi_H^V)(T^V - T^I) + N_1 \Delta H_{\mathrm{vap}1}$$

Step 5: If $(E^W/E^V) - 1 > 0.001$ (say) reestimate T^I.
Return to Step 2

Example 15.1.3 Condensation of Methanol in the Presence of Nitrogen

A mixture of methanol vapor and nitrogen gas (40% CH_3OH, 60% N_2) enters inside the top of a single vertical tubular condenser at a temperature of 70°C. The tube is cooled on the outside by means of cooling water that enters the top section of the annular section at a temperature of 15°C. The condensation takes place at atmospheric pressure (101.3 kPa).

Find

1. The rate of condensation in moles per meter square per second (mol/m²s) of methanol at the top of the tube.
2. The energy flux in watts per meter square (W/m²) at the top of the tube.

DATA

The mass transfer coefficient in the vapor phase (obtained under low rates of transfer): $k^V = 0.0091$ ms^{-1}.

The vapor-phase heat transfer coefficient (uncorrected for finite transfer rates): $h^V = 11.6$ W/m²K.

The overall heat transfer coefficient including resistance of condensate film, wall and coolant: $h_0 = 400$ W/m²K.

Molar heat capacity of methanol: $C_{p1} = 45$ J/mol K.

Heat capacity of nitrogen: $C_{p2} = 29$ J/mol K.

Heat of vaporization of methanol: $\Delta H_{vap1} = 36.4$ kJ/mol.

The vapor pressure of methanol in millimeters of mercury is given by the Antoine equation as

$$\ln P_1^s = 18.587 - 3626.55/[T(K) - 34.29]$$

(760 mmHg = 1 atm = 101.3 kPa).

SOLUTION Let us follow the steps of Algorithm 15.3.

Step 1: We begin by making an initial estimate of the interface temperature T^I at a value midway between the vapor and coolant temperatures 42.5°C in this case.

Step 2: The vapor pressure of methanol is calculated from the Antoine equation given above as

$$P_1^s = \exp(18.587 - 3626.55/(42.5 + 273.15 - 34.29)) \times 101.3/760$$
$$= 39.77 \text{ kPa}$$

The interface concentration of methanol may be estimated as follows:

$$y_1^I = P_1^S/P$$
$$= 39.77/101.3$$
$$= 0.3925$$

Step 3: The flux of methanol follows from Eq. 8.2.19 as written above after computing the molar density c_t^V at the average of T^V and T^I ($c_t^V = 37$ mol/m³)

$$N_1 = c_t^V k^V \ln\left\{\frac{1 - y_1^I}{1 - y_1^V}\right\}$$
$$= 37 \times 0.0091 \times \ln\left\{\frac{1 - 0.3925}{1 - 0.40}\right\}$$
$$= 4.173 \times 10^{-3} \text{ mol/m}^2\text{s}$$

Step 4: The energy flux through the wall is

$$E^W = h_0(T^I - T^C)$$
$$= 400 \times (42.5 - 15)$$
$$= 11{,}000 \text{ W/m}^2$$

The heat transfer rate factor Φ_H is given by

$$\Phi_H^V = N_1 C_{p1}/h^V$$
$$= 4.173 \times 10^{-3} \times 45/400$$
$$= 0.0162$$

The heat transfer coefficient correction factor Ξ_H is calculated next.

$$\Xi_H = \Phi_H^V/(\exp \Phi_H^V - 1)$$
$$= 0.0162/[\exp(0.0162) - 1]$$
$$= 0.9939$$

The energy flux in the vapor film is found from the second part of Eq. 15.1.18 as

$$E^V = h^V(\Xi_H + \Phi_H^V)(T^V - T^I) + N_1 \Delta H_{vap1}$$
$$= 400 \times (0.9939 + 0.0162)(70 - 42.5) + 4.173 \times 10^{-3} \times 36.4 \times 10^3$$
$$= 473.5 \text{ W/m}^2$$

Clearly, there is a substantial difference between E^V and E^W so we must reestimate T^I. A rather simple way to do this is to rearrange Eq. 15.1.18 as an explicit expression for T^I and using the energy flux in the vapor phase in place of the energy flux in the coolant

$$T^I = E^V/h_0 + T^C$$
$$= 473.5/400 + 15$$
$$= 16.18°C$$

We now return to Step 2 and carry out further iterations. The final results are summarized as follows:

$$T^I = 26.4°C$$
$$y_1^I = P_1^S/P = 0.1794$$
$$N_1 = 0.108 \text{ mol/m}^2\text{s}$$
$$\Phi_H = 0.419$$
$$E^V = E^W = 4553.3 \text{ W/m}^2\text{s}$$

The numerical value of the high flux correction factor is 0.805. Thus, a calculation of the heat flux q^V from Eq. 11.5.8 would be some 20% in error if the flux correction factor were ignored.

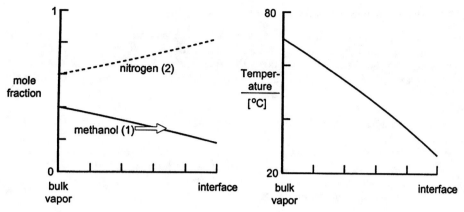

Figure 15.8. Composition and temperature profiles in the vapor phase at the inlet to the condenser tube.

The simple way of reestimating T^I used here works because E^W is a linearly increasing function of T^I, whereas E^V is mildly nonlinear but decreasing function of T^I. Approximately 10 iterations were needed to reach 0.1% accuracy in the energy balance. Applying Newton's method to the interface energy balance does not improve things much unless N_1 and y_1^I are regarded as functions of T^I and that makes the method much more complicated than the simple and entirely workable method described above.

Composition and temperature profiles in the vapor phase at the top of the condenser tube are shown in Figure 15.8. Note how the concentration of inert gas increases towards the interface. The noncondensing nitrogen is acting as a blanket, making it harder for the methanol to condense.

15.2   CONDENSER DESIGN

Existing models for designing heat exchangers to condense multicomponent mixtures are of two basic kinds: equilibrium models, such as those of Kern (1950), Silver (1947) and Bell and Ghaly (1972), and the differential or nonequilibrium models that have developed following the original work of Colburn and Drew (1937). In the latter class of models a set of one-dimensional differential material and energy balances is integrated numerically along the length of the condenser. Mass and energy transfer rates usually are calculated using equations based on a film model (Colburn and Drew, 1937; Schrodt, 1973; Price and Bell, 1974; Krishna et al., 1976, Krishna and Panchal 1977; Krishna, 1979a–d; Röhm, 1980; Bandrowski and Kubaczka, 1981; Webb and McNaught, 1980; Webb and Sardesai, 1981; Webb, 1982; Taylor and Noah, 1982; Shah and Webb, 1983; McNaught, 1983a, b, Webb and Panagoulias, 1987). Still more sophisticated nonequilibrium models based on boundary layer theory are limited primarily to describing the condensation of binary vapors or of one vapor in the presence of a noncondensing gas. Extensions of the boundary layer models to multicomponent systems are few in number (see, however, Taitel and Tamir, 1974; Tamir and Merchuk, 1979; Sage and Estrin, 1976) but have not been developed to the point where they could be used in the design of heat exchangers of complex geometry. While the equilibrium methods are widely used (the reasons being their simplicity, rapidity in computation and because there is no need to compute intermediate vapor compositions or to obtain diffusivity data), the one-dimensional nonequilibrium methods are more soundly based. We develop the design relations for both vertical and horizontal condensers below.

15.2.1   Material Balance Relations

Consider the condensation of a vapor mixture inside the vertical tube as shown in Figure 15.1. Vapor enters the top of the tube and flows downwards. The condensate flows cocurrently with the remaining vapor down the tube. The coolant may flow cocurrently with or countercurrently to the vapor and liquid streams. Condensation on a horizontal tube was illustrated in Figure 15.2. In this case the condensed liquid drips off the tube and eventually collects at the bottom of the heat exchanger. The vapor flow is directed along or, more likely, across bundle of tubes. The design equations are developed in the same way for both types of condenser.

We start by drawing up the appropriate material balances which, here, are written around a section of condenser tube of differential area shown in Figure 15.9. This figure also serves to introduce the notation we shall use here.

For the vapor phase the component material balance reads

$$v_i|_{A+\Delta A} - v_i|_A + N_i \, \Delta A = 0 \tag{15.2.1}$$

Dividing by ΔA and taking the limit as ΔA goes to zero gives

$$\frac{dv_i}{dA} = -N_i \qquad i = 1, 2, \ldots, n \tag{15.2.2}$$

The total material balance is obtained simply by summing Eqs. 15.2.2 over all n species.

$$\frac{dV}{dA} = -N_t \tag{15.2.3}$$

The differential material balances for the liquid phase are obtained in a similar way: For each component we have

$$\frac{dl_i}{dA} = N_i \qquad i = 1, 2, \ldots, n \tag{15.2.4}$$

and the total material balance is

$$\frac{dL}{dA} = N_t \tag{15.2.5}$$

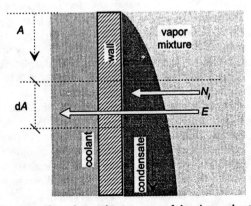

Figure 15.9. Differential section of condenser for purposes of drawing up heat and material balances.

The terms on the right-hand sides of Eqs. 15.2.2 and 15.2.4 are the molar fluxes of species i in the vapor and liquid phases, respectively; we assume that transfers from the vapor phase to the liquid phase are positive.

For a horizontal condenser where the condensed liquid drains off as soon as it forms the component and total material balances for the liquid (Eqs. 15.2.4 and 15.2.5), do not apply. The condensed liquid-phase composition is dictated by the *local* rates of condensation N_i and we have (cf. Eqs. 15.1.1 and 15.1.2)

$$x_i^I = x_i^L = N_i/N_t \tag{15.2.6}$$

15.2.2 Energy Balance Relations

The energy balance for the vapor phase is

$$\frac{d(VH^V)}{dA} = -E^V \tag{15.2.7}$$

where H^V is the enthalpy of the bulk vapor.

$$H^V = \sum_{i=1}^{n} y_i \overline{H}_i^V \tag{15.2.8}$$

The product VH^V may be expressed as

$$VH^V = \sum_{i=1}^{n} v_i \overline{H}_i^V \tag{15.2.9}$$

where we have made use of the relation $v_i = Vy_i$. On differentiating Eq. 15.2.9 we obtain

$$\begin{aligned}
\frac{d(VH^V)}{dA} &= \sum_{i=1}^{n} \frac{d(v_i \overline{H}_i^V)}{dA} \\
&= \sum_{i=1}^{n} \left\{ v_i \frac{d\overline{H}_i^V}{dA} + \overline{H}_i^V \frac{dv_i}{dA} \right\} \\
&= \sum_{i=1}^{n} v_i \frac{d\overline{H}_i^V}{dA} - \sum_{i=1}^{n} N_i \overline{H}_i^V
\end{aligned} \tag{15.2.10}$$

The component material balance (Eq. 15.2.2) was used to eliminate dv_i/dA. Substituting this result into Eq. 15.2.7, together with Eq. 15.1.4 for E^V, gives

$$\sum_{i=1}^{n} v_i \frac{d\overline{H}_i^V}{dA} = -q^V \tag{15.2.11}$$

where q^V is the conductive heat flux out of the bulk vapor.

To proceed further we make use of the following expression to calculate the partial molar enthalpies

$$\overline{H}_i^V(T) = C_{pi}^L(T_{b,i} - T_{\text{ref}}) + \Delta H_{\text{vap},i} + C_{pi}^V(T - T_{b,i})$$

where $T_{b,i}$ is the normal boiling point of species i and T_{ref} is a reference temperature. With the help of this model for the enthalpies, the vapor-phase energy balance may be expressed as

$$VC_p^V \frac{dT^V}{dA} = -q^V \qquad (15.2.12)$$

where C_p^V is the molar heat capacity of the vapor mixture.

If we neglect subcooling of the liquid condensate, the conductive energy flux through the condensed liquid film E^l will also equal the energy flux through the wall of the condenser and carried away by the coolant on the other side of the wall. The energy balance for the coolant is

$$L_C G_p^C \frac{dT^C}{dA} = \pm q^W + = \text{cocurrent}$$

$$- = \text{countercurrent} \qquad (15.2.13)$$

where q^W is the heat flux into the coolant. The sign in Eq. 15.2.13 is determined by whether the coolant is flowing counter to the vapor phase or not. Differences in the interfacial area at the vapor-condensate and condensate-tube wall interfaces are ignored here; it would, however, be easy to allow for such variations.

The equations presented above can be used (with or without modifications) to describe mass transfer processes in cocurrent flow. See, for example, the work of Modine (1963), whose wetted wall column experiments formed the basis for Example 11.5.3 and are the subject of further discussion in Section 15.4. The coolant energy balance is not needed to model an adiabatic wetted wall column and must be replaced by an energy balance for the liquid phase. Readers are asked to develop a complete mathematical model of a wetted wall column in Exercise 15.2.1.

15.2.3 Solving the Model Equations

The set of differential and algebraic equations that make up the model must be solved numerically in general; their complexity and nonlinearity precludes analytical solution in the majority of cases of practical importance. It is beyond the scope of this book to discuss in any detail the multitude of numerical methods that have been developed for solving systems of differential equations. The equations developed above can be solved perfectly satisfactorily with nothing more complicated than a fourth-order Runge–Kutta method (used by, e.g., Webb and Sardesai, 1981) and the first-order Euler method has also been used on occasions (by, e.g., Krishna and Panchal, 1977). If an implicit Euler method is used, the derivatives are, in effect, replaced by finite difference approximations. The resulting set of algebraic material and energy balance equations can be combined with the algebraic equations that model the mass and energy transfer rate calculation and solved simultaneously using Newton's method (Taylor et al., 1986).

Whichever method is selected, the calculations start at one end of the condenser, the vapor inlet presumably, where the vapor temperature, pressure, flow rate, and composition is known and proceed until either a specified area has been reached (a simulation problem) or until a specified amount of vapor has been condensed (a design problem). To simplify the calculations it is usual to specify the temperature of the coolant at the vapor inlet end of the condenser and determine the coolant entry temperature as part of the solution. Multiple pass condensers (where the coolant and/or gas vapor streams make more than one pass through the heat exchanger) pose additional computational challenges that are discussed by, for example, Webb and McNaught (1980) and by Taylor et al. (1986).

15.3   DESIGN STUDIES

In this section we present a number of examples designed to demonstrate the use of the models presented in the preceding section as a condenser simulation and design tool. Four example problems, taken from the literature on multicomponent condensation, are used as a basis for illustrating the features of the model and calculation procedure. As was the case in Chapter 14, it is not practical to present illustrative examples of the condensation model that are as detailed as the examples in prior sections. Only a summary of the problem specifications and a selection of results is provided here. The discussion that follows is based on a paper by Taylor et al. (1986).

15.3.1   Example Specifications

Example 1 involves the condensation of methanol and water in the presence of air, a system studied earlier by Schrodt (1973), by Krishna and Panchal (1977), and by Krishna (1979c). Example 2 is similar to Example 1 but involves helium instead of air as the noncondensing species. This change serves to increase the importance of diffusional interaction effects which are quite considerable in this system (Krishna, 1979b, c). Example 3 involves the condensation of a mixture of straight-chain hydrocarbon vapors taken originally from Kern (1950) and reconsidered by Webb and McNaught (1980) who used an effective diffusivity model to calculate the vapor phase mass transfer rates. Example 4 is a variation on an example considered by Krishna et al. (1976); it involves the same hydrocarbons as Example 3 with the addition of hydrogen (which, in fact, has a significant effect on the results). Unlike Krishna et al., in the results shown here, hydrogen is not considered to be a noncondensing gas; it is, however, only sparingly soluble in the condensate. The problem specifications are summarized in Tables 15.1 and 15.2.

TABLE 15.1   Specification of Condenser Design Examples 1 and 2

Example		1	2
Components	1	Methanol	Methanol
	2	Water	Water
	3	Air	Helium
Tube orientation		Vertical	Vertical
Tube diameter (m)		0.0254	0.0254
Surface area per meter length of one tube (m^2)		0.08	0.08
Inlet vapor flow rates ($mol/s \times 10^3$ per tube)			
Component	1	128.9	92.1
	2	36.8	46.0
	3	18.4	46.0
Vapor inlet temperature (K)		360.0	350.0
Coolant temperature at top (K)		308.15	308.15
Coolant flow rate (kg/s)		0.06	0.06
Coolant flow direction		Counter	Counter
Heat transfer coefficients			
h_0 (W/m^2K)		1700	1700

References: Schrodt (1973); Krishna and Panchal (1977); Krishna (1979c); Taylor and Noah (1982); Taylor et al. (1986).

TABLE 15.2   Specification of Condenser Design Examples 3 and 4

Example		3	4
Components	1	n-Octane	n-Octane
	2	n-Heptane	n-Heptane
	3	n-Hexane	n-Hexane
	4	n-Butane	n-Butane
	5	Propane	Propane
	6		Hydrogen
Tube orientation		Vertical	Vertical
Tube diameter (m)		0.0254	0.0254
Surface area per meter length of one tube (m^2)		0.08	0.08
Inlet vapor flow rates (mol/s $\times$ 10^3 per tube)			
Component	1	204.4	8.4
	2	245.5	4.2
	3	40.9	12.6
	4	204.6	6.3
	5	122.7	10.5
	6		28.0
Vapor inlet temperature (K)		413.06	345.4
Coolant temp. at top (K)		300.0	283.15
Coolant flow rate (kg/s)		NA	0.04376
Coolant flow direction		NA	Counter
Heat transfer coefficients			
h_0 (W/m^{2}K)			1700
h^C (W/m^{2}K)		1700	

References: Krishna et al., (1976); Kern (1950); Webb and McNaught (1980).

In the discussion that follows we focus attention on the following:

1. A comparison of the film models that ignore diffusional interaction effects (the effective diffusivity methods) with the film models that take multicomponent interaction effects into account (Krishna–Standart (1976), Toor–Stewart–Prober (1964), Krishna, (1979b, c) and Taylor–Smith, 1982).
2. A comparison of the interactive film models that use the Chilton–Colburn analogy to obtain the heat and mass transfer coefficients with the turbulent eddy diffusivity models.
3. The influence of the model used to approximate the mass transfer behavior in the liquid phase (i.e., mixed, unmixed or finite mass transfer rate model).

15.3.2   Significance of Interaction Effects

In considering very many condenser simulations (not just those reviewed here) we have yet to find an application where the differences between any of the multicomponent film models that account for interaction effects (Krishna–Standart, 1976; Toor–Stewart–Prober, 1964; Krishna, 1979a–d; Taylor–Smith, 1982) are significant. There is also very little difference between the turbulent eddy diffusivity model and the film models that use the Chilton–Colburn analogy (Taylor et al., 1986). This result is important because it indicates that the Chilton–Colburn analogy, widely used in design calculations, is unlikely to lead to large

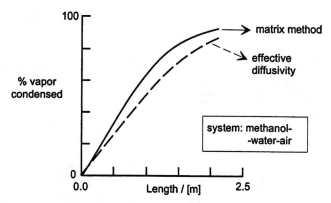

Figure 15.10. Percentage condensed as a function of condenser length for condensation of methanol and water in the presence of air.

errors in design calculations. However, effective diffusivity methods may yield results that can differ quite markedly from the results obtained using an interactive model.

Figures 15.10–15.13 can be used to determine the relative amounts of vapor that would condense in a device of given size for each of the four examples. Consider Example 2: from Figure 15.11 we find that the tube needed to condense 60% of the incoming vapor is 1.16 m predicted using an interactive model but as much as 1.66 m predicted using the effective diffusivity model, an increase in tube length of 43%! The temperature of the vapor leaving a condenser of either size (or any other size for that matter) can be determined directly from Figure 15.14. The condensate composition as a function of tube length is shown, for the same example, in Figure 15.15. For this particular example the models that account for interaction effects are in good agreement on all these quantities but the effective diffusivity method predicts a mole fraction of methanol in the condensate quite different to that predicted by the interactive models (Fig. 15.15). The large differences between the percentage condensed predicted by the matrix and the effective diffusivity methods is indicative of large differences between the respective predictions of the total fluxes, the disagreement between the two predictions of the condensate composition is a measure of the disagreement between the predictions of the component fluxes.

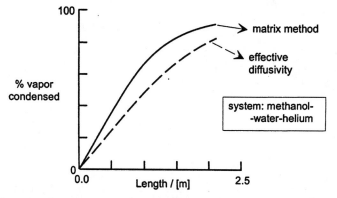

Figure 15.11. Percentage condensed as a function of condenser length for condensation of methanol and water in the presence of helium.

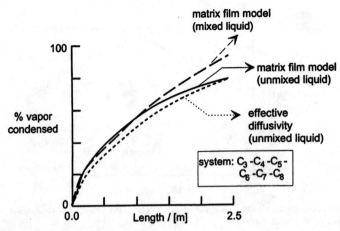

Figure 15.12. Percentage condensed as a function of condenser length for condensation of a mixture of normal paraffins (propane to *n*-octane). Note differences between condensate mixing models.

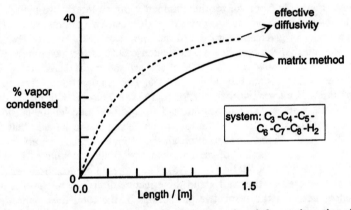

Figure 15.13. Percentage condensed as a function of condenser length for condensation of a mixture of normal paraffins (propane to *n*-octane) in the presence of hydrogen.

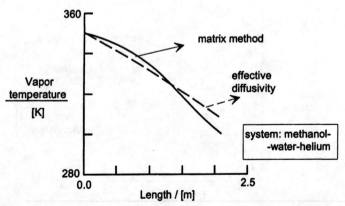

Figure 15.14. Vapor temperature as a function of condenser length for condensation of methanol and water in the presence of helium.

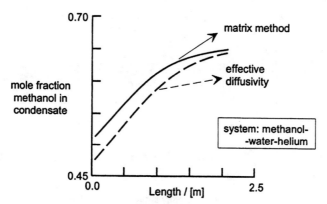

Figure 15.15. Condensate composition a function of condenser length for condensation of methanol and water in the presence of helium.

For Example 1, we find that the matrix models suggest that tubes 1.41 m long will condense 65% of the incoming vapor, whereas the effective diffusivity method suggests that 1.84 m would be needed, a 30% larger condenser (Fig. 15.10). The same relative difference in the size of device needed to condense 50% of the inlet vapor is found for Example 3 (Fig. 15.12). In each of Examples 1–3, the effective diffusivity model gave the most conservative design (larger condenser surface area). That this is not always the case is demonstrated in Example 4, where the effective diffusivity method requires some 44% *less* area to condense 30% of the inlet stream (Fig. 15.13). It is now clear from these few results that the effective diffusivity methods may result in either a *significantly over designed* or an *under designed* condenser and it is not possible to forecast whether the effective diffusivity method will be conservative or not!

Interaction effects are most important in systems containing species whose molecular size and nature differ widely, as is the case in Examples 2 and 4. Temperature, composition, and flux profiles for Example 4 are shown in Figures 15.16–15.18. There are significant differences between the matrix methods and the effective diffusivity methods. Without hydrogen in the mixture (Example 3) all of the models give very similar results. This result

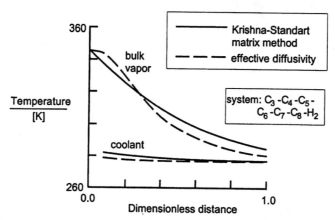

Figure 15.16. Temperature profiles during condensation of a mixture of normal paraffins (propane to *n*-octane) in the presence of hydrogen.

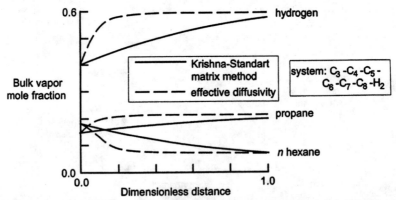

Figure 15.17. Mole fractions in the vapor phase as a function of condenser length for condensation of a mixture of normal paraffins (propane to n-octane) in the presence of hydrogen.

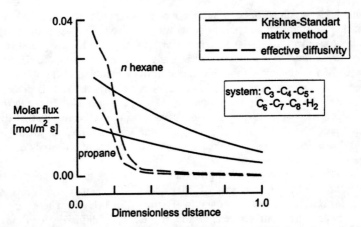

Figure 15.18. Molar fluxes as a function of tube length for condensation of a mixture of normal paraffins (propane to n-octane) in the presence of hydrogen.

occurs because the diffusivities of all of the binary pairs cover a much narrower range of values. For the same reasons, interaction effects are more pronounced in system 2 than in system 1.

Further evidence for the significance of interaction effects in condensation can be found in the papers by Krishna et al. (1976), Krishna (1979a–d), Röhm (1980), and Webb et al. (1981). We shall return to this topic in Section 15.4, where we provide some experimental support for our recommendation that an effective diffusivity model should not be used. It might be worth pointing out that, if the calculations are done as we have suggested in preceding sections, there are no computational advantages to using an effective diffusivity approach.

15.3.3 Liquid-Phase Models

There is no significant difference between the two extremes of condensate mixing (i.e., completely mixed or completely unmixed) if there is a noncondensing or sparingly soluble gas present in the vapor stream (Examples 1, 2, and 4). This conclusion was also reached by Krishna et al. (1976), Webb and Saredesai (1981), and Taylor and Noah (1982). However, in Example 3, in which all of the components are condensable, there is a very considerable difference between these two extremes (Fig. 15.12). McNaught (1983a) writes that it is common practice to assume complete mixing of the condensate when applying the Silver (1947) (equilibrium) method and "it is well established that this can lead to under design if significant separation of the phases occurs" (as in a horizontal condenser). The same thing can be said if complete mixing of the condensate is assumed in the models described in this paper. A design based on the well-mixed condensate model might lead to a significant under design. We conclude from these and other results that the more conservative design is, in general, obtained with the condensate assumed to be completely unmixed. We would emphasize once more that there is no computational penalty for adopting either one of the two extreme cases. Interestingly enough, it is impossible to predict situations where some components condense while others evaporate using the no mixing option. It is quite possible to predict this situation with a rate model or with the well-mixed condensate; indeed, computations indicate that this happens in Example 3 in some of the sections closer to the vapor inlet.

15.4 EXPERIMENTAL STUDIES

In this section we review the experimental studies that have been carried out with a view to testing models of multicomponent condensation. There is a great shortage of experimental data on mass transfer in multicomponent vapor (plus inert gas)–liquid systems. Most published works deal with absorption (or condensation or evaporation) of a single species in the presence of a nontransferring component. Thus, this review is necessarily brief.

15.4.1 Multicomponent Condensation

The most comprehensive experimental studies of multicomponent condensation have been carried out by Webb and co-workers. Webb and Sardesai (1981) report the results of a number of experiments involving the condensation of 2-propanol and water in the presence of a noncondensing gas. The experiments were carried out in a vertical condenser tube 1 m long and 0.023-m internal diameter. Eight runs were carried out in which 2-propanol and water were condensed in the presence of nitrogen as inert gas and another seven runs with freon-12 as the noncondensable component. These experiments cover a somewhat larger range of Reynolds numbers (7000–20,000) and vapor composition. Complete details of the experiments are available in the thesis of Sardesai (1979); a summary of the results is given by Webb and Sardesai (1981). Webb (1982) reviewed other data obtained by his co-workers, including the results of Deo (1979) obtained in a condenser of annular geometry and Shah and Webb (1983) who condensed water and methanol in the presence of a variety of inert gases in a condenser consisting of 50 horizontal tubes mounted in a rectangular duct.

Numerical simulations of Sardesai's experiments are discussed by Webb and Sardesai (1981) and Webb (1982) (who used the Krishna–Standart (1976), Toor–Stewart–Prober (1964) and effective diffusivity methods to calculate the condensation rates), McNaught (1983a, b) (who used the equilibrium model of Silver, 1947), and Furno et al. (1986) (who used the turbulent diffusion models of Chapter 10 in addition to methods based on film theory). It is the results of the last named that are presented here.

Webb and Sardesai (1981) used a fourth-order Runge-Kutta method to integrate the differential equations modeling the condenser. Furno (1986) divided the condenser tube into 40 sections. Both investigators used the same method to calculate physical properties. The Fanning friction was calculated from a correlation obtained by Webb and Sardesai

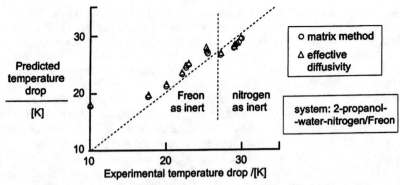

Figure 15.19. Comparison between predicted and measured temperature drop over a vertical tube condensing 2-propanol and water in the presence of nitrogen or freon-12. Experimental data of Webb and Sardesai (1981); predictions by Furno et al. (1986).

from dry gas cooling experiments carried out in their condenser. Furno's simulations of Sardesai's experiments were made easier by employing the measured wall temperature profile. This obviates the need to consider the energy balance for the coolant and the uncertainties in the estimation of the coolant heat transfer coefficient (the dimensions of the annular space occupied by the coolant and the coolant flow rate are not specified in the paper by Webb and Sardesai). The wall temperatures at positions located between the thermocouples were estimated by interpolation.

The results of Furno's simulations of Sardesai's experiments are summarized in Figure 15.19 (predicted versus measured overall temperature drop), Figure 15.20 (predicted versus measured total condensation rate), and Figure 15.21 (predicted versus measured condensate composition). The condensate composition is a good indicator of the individual condensation rates.

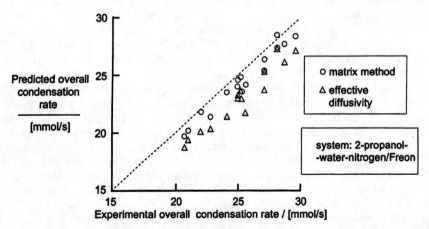

Figure 15.20. Comparison between predicted and measured overall rates of condensation in a vertical tube condensing isopropanol and water in the presence of nitrogen or freon-12. Experimental data of Webb and Sardesai (1981); predictions by Furno et al. (1986).

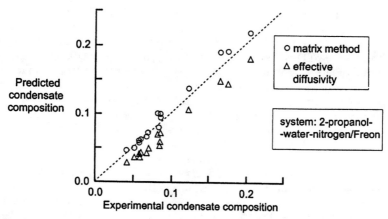

Figure 15.21. Comparison between predicted and measured mole fractions of 2-propanol in a condensate of 2-propanol and water. Experimental data of Webb and Sardesai (1981); predictions by Furno et al. (1986).

Both extremes of condensate mixing yield almost identical results; those shown in Figures 15.19–15.21 were obtained using the no-mixing option. The use of the Chilton–Colburn analogy to estimate the heat and binary pair mass transfer coefficients consistently underpredicted the total condensation rates. Figures 15.19–15.21 show the results of simulations in which the Gilliland–Sherwood correlation (Eq. 8.8.5) was used for estimating only the mass transfer coefficients (heat transfer coefficients were calculated from the analogy). The model using the Gilliland–Sherwood correlation still tends to underpredict the total condensation rates, but not by as much as was found using the Chilton–Colburn analogy to estimate the mass transfer coefficients).

We see from these figures that the mass transfer models that take diffusional interactions into account are quite a lot better than the effective diffusivity model, which underpredicts the rate of condensation of 2-propanol in every case. However, the effective diffusivity methods give good predictions of the overall temperature drops (Fig. 15.19) although there is little to distinguish any of the models here on this basis.

15.4.2 Ternary Mass Transfer in a Wetted Wall Column

A set of ternary mass transfer experiments was carried out by Toor and Sebulsky (1961b) and Modine (1963) in a wetted-wall column and also in a packed column. These authors measured the simultaneous rates of transfer between a vapor–gas mixture containing acetone, benzene, and nitrogen or helium, and a binary liquid mixture of acetone and benzene. Vapor and liquid streams were in cocurrent flow in the wetted-wall column and in countercurrent flow in the packed column. Their experimental results show that diffusional interaction effects were significant in the vapor phase, especially for the experiments with helium in the wetted wall column.

Modine (1963) carried out his experiments in an adiabatic wetted-wall column 0.6096 m in length and 0.025019-m inside diameter. A gas–vapor stream containing acetone (1)–benzene(2)–nitrogen(3) or helium(3) was contacted with a cocurrently flowing liquid stream containing only the first two components. Seven runs were carried out with nitrogen as inert gas and six runs with helium. The experimental results are most accessible in a paper by

Krishna (1981); for this reason, they are not repeated here. It is worth noting, however, that most of the experiments involve the simultaneous evaporation of benzene and condensation of acetone. The experiments cover a rather narrow range of liquid compositions (mole fraction acetone ~ 0.1) and vapor flow rates (Re numbers ~ 9000).

A number of investigators used the wetted-wall column data of Modine to test multicomponent mass transfer models (Krishna, 1979, 1981; Furno et al., 1986; Bandrowski and Kubaczka, 1991). Krishna (1979b, 1981a) tested the Krishna–Standart (1976) multicomponent film model and also the linearized theory of Toor (1964) and Stewart and Prober (1964). Furno et al. (1986) used the same data to evaluate the turbulent eddy diffusion model of Chapter 10 (see Example 11.5.3) as well as the explicit methods of Section 8.5. Bandrowski and Kubaczka (1991) evaluated a more complicated method based on the development in Section 8.3.5. The results shown here are from Furno et al. (1986).

To simulate Modine's experiments, Krishna (1981a) and Furno et al. (1986) divided the column into 24 sections, each of which was modeled by a nonequilibrium section as described in Section 15.2.3. Thus, the model of the wetted wall column is simpler than the model of a condenser in that the coolant energy balance is not needed and the interphase energy balance does not include a term for heat transfer through the outer wall (the column is assumed to be adiabatic). In all other respects, however, the model of a wetted wall column is the same as the model of a condenser tube. Physical properties like density, viscosity, heat capacity and thermal conductivity were evaluated using the same methods employed by Krishna (1981a). The Fanning friction factor (needed in the calculation of the heat and mass transfer coefficients) was estimated using an expression obtained by Modine in his column (see Example 11.5.3). K values were estimated using the Antoine equation for the vapor pressures and the Wilson equation for the liquid-phase activity coefficients. Latent heats of vaporization were calculated using the Watson equation (Reid et al., 1977). All physical and thermodynamic properties were evaluated separately in each section of the condenser.

Figure 15.22 provides a comparison between the mass transfer rates measured by Modine (1963) and the rates predicted using the different classes of model for the vapor-phase mass transfer process: methods based on the multicomponent film model (all of

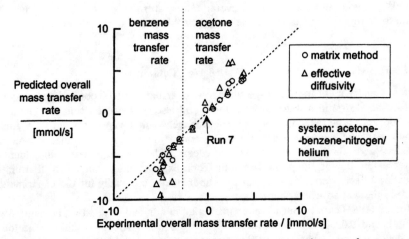

Figure 15.22. Comparison between predicted and measured mass transfer rates for mass transfer between a falling liquid film of acetone and benzene and a gas–vapor mixture containing, in addition, nitrogen or helium. Experimental data of Modine (1963); predictions by Krishna (1981a) and by Furno et al. (1986).

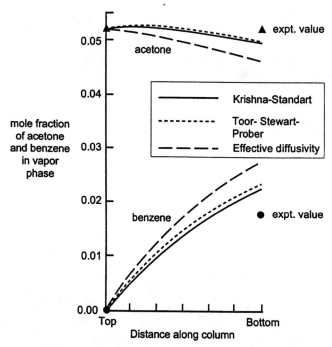

Figure 15.23. Acetone and benzene composition profiles along the length of a wetted wall column for Run 7 of Modine (1963). After Krishna (1981a).

them yield essentially identical results), and the effective diffusivity models. It is clear from this figure that the models that account for interactive effects are in better agreement with the experimental data than is the effective diffusivity method.

Run 7 of the simulations of Modine's experiments is particularly sensitive to the mass transfer model used. This experiment was used as the basis for the flux calculation in Example 10.4.1. For this experiment none of the models does well at predicting the total amount of acetone transfer. The Krishna–Standart method, the linearized theory and the effective diffusivity model predict the wrong direction of mass transfer; while the experimental data show that there is net vaporization of acetone, the models predict net condensation. This erroneous prediction comes about in part because of the extremely small magnitude of the acetone flux relative to the total flux.

Predicted vapor-phase composition profiles for Run 7 are shown in Figure 15.23 and the fluxes calculated by three different mass transfer models are shown in Figure 15.24. The latter figure shows that the Krishna–Standart and linearized theory predict that the flux of acetone undergoes a change in sign somewhere in the column from negative values at the top of the column (corresponding to vaporization of acetone) to positive in the lower section of the column (corresponding to evaporation of acetone). The effective diffusivity method, on the other hand, predicts that acetone will condense everywhere in the column; completely opposite to the data. These results must be considered as evidence that reverse mass transfer takes place over part of the length of the column. As has been noted elsewhere in this book, models of the effective diffusivity type cannot account for this behavior, even qualitatively. The detailed calculation of the mass fluxes provided in Example 11.5.3 provides further evidence for the reverse mass transfer. Similar calculations using the Krishna–Standart (1976), Toor–Stewart–Prober (1964) and effective diffusivity models are given by Krishna (1981a).

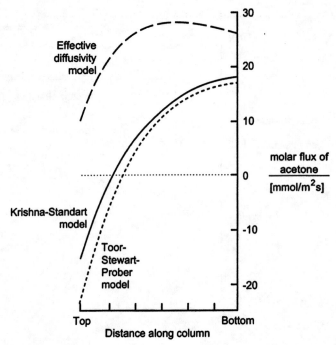

Figure 15.24. Acetone flux along the length of a wetted wall column for Run 7 of Modine (1963). After Krishna (1981a).

15.5   CONCLUSIONS AND RECOMMENDATIONS

In this chapter we have considered models of multicomponent condensation. In particular, we have considered various approaches to calculating the rates of mass and energy transfers in the vapor and condensate, respectively. Methods of solving the model equations have also been discussed.

With regard to the various models of vapor-phase mass transfer we conclude that

1. Effective diffusivity models should not be used in the determination of the rates of mass transfer in the vapor phase. These models are not justified on theoretical grounds, nor on experimental grounds and their use offers no reduction in the cost of obtaining a solution or any increase in the ease by which that solution is obtained.

2. The film models that take interaction effects into account (Krishna–Standart, 1976; Toor–Stewart–Prober, 1964; Krishna, 1979; Taylor–Smith, 1982) yield temperature and composition profiles that, for all practical purposes, are indistinguishable.

3. There is little to choose between the film models that use the Chilton–Colburn analogy to obtain the heat and mass transfer coefficients and the turbulent eddy diffusivity methods when they are used to predict the performance of multicomponent condensers.

With regard to the liquid phase, there is very little to distinguish between the results obtained using the two extremes of condensate mixing if noncondensing or sparingly soluble gases are present in the vapor phase. There can be a very considerable difference if all species condense. More conservative design is obtained with the liquid phase assumed to be completely unmixed.

The simple one-dimensional models of multicomponent condensation and cocurrent separation processes described in this chapter are well able to model the performance of a wetted-wall column operated by Modine (1963) and a vertical tube condenser operated by Sardesai (1979). The results obtained with the one-dimensional model are probably good enough for design purposes; it is doubtful if a more sophisticated boundary layer analysis could yield any better results.

Postface

In this textbook we have concentrated our attention on mass transfer in mixtures with three or more species. The rationale for doing this should be apparent to the reader by now; *multicomponent mixtures have characteristics fundamentally different from those of two component mixtures*. In fact, a binary system is peculiar in that it has none of the features of a general multicomponent mixture. We strongly believe that treatments of even binary mass transfer are best developed using the Maxwell–Stefan equations. We hope that this text will have the effect of persuading instructors to use the Maxwell–Stefan approach to mass transfer even at the undergraduate level.

We have emphasized the proper modeling of *thermodynamic nonideality* both with regard to molecular diffusion and interphase mass transfer. The benefits of adopting the irreversible thermodynamic approach are particularly apparent here; it would not be possible otherwise to explain the "peculiar" behavior of the Fick diffusivities. The practical implications of this behavior in the design of separation equipment operating close to the phase transition or critical point (e.g., crystallization, supercritical extraction, and zone refining) are yet to be explored. In any case, the theoretical tools are available to us.

We anticipate that in the future, most separation process design will be carried out by a "head-on" approach using nonequilibrium stage models rather than by the time honored method of equilibrium stage simulations followed by efficiency estimation. After all, designers of chemical reactors never resort to equilibrium stage approaches to packed bed reactors. There is going to be a computational penalty for incorporating matrix formulations of multicomponent mass transfer into equipment design procedures. We have tried to indicate how these models may most effectively be used in design. With the help of the worked out design examples we hope we have convinced the process designer of the need to exercise particular caution when dealing with nonideal multicomponent mixtures.

An important application of multicomponent mass transfer theory that we have not considered in any detail in this text is diffusion in porous media with or without heterogeneous reaction. Such applications can be handled with the "dusty gas" (Maxwell–Stefan) model in which the porous matrix is taken to be the $n + 1$th component in the mixture. Readers are referred to monographs by Jackson (1977), Cunningham and Williams (1980), and Mason and Malinauskas (1983) and a review by Burghardt (1986) for further study. Krishna (1993a) has shown the considerable gains that accrue from the use of the Maxwell–Stefan formulation for the description of surface diffusion within porous media.

One of the most exciting possibilities afforded by a proper appreciation of the theory of multicomponent mass transfer is that we are able to effect separations that would otherwise not be possible using "simple-minded" binary-like approaches. An example of this is diffusional distillation where deliberate use is made of multicomponent effects to separate azeotropic mixtures. The trick of deliberately adding a third component, in the case of diffusional distillation of a noncondensing gas, to act as a selective "filter" or "membrane" can also be used to separate gaseous mixtures that are noncondensable; the third components can be a condensable vapor, such as steam. This technique has been used to separate mixtures as helium–neon and isotopes of helium. Conceptually speaking, there is no difference between diffusional distillation and membrane pervaporation; the inert gas in the former case plays the same role as the membrane in the latter case. In fact, gaseous membrane separation processes lend themselves very nicely to analysis using the

Maxwell–Stefan (dusty gas) approach by taking the membrane to be the additional "component" in the mixture. When the model is extended to account for thermodynamic nonidealities (what may be considered to be a dusty fluid model) almost all membrane separation processes can be modeled systematically. Put another way, the Maxwell–Stefan approach is the most promising candidate for developing a generalized theory of separation processes (Lee et al., 1977; Krishna, 1987).

Exercises

To complete this text we offer a selection of exercises. Some of the exercises are of a theoretical nature and we ask you to verify some of the equations presented in the text or to derive established results in other ways. In other exercises we invite you to extend the theoretical treatments in ways that we have not considered in detail (although in some cases the results are available in the literature). Still other exercises are of a computational nature and we invite you to compute mass transfer coefficients, molar fluxes, composition profiles, and other quantities. In order to gain some familiarity with the various methods described in the text we strongly recommend solving the computational problems. Hand calculation is extremely instructive but does get to be a bit tiresome after a while. There are also exercises that require you to evaluate papers in the literature. Finally, we have provided exercises that can be assigned as term projects for students, perhaps as a replacement for a final examination.

CHAPTER 1

1.1 Starting from the definitions of the molar diffusion fluxes in the molar average velocity reference frame and in volume average velocity reference frame (Table 1.3), verify Eqs. 1.2.20 and 1.2.22. These equations may be written in $n - 1$ dimensional matrix form as

$$(J) = [B^{uV}](J^V)$$
$$(J^V) = [B^{Vu}](J)$$

Show that $[B^{Vu}] = [B^{uV}]^{-1}$.
Hint: Use the Sherman–Morrison formula (Eqs. A.3.22 and A.3.23).

1.2 Starting from the definitions of the mass diffusion fluxes in the mass average velocity reference frame and in molar average velocity reference frame (Table 1.3), verify Eqs. 1.2.24 and 1.2.26. These equations may be written in $n - 1$ dimensional matrix form as

$$(j^u) = [B^{uo}](j)$$
$$(j) = [B^{ou}](j^u)$$

Show that $[B^{uo}] = [B^{ou}]^{-1}$.

1.3 Obtain an equation similar to Eqs. 1.2.20 or 1.2.22 that transforms the molar diffusion flux relative to an arbitrary reference velocity u^a to the molar diffusion flux relative to some other arbitrary reference velocity u^b. Verify that Eqs. 1.2.20 and 1.2.22 are special cases of the general result.

1.4 Repeat Exercise 1.3 for transforming mass diffusion fluxes between reference velocities u^a and u^b.

1.5 Derive an expression for transforming the gradient mole fractions to gradient of mass fractions (and vice versa) analogous to Eqs. 1.2.20 et seq. That is, find the matrix $[B^{x\omega}]$

in the equation

$$(\nabla x) = [B^{x\omega}](\nabla\omega)$$

and find the inverse transformation

$$(\nabla\omega) = [B^{\omega x}](\nabla x)$$

What is the relationship between $[B^{\omega x}]$ and $[B^{\mu o}]$?

CHAPTER 2

2.1 The chemical potential of species i in a mixture is given by

$$\mu_i = \mu_i^\circ + RT \ln(a_i)$$

where μ_i° is the standard state chemical potential, a function only of temperature. The activity a_i of species i is defined by

$$a_i = f_i/f_i^\circ$$

where f_i is the fugacity of species i and f_i° is the standard state fugacity. For ideal gases the fugacity is equal to the partial pressure

$$f_i = p_i$$

Use the above thermodynamic definitions to show that the driving force d_i defined by

$$d_i \equiv (x_i/RT)\nabla_{T,P}\mu_i \qquad (2.2.2)$$

reduces, for ideal gases, to

$$d_i = (1/P)\nabla p_i$$

2.2 Experiments in ultracentrifugation of binary solutions have been reported by Cullinan and Lenczyk (1969). In their experiments an equimolar mixture of hexane (1) and carbon tetrachloride (2) was placed in a sedimentation cell in an ultracentrifuge and rotated at 30,000 rpm. The temperature of the cell was held constant at 20°C. After sedimentation equilibrium had been obtained, samples were withdrawn from several radial positions and analyzed in a gas chromatograph with accuracy better than 1%. The composition profile at equilibrium is given in the following table.

r^2 [10^3 mm^2]	x_1	r^2 [10^3 mm^2]	x_1	r^2 [10^3 mm^2]	x_1
2.91	0.651	4.05	0.592	6.34	0.467
3.08	0.642	4.46	0.574	6.86	0.439
3.37	0.629	4.92	0.548	7.37	0.413
3.61	0.612	5.36	0.524	7.95	0.378
		5.88	0.498		

Estimate the thermodynamic derivative $\partial\mu_1/\partial x_1$ for an equimolar mixture of hexane and carbon tetrachloride, at 20°C. Assume that the partial molar volumes at the constituents are independent of pressure and composition. Compare your result with the one obtained from the equation given by Christian et al. (1960) for the activity coefficient of hexane in the same system at 20°C.

$$\ln \gamma_1 = 0.18(1 - x_1)^2$$

Data

Molar masses: $M_1 = 0.0862$ kg/mol $M_2 = 0.1538$ kg/mol
Densities at 20°C: $\rho_1 = 659$ kg/m^3 $\rho_2 = 1595$ kg/m^3

2.3 Laity (1963) reported values of "friction coefficients" $f_{ij} = RT/Đ_{ij}$ for aqueous solutions of NaCl at 25°C, at various salt concentrations. The data, recalculated in terms of the $Đ_{ij}$, is given below. Numbering is $1 = Na^+$; $2 = Cl^-$; $3 = H_2O$

c_s [mol/m^3]	$Đ_{13}$	$Đ_{23}$ [10^{-9} m^2/s]	$Đ_{12}$
0.0	1.333	2.032	0
10	1.333	2.065	0.0078
20	1.333	2.083	0.0113
50	1.325	2.083	0.0197
100	1.312	2.083	0.0302
200	1.298	2.101	0.0468

a. Determine the matrix $[B^n]^{-1}$ at a salt concentration of 200 mol/m^3.

b. Determine the salt concentration up to which the Nernst–Planck equation can "safely" be used.

CHAPTER 3

3.1 Show that Eq. 3.2.11 relates the matrix of Fick diffusion coefficients in the mass average reference velocity frame to the matrix of Fick diffusion coefficients in the molar average reference velocity frame.

3.2 Show that Eq. 3.2.12 relates the matrix of Fick diffusion coefficients in the molar average reference velocity frame to the matrix of Fick diffusion coefficients in the volume average reference velocity frame.

3.3 Continue the calculations of Example 3.2.1 for the remainder of the data in Table 3.2 (transformation of Fick diffusion coefficients in the volume average reference velocity frame to the molar average reference velocity frame).

3.4 For the system dodecane(1)–hexadecane(2)–hexane(3), Kett and Anderson (1969) reported the values of the matrix of Fick diffusion coefficients in the volume average reference velocity frame. At a temperature of 25°C and at the composition: $x_1 = 0.35$, $x_2 = 0.317$ the matrix $[D^V]$ is

$$[D^V] = \begin{bmatrix} 0.968 & 0.266 \\ 0.225 & 1.031 \end{bmatrix} \times 10^{-9} \text{ m}^2/\text{s}$$

Determine the values of the coefficients of $[D]$ in the molar average reference velocity frame.

Data

$$\bar{V}_1 = 229 \times 10^{-6} \text{ m}^3/\text{mol}$$

$$\bar{V}_2 = 293 \times 10^{-6} \text{ m}^3/\text{mol}$$

$$\bar{V}_3 = 132 \times 10^{-6} \text{ m}^3/\text{mol}$$

3.5 Carry out a review of the literature on diffusion in the vicinity of critical points and spinodal lines. Your review should concentrate on summarizing the data that has been measured as well as on contrasting the various theories that have been proposed to explain diffusion phenomena near critical conditions. Consider the question Can the fact that the matrix of Fick diffusion coefficients is close to being singular near the spinodal be exploited in a commercially useful separation process?

CHAPTER 4

4.1 Carry out a review of methods for estimating infinite dilution diffusion coefficients in binary liquid mixtures. Your review should include calculations to test the accuracy of the methods that have been proposed. Fundamental data for computing the coefficients should be obtained from a single source as far as is possible; we recommend the compilation by Daubert and Danner (1985).

4.2 Tyn and Calus (1975b) measured the Fick diffusion coefficient D in mixtures of ethanol and water at 40°C. Their data are tabulated below.

x_1	$D[10^{-9} \text{ m}^2/\text{s}]$
0.240	0.151
0.100	0.100
0.144	0.780
0.200	0.680
0.254	0.635
0.300	0.610
0.400	0.640
0.500	0.730
0.590	0.850
0.600	0.865
0.680	1.020
0.700	1.060
0.792	1.260
0.800	1.270
0.880	1.440
0.900	1.470
0.960	1.570

These data were shown in Figure 4.1 and used as the basis for Example 4.1.3. Carry out your own check of the accuracy of the Vignes correlation using the NRTL model for the activity coefficients and the parameters in Example 4.1.3. Repeat the calculations using the Van Laar model with parameters $A_{12} = 1.5922$ and $A_{21} = 0.9836$.

4.3 Look up activity coefficient model parameters for the system methanol–n-hexane in the collection of Gmehling and Onken (1977ff, Vol. I/2c). Hence, compute the thermodynamic factor Γ for a range of concentrations for each of the models in Appendix D.1. What conclusions can be drawn from a comparison of your results with those in Figure 4.2.

4.4 Sanni and Hutchison (1973) presented data on the binary Fick diffusivity for the systems benzene–chloroform, cyclohexane–carbon tetrachloride, cyclohexane–toluene, benzene–cyclohexane, benzene–toluene, and diethyl ether–chloroform. Calculate the thermodynamic factor Γ for these systems using parameters from Gmehling and Onken (1977ff). Hence, estimate the Maxwell–Stefan diffusion coefficients and test the applicability of the Vignes model.

4.5 Carry out a comparitive review of methods for estimating diffusivities in binary liquid mixtures as a function of composition. Your review should include sample calculations to support any conclusions you may have reached regarding the efficacy of any particular method. Write up your work in the format of a paper for publication in *Industrial and Engineering Chemistry Research*.

4.6 Myerson and Senol (1984) provided data for the Fick diffusivity of urea in aqueous solutions along with thermodynamic data. Determine the Maxwell–Stefan diffusion coefficient for this system and satisfy yourself that this diffusivity is better "behaved" than the Fick diffusivity, which approaches zero near the spinodal curve.

4.7 Explore the structure of the matrix of Fick diffusion coefficients for a gas–vapor mixture of acetone (1)–benzene(2)–helium(3). The Maxwell–Stefan diffusion coefficients in the vapor phase are $Ð_{12} = 2.93 \times 10^{-6}$ m^2/s, $Ð_{13} = 31.8 \times 10^{-6}$ m^2/s, and $Ð_{23} = 29.0 \times 10^{-6}$ m^2/s.

4.8 Change the order in which the components in Exercise 4.7 are numbered and recalculate the matrix of Fick diffusion coefficients at the same compositions you used in that exercise.

4.9 Compute the eigenvalues of the matrices of Fick diffusion coefficients you computed in Exercises 4.7 and 4.8. Convince yourself that at the same composition, the eigenvalues of $[D]$ are the same regardless of the order in which the components are numbered. Prove that this must always be the case.

4.10 Calculate the matrix of vapor-phase Fick diffusion coefficients in the four component mixture: acetone(1)–methanol(2)–2-propanol(3)–water(4). The composition of the vapor is

$$y_1 = 0.003157 \qquad y_2 = 0.164701$$
$$y_3 = 0.394518 \qquad y_4 = 0.437623$$

Data: The Maxwell–Stefan diffusivities in the vapor phase [units are 10^{-5} m^2/s].

$$Ð_{12}^V = 1.258 \qquad Ð_{13}^V = 0.8084$$
$$Ð_{14}^V = 1.944 \qquad Ð_{23}^V = 1.220$$
$$Ð_{24}^V = 2.896 \qquad Ð_{34}^V = 1.883$$

4.11 Calculate the matrix of liquid-phase Fick diffusion coefficients in the four component mixture: acetone(1)–methanol(2)–2-propanol(3)–water(4). The composition of the liquid is

$$x_1 = 0.003157 \quad x_2 = 0.164701$$
$$x_3 = 0.394518 \quad x_4 = 0.437623$$

The Maxwell–Stefan diffusivities in the liquid phase [units are 10^{-9} m^2/s].

$$Ð_{12}^L = 3.423 \quad Ð_{13}^L = 2.925$$
$$Ð_{14}^L = 3.742 \quad Ð_{23}^L = 3.637$$
$$Ð_{24}^L = 5.029 \quad Ð_{34}^L = 4.358$$

Activity coefficients may be calculated with the help of the NRTL model. The NRTL parameters are taken from the Gmehling and Onken (1977ff) [units of Δg are cal/mol K].

$$\Delta g_{12} = 1332.548 \quad \Delta g_{21} = -519.890 \quad \alpha_{12} = \alpha_{21} = 0.2840$$
$$\Delta g_{13} = 536.728 \quad \Delta g_{31} = -97.8216 \quad \alpha_{13} = \alpha_{31} = 0.3018$$
$$\Delta g_{14} = 189.113 \quad \Delta g_{41} = 1581.274 \quad \alpha_{14} = \alpha_{41} = 0.3010$$
$$\Delta g_{23} = 1520.984 \quad \Delta g_{32} = -961.576 \quad \alpha_{23} = \alpha_{32} = 0.2820$$
$$\Delta g_{24} = 263.258 \quad \Delta g_{24} = 115.782 \quad \alpha_{24} = \alpha_{42} = 0.3550$$
$$\Delta g_{34} = -51.824 \quad \Delta g_{43} = 1495.643 \quad \alpha_{34} = \alpha_{43} = 0.2700$$

4.12 Compute the matrix of Fick diffusion coefficients in the mass average reference velocity frame for a mixture of MEK(1)–2-butanol(2)–hydrogen(3). The vapor composition (mole fraction) is $y_1 = 0.4737$ and $y_2 = 0.0579$. The Maxwell–Stefan diffusion coefficients in the vapor phase are $Ð_{12} = 6.39 \times 10^{-6}$ m^2/s, $Ð_{13} = 54.5 \times 10^{-6}$ m^2/s, and $Ð_{23} = 53.4 \times 10^{-6}$ m^2/s and the molar masses are $M_1 = 0.072107$, $M_2 = 0.07412$, and $M_3 = 0.002016$ kg/mol.

CHAPTER 5

5.1 A set of multicomponent diffusion experiments in a two bulb diffusion cell apparatus was carried out by Duncan and Toor (1962) in an investigation of diffusional interaction effects. In Experiment 1 the initial concentration in each cell is

$$x_{10} = 0.50199 \quad x_{20} = 0.00000 \quad x_{30} = 0.49801$$
$$x_{1\ell} = 0.00000 \quad x_{2\ell} = 0.49930 \quad x_{3\ell} = 0.50070$$

where x_{i0} is the mole fraction of species i in bulb 1 and $x_{i\ell}$ is the mole fraction of species i in bulb 2. The three components are hydrogen (1)–nitrogen(2)–carbon dioxide(3). The binary diffusivities are given in Example 5.3.1. Geometric details of the two bulb cell used by Duncan and Toor and operating conditions are given in Example 5.4.1.

Compute the mole fractions in each bulb after 1 h and after 3 h. Plot the change in concentration with time.

5.2 Repeat Exercise 5.1 for Experiment 3 of Duncan and Toor (1962). The initial mole fractions in each bulb for this experiment were

$$x_{10} = 0.50056 \quad x_{20} = 0.49954 \quad x_{30} = 0.00000$$
$$x_{1\ell} = 0.49923 \quad x_{2\ell} = 0.00000 \quad x_{3\ell} = 0.50077$$

5.3 Arnold and Toor (1967) investigated diffusional interaction effects in a Loschmidt tube with the system methane (1)–argon(2)–hydrogen(3). The composition in each tube at the start of their experiment $2T$ was

$$x_{1-} = 0.515 \qquad x_{2-} = 0.485 \qquad x_{3-} = 0.0$$
$$x_{1+} = 0.488 \qquad x_{2+} = 0.000 \qquad x_{3+} = 0.512$$

Calculate the average concentration in the bottom tube after 180 min and plot the change in concentration with time.

  The experiments were carried out at 34°C and 101.3 kPa. The diffusion tube had a length of $(\pi/\ell)^2 = 60 \text{ m}^{-2}$. The binary diffusion coefficients of the three ternary gas pairs are given in Example 5.5.1.

5.4 The composition in the two tubes of a Loschmidt diffusion apparatus are

$$x_{1-} = 0.000 \qquad x_{2+} = 0.509 \qquad x_{3+} = 0.491$$
$$x_{1+} = 0.488 \qquad x_{2-} = 0.000 \qquad x_{3-} = 0.512$$

Repeat the calculations of Exercise 5.3 for this set of data, which corresponds to the start of experiment $3T$ of Arnold and Toor (1967).

5.5 McKay (1971) carried out experiments in a Loschmidt-type diffusion cell with a *nine* component system including nitrogen, carbon dioxide, and a number of paraffins. The experiments were carried out at the high temperatures and pressures typical of underground oil and gas reservoirs. McKay's paper includes a dimensioned drawing of the Loschmidt-type cell used for the experiment and the initial conditions for some of his experiments. Use the theory of Section 5.5 to predict the composition profiles in this system. State clearly any assumptions made in your calculations. You will need to estimate diffusion coefficients at high pressure. Consult Reid et al. (1987) for details of methods that can be used for this purpose.

5.6 Consider interphase mass transfer in the system glycerol(1)–water(2)–acetone(3). In a particular equilibration experiment carried out by Krishna et al. (1985) the initial composition in the acetone-rich phase is

$$x_{10} = 0.85 \qquad x_{20} = 0.15 \qquad x_{30} = 0.0$$

The composition of the interface is

$$x_{1I} = 0.5480 \qquad x_{2I} = 0.2838 \qquad x_{3I} = 0.1682.$$

The matrix of volumetric transfer coefficients $[K]$, averaged over the entire equilibration process is

$$[K] = \begin{bmatrix} 0.4948 & -0.0385 \\ -0.6552 & 0.2852 \end{bmatrix} \text{h}^{-1}$$

Determine the equilibration trajectory for the acetone-rich phase. Draw this path on a triangular diagram together with the eigenvectors passing through the initial state.

5.7 In a diaphragm cell diffusion occurs through a horizontal porous diaphragm separating two chambers containing liquid mixtures of different compositions. In the mathematical analysis of diffusion cell data, the following assumptions normally are made.

a. Diffusion through the diaphragm is unidirectional.

b. Quasisteady state is obtained.

c. The contents of the cell chambers are well mixed.

d. Partial molar volumes are constant and there is no net volume change on mixing. With the above assumptions, the diffusion coefficient in a binary liquid mixture D_{12} is obtained from

$$\ln(\Delta c_1/\Delta c_{10}) = -\beta D_{12}t$$

where Δc_{10} is the initial difference in concentration of species 1 in the two chambers; Δc_1 is the difference in concentration of species 1 in the two chambers at any time t after the start of the experiment, and β is the cell constant. Use the linearized theory of multicomponent diffusion to derive the multicomponent generalization of the above diaphragm cell relation. Suggest a procedure for calculating the four elements of $[D]$ from experimental data obtained with a ternary system.

For more information (and a derivation) see Cussler (1976). A more general analysis is given by DeLancey (1969).

CHAPTER 6

6.1 Using the Maxwell–Stefan equations for nonideal fluids, Eqs. 2.2.1, as a basis, develop a general expression for an effective diffusivity defined in terms of the generalized driving force as

$$J_i = -c_t D_{i,\text{eff}} d_i$$

With the help of Eq. 2.2.4, develop a more practical expression for the effective diffusivity in nonideal fluids defined in terms of the gradient of mole fraction.

$$J_i = -c_t D_{i,\text{eff}} \nabla x_i$$

6.2 Repeat Example 6.4.1 for the conditions at the start of Experiments 1 and 3 of Duncan and Toor (1962). See Exercises 5.1 and 5.2 for data.

6.3 Show that the effective diffusivity method leads to composition profiles in ternary systems that are straight lines when plotted on triangular diagrams.

6.4 Repeat Exercise 5.5 using an effective diffusivity method.

CHAPTER 8

8.1 In Section 8.3 we presented a derivation of an exact matrix solution of the Maxwell–Stefan equations for diffusion in ideal gas mixtures. Although the final expression for the composition profiles (Eq. 8.3.12), is valid whatever relationship exists between the fluxes (i.e., bootstrap condition), the derivation given in Section 8.3.1 cannot be used when the total flux is zero. Why not? Use the method of solving coupled differential equations developed in Appendix B.3 to develop Eq. 8.3.12. Does the same limitation apply to this alternative method of obtaining the composition profiles. If not, are there any other limitations that do apply.

In the course of carrying out Exercise 8.1 you should obtain the following expressions

$$\ln\left\{\frac{\hat{\Phi}_i \hat{y}_{i\delta} + \hat{\phi}_i}{\hat{\Phi}_i \hat{y}_{i0} + \hat{\phi}_i}\right\} = \hat{\Phi}_i \qquad \text{when} \qquad \hat{\Phi}_i \neq 0$$

and

$$\hat{y}_{i\delta} - \hat{y}_{i0} = \hat{\phi}_i \quad \text{when} \quad \hat{\Phi}_i = 0$$

where the $\hat{\Phi}_i$ are the eigenvalues of the matrix of mass transfer rate factors defined by Eqs. 8.3.4 and 8.3.5. In addition,

$$(\hat{y}) = [P]^{-1}(y)$$

and

$$(\hat{\phi}) = [P]^{-1}(\phi)$$

where $[P]$ is the modal matrix of $[\Phi]$.

These equations, which sometimes are referred to as parametric solutions, are quite general in the sense that they hold for mixtures with any numbers of constituents and for any relationship between the fluxes. In fact, for certain special cases these equations have been available in the literature for a long time. The next five exercises explore these special cases.

8.2 Slattery (1981) presents a solution of the Maxwell–Stefan equations for the special case when two molar fluxes are zero, $N_1 = N_2 = 0$. Write down expressions for $\hat{\Phi}_i$, $[P]$, $(\hat{y})$, and show that, for this special case, the eigenvalue solutions are equivalent to the expressions given by Slattery.

8.3 Toor (1957) derived a solution of the Maxwell–Stefan equations for ternary systems when the total molar flux is zero, $N_t = 0$. Write down expressions for $\hat{\Phi}_i$, $[P]$, $(\hat{y})$, and show that, for $N_t = 0$, the eigenvalue solutions are equivalent to the expressions given by Toor.

8.4 Gilliland (1937) derived a solution of the Maxwell–Stefan equations for ternary systems when the flux of component 3 is zero, $N_3 = 0$. Write down expressions for $\hat{\Phi}_i$, $[P]$, $(\hat{y})$, and show that, for this special case, the eigenvalue solutions are equivalent to the expressions obtained by Gilliland.

8.5 Bröcker and Schulze (1991) expounded on Gilliland's solution to the Maxwell–Stefan equations. It has long been known that Gilliland's equations are satisfied by more than one set of numerical values for the molar fluxes. Bröcker and Schulze present criteria for choosing the "correct" set. Use the Bröcker and Schulze algorithm and determine the three sets of molar fluxes that satisfy Gilliland's equations for, say, Examples 8.3.2 and 8.6.1. Is the exact matrix solution satisfied by more than one set of numerical values of the fluxes?

8.6 The eigenvalue solutions are also valid for two component systems where all matrices are of order one. Simplify the eigenvalue solutions for mass transfer in binary systems when

a. $N_t = 0$

b. $N_2 = 0$

c. $N_1/(N_1 + N_2) = z_1$

See Section 8.2 for the answers you should get.

8.7 Verify that for three component systems the matrix $[R]$, with elements defined by Eqs. 8.3.25, can be inverted explicitly to give Eqs. 8.3.30.

8.8 Do the algorithms of Section 8.3 work if one component, No. 2 say, is totally absent from the mixture (in other words, $y_2 = 0$ at both ends of the film)? *Hint*: Using a ternary system as a basis, write expressions (not numbers) for every calculation required by Algorithm 8.3.1.

8.9 Johns and DeGance (1975) presented an exact solution of the Maxwell–Stefan equations. The starting point for their analysis is the n-dimensional matrix equation

$$\frac{d(y)}{d\eta} = [T](y)$$

where $[T]$ is a matrix with elements defined by

$$T_{ii} = \sum_{\substack{k=1 \\ i \neq k}}^{n} \frac{N_k}{c_t \mathcal{D}_{ik}/\ell} \qquad i = 1, \ldots, n$$

$$T_{ij} = -N_i/(c_t \mathcal{D}_{ij}/\ell) \qquad i \neq j = 1, \ldots, n$$

The solution to this equation, subject to the initial condition $\eta = 0$ and $y_i = y_{i0}$ is

$$(y) = \exp[[T]\eta](y_0)$$

Show that this expression is equivalent to Eq. 8.3.9 when reduced to $n - 1$ dimensional matrix form. What, if any, is the relationship between the eigenvalues of $[\Phi]$ and $[T]$? See Taylor (1982c) for help if needed.

8.10 In some cases it is possible (although not for the bootstrap conditions $N_t = 0$ and $N_3 = 0$) for the eigenvalues of $[\Phi]$ to be complex (see, e.g., the paper of Johns and DeGance cited in Exercise 8.9). Develop alternatives to Eqs. 8.3.46 for the evaluation of the correction factor matrix when $[\Phi]$ has complex eigenvalues. Show that any imaginary part of the composition profiles is identically zero (whoever heard of an imaginary mole fraction)! This exercise can be extremely difficult or trivially simple depending on how you choose to go about it.

8.11 Repeat Example 8.4.1 (dehydrogenation of ethanol) using an exact matrix method of determining the fluxes.

8.12 Repeat Example 8.6.1 (diffusion in a Stefan tube) using an exact matrix method of determining the fluxes.

8.13 Repeat Example 8.5.1 (evaporation into two inert gases) using an exact solution of the Maxwell–Stefan equations for determining the fluxes.

8.14 A condenser operates with a feed vapor consisting of ammonia(1)–water vapor(2)–hydrogen(3) at a pressure of 340 kPa. At one point in the condenser the mole fractions in the bulk vapor are $y_{10} = 0.30$, $y_{20} = 0.40$, and $y_{30} = 0.30$. The liquid on the condensing surface at this point is at 93.3°C and contains 10 mol% ammonia and 90 mol% water, with negligible hydrogen. The composition of the vapor–gas mixture at the liquid surface, assumed to be in equilibrium with the liquid surface of the stated composition, is $y_{1\delta} = 0.455$, $y_{2\delta} = 0.195$, and $y_{3\delta} = 0.35$. Employ the exact matrix solution of the Maxwell–Stefan equations to estimate the rate of condensation of water relative to that of ammonia.

The diffusivities of the three binary pairs at 101 kPa are

$$\mathcal{D}_{12} = 29.4 \text{ mm}^2/\text{s}$$

$$\mathcal{D}_{13} = 114 \text{ mm}^2/\text{s}$$

$$\mathcal{D}_{23} = 130 \text{ mm}^2/\text{s}$$

Molar density of vapor mixture: $c_t = 30$ mol/m^3. This exercise is worked out in detail by Sherwood et al. (1975) who used the eigenvalue expressions derived in Exercise 8.4 as a basis for computing the fluxes.

8.15 Taylor and Krishnamurthy (1982) provided the following data for diffusion in the gaseous system He(1), SF$_6$(2), O$_2$(3):

Bulk vapor-phase composition: $y_{10} = x$, $y_{20} = 0.0$, $y_{30} = 1 - x$

Interface vapor composition: $y_{1\delta} = 0$, $y_{2\delta} = x$, $y_{3\delta} = 1 - x$

where x is to be varied in the range 0.0–1.0.

The diffusivities are

$$Ð_{12} = 43.54 \text{ mm}^2/\text{s}$$

$$Ð_{13} = 79.15 \text{ mm}^2/\text{s}$$

$$Ð_{23} = 9.993 \text{ mm}^2/\text{s}$$

The diffusion process may be assumed to be equimolar at a temperature of 310 K and a pressure of 100 kPa across a film of thickness $\delta = 0.3$ mm.

Use an exact matrix method to estimate the fluxes and plot the composition profiles on a triangular diagram for selected values of x.

8.16 Hydrogen is being oxidized on a solid catalyst surface at steady state according to the reaction

$$2H_2 + O_2 \rightarrow 2H_2O$$

At a given point in the reactor at 473 K and 101.35-kPa total pressure the composition (mole fraction) of the bulk gas phase is for H$_2$(1), $y_1 = 0.40$; O$_2$, $y_2 = 0.20$; and for H$_2$O(3), $y_3 = 0.40$. Assume that H$_2$ and O$_2$ diffuse through a gas film 1 mm thick to the catalyst surface, react instantly, and the H$_2$O diffuses back through the same film.

Estimate the rate of reaction in kilomoles of hydrogen per second per meters squared of catalyst surface (kmol H$_2$/s m^2 catalyst surface) that can be obtained if the overall reaction rate is diffusion controlled.

The diffusivities are

$$Ð_{12} = 180 \text{ mm}^2/\text{s}$$

$$Ð_{13} = 212 \text{ mm}^2/\text{s}$$

$$Ð_{23} = 59 \text{ mm}^2/\text{s}$$

8.17 The development of the Toor–Stewart–Prober method in Section 8.4 is based on the assumptions that the molar density and the matrix of Fick diffusion coefficients in the molar average reference velocity frame can be assumed constant along the diffusion path. Develop the theory anew in the mass average reference velocity frame; that is, assuming ρ_t and $[D^\circ]$ can be considered constant. You will need to work with mass fluxes n_i and mass diffusion fluxes j_i.

8.18 Repeat Example 8.3.1 (equimolar diffusion in a ternary system) using the linearized equations for determining the fluxes and composition profiles.

8.19 Repeat Example 8.3.2 (diffusional distillation) using the linearized equations for determining the fluxes and composition profiles. Compare your results to those given in Example 8.3.2.

8.20 Repeat Example 8.6.1 (diffusion in a Stefan tube) using the Toor–Stewart–Prober method of determining the fluxes. Compare the profiles computed from the linearized equations to the profiles obtained with the exact method and the experimental data given in Example 2.2.1.

8.21 Repeat Example 8.5.1 (evaporation into two inert gases) using the Toor–Stewart–Prober method of determining the fluxes.

8.22 Repeat Exercise 8.14 using the Toor–Stewart–Prober method of determining the fluxes.

8.23 Show that the explicit method of Taylor and Smith (1982) (discussed in Section 8.5) is an exact solution of the Maxwell–Stefan equations if all the binary diffusion coefficients are equal. Solutions are given by Burghardt (1984) and Taylor (1984).

8.24 The explicit method of Taylor and Smith (1982) discussed in Section 8.5 makes use of the mass transfer rate factor

$$\Phi = \ln(\bar{\nu}_\delta / \bar{\nu}_0) \tag{8.5.25}$$

where

$$\bar{\nu} = \sum_{j=1}^{n} \nu_j y_j$$

Show that Φ defined in Eq. 8.5.25 is an exact eigenvalue of the matrix $[\Phi]$ defined by Eqs. 8.3.4 and 8.3.5 for three cases.

a. All diffusivities equal (regardless of the values of the ν_i).
b. Stefan diffusion ($N_n = 0$) (regardless of the values of $Đ_{ij}$ and ν_i).
c. Equimolar countertransfer ($N_t = 0$, $\Phi = 0$) (regardless of the values of the $Đ_{ij}$ and ν_i).
Is Φ always an eigenvalue of $[\Phi]$?

8.25 A matrix $[A]$ that arises in the explicit methods of Section 8.5 has elements defined by

$$A_{ii} = \frac{y_i(\nu_i/\nu_n)}{Đ_{in}} + \sum_{\substack{k=1 \\ i \neq k}}^{n} \frac{y_k}{Đ_{ik}} \tag{8.5.21}$$

$$A_{ij} = -y_i\left(\frac{1}{Đ_{ij}} - \frac{(\nu_j/\nu_n)}{Đ_{in}}\right) \tag{8.5.22}$$

Invert $[A]$ for the three component case. If $[W]$ denotes $[A]^{-1}$, what are the elements of $[W]$ when

a. $N_t = 0$
b. $N_3 = 0$
What is the determinant of $[A]$?

8.26 Prove that

$$[A] = [B][\beta]^{-1} \quad \text{or} \quad [A]^{-1} = [\beta][B]^{-1} \tag{8.5.23}$$

8.27 Repeat Example 8.3.2 (diffusional distillation) using the explicit method for determining the fluxes and composition profiles. Compare your results to those given in Example 8.3.2.

8.28 Repeat Example 8.4.1 (dehydrogenation of ethanol) using an explicit method of determining the fluxes.

8.29 Repeat Example 8.6.1 (diffusion in a Stefan tube) using an explicit method of determining the fluxes. Plot the composition profiles and compare them to the results of an exact solution and to the experimental data given in Example 2.2.1.

8.30 Repeat Example 8.6.1 (diffusion in a Stefan tube) using Wilke's effective diffusivity formula (Eq. 6.1.14).

8.31 Repeat Example 8.6.1 (diffusion in a Stefan tube) using the effective diffusivity formula of Kato et al. (1981), Eq. 6.1.16.

8.32 Repeat Example 8.6.1 (diffusion in a Stefan tube) using the effective diffusivity formula of Burghardt and Krupiczka (1975), Eq. 6.1.15.

8.33 Repeat Example 8.3.2 (diffusional distillation) using an effective diffusivity method for determining the fluxes and composition profiles. Compare your results to those given in Example 8.3.2.

8.34 Repeat Example 8.4.1 (dehydrogenation of ethanol) using an effective diffusivity method of determining the fluxes.

8.35 Repeat Example 8.3.1 (equimolar diffusion in a ternary system) using an effective diffusivity method for determining the fluxes and composition profiles. Compare the fluxes calculated with the effective diffusivity model to those obtained in Example 8.3.1.

8.36 Simplify the definition of the effective diffusivity given by Eq. 6.1.7 for the special case when two molar fluxes are zero, $N_1 = N_2 = 0$. What is the relationship between the effective diffusivity method and the method of Section 8.5.2 for this special case.

8.37 An exact solution of the Maxwell–Stefan equations for diffusion in nonideal fluids may be obtained using, as a basis, the method developed in Section 8.3.5. All of the results given in that section are valid with the proviso that the matrix $[D]$ is given by $[B]^{-1}[\Gamma]$.

  As an alternative to this approach we may cast the Maxwell–Stefan equations in $n-1$ dimensional matrix form as

$$[\Gamma]\frac{d(x)}{d\eta} = [\Phi](x) + (\phi) \tag{8.7.3}$$

where $[\Phi]$ and (ϕ) are as defined for the ideal gas case.

  For nonideal fluids $[\Gamma]$, $[\Phi]$, and (ϕ) are functions of composition (and, hence, of position). Develop an exact solution to this $n-1$ dimensional matrix equation using the method of repeated solution (see Appendix B.2 for details of the method). Note that at this level of analysis it is not necessary to know exactly how $[\Gamma]$, $[\Phi]$, and (ϕ) depend on position; it suffices to know only that they are functions of η.

8.38 Kubaczka and Bandrowski (1991) presented a method for calculating fluxes in nonideal fluid mixtures. Examine their paper and prepare a critical essay discussing any novel features of their paper. Back up your conclusions with sample calculations based on the problems considered in the paper by Kubaczka and Bandrowski.

8.39 Devise an efficient computational algorithm for computing the molar fluxes using the models developed by Kubaczka and Bandrowski (1991). Present your algorithm in the form of a paper suitable for submission to *Computers and Chemical Engineering*.

8.40 A generalized derivation of the explicit methods of Section 8.5 for diffusion in ideal gas mixtures is given by Taylor and Smith (1982). The starting point for their analysis is the set of Maxwell–Stefan equations written in n-dimensional matrix form. By making use of a matrix transformation, they eventually obtain a generalized explicit method. Starting from the Maxwell–Stefan equations for nonideal fluids, Eq. 2.2.1, use the method of solution employed by Taylor and Smith to develop an explicit method for nonideal fluids. *Hint*: You will need to develop an n-dimensional analog of Eq. 2.2.4 for the driving forces d_i (Appendix D can be helpful here). Write up your work in a format suitable for submission as a paper in *Chemical Engineering Communications*. However, before you submit it, look up articles by Kubaczka and Bandrowski (1990) and by Taylor (1991).

8.41 Repeat Example 8.7.1 using the linearized method of Toor–Stewart–Prober (1964) discussed in Section 8.4.

8.42 The explicit method of Taylor and Smith (1982) for mass transfer in ideal gas mixtures is an exact solution of the Maxwell–Stefan equations for two component systems where all matrices are of order 1. Does the generalized explicit method derived in Exercise 8.40 reduce to the expressions given in Section 8.2 for a film model of mass transfer in binary systems?

8.43 Repeat Example 8.7.1 using the explicit methods derived in Exercise 8.40.

8.44 Repeat Example 8.7.1 using the effective diffusivity methods you derived in Exercise 6.1. Compare the results with those obtained in Example 8.7.1 and in Exercise 8.43.

8.45 Repeat Example 8.8.1 (ternary distillation in a wetted wall column) using the linearized equations to calculate the molar fluxes but following the suggestion given in the second paragraph of Section 8.8.3 for estimating the multicomponent mass transfer coefficients in terms of the binary mass transfer coefficients.

8.46 Repeat Example 8.8.1 (ternary distillation in a wetted wall column) using the method of calculating the fluxes developed in Section 8.3 but following the suggestion of Krishna and Standart as described in Section 8.8.3 for estimating the binary mass transfer coefficients.

8.47 In Example 8.4.1 we considered multicomponent diffusion with heterogeneous chemical reaction. In this example the reaction was considered to be first order in the mole fraction of ethanol. For the catalytic reaction $A + 2B \rightarrow C$, conforming to Langmuir–Hinshelwood kinetics, it has been demonstrated by Löwe and Bub (1976) that multiple steady states are possible. The analysis of Löwe and Bub assumes equal diffusivities for the three reacting species A, B, and C. Extend their analysis for the general case of unequal diffusivities using the Maxwell–Stefan equations. In addition to the possibility of multiple steady states show that it is possible to obtain diffusion of a species against its gradient.

CHAPTER 9

9.1 Extend the film-penetration model of mass transfer developed by Toor and Marchello (1958) to multicomponent mixtures. See also, Krishna (1978a).

9.2 A droplet containing a mixture of acetone(1)–benzene(2)–methanol(3) has a diameter of 8 mm and attains a velocity of 0.1 m/s in a sieve tray extraction column when it is dispersed in a continuous hydrocarbon phase. Use the penetration model to estimate the matrix of low flux mass transfer coefficients $[k]$ inside the droplet.

The matrix of Fick diffusion coefficients is

$$[D] = \begin{bmatrix} 3.860 & 0.890 \\ -0.30 & 2.240 \end{bmatrix} \times 10^{-9} \text{ m}^2/\text{s}$$

9.3 For the system considered in Exercise 9.2 calculate the unaccomplished composition change

$$(x_i - x_{i\infty})/(x_{i0} - x_{i\infty})$$

as a function of the dimensionless parameter

$$z/\sqrt{4D_{11}t}$$

using Eq. 9.3.28 valid for low transfer fluxes. For these calculations assume that the ratio of driving forces

$$(x_{10} - x_{1\infty})/(x_{20} - x_{2\infty})$$

for components acetone(1) and benzene(2) to have the following set of values: (a) 4.5 and (b) 0.1.

Examine the behavior of the composition profiles for acetone and benzene in the two cases (a) and (b) and see if you detect any unusual, unbinary-like behavior. After completing your calculations have a look at the paper by Krishna (1978b) and compare your results with his. How would you go about tackling the same problem for finite mass transfer rates?

CHAPTER 10

10.1 Your objective here is to examine the influence of turbulence on the interfacial fluxes of acetone(1)–benzene(2)–helium(3) for interphase mass transfer in a wetted wall column at the top of the column where the incoming vapor–gas mixture first comes into contact with the downflowing liquid mixture of acetone and benzene. The details of the column and the operating conditions are given in Example 11.5.3. For the purposes of this exercise you may ignore thermal effects. Take the vapor-phase compositions, expressed in mass fractions, in the bulk vapor and at the interface to be

$$\omega_1^V = 0.44512 \qquad \omega_2^V = 0.0 \qquad \omega_3^V = 0.55488$$

$$\omega_1^I = 0.13770 \qquad \omega_2^I = 0.64140 \qquad \omega_3^I = 0.22089$$

Carry out calculations of the mass fluxes at Reynolds numbers of 10^4, 1.5×10^4, 2×10^4, 3×10^4, 4×10^4, and 5×10^4 using the following models: (1) Turbulent kinematic viscosity determined from the Von Karman universal velocity profile (Eqs. 10.2.17 and 10.2.18). (2) Turbulent kinematic viscosity determined from the Prandtl mixing length model (Eq. 10.2.19), along with the Van Driest damping factor (Eq. 10.2.22). (3) Chilton–Colburn analogy (Eq. 10.3.28).

Note that the sign and magnitude of the flux of acetone is sensitive to the value of the Re number (why?) and to the choice of the turbulent mass transfer model (again, why?). Rationalize your results in terms of the results portrayed in Figure 10.8.

10.2 2-Propanol(1) and water(2) are condensing in the presence of nitrogen(3) inside a vertical tube. At the vapor inlet, the gas–vapor phase has the composition

$$y_1 = 0.1123 \qquad y_2 = 0.4246$$

The composition of the vapor in equilibrium with the condensate at that point is

$$y_{10} = 0.1457 \qquad y_{20} = 0.1640$$

Estimate the rates of condensation using the Von Karman model for calculating the mass transfer coefficients.

The diffusivities of the three binary pairs are

$$Đ_{12} = 15.99 \text{ mm}^2/\text{s}$$
$$Đ_{13} = 14.44 \text{ mm}^2/\text{s}$$
$$Đ_{23} = 38.73 \text{ mm}^2/\text{s}$$

Density of vapor mixture: $\rho_t^V = 0.882 \text{ kg/m}^3$.
Viscosity of vapor mixture: $\mu^V = 1.606 \times 10^{-5}$ Pa s.
Vapor phase Reynolds number: $\text{Re}_V = 9574$.
Friction factor: $f/2 = 0.023 \text{ Re}_V^{-0.17}$.

10.3 Repeat Exercise 10.2 using the multicomponent generalization of the Chilton-Colburn analogy to estimate the mass transfer coefficients.

10.4 Repeat Example 10.4.1 using the multicomponent generalization of the Chilton-Colburn analogy to estimate the mass transfer coefficients.

10.5 Von Behren et al. (1972) analyzed multicomponent mass transfer in turbulent flow in a pipe. Show that their model is fundamentally incorrect. You may also refer to the paper by Stewart (1973).

10.6 Reijnhart et al. (1980) analyzed mass transfer from spills of volatile, single component, liquids into turbulent air streams. As the temperature increases the high flux correction factors become significant due to the increased mass transfer rates. Reconcile the analysis of Reijnhart et al. with the material presented in Chapter 10. Attempt to simulate their experimental results for the evaporation of toluene into air. Note that for spills of multicomponent liquid mixtures, the analysis of Chapter 10 really comes into its own.

CHAPTER 11

11.1 One of Modine's (1963) experiments involving mass transfer in a wetted wall column was the basis for Example 11.5.3. Here we provide data for another of Modine's experiments and ask you to compute the mass fluxes at the top of the column. The components for this experiment were acetone (1)–benzene (2)–nitrogen (3).

Diameter of column, $d = 0.025019$ m.
Nitrogen flow rate = 0.0891404 mol/s.
Vapor inlet temperature = 37.4°C.
Liquid flow rate = 5.8223×10^{-3} kg/s.
Liquid inlet temperature = 38.05°C.
Inlet pressure = 154.99 kPa.
Inlet vapor composition (mole fraction): $y_1 = 0.123570 \qquad y_2 = 0.0$.
Inlet liquid composition (mole fraction): $x_1 = 0.076529$.

Estimate the mass fluxes at the top of the wetted wall column. Physical property data (where it differs from that provided in Example 11.5.3) is given below.

Maxwell–Stefan diffusion coefficients in the vapor phase

$$\mathcal{D}_{12} = 3.54 \times 10^{-6} \text{ m}^2/\text{s}$$

$$\mathcal{D}_{13} = 7.65 \times 10^{-6} \text{ m}^2/\text{s}$$

$$\mathcal{D}_{23} = 6.49 \times 10^{-6} \text{ m}^2/\text{s}$$

Molar masses are $M_1 = 0.05808$, $M_2 = 0.0781$, and $M_3 = 0.028$ kg/mol.

Vapor viscosity and thermal conductivity

$$\mu^V = 1.400 \times 10^{-5} \text{ Pa s}$$

$$\lambda^V = 0.025 \text{ W/m K}$$

The liquid-phase mass transfer coefficient is taken to be

$$k^L = 2.26 \times 10^{-4} \text{ m/s}$$

The molar density of the liquid phase has been estimated as

$$c_t^L = 11.34 \text{ kmol/m}^3$$

The vapor-phase molar density may be calculated from the ideal gas law.

The liquid-phase heat transfer coefficient may be estimated using a correlation provided by Modine et al. (1963). The numerical value used in this example is

$$h^L = 2063 \text{ W/m}^2\text{K}$$

The pure component vapor-phase heat capacities have been estimated at the bulk vapor temperature as

$$C_{p1}^V = 75.13 \qquad C_{p2}^V = 85.98 \qquad C_{p3}^V = 29.15 \text{ J/kmol K}$$

and the pure component liquid heat capacities at the bulk liquid temperature are

$$C_{p1}^L = 128.8 \qquad C_{p2}^L = 138.8 \text{ J/kmol K}$$

11.2 Repeat Exercise 11.1 using the Chilton–Colburn analogy to estimate the mass transfer coefficients.

11.3 Revise the analysis of Example 11.5.3 and show how a method based on the film models of Chapter 8 could be used to compute the rates of mass transfer. Then use the Krishna–Standart method (of Sections 8.3 and 8.8.3) and compute the molar fluxes. Binary pair mass transfer coefficients may be estimated using the Chilton–Colburn analogy.

11.4 Repeat the calculations of Exercise 11.3 using the linearized theory of multicomponent mass transfer to compute the molar fluxes.

11.5 Calculate the molar fluxes in the system methanol(1)–ethanol(2)–water(3) under the following conditions:

The bulk vapor composition.

$$y_1 = 0.94200 \qquad y_2 = 0.02800 \qquad y_3 = 0.03000$$

The bulk liquid composition.

$$x_1 = 0.94199 \qquad x_2 = 0.02801 \qquad x_3 = 0.03000$$

Bulk vapor temperature: $T^V = 67°C$.

Bulk liquid temperature: $T^L = 64°C$.

Pressure: $P = 101.3$ kPa.

Binary mass transfer coefficients in the vapor phase where $c_t^V \kappa_{12}^V = 0.4054$ mol/m²s, $c_t^V \kappa_{13}^V = 0.6435$ mol/m²s, and $c_t^V \kappa_{23}^V = 0.5397$ mol/m²s.

Binary mass transfer coefficients in the liquid phase where $c_t^L \kappa_{12}^L = 0.2141$ mol/m²s, $c_t^L \kappa_{13}^L = 0.5841$ mol/m²s, and $c_t^L \kappa_{23}^L = 0.4561$ mol/m²s.

Vapor–liquid equilibrium data for this system can be found in Gmehling and Onken (1977ff, Vol. I/1 pp. 562–568).

This exercise is adapted from an example worked out by Röhm and Vogelpohl (1980).

11.6 Repeat Exercise 11.5 but use a pseudobinary (effective diffusivity) approach for the diffusion fluxes.

11.7 Develop the film model for simultaneous mass and energy transfer including Soret and Dufour effects. Use the Toor–Stewart–Prober linearized theory in developing the model. An example of a process where thermal diffusion effects cannot be ignored is chemical vapor deposition. Use the model to perform some sample calculations for a system of practical interest. You will have to search the literature to find practical systems. To get an idea of the numerical values of the transport coefficients consult the book by Rosner (1986).

CHAPTER 12–13

12.1 Lockett and Ahmed (1983) studied the distillation of methanol and water at total reflux in a sieve tray column with the following dimensions:

Column diameter: 0.59 m

Weir length: 0.457 m

Flow path length: 0.374 m

Active tray bubbling area: 0.200 m²

Downcomer area: 0.034 m²

Total hole area: 0.0185 m²

Tray spacing: 0.6 m

Hole diameter: 4.8 mm

Hole pitch: 12.7 mm

Exit weir height: 50 mm

Below we provide the following data for one of the experiments.

F factor based on active area of tray.

$$F_s = 1.06 \ (m/s)(kg/m^3)^{0.5}$$

Composition of the liquid leaving the tray is

$$x_1 = 0.154$$

The pressure is 101 kPa and the temperature is 357.3 K.

The fractional approach to flooding was estimated to be

$$F_f = 0.241$$

Use the Chan and Fair method to estimate the numbers of transfer units and, hence, the point efficiency for this system. The experimental value of the point efficiency for this experiment was 0.696.

Physical properties of the mixture can be estimated using Figure 3 in the paper Lockett and Plaka (1983).

12.2 Use the fundamental model of mass transfer on distillation trays in Section 12.1.7 to investigate the performance of the system methanol(1)–water(2) considered in Exercise 12.1.

Assume a bimodal bubble population with the following parameters:

The small bubble diameter: $d_{II,1} = 5$ mm.

The small bubble rise velocity: $U_{II,1} = 0.6$ m/s.

The fractional holdup of vapor in the small bubbles: $\varepsilon_{II,1} = 0.2$.

The large bubble diameter: $d_{II,2} = 50$ mm.

The fractional holdup of vapor in the small bubbles: $\varepsilon_{II,2} = 0.5$.

The height of the bubbling zone: $h_{II} = 0.11$ m.

Neglect the formation zone.

Compare your results with those of Lockett and Plaka (1983). In what ways does the model of Section 12.1.7 differ from theirs?

12.3 Use the fundamental model of mass transfer on distillation trays in Section 12.1.7 to investigate the performance of the system toluene(1)–methylcyclohexane (2). Tray dimensions and physical properties are provided in Example 12.1.1.

Assume a bimodal bubble population with the following parameters:

The small bubble diameter: $d_{II,1} = 5$ mm.

The small bubble rise velocity: $U_{II,1} = 0.3$ m/s.

The fractional holdup of vapor in the small bubbles: $\varepsilon_{II,1} = 0.2$.

The large bubble diameter: $d_{II,2} = 50$ mm.

The fractional holdup of vapor in the small bubbles: $\varepsilon_{II,2} = 0.5$.

The height of the bubbling zone (neglect the formation zone so that $h_{II} = h_f$) and superficial vapor velocity was calculated in Example 12.1.1.

How does the mass transfer performance of the tray change if the liquid-phase mass transfer resistance is ignored?

12.4 Biddulph and Kalbassi (1988) investigated the distillation at total reflux of the ternary system methanol(1)–1-propanol(2)–water(3). Their experimental data is tabulated below.

Run	x_1	x_2	x_3
WA	0.1533	0.5231	0.3255
WB	0.2847	0.4126	0.3027
WC	0.3611	0.3557	0.2832
WD	0.2474	0.3927	0.3098
WE	0.1460	0.2015	0.6525
WF	0.0785	0.1343	0.7870
WG	0.2127	0.2103	0.5769
WI	0.1572	0.2388	0.6041

In Example 12.2.1 we computed the matrices of numbers of transfer units for run WA. Estimate the number of transfer units for each of the remaining experiments.

Hence, compute the Murphree point efficiencies for each data set (see Example 13.3.1). Note that it will be necessary to compute the composition of the vapor that would be in equilibrium with the liquid for each experiment. Physical property data is provided in Example 12.2.1.

12.5 Correlations of numbers of transfer units developed for binary systems may be used to compute numbers of transfer units for multicomponent systems as described in Section 12.1.5. An alternative method that follows the ideas put forward by Toor in his development of the linearized theory of mass transfer is to generalize binary correlations by replacing the binary diffusivity with the matrix of Fick diffusion coefficients (in much the same way that we generalized correlations of binary mass transfer coefficients in Section 8.8.2). Let the number of transfer units in a binary system be expressed as

$$\mathbb{N}_V = \alpha (D^V)^\beta$$

where α and β are constants or functions of physical properties, equipment design, and operational parameters. The matrix of transfer units for a multicomponent vapor mixture would be expressed as

$$[\mathbb{N}_V] = \alpha [D^V]^\beta$$

Sylvester's formula could be used to evaluate $[D^V]^\beta$. A similar expression could be written for the liquid phase.

Based on these ideas, develop multicomponent forms of the AIChE, Chan and Fair, and Zuiderweg methods of computing numbers of transfer units.

12.6 A mixture of acetone(1)–methanol(2)–water(3) is being distilled in a tray column. You are asked to examine the mass transfer behavior of the system on a particular tray operating at 101.3 kPa and 80°C. The composition of the vapor entering the tray is $y_{1E} = 0.257$, $y_{2E} = 0.518$, and $y_{3E} = 0.225$. The composition of the vapor in equilibrium with the liquid leaving the tray has been determined to be $y_1^* = 0.480$, $y_2^* = 0.445$, and $y_3^* = 0.075$. Measurements of the tray hydrodynamics have shown that the bubble size distribution on the tray is bimodal. The "small" bubble population consists of bubbles of 5-mm diameter having a rise velocity of 0.3 m/s. The "large" bubble population consists of bubbles of 12.5-mm diameter and having a rise velocity of 1.5 m/s. The froth height is 75 mm and 90% of the vapor passes through the froth in the large bubbles.

Estimate the mass transfer rates in each bubble population. Hence, the point efficiencies for the three components. The mass transfer process may be assumed to be equimolar and to be controlled by the vapor-phase resistance. The Maxwell–Stefan diffusion coefficients are

$$Ð_{12}^V = 8.7 \text{ mm}^2/\text{s}$$

$$Ð_{13}^V = 12.1 \text{ mm}^2/\text{s}$$

$$Ð_{23}^V = 18.4 \text{ mm}^2/\text{s}$$

12.7 Distillation test runs on the following binary mixtures:

ethanol–water   ethanol–*t*-butyl alcohol   *t*-butyl alcohol–water were carried out in a 3 in. diameter sieve tray column under the same hydrodynamic conditions. It was found that the overall number of transfer units N_{OV} could be correlated by

$$N_{OV} = 3.1 \, D^{0.5}$$

where D is the vapor-phase Fick diffusion coefficients (in cm^2/s) of the corresponding binary pair.

In a test run with the ternary mixture ethanol–t-butyl alcohol–water, the following results were obtained for a particular tray.

Composition of vapor entering tray (mole fractions)

$$y_1 = 0.2720 \qquad y_2 = 0.2874 \qquad y_3 = 0.4405$$

Composition of vapor in equilibrium with liquid on the tray

$$y_1^* = 0.3106 \qquad y_2^* = 0.3580 \qquad y_3^* = 0.3292$$

Using the matrix generalization technique developed in Exercise 12.2, calculate the matrix of overall numbers of transfer units. Hence, estimate the composition of the vapor above the tray and the Murphree efficiencies of the three components. Comment on the relative values of the efficiencies. Under what circumstances would you expect them to be equal?

Data: The matrix of Fick diffusion coefficients is

$$[D] = \begin{bmatrix} 0.151 & 0.043 \\ 0.035 & 0.131 \end{bmatrix} \times cm^2/s$$

12.8 Distillation test runs on the following binary mixtures

$$\text{acetone–methanol} \qquad \text{methanol–water} \qquad \text{acetone–water}$$

were performed in a rectangular (5 in. $\times$ 6 in.) split flow sieve tray column under identical hydrodynamic conditions.

In a test run at atmospheric pressure with the ternary mixture acetone(1)–methanol(2)–water(3), the following results were obtained for a particular tray.

Composition of vapor entering tray (mole fractions)

$$y_1 = 0.257 \qquad y_2 = 0.518 \qquad y_3 = 0.225$$

Composition of vapor in equilibrium with liquid on the tray

$$y_1^* = 0.480 \qquad y_2^* = 0.445 \qquad y_3^* = 0.075$$

If the distillation operation is assumed to be controlled by the vapor-phase resistance to mass transfer, indicate which of the following inequalities will hold

a. $E_{OV1} < E_{OV3} < E_{OV2}$

b. $E_{OV3} > E_{OV1} > E_{OV2}$

c. $E_{OV1} \simeq E_{OV3} \simeq E_{OV2}$

where E_{OVi} are the Murphree point efficiencies of the components acetone, methanol, and water. Support your conclusions by performing preliminary calculations using the data supplied below.

Data: The Maxwell–Stefan diffusion coefficients are

$$Đ_{12} = 8.7 \text{ mm}^2/\text{s}$$

$$Đ_{13} = 12.1 \text{ mm}^2/\text{s}$$

$$Đ_{23} = 18.4 \text{ mm}^2/\text{s}$$

12.9   Hofer (1983) has analyzed the influence of gas-phase dispersion on the tray efficiency for binary systems. Extend the analysis for multicomponent mixtures. Include some numerical calculations and write up your work in the form of a paper for possible publication in the *AIChEJ*.

12.10   Mora and Bugarel (1976) used the theory of irreversible thermodynamics to analyze interphase mass transfer on a distillation tray. Compare their treatment with the analysis in Chapters 12 and 13; which is more general, theirs or that given in this book? Write up your comparison in the form of a paper to *Entropie*.

12.11   A packed distillation column for the separation of ethylbenzene(1) and styrene(2) was designed in Example 14.3.3. The packing is Koch Flexipac II with characteristic dimensions in Table 14.6. Estimate the mass transfer coefficients and the heights of transfer units for the rectifying and stripping sections in this column.

Physical properties of the system have been estimated as follows:

Viscosity of vapor mixture: 7.3×10^{-6} Pa s.

Viscosity of liquid mixture: 4.2×10^{-4} Pa s.

Liquid density: 790 kg/m³.

Molar mass of ethylbenzene: 0.106 kg/mol.

Molar mass of styrene: 0.104 kg/mol.

Surface tension: $\sigma = 0.024$ N/m.

Use the methods of Chapter 4.1 to estimate the vapor and liquid-phase diffusion coefficients. *K* values may be estimated using Raoult's law.

12.12   Gorak (1991) conducted a number of distillation experiments with the four component system acetone(1)–methanol(2)–2-propanol(3)–water(4). The data for one of Gorak's experiments were used as the basis for Example 12.3.3. Here we ask you to estimate the mass transfer coefficients and the numbers of transfer units for the conditions existing in a different experiment:

Flow rate of the vapor: 38.1557 mol/(m²s).

Pressure: 101 kPa.

Composition of the vapor below the packing is $y_1 = 0.001899$, $y_2 = 0.128267$, $y_3 = 0.440088$, and $y_4 = 0.429745$.

The experiments were carried out at total reflux.

Data: Physical properties of the vapor and liquid phases at the composition reported above and at the bubble point temperature of the liquid have been estimated as follows:

Viscosity of vapor mixture: $\mu^V = 1.083 \times 10^{-5}$ Pa s.

Viscosity of liquid mixture: $\mu^L = 4.112 \times 10^{-4}$ Pa s.

Vapor density: $\rho_t^V = 1.319$ kg/m³.

Liquid density: $\rho_t^L = 822.0$ kg/m³.

Average molar mass of vapor: $M^V = 0.0384$ kg/mol.

Surface tension: $\sigma = 0.0481$ N/m.

The Maxwell–Stefan diffusivities in the vapor phase [units are 10^{-5} m^2/s].

$$\mathcal{D}_{12}^V = 1.223$$
$$\mathcal{D}_{13}^V = 0.7863$$
$$\mathcal{D}_{14}^V = 1.891$$
$$\mathcal{D}_{23}^V = 1.186$$
$$\mathcal{D}_{24}^V = 2.817$$
$$\mathcal{D}_{34}^V = 1.832$$

The Maxwell–Stefan diffusivities in the liquid phase [units are 10^{-9} m^2/s].

$$\mathcal{D}_{12}^L = 3.219$$
$$\mathcal{D}_{13}^L = 2.735$$
$$\mathcal{D}_{14}^L = 3.507$$
$$\mathcal{D}_{23}^L = 3.452$$
$$\mathcal{D}_{24}^L = 4.667$$
$$\mathcal{D}_{34}^L = 4.169$$

Equimolar countertransfer may be assumed to prevail.

The K values determined at the bubble point of the liquid with a composition equal to that of the vapor are $K_1 = 3.6750$, $K_2 = 1.5678$, $K_3 = 1.0820$, and $K_4 = 0.7348$.

The column had an internal diameter of 0.1 m and was 0.8 m high. The column was fitted with Sulzer BX structured packing. The characteristics of Sulzer BX packing are given in Example 12.3.3.

CHAPTER 14

There are many excellent texts that discuss the design of distillation columns using equilibrium stage calculations. Some of them were cited in Chapters 12–14. These texts provide a wealth of examples that could be used as the basis for a design using the nonequilibrium model described in Chapter 14. We adapt one such example below (Exercise 14.1) in order to indicate how this might be done.

The worked examples and simulation results in Chapter 14 were, for the most part, obtained with a simulation program known as *ChemSep* (Kooijman and Taylor, 1992). You will need to obtain this program (or something equivalent) in order to carry out the numerical exercises for this chapter. Contact R. Taylor for more information on the availability of *ChemSep*.

14.1 Fair (1987) presents the results of a design calculation for a six component de-ethanizer (see Table 5.3.4 of this reference). Use a nonequilibrium model to determine the number of trays required to reach the purity requirements.

14.2 How would the performance of the ethylbenzene–styrene column in Example 14.3.3 differ if the column were filled with: (a) 50-mm metallic Pall rings, (b) Sulzer BX packing, and (c) Sieve trays.

14.3 In Section 12.2.2 we derived an expression that allows us to calculate the average molar fluxes in a vertical slice of froth on a tray under the assumptions that the vapor rises through the froth in plug flow and the liquid in the vertical slice is well mixed. Extend the treatment and derive an expression for the average mass transfer rates for the entire tray if the liquid is in plug flow. Some clues as to how to proceed may be found in Section 13.3.3.

14.4 Discuss how the fundamental models of mass transfer in Sections 12.1.7 (binary systems) and 12.2.4 (multicomponent systems) may be used to estimate mass transfer rates for use in a nonequilibrium simulation of an existing distillation column. Your essay should address the important question of how the model parameters are to be estimated.

14.5 Most experimental studies of distillation are carried out at total reflux to prevent loss of materials. How would the nonequilibrium model of Chapter 14 be used to simulate a distillation column operating under total reflux conditions?

14.6 Discuss the simplifications of the nonequilibrium model that are possible when: (a) constant pressure operation is assumed and (b) the resistance to mass transfer in the liquid phase is ignored.

14.7 Gorak and Vogelpohl (1985) present an experimental study of ternary distillation in a packed column. The system used was methanol(1)–2-propanol(2)–water(3) and the column was 0.1 m wide and filled with Sulzer CY packing. Use the nonequilibrium model to simulate their experiments. Investigate the sensitivity of the simulation results to the thermodynamic model parameters. Write an article in the format required by *Separation Science Technology* that summarizes your calculations.

14.8 Burghardt et al. (1983) present data for the distillation of methanol(1), 2-propanol(2), water(3) and acetone(1), ethanol(2), water(3) in a sieve tray column (0.25-m diameter). Use the nonequilibrium model of Chapter 14 to simulate their experiments. Write up your work in an article in the format required by the *Chemical Engineering Journal*.

14.9 In a paper entitled "A new simulation model for a real trays absorption column," Grottoli et al. (1991) present their version of a nonequilibrium model for computer simulation of absorption columns. Using the material presented in Chapter 14 as a basis, critically examine any novel aspects of their paper.

CHAPTER 15

15.1 Repeat Example 15.1.2 using the Toor–Stewart–Prober method of estimating the multicomponent mass transfer coefficients.

15.2 Repeat Example 15.1.2 using an effective diffusivity formulation for the diffusion fluxes. Compare your results with those of Example 15.1.2, Exercise 15.1, and the results obtained by Austin and Jeffreys (1979).

15.3 Webb and McNaught (1980) present in detail the design of a heat exchanger for condensing a mixture of five hydrocarbons [propane(1)–n-butane(2)–n-hexane(3)–n-heptane(4)–n-octane(5)]. Using the data presented in their article, calculate the molar rates of condensation at the entrance to the condenser.

15.4 Ketene (CH_2CO) is an industrially important intermediate that has a number of uses, particularly for acetylation purposes. Nielsen and Villadsen (1984) provide a nice example of the theories discussed in this book to the design of a condenser for treating a vapor mixture consisting of 10% acetic anhydride, 45% acetic acid, and 45% ketene. One added complication in this condensation system is that when it is condensed the ketene and acetic acid will react with each other. At the conditions chosen for design this reaction may be assumed to be a surface reaction. With the information given in the paper by Nielsen and Villadsen (1984) you are required to simulate the condenser. Other thermodynamic and transport properties can be estimated by reference to Reid et al. (1987).

15.5   Krishna et al. (1976) presented a condenser design problem in which the condenser area requirements are estimated for condensing 50% of a vapor mixture containing propane(1), *n*-butane(2), *n*-hexane(3), *n*-heptane(4), *n*-octane(5), and hydrogen(6). Repeat their calculations using the design algorithms outlined in Chapter 15. This example should provide a strong reason for taking multicomponent diffusion interactions into account in design calculations.

15.6   Develop a mathematical model of an adiabatic wetted wall column. *Hint*: The model will be similar to that presented in Section 15.2. Use your model to generate composition profiles for the wetted wall column of Modine as discussed in Example 11.5.3. This is a lengthy exercise and will require you to calculate physical properties as a function of temperature, pressure, and composition. Modine's experimental data are most accessible in a paper by Krishna (1981a).

MISCELLANEOUS TOPICS

M.1   The fathers of multicomponent diffusion are James Clerk Maxwell and Josef Stefan. It is remarkable how several of the multicomponent diffusional interaction phenomena introduced in Chapter 5 were anticipated by these two scientists. Study the classic contributions of Maxwell (1866, 1868) (see his collected papers; Maxwell, 1952) and Stefan (1871) and draw your own conclusions as to how much (or how little) extra insight has been gained in the last century.

M.2   The Dusty Gas Model (DGM) (Mason and Malinauskas, 1983) is commonly used for describing diffusion in porous media. In this model the medium is modeled as giant molecules (dust) held motionless in space ($u_{dust} = 0$). Derive the DGM equations using the treatment given in Chapter 2 by taking the medium as the $(n + 1)$th species in the mixture. Compare the results of your derivations with that in Mason and Malinauskas (1983). You may also refer to Wesselingh and Krishna (1990) for more information.

M.3   Carry out a literature survey on multicomponent diffusion in inorganic ceramic materials and examine whether it would be beneficial for researchers in this area to adopt the Maxwell-Stefan approach instead of the generalized Fick's law. A paper by Cooper (1974) is a good starting point for your survey.

M.4   Frey (1986) has analyzed mass transfer within the liquid phase in an ion exchanger. He presents a detailed comparison of the approaches of Krishna and Standart (1976a), Toor (1964a) and Stewart and Prober (1964), Van Brocklin and David (1972), and Tunison and Chapman (1976). Try to reproduce the results of Frey and satisfy yourself of the need for rigorous modeling of mass transfer in multicomponent ionic systems.

M.5   The process of azeotropic distillation is widely used to separate mixtures that are difficult or impossible to separate by simple fractional distillation. An example of an azeotropic distillation process is the separation of ethanol from water using, as entrainer, benzene or n-pentane. A potential disadvantage of the use of benzene (depending on the intended use for the ethanol) is its toxicity. Consider an alternative process based on diffusional distillation as illustrated in Example 8.3.2 in which an inert gas is introduced in the condenser in order to effect a separation of the azeotropic mixture that can be further enhanced in a second distillation column. Develop a process for separating ethanol from water using these idea. You may consult Fullarton and Schlünder (1986a, b) for further details of the use of diffusional distillation in process development.

M.6 One topic that has not been covered in this book is that of multicomponent diffusion with simultaneous homogenous reaction. Carry out a literature survey of this field (the number of papers is not very large) and determine whether there is scope for further analysis using the Maxwell-Stefan formulation of multicomponent diffusion as opposed to the generalized Fick's law. A paper by Şentarli and Hortaçsu (1987) can serve as a starting point for your survey.

APPENDIX A
Review of Matrix Analysis

A.1 INTRODUCTION

Diffusion and mass transfer in multicomponent systems are described by systems of differential equations. These equations are more easily manipulated using matrix notation and concepts from linear algebra. We have chosen to include three appendices that provide the necessary background in matrix theory in order to provide the reader a convenient source of reference material. Appendix A covers linear algebra and matrix computations. Appendix B describes methods for solving systems of differential equations and Appendix C briefly reviews numerical methods for solving systems of linear and nonlinear equations. Other books cover these fields in far more depth than what follows. We have found the book by Amundson (1966) to be particularly useful as it is written with chemical engineering applications in mind. Other books we have consulted are cited at various points in the text.

A.1.1 Definition

A *matrix* is a rectangular array of elements arranged in horizontal and vertical columns. For example,

$$[A] = \begin{bmatrix} A_{11} & A_{12} & A_{13} & \cdots & A_{1m} \\ A_{21} & A_{22} & A_{23} & \cdots & A_{2m} \\ \vdots & & & & \vdots \\ A_{p1} & A_{p2} & A_{p3} & \cdots & A_{pm} \end{bmatrix} \tag{A.1.1}$$

The square brackets are used to denote a matrix. The A_{ij} are the *elements* of the matrix. They may be real or complex numbers. Quite often the elements of a matrix will be functions or operators. The matrix $[A]$ above has p rows and m columns and is said to be a $p \times m$ *matrix*; alternatively, the matrix $[A]$ is said to be of *order* $p \times m$. (It is common to use the letter n to denote the number of rows a matrix has. To avoid confusion with the rest of this book, we have reserved the letter n for denoting the number of components in a mixture.) The order of the matrices we will encounter in this book usually is obvious from the context but, in the event that it becomes necessary to distinguish between matrices of different order, this will be done by appending a subscript to the bracket notation. Thus, the $p \times m$ matrix $[A]$ would be written $[A]_{p \times m}$.

The *transpose*, denoted by $[A]^T$, of a matrix is the matrix formed by interchanging the rows and columns of the matrix. That is, the first row becomes the first column, the second row becomes the second column, and so on.

$$[A]^T = \begin{bmatrix} A_{11} & A_{21} & A_{31} & \cdots & A_{m1} \\ A_{12} & A_{22} & A_{32} & \cdots & A_{m2} \\ \vdots & & & & \vdots \\ A_{1p} & A_{2p} & A_{3p} & \cdots & A_{mp} \end{bmatrix} \tag{A.1.2}$$

The transpose of a $p \times m$ matrix is an $m \times p$ matrix.

A.1.2 Principal Types of Matrix

Besides the rectangular matrix discussed above, the following principal types of matrices occur in matrix algebra.

Square Matrix If the number of columns is equal to the number of rows, that is, $p = m$, the matrix $[A]$ is said to be a square matrix of order m.

$$[A] = \begin{bmatrix} A_{11} & A_{12} & A_{13} & \cdots & A_{1m} \\ A_{21} & A_{22} & A_{23} & \cdots & A_{2m} \\ \vdots & & & & \vdots \\ A_{m1} & A_{m2} & A_{m3} & \cdots & A_{mm} \end{bmatrix} \tag{A.1.3}$$

If necessary, the order of a square matrix will be denoted by a single subscript; for example, $[A]_m$.

The diagonal of the square matrix that contains the elements A_{ii} (i.e., $A_{11}, A_{22}, \ldots, A_{mm}$) is called the *principal diagonal*.

Column Matrix If the matrix has only one column ($m = 1$), the m elements are arranged in a single column; such a matrix is a column matrix (sometimes called a column vector) and is of order $p \times 1$. A column matrix is denoted by parentheses

$$(b) = \begin{pmatrix} b_1 \\ b_2 \\ \vdots \\ b_p \end{pmatrix} \tag{A.1.4}$$

Row Matrix A set of n elements in a row ($p = 1$) is a matrix of order $1 \times m$; such a matrix is a row matrix or row vector. Since the transpose of a column matrix is a row matrix, this helps determine the way in which row matrices are denoted.

$$(c)^T = (c_1 \quad c_2 \quad \cdots \quad c_m) \tag{A.1.5}$$

Diagonal Matrix A square matrix with nonzero elements only on the principal diagonal and zeroes everywhere else is called a diagonal matrix.

$$[\lambda] = \begin{bmatrix} \lambda_1 & 0 & 0 & \cdots & 0 \\ 0 & \lambda_2 & 0 & \cdots & 0 \\ \vdots & & & & \vdots \\ 0 & 0 & 0 & \cdots & \lambda_m \end{bmatrix} \tag{A.1.6}$$

Identity Matrix If the nonzero elements of a diagonal matrix are all unity, it is called an *identity*, idem, or unit matrix and given the symbol $[I]$

$$[I] = \begin{bmatrix} 1 & 0 & 0 & \cdots & 0 \\ 0 & 1 & 0 & \cdots & 0 \\ \vdots & & & & \vdots \\ 0 & 0 & 0 & \cdots & 1 \end{bmatrix} \tag{A.1.7}$$

Sparse Matrices   Any matrix that has a high proportion of its elements equal to zero is a sparse matrix. Thus, any diagonal matrix is sparse. Another common sparse matrix is the *tridiagonal* matrix.

$$[A] = \begin{bmatrix} B_1 & C_1 & & & & \\ A_2 & B_2 & C_2 & & & \\ & \cdots & & & & \\ & & & A_{m-1} & B_{m-1} & C_{m-1} \\ & & & & A_m & B_m \end{bmatrix} \qquad (A.1.8)$$

where we have omitted the zeroes for clarity.

Partitioned Matrix   A matrix of matrices is a partitioned matrix. For example, the *block-tri-diagonal* matrix

$$[ABC] = \begin{bmatrix} [B_1] & [C_1] & & & & \\ [A_2] & [B_2] & [C_2] & & & \\ & [A_3] & [B_3] & [C_3] & & \\ & & \cdots & & & \\ & & & \cdots & & \\ & & & [A_{m-1}] & [B_{m-1}] & [C_{m-1}] \\ & & & & [A_m] & [B_m] \end{bmatrix} \qquad (A.1.9)$$

in which each entry $[A]$, $[B]$, $[C]$ is a submatrix, is a partitioned matrix. Matrices with this structure arise when the equations used in the modeling of separation processes like distillation, absorption, and extraction are solved numerically. The matrix in Eq. A.1.9 is also a sparse matrix.

Symmetric Matrix   A symmetric matrix is a square matrix that is equal to its transpose. That is,

$$[A] = [A]^T \qquad \text{or} \qquad A_{ij} = A_{ji} \qquad (A.1.10)$$

Diagonal matrices are symmetric.

Skew or Antisymmetric Matrix   A skew or antisymmetric matrix has the property that

$$[A] = -[A]^T \qquad \text{or} \qquad A_{ij} = -A_{ji} \qquad (A.1.11)$$

Any matrix that has this property must have zeroes on the principal diagonal.

A.2   ELEMENTARY OPERATIONS AND PROPERTIES OF MATRICES

The simplest relationship between two matrices is equality. Intuitively, one feels that two matrices should be equal if their corresponding elements are equal. This is the case, providing the matrices are of the same order. Thus, the matrices $[A]_{p \times m}$ and $[B]_{p \times m}$ are equal if they have the same order and if

$$A_{ij} = B_{ij} \qquad i = 1, 2, \ldots, p; \ j = 1, 2, \ldots, m \qquad (A.2.1)$$

A.2.1 Addition

Two matrices $[A]$ and $[B]$ can be added together only if they are of the same order, and then the elements in the corresponding positions are added, that is, if $[A] + [B] = [C]$, then

$$C_{ij} = A_{ij} + B_{ij} \tag{A.2.2}$$

It is not difficult to show that the addition of matrices is both *commutative* and *associative*.

$$[A] + [B] = [B] + [A] \tag{A.2.3}$$

$$[A] + ([B] + [C]) = ([A] + [B]) + [C] \tag{A.2.4}$$

If we define a *null matrix* [0] to be a matrix consisting of only zero elements, then

$$[A] + [0] = [A] \tag{A.2.5}$$

A.2.2 Multiplication by a Scalar

If $[A]$ is a $p \times m$ matrix and λ is a scalar, then $\lambda[A]$ is a $p \times m$ matrix $[B]$ where

$$B_{ij} = \lambda A_{ij} \qquad i = 1, 2, \ldots, p \qquad j = 1, 2, \ldots, m \tag{A.2.6}$$

The related operation of division by the scalar λ will be denoted by $[A]/\lambda$. The elements of this matrix are A_{ij}/λ. This operation is equivalent to multiplication of $[A]$ by $1/\lambda$.

A.2.3 Multiplication of Two Matrices

Many of the important uses of matrices depend on the definition of multiplication. Multiplication is defined only for *conformable* matrices. Two matrices $[A]$ and $[B]$ are said to be conformable in the order $[A][B]$ if $[A]$ has the same number of columns as $[B]$ has rows.

Given a matrix $[A]$ with p rows and m columns and a matrix $[B]$ with m rows and q columns, the product matrix $[C]$ is a matrix of p rows and q columns in which the elements in the ith row and jth column of $[C]$ is obtained as the sum of the products of the corresponding elements in the ith row of $[A]$ and the jth column of $[B]$. That is, if $[C] = [A][B]$, then

$$C_{ij} = \sum_{k=1}^{n} A_{ik} B_{kj} \tag{A.2.7}$$

For example, two square matrices of order 2, $[A]$ and $[B]$, when multiplied together give a third matrix $[C]$

$$[C] = [A][B] = \begin{bmatrix} A_{11} & A_{12} \\ A_{21} & A_{22} \end{bmatrix} \begin{bmatrix} B_{11} & B_{12} \\ B_{21} & B_{22} \end{bmatrix}$$

$$= \begin{bmatrix} A_{11}B_{11} + A_{12}B_{21} & A_{11}B_{12} + A_{12}B_{22} \\ A_{21}B_{11} + A_{22}B_{21} & A_{21}B_{12} + A_{22}B_{22} \end{bmatrix}$$

A square matrix $[A]$ of order 2 and a column matrix (b) of order 2 when multiplied together yield a column matrix (c), also of order 2

$$(c) = [A](b) = \begin{bmatrix} A_{11} & A_{12} \\ A_{21} & A_{22} \end{bmatrix} \begin{pmatrix} b_1 \\ b_2 \end{pmatrix}$$

$$= \begin{pmatrix} A_{11}b_1 + A_{12}b_2 \\ A_{21}b_1 + A_{22}b_2 \end{pmatrix}$$

A row matrix of order m can be multiplied by a column matrix of order m in two different ways. The *inner product* is a scalar defined by

$$c = (a)^T(b) = (a_1 \quad a_2 \quad \cdots \quad a_m) \begin{pmatrix} b_1 \\ b_2 \\ \vdots \\ b_m \end{pmatrix} = \sum_{i=1}^{m} a_i b_i \qquad \text{(A.2.8)}$$

The *outer product* of a row and column matrix, both of order m, is a *square* matrix of order m and is written as $[C] = (a)(b)^T$ with

$$C_{ij} = a_i b_j \qquad i, j = 1, 2, \ldots, m \qquad \text{(A.2.9)}$$

For example, if (a) and (b) are of order 2 then

$$[C] = \begin{pmatrix} a_1 \\ a_2 \end{pmatrix} (b_1 \quad b_2) = \begin{bmatrix} a_1 b_1 & a_1 b_2 \\ a_2 b_1 & a_2 b_2 \end{bmatrix}$$

When the product is written $[A][B]$, $[A]$ is said to premultiply $[B]$ while $[B]$ is said to postmultiply $[A]$.

In general, it can be shown that matrix multiplication is associative

$$[A][[B][C]] = [[A][B]][C] \qquad \text{(A.2.10)}$$

and distributive

$$[A][[B] + [C]] = [A][B] + [A][C] \qquad \text{(A.2.11)}$$

In contrast with scalar multiplication, the one basic property that matrix multiplication does not possess is commutativity; that is, in general

$$[A][B] \neq [B][A] \qquad \text{(A.2.12)}$$

If we have the equality $[A][B] = [B][A]$, the matrices $[A]$ and $[B]$ are said to *commute* or to be *permutable*. The identity matrix $[I]$ commutes with any square matrix of the same order,

$$[A][I] = [I][A] = [A] \qquad \text{(A.2.13)}$$

and any matrix commutes with itself, of course.

One "unfortunate" property of matrix multiplication is that the equation $[A][B] = [A][C]$ does *NOT* imply that $[B] = [C]$. In the same vein, the equation $[A](c) = [B](c)$ does not imply equality of $[A]$ and $[B]$.

If a square matrix is multiplied by itself k times, the resultant matrix is defined as $[A]^k$

$$[A]^k = \underbrace{[A][A]\ldots[A]}_{(k \text{ factors})} \qquad \text{(A.2.14)}$$

A.2.4   Differentiation and Integration of Matrices

The derivative of a $p \times m$ matrix $[A]$ is defined by

$$\frac{d[A]}{dt} = \begin{bmatrix} \dfrac{dA_{11}}{dt} & \dfrac{dA_{12}}{dt} & \cdots & \dfrac{dA_{1m}}{dt} \\ \dfrac{dA_{21}}{dt} & \dfrac{dA_{22}}{dt} & \cdots & \dfrac{dA_{2m}}{dt} \\ \cdots & & & \vdots \\ \dfrac{dA_{p1}}{dt} & \dfrac{dA_{p2}}{dt} & \cdots & \dfrac{dA_{pm}}{dt} \end{bmatrix} \qquad (A.2.15)$$

For example, the derivative of the matrix $[A]t$ where $[A]$ is a constant matrix is

$$\frac{d[[A]t]}{dt} = [A]$$

and the gradient of a column matrix $\nabla(x)$ is

$$\nabla(x) = \begin{pmatrix} \nabla x_1 \\ \nabla x_2 \\ \vdots \\ \nabla x_m \end{pmatrix}$$

The chain rule for the differentiation of products has the matrix generalization

$$\frac{d[[A][B]]}{dt} = [A]\frac{d[B]}{dt} + \frac{d[A]}{dt}[B] \qquad (A.2.16)$$

(note the order of multiplication). For example, the gradient of $[P](x)$, where $[P]$ is a constant matrix, is

$$\nabla([P](x)) = [P]\nabla(x) \qquad (\nabla[P] = [0])$$

Integration is defined in a way analogous to differentiation.

$$\int [A]dt = \begin{bmatrix} \int A_{11}dt & \int A_{12}dt & \cdots & \int A_{1m}dt \\ \int A_{21}dt & \int A_{22}dt & \cdots & \int A_{2m}dt \\ \vdots & \vdots & \cdots & \vdots \\ \int A_{p1}dt & \int A_{p2}dt & \cdots & \int A_{pm}dt \end{bmatrix} \qquad (A.2.17)$$

For example, the integral of $[A]t$, where $[A]$ is constant is

$$\int [A]t\,dt = \tfrac{1}{2}[A]t^2$$

A.3   THE INVERSE

The inverse of an $m \times m$ matrix $[A]$ is an $m \times m$ matrix $[B]$ having the property that

$$[A][B] = [B][A] = [I] \tag{A.3.1}$$

Here $[B]$ is called the inverse of $[A]$ and is usually denoted by $[A]^{-1}$.

One method for finding inverses requires calculating the *cofactor* and *adjoint* matrices, as explained below.

The *cofactor* matrix associated with an $m \times m$ matrix $[A]$ is an $m \times m$ matrix $[A^c]$ obtained from $[A]$ by replacing each element of $[A]$ by its signed cofactor A_{ij}^c where

$$A_{ij}^c = (-1)^{i+j} M_{ij} \tag{A.3.2}$$

here M_{ij} is the *minor* and is the *determinant* of the $m - 1 \times m - 1$ submatrix of $[A]$, formed by deleting the ith row and jth column. The determinant of $[A]$ is then given by

$$|A| = \sum_{i=1}^{m} A_{ij}^c A_{ij} = \sum_{j=1}^{m} A_{ij}^c A_{ij} \tag{A.3.3}$$

where i and j may take any value (from 1 to m). (There are many other ways to defining and computing determinants; the recursive definition given here suffices for our needs.)

The adjoint of an $m \times m$ matrix $[A]$ is the transpose of the cofactor matrix of $[A]$. Thus, if we denote the adjoint of $[A]$ by $[A^a]$, we have

$$[A^a] = [A^c]^T \tag{A.3.4}$$

The adjoint matrix $[A^a]$ has the following important property

$$[A][A^a] = [A^a][A] = |A|[I] \tag{A.3.5}$$

If $|A| = 0$, $[A]$ is said to be *invertible* or *nonsingular*. If $|A| = 0$, $[A]$ is said to be *singular* and its inverse is not defined (in this way at least). In particular, the identity matrix $[I]$ is invertible and is its own inverse, that is,

$$[I][I] = [I] \tag{A.3.6}$$

while the null matrix [0] is singular.

We divide Eq. A.3.5 by $|A|$ to obtain

$$[A][A^a]/|A| = [A^a][A]/|A| = [I] \tag{A.3.7}$$

Thus, using the definition of the inverse, we have

$$[A]^{-1} = [A^a]/|A| \qquad \text{if } |A| \neq 0 \tag{A.3.8}$$

That is, the inverse of $[A]$ may be obtained by dividing the adjoint of $[A]$ by the determinant $|A|$.

For a square matrix of order 2, the inverse can be computed from Eq. A.3.8 as follows: The determinant of a matrix $[A]$ of order 2 is

$$|A| = A_{11}A_{22} - A_{12}A_{21} \tag{A.3.9}$$

The cofactor matrix is

$$[A^c]^T = \begin{bmatrix} A_{22} & -A_{21} \\ -A_{12} & A_{11} \end{bmatrix}^T = \begin{bmatrix} A_{22} & -A_{12} \\ -A_{21} & A_{11} \end{bmatrix} \qquad (A.3.10)$$

Thus, the elements of $[A]^{-1}$, denoted below by A_{ij}^{-1} are given by

$$\begin{aligned} A_{11}^{-1} &= A_{22}/|A| \\ A_{12}^{-1} &= -A_{12}/|A| \\ A_{21}^{-1} &= -A_{21}/|A| \\ A_{22}^{-1} &= A_{11}/|A| \end{aligned} \qquad (A.3.11)$$

One of the uses of matrix inversion is in the solution of systems of simultaneous linear equations

$$[A](x) = (b) \qquad (A.3.12)$$

If $[A]$ is invertible, then the solution is given by

$$(x) = [A]^{-1}(b) \qquad (A.3.13)$$

This method of solving systems of linear equations (or even of inverting a matrix) is *NOT* recommended for systems of order greater than 2. Better methods are discussed in texts on numerical methods (e.g., Press et al., 1992) (see, also, Appendix C).

A.3.1 Properties of the Inverse and Transpose

1. $$\left[[A]^{-1}\right]^{-1} = [A] \qquad (A.3.14)$$

2. $$[[A][B]]^{-1} = [B]^{-1}[A]^{-1} \qquad (A.3.15)$$

3. $$[[A][B][C]]^{-1} = [C]^{-1}[B]^{-1}[A]^{-1} \qquad (A.3.16)$$

4. $$[[A][B]]^T = [B]^T[A]^T \qquad (A.3.17)$$

5. $$\left[[A]^T\right]^{-1}\left[[A]^{-1}\right]^T \qquad (A.3.18)$$

6. The inverse of a nonsingular diagonal matrix is diagonal. One can readily check that the inverse of $[\lambda]$

$$[\lambda] = \begin{bmatrix} \lambda_1 & 0 & 0 & \cdots & 0 \\ 0 & \lambda_2 & 0 & \cdots & 0 \\ \vdots & & & & \vdots \\ 0 & 0 & 0 & \cdots & \lambda_m \end{bmatrix} \qquad (A.3.19)$$

is

$$[\lambda]^{-1} = \begin{bmatrix} \lambda_1^{-1} & 0 & 0 & \cdots & 0 \\ 0 & \lambda_2^{-1} & 0 & \cdots & 0 \\ \vdots & & & & \vdots \\ 0 & 0 & 0 & \cdots & \lambda_m^{-1} \end{bmatrix} \qquad (A.3.20)$$

Note that no λ_i ($i = 1, 2, \ldots, n$) can be equal to zero.

7. The inverse of a nonsingular symmetric matrix is also symmetric.
8. The inverse of a matrix that can be written in the form

$$[B] = [A] + (u)(v)^T \qquad (A.3.21)$$

where (u) and (v) are column matrices of the same order as $[A]$, can be computed using the Sherman–Morrison formula (Ortega and Rheinbolt, 1970, p. 50)

$$[B]^{-1} = [A]^{-1} - \alpha[A]^{-1}(u)(v)^T[A]^{-1} \qquad (A.3.22)$$

The scalar α is defined by

$$\alpha = 1 + (v)^T[A]^{-1}(u) \qquad (A.3.23)$$

Equation (A.3.22) is especially useful if $[A]$ is diagonal.

A.4   EIGENVALUES AND EIGENVECTORS

Consider a square matrix $[A]$ of order m. Let (x) be a column matrix of the same order, that is, with m rows and 1 column. From the definition of matrix multiplication it is known that the premultiplication of the matrix (x) by the matrix $[A]$ generates a new column matrix (y) so that

$$(y) = [A](x) \qquad (A.4.1)$$

The matrix (y) can be considered to be a transformation of the original matrix (x). The question that will now be asked is Is it possible for the matrix (y) to be proportional to (x), that is, $(y) = \lambda(x)$, where λ is a scalar multiplier? That is, for (y) to have the same "direction" as the matrix (x). For the case of a collinear transformation we have

$$(y) = [A](x) = \lambda(x) \qquad (A.4.2)$$

or

$$[[A] - \lambda[I]](x) = (0) \qquad (A.4.3)$$

where (0) is a null column matrix. Equation A.4.3 represents a system of linear homogeneous equations. If $[[A] - \lambda[I]]$ is nonsingular, that is, $\|[A] - \lambda[I]\| \neq 0$, then we may formally write

$$(x) = [[A] - \lambda[I]]^{-1}(0) = (0) \qquad (A.4.4)$$

This result is, however, trivial for $(x) = (0)$ represents a trivial solution. But we may ask the question Are there values of λ that will produce nontrivial solutions? The necessary and sufficient condition that there be nontrivial solutions is that the determinant of the coefficients vanishes, that is,

$$\|[A] - \lambda[I]\| = 0 \qquad (A.4.5)$$

which is called the *characteristic equation* of $[A]$. The determinantal Eq. A.4.5, when expanded, will be of the form

$$\|[A] - \lambda[I]\| = P_m(\lambda) = 0 \qquad (A.4.6)$$

where

$$P_m(\lambda) = \lambda^m + C_1\lambda^{m-1} + C_2\lambda^{m-2} \cdots + C_{m-1}\lambda + C_m \qquad (A.4.7)$$

will be a polynomial of mth degree in λ. From a fundamental theorem in algebra it is known that a polynomial of the mth degree has m roots. In other words, there are m values λ_i such that

$$P_m(\lambda_i) = 0 \qquad i = 1, 2, \ldots, m \qquad (A.4.8)$$

The λ_i that satisfy this Eq. A.4.8 may be complex, since an algebraic equation with real coefficients may have complex conjugate pairs of roots, or they may be complex since the polynomial may have complex coefficients if the matrix $[A]$ has nonreal elements.

The values (roots) $\lambda_1, \lambda_2, \lambda_3, \ldots, \lambda_m$ of the characteristic equation $P_m(\lambda) = 0$ are called *eigenvalues* of the matrix $[A]$.

A.4.1 Properties of Eigenvalues

1. The sum of the eigenvalues of a matrix equals the *trace* of the matrix: the trace of a matrix being the sum of the elements on the principal, or main, diagonal.

$$\lambda_1 + \lambda_2 + \lambda_3 + \cdots + \lambda_n = A_{11} + A_{22} + A_{33} + \cdots + A_{nn}$$
$$= \mathrm{tr}[A] \qquad (A.4.9)$$

2. The product of the eigenvalues of a matrix equals the determinant of the matrix, thus,

$$\lambda_1\lambda_2\lambda_3 \cdots \lambda_m = |A| \qquad (A.4.10)$$

3. A matrix is singular if and only if it has a zero eigenvalue.
4. If λ is an eigenvalue of $[A]$, then $1/\lambda$ is the corresponding eigenvalue of $[A]^{-1}$.
5. If λ is an eigenvalue of $[A]$, then $c\lambda$ is an eigenvalue of $c[A]$ where c is any arbitrary scalar.
6. The eigenvalues of a real symmetric matrix are real.
7. For a matrix of order 2,

$$[A] = \begin{bmatrix} A_{11} & A_{12} \\ A_{21} & A_{22} \end{bmatrix}$$

where the two eigenvalues $\hat{A}_1$ and $\hat{A}_2$ are the roots of the quadratic equation

$$\hat{A}_i^2 - (A_{11} + A_{22})\hat{A}_i + (A_{11}A_{22} - A_{12}A_{21}) = 0 \qquad (A.4.11)$$

That is,

$$\hat{A}_1 = \tfrac{1}{2}\left\{\mathrm{tr}[A] + \sqrt{\mathrm{disc}[A]}\right\}$$
$$\hat{A}_2 = \tfrac{1}{2}\left\{\mathrm{tr}[A] - \sqrt{\mathrm{disc}[A]}\right\} \qquad (A.4.12)$$

where

$$\mathrm{tr}[A] = A_{11} + A_{22}$$
$$|A| = A_{11}A_{22} - A_{12}A_{21}$$
$$\mathrm{disc}[A] = (\mathrm{tr}[A])^2 - 4|A|$$

are the trace, determinant and discriminant of the two dimensional matrix $[A]$.

A.4.2 Eigenvectors

For each possible value of $\lambda = \lambda_i$, $i = 1, 2, \ldots, m$, a solution of the homogeneous equation

$$[[A] - \lambda_i[I]](x) = (0) \qquad i = 1, 2, \ldots, m \tag{A.4.13}$$

can be found.

Let $(x) = (e_i)$ be the column matrix associated with $\lambda = \lambda_i$. We may write

$$[[A] - \lambda_i[I]](e_i) = (0) \qquad i = 1, 2, \ldots, m \tag{A.4.14}$$

provided the characteristic equation has m distinct roots. The column matrices (e_i) are called *eigenvectors* or principal axes of the matrix $[A]$.

For a matrix of order 2, Eq. A.4.14 simplifies to

$$\begin{bmatrix} A_{11} - \hat{A}_i & A_{12} \\ A_{21} & A_{22} - \hat{A}_i \end{bmatrix} \begin{pmatrix} e_{i_1} \\ e_{i_2} \end{pmatrix} = \begin{pmatrix} 0 \\ 0 \end{pmatrix} \tag{A.4.15}$$

Carrying out the multiplications required by Eq. A.4.15 and solving for e_2 in terms of e_1 gives

$$e_{i_2} = -\left(A_{11} - \hat{A}_i\right)e_{i_1}/A_{12} \tag{A.4.16}$$

which has been derived from the first row or

$$e_{i_2} = -A_{21}e_{i_1}/\left(A_{22} - \hat{A}_i\right) \tag{A.4.17}$$

which is obtained from the second row.

Since the eigenvectors are solutions of a homogeneous system of equations, the solution is determined only up to a constant factor and only the ratios of the elements in the columns (e_i) are uniquely determined. The geometrical interpretation of this is that the eigenvectors are uniquely determined only in their direction, but their length or absolute value is arbitrary.

A.4.3 Similar Matrices

A matrix $[A]$ is said to be *similar* to a matrix $[B]$ if there exists an invertible matrix $[P]$ such that

$$[A] = [P]^{-1}[B][P] \tag{A.4.18}$$

If $[A]$ is similar to $[B]$ then $[B]$ is similar to $[A]$, that is,

$$[B] = [P][A][P]^{-1} \tag{A.4.19}$$

The above transformations are called *similarity* transformations and are very important in developing solutions to systems of coupled differential equations.

Similar matrices have the same characteristic equation and, therefore, the same eigenvalues.

A.4.4 Diagonalizable Matrices

A matrix is diagonalizable if it is *similar* to a diagonal matrix. Diagonalizable matrices are of particular interest since their matrix functions can be easily computed. Let us see how a square matrix $[A]$ of order $m \times m$ can be diagonalized. Let $\lambda_i, \lambda_2, \ldots, \lambda_m$ represent the m eigenvalues of $[A]$ and let us assume that these are real and distinct. There will be m eigenvectors (e_i) corresponding to these eigenvectors, placed in sequence, thus,

$$[E] = [(e_1)(e_2)(e_3) \cdots (e_m)] \tag{A.4.20}$$

that is, the columns of $[E]$ are the eigenvectors of $[A]$. The matrix $[E]$ is called the *modal matrix* of $[A]$. Since the eigenvectors are determined only to within an arbitrary multiplying factor, the modal matrix $[E]$ is also not uniquely determined.

Consider the matrix product $[A][E]$

$$
\begin{aligned}
[A][E] &= [A][(e_1)(e_2)(e_3) \cdots (e_n)] \\
&= [[A](e_1)\ [A](e_2)\ [A](e_3) \cdots [A](e_m)]
\end{aligned}
\tag{A.4.21}
$$

But, by definition

$$[A](e_i) = \lambda_i(e_i) \tag{A.4.22}$$

and so

$$
\begin{aligned}
[A][E] &= [\lambda_1(e_1)\lambda_2(e_2)\lambda_3(e_3) \cdots \lambda_m(e_m)] \\
&= [E][\lambda]
\end{aligned}
\tag{A.4.23}
$$

where $[\lambda]$ is a diagonal matrix containing the eigenvalues on the principal diagonal.

If the eigenvalues of $[A]$ are distinct so that the characteristic equation of $[A]$ does not have multiple roots, it can be shown that the modal matrix $[E]$ is nonsingular. The matrix $[E]^{-1}$ is, therefore, defined and exists. Premultiplying the foregoing equation by $[E]^{-1}$ we get

$$[E]^{-1}[A][E] = [E]^{-1}[E][\lambda] = [\lambda] \tag{A.4.24}$$

or

$$[A] = [E][\lambda][E]^{-1} \tag{A.4.25}$$

For a matrix of order 2, a possible modal matrix formed from the eigenvectors (Eq. A.4.17 and A.17.18) is

$$
\begin{aligned}
[E([A])] &= \left[e_1(\hat{A}_i)\ e_2(\hat{A}_2) \right] \\
&= \begin{bmatrix} \dfrac{1}{\hat{A}_1 - A_{11}} & \dfrac{1}{-A_{21}} \\[2mm] \dfrac{}{A_{12}} & \dfrac{}{A_{22} - \hat{A}_2} \end{bmatrix}
\end{aligned}
\tag{A.4.26}
$$

Other structures for $[E]$ are possible; it all depends on what values we choose for E_{11} and E_{12}.

A.5 MATRIX CALCULUS

Polynomials and exponentials play an important role in matrix calculus and matrix differential equations. It is, therefore, necessary to develop techniques for calculating these functions.

Let $P_k(x)$ denote an arbitrary polynomial of degree k in x:

$$P_k(x) = a_k x^k + a_{k-1} x^{k-1} + a_{k-2} x^{k-2} + \cdots + a_1 x + a_0 \qquad (A.5.1)$$

where $a_0, a_1, a_2, \ldots, a_k$ are real numbers. The matrix $P_k[[A]]$, a polynomial of the square matrix $[A]$ of the kth degree, is defined as

$$P_k[[A]] = a_k[A]^k + a_{k-1}[A]^{k-1} + \cdots + a_1[A] + a_0[I] \qquad (A.5.2)$$

which represents a sum of $(k + 1)$ matrices, all of order $m \times m$.

If we recall from calculus that many functions can be written as a Maclaurin series, then we can define functions of matrices quite easily. For instance, the Maclaurin series for e^x is

$$e^x = 1 + x + \frac{x^2}{2!} \cdots \qquad (A.5.3)$$

We define a matrix exponential $\exp[A]$ in a similar manner.

$$\exp[A] = [I] + [A] + \frac{1}{2!}[A]^2 + \cdots$$

$$= \sum_{k=0}^{\infty} \frac{1}{k!}[A]^k \qquad (A.5.4)$$

(Note that when we change a scalar equation to a matrix equation the unity element 1 is replaced by the identity matrix $[I]$.)

Other functions like $\sin[A]$ or $\cos[A]$ are also defined in a manner analogous to that for scalars. For example,

$$\sin[A] = \sum_{k=1}^{\infty} \frac{(-1)^k}{k!}[A]^{2k+1}$$

$$= [A] - \frac{1}{3!}[A]^3 + \frac{1}{5!}[A]^5 - \frac{1}{7!}[A]^7 + \cdots \qquad (A.5.5)$$

The question of convergence now arises. For an infinite series of matrices we define convergence as follows. A sequence $[[B]^k]$ of matrices with elements B_{ij}^k is said to converge to a matrix $[B]$ with elements B_{ij} if the elements B_{ij}^k converge to B_{ij} for every i and j.

A.5.1   The Cayley–Hamilton Theorem

The Cayley–Hamilton theorem is one of the most powerful theorems of matrix theory. It states "A matrix satisfies its own characteristic equation." That is, if the characteristic equation of an $m \times m$ matrix $[A]$ is

$$\lambda^m + a_{m-1}\lambda^{m-1} + a_{m-2}\lambda^{m-2} + \cdots + a_1\lambda + a_0 = 0 \qquad (A.5.6)$$

then

$$[A]^m + a_{m-1}[A]^{m-1} + a_{m-2}[A]^{m-2} + \cdots + a_1[A] + a_0[I] = [0] \qquad (A.5.7)$$

In many cases it is very difficult to compute functions of matrices from their definitions as infinite series (an exception is the exponential matrix). The Cayley–Hamilton theorem, however, provides a starting point for the development of a straightforward method for calculating these functions. Let us first consider how to calculate polynomials of matrices.

Let $[A]$ represent an $m \times m$ matrix and let us define $d(\lambda) = |[A] - \lambda[I]|$. Thus, $d(\lambda)$ is an mth degree polynomial in λ and the characteristic equation of $[A]$ is $d(\lambda) = 0$. If λ_i is an eigenvalue of $[A]$ then λ_i is a root of the characteristic equation and

$$d(\lambda_i) = 0 \tag{A.5.8}$$

Let $P[[A]]$ be a matrix polynomial of arbitrary degree that we wish to compute. Then $P(\lambda)$ represents the corresponding polynomial of λ. A theorem of algebra states that there exist polynomials $q(\lambda)$ and $r(\lambda)$ such that

$$P(\lambda) = d(\lambda)q(\lambda) + r(\lambda) \tag{A.5.9}$$

where $r(\lambda)$ is called the remainder and is of degree $m - 1$. The degree of $r(\lambda)$ is one less than that of $d(\lambda)$, which is m, and must be less than or equal to the degree of $P(\lambda)$. Since a matrix commutes with itself, many of the properties of polynomials are still valid for polynomials of a matrix. Therefore, we may write

$$P[[A]] = d[[A]]q[[A]] + r[[A]] \tag{A.5.10}$$

But $d[[A]] = [0]$. Therefore $P[[A]] = r[[A]]$. It follows that any polynomial on an $m \times m$ matrix $[A]$ may be written as a polynomial of degree $m - 1$. The latter polynomial may be much easier to compute.

If $[A]$ is a 2×2 matrix, $r(\lambda)$ will be a polynomial having the form

$$r(\lambda) = \alpha_1 \lambda + \alpha_0 \tag{A.5.11}$$

where α_0 and α_1 are coefficients to be determined.

If λ_1 and λ_2 are the two eigenvalues of $[A]$, then we must have

$$f(\lambda_1) = \alpha_1 \lambda_1 + \alpha_0 \tag{A.5.12}$$

$$f(\lambda_2) = \alpha_1 \lambda_2 + \alpha_0 \tag{A.5.13}$$

which represent two simultaneous linear equations in the two unknowns α_1 and α_0. The two equations can be solved to yield

$$\alpha_1 = \frac{f(\lambda_1) - f(\lambda_2)}{\lambda_1 - \lambda_2}$$
$$\alpha_0 = \frac{\lambda_1 f(\lambda_2) - \lambda_2 f(\lambda_1)}{\lambda_1 - \lambda_2} \tag{A.5.14}$$

We are now in a position to evaluate $P[[A]]$.

$$f[[A]] = r[[A]] = \alpha_1[A] + \alpha_0[I]$$
$$= \frac{(f(\lambda_1) - f(\lambda_2))[A] + (\lambda_1 f(\lambda_2) - \lambda_2 f(\lambda_1))[I]}{\lambda_1 - \lambda_2} \tag{A.5.15}$$

or

$$f[[A]] = \frac{[f(\lambda_1)[A] - \lambda_2[I]]}{\lambda_1 - \lambda_2} + \frac{[f(\lambda_2)[A] - \lambda_1[I]]}{\lambda_2 - \lambda_1} \tag{A.5.16}$$

The procedure outlined above for a 2×2 matrix may be extended to $m \times m$ matrices. This involves the solution of m simultaneous linear equations in m unknowns, $\alpha_{m-1}, \alpha_{m-2}, \cdots \alpha_1, \alpha_0$. The final result for $P[[A]]$ can be written as

$$P[[A]] = \sum_{i=1}^{m} P(\lambda_i) \left\{ \prod_{\substack{j=1 \\ j \neq i}}^{m} [[A] - \lambda_j[I]] \middle/ \prod_{\substack{j=1 \\ j \neq i}}^{m} (\lambda_i - \lambda_j) \right\} \qquad (A.5.17)$$

and is Sylvester's expansion formula for a matrix *polynomial*. The symbol Π represents a product over $m - 1$ factors.

A.5.2   Functions of a Matrix

Suppose we wish to compute $f[[A]]$, $f(\lambda)$ being an arbitrary *function* of λ. It can be shown that, for a large class of problems, there exists a function $q(\lambda)$ and a polynomial of degree $m - 1, r(\lambda)$, such that

$$f(\lambda) = d(\lambda)q(\lambda) + r(\lambda) \qquad (A.5.18)$$

where

$$d(\lambda) = |[A] - \lambda[I]|$$

It follows that

$$\begin{aligned} f[[A]] &= d[[A]]q[[A]] + r[[A]] \\ &= r[[A]] \qquad \text{(since } d[[A]] = [0]) \end{aligned} \qquad (A.5.19)$$

and $f[[A]]$ can be computed using Sylvester's theorem.

$$f[[A]] = \sum_{i=1}^{m} f(\lambda_i) \left\{ \prod_{\substack{j=1 \\ j \neq i}}^{m} [[A] - \lambda_j[I]] \middle/ \prod_{\substack{j=1 \\ j \neq i}}^{m} (\lambda_i - \lambda_j) \right\} \qquad (A.5.20)$$

Equations A.5.17 and A.5.20 may be used for computing polynomials and functions of diagonalizable matrices of any order $(> m)$ with m distinct eigenvalues $(\lambda_i \neq \lambda_j)$ (Lancaster and Tismenetsky, 1985).

A.5.3   Functions of Diagonalizable Matrices

Functions of a diagonalizable matrix can be evaluated by an alternative procedure. If $[\lambda]$ represents a diagonal matrix

$$[\lambda] = \begin{bmatrix} \lambda_1 & 0 & 0 & \cdots & 0 \\ 0 & \lambda_2 & 0 & \cdots & 0 \\ \vdots & & & & \vdots \\ 0 & 0 & 0 & \cdots & \lambda_m \end{bmatrix} \qquad (A.5.21)$$

then $[\lambda]^k$ is

$$[\lambda]^k = \begin{bmatrix} \lambda_1^k & 0 & 0 & \cdots & 0 \\ 0 & \lambda_2^k & 0 & \cdots & 0 \\ \vdots & & & & \vdots \\ 0 & 0 & 0 & \cdots & \lambda_m^k \end{bmatrix} \qquad (A.5.22)$$

An arbitrary polynomial $P_k[[\lambda]]$ is given by

$$P_k[[\lambda]] = \begin{bmatrix} P_k(\lambda_1) & 0 & 0 & \cdots & 0 \\ 0 & P_k(\lambda_2) & 0 & \cdots & 0 \\ \vdots & & & & \vdots \\ 0 & 0 & 0 & \cdots & P_k(\lambda_m) \end{bmatrix} \qquad (A.5.23)$$

An arbitrary function $f[[\lambda]]$ of $[\lambda]$ is calculable from

$$f[[\lambda]] = \begin{bmatrix} f(\lambda_1) & 0 & 0 & \cdots & 0 \\ 0 & f(\lambda_2) & 0 & \cdots & 0 \\ \vdots & & & & \vdots \\ 0 & 0 & 0 & \cdots & f(\lambda_m) \end{bmatrix} \qquad (A.5.24)$$

For example, $\exp[\lambda]$ is

$$\exp[[\lambda]] = \begin{bmatrix} \exp(\lambda_1) & 0 & 0 & \cdots & 0 \\ 0 & \exp(\lambda_2) & 0 & \cdots & 0 \\ \vdots & & & & \vdots \\ 0 & 0 & 0 & \cdots & \exp(\lambda_m) \end{bmatrix}$$

Consider a diagonalizable matrix $[A]$

$$[A] = [E][\lambda][E]^{-1} \qquad (A.5.25)$$

It is easy to see that

$$[A]^2 = \left[[E][\lambda][E]^{-1}\right]\left[[E][\lambda][E]^{-1}\right]$$
$$= [E][\lambda][\lambda][E]^{-1}$$
$$= [E][\lambda]^2[E]^{-1} \qquad (A.5.26)$$

and, in general,

$$[A]^k = [E][\lambda]^k[E]^{-1} \qquad (A.5.27)$$

which provides a straightforward procedure for evaluating $[A]^k$, since $[\lambda]^k$ can be evaluated simply.

Let us now consider the evaluation of the matrix exponential $\exp[A]$.

$$\exp[A] = \sum_{k=0}^{\infty} \frac{1}{k!}[A]^k$$
$$= \sum_{k=0}^{\infty} \frac{1}{k!}[E][\lambda]^k[E]^{-1}$$
$$= [E]\left(\sum_{k=0}^{\infty} \frac{1}{k!}[\lambda]^k\right)[E]^{-1}$$
$$= [E][\exp[\lambda]][E]^{-1} \qquad (A.5.28)$$

In general, we can evaluate any function of a diagonalizable matrix $[A]$ by the above procedure. Thus,

$$f[[A]] = [E]f[[\lambda]][E]^{-1} \tag{A.5.29}$$

A.6   MATRIX COMPUTATIONS

Matrix computations can be very time consuming, particularly for matrices of large size. By performing the calculations in the most efficient way it is often possible to save a lot of time. Here, we provide some guidelines as to how certain matrix computations can most expediently be carried out. For definiteness, we consider the problem of computing the molar diffusion fluxes J_i from the equation

$$(J) = c_t[k][\Phi][\exp[\Phi] - [I]]^{-1}(\Delta x) \tag{A.6.1}$$

Equation A.6.1 arises when the Maxwell–Stefan equations are solved for the case of steady-state, one-dimensional mass transfer, as discussed in Chapter 8. The matrices $[k]$ and $[\Phi]$ are as defined in Chapter 8, c_t is the molar density of the mixture and a scalar, and (Δx) is a column matrix of mole fraction differences. All matrices in Eq. A.6.1 are of order $n - 1$ where n is the number of components in the mixture. For the purposes of this discussion we shall assume that the matrices $[k]$ and $[\Phi]$ have already been calculated. The matrix function $[\Phi][\exp[\Phi] - [I]]^{-1}$, denoted by $[\Xi]$, can be computed using Sylvester's expansion formula (see, however, below) so the immediate problem is the calculation of the column matrix (J) from

$$(J) = c_t[k][\Xi](\Delta x) \tag{A.6.2}$$

A.6.1   Arithmetic Operations

Now, the multiplication of two square matrices of order m involves m^3 multiplication operations and $m^2(m - 1)$ additions. However, the multiplication of a square and a column matrix involves m^2 multiplications and $m(m - 1)$ additions and the multiplication of a square matrix with a scalar involves just m^2 multiplications. With regard to Eq. A.6.2, there is, therefore, some incentive to computing (J) from right to left rather than from left to right, as written.

A.6.2   Matrix Functions

The most time consuming step in the calculation of the fluxes is the evaluation of the matrix $[\Xi]$. For problems involving only a few (three or four) components the use of Sylvester's theorem is recommended. Unfortunately, Sylvester's expansion formula demands an increasing percentage of computer time as the number of components and, therefore, the size of the matrices involved increases. An alternative to the use of Sylvester's formula is to evaluate a matrix function from its power series definition. Thus, for example, the exponential matrix in Eq. A.6.1 could be evaluated directly from Eq. A.5.4. The number of terms needed to evaluate $\exp[\Phi]$ in this way depends on the magnitude of the dominant eigenvalue $\hat{\Phi}_d$ of $[\Phi]$ (the eigenvalue with the largest absolute value). Convergence will be rapidly obtained for small $\hat{\Phi}_d$. However, for large $\hat{\Phi}_d$ we may need a great many terms of the series to evaluate the exponential.

Moler and Van Loan (1978) discuss at some length "Nineteen dubious ways to compute the exponential of a matrix." They consider three or four methods as candidates for the "best" method. Included among this small group of "best" methods is one that they call "scaling and squaring," which is attributed to many, but which we first came across in a paper by Buffham and Kropholler (1971). They write e^x in the form

$$e^x = \left\{\exp\{x/2^q\}\right\}^{2^q} \tag{A.6.3}$$

where q is a positive integer. By choosing a suitable value for q the magnitude of the exponential argument will be of order 1. Convergence of $\exp\{x/2^q\}$ from the power series (Eq. A.5.3) will be very rapidly obtained. It only remains to multiply this quantity by itself q times to obtain e^x. Equation A.6.3 may be generalized in matrix form as

$$\exp[\Phi] = \left[\exp[\Phi]/2^q\right]^{2^q} \tag{A.6.4}$$

A suitable value of q may be estimated from (Buffham and Kropholler, 1971)

$$q = \text{Int}\left(\frac{\ln|\hat{\Phi}_d|}{\ln 2} + 1\right) \tag{A.6.5}$$

If q calculated from Eq. A.6.5 is negative, it should be assigned a value of zero. Subtraction of the identity matrix $[I]$ from $\exp[\Phi]$, followed by inversion and premultiplication by $[\Phi]$ gives the matrix $[\Xi]$. Using Eq. A.6.4 to calculate $[\Xi]$ can be several times faster than Sylvester's formula (Taylor and Webb, 1981).

Since the computation of eigenvalues can be quite time consuming, it would be a good idea to replace an exact determination of the dominant eigenvalue with an approximation. Gershgorin's circle theorem (see, e.g., Wilkinson, 1965) may be used to estimate the limits of the eigenvalues. This theorem states that the ith eigenvalue of $[\Phi]$ lies within a circle with center Φ_{ii} and radius given by the sum of $|\Phi_{ik}|$, excluding the diagonal element. Thus, $\hat{\Phi}_d$ may be estimated quite readily from

$$|\hat{\Phi}_d| \leq \max_i \left(\sum_{k=1}^{n-1} |\Phi_{ik}|\right) \tag{A.6.6}$$

Unfortunately, the above procedure cannot be used when $[\Phi]$ is singular. In this case the matrix $[\exp[\Phi] - [I]]$ is singular and cannot be inverted. Hence, the matrix function $[\Xi]$, though finite, cannot be obtained in this way. A power series expansion of $[\Xi]$ is not convergent for all $[\Phi]$. On the other hand, an expansion of the inverse, $[\Xi]^{-1}$ is convergent and can be calculated even when $[\Phi]$ is singular. The series $[\Xi]^{-1}$ may be expressed as (Taylor and Webb, 1981)

$$[\Xi]^{-1} = [I] + \sum_{k=1}^{\infty} \frac{1}{(k+1)!}[\Phi]^k \tag{A.6.7}$$

The matrix $[\Xi]$ is then obtained by inversion of the result of this series. The series representation (Eq. A.6.6) is preferred to Sylvester's formula (especially when the order of the matrix is > 3 or 4) but is not as fast as the truncated power series (Eq. A.6.4) (Taylor and Webb, 1981). For problems involving a singular, or nearly singular $[\Phi]$, the series (Eq. A.6.7) is the best alternative to Sylvester's formula.

APPENDIX B
Solution of Systems
of Differential Equations

B.1 GENERALIZATION OF THE SOLUTIONS
OF SCALAR DIFFERENTIAL EQUATIONS

In many cases, the solution to a matrix differential equation can be obtained as the matrix generalization of the equivalent scalar differential equation. For example, the first-order differential equation

$$\frac{dx}{dt} = Ax \tag{B.1.1}$$

with "initial" condition

$$x = x_0 \quad \text{at} \quad t = 0 \tag{B.1.2}$$

has the solution

$$x = e^{At}x_0 \tag{B.1.3}$$

The matrix generalization of Eq. B.1.1

$$\frac{d(x)}{dt} = [A](x) \tag{B.1.4}$$

with initial conditions

$$(x) = (x_0) \quad \text{at} \quad t = 0 \tag{B.1.5}$$

has the solution

$$(x) = [\exp[A]t](x_0) \tag{B.1.6}$$

Using the definition of the exponential matrix (Eq. A.5.4), we may rewrite Eq. B.1.6 as

$$(x) = \left[[I] + [A]t + \tfrac{1}{2}[A]^2t^2 + \tfrac{1}{6}[A]^3t^3 + \cdots\right](x_0) \tag{B.1.7}$$

and differentiate term by term to get

$$
\begin{aligned}
\frac{d(x)}{dt} &= \left[[A] + \tfrac{2}{2}[A]^2t + \tfrac{3}{6}[A]^3t^2 + \cdots\right](x_0) \\
&= [A]\left[[I] + [A]t + \tfrac{1}{2}[A]^2t^2 + \cdots\right](x_0) \\
&= [A][\exp[A]t](x_0) \\
&= [A](x) \tag{B.1.8}
\end{aligned}
$$

which completes the proof.

Let us now consider the matrix differential equation

$$\frac{d(x)}{dt} = [A](x) + (b) \tag{B.1.9}$$

where both $[A]$ and (b) are constant matrices. The "initial" or boundary condition is, again, given by Eq. B.1.5. This system of equations can be solved as follows. First, we rewrite Eq. B.1.9 as

$$\frac{d(x)}{dt} = [A]\big((x) + [A]^{-1}(b)\big) \tag{B.1.10}$$

Since $[A]$ and (b) are constant matrices, it follows that

$$\frac{d(\omega)}{dt} = \frac{d(x)}{dt} \tag{B.1.11}$$

where we have defined (ω) by

$$(\omega) = (x) + [A]^{-1}(b) \tag{B.1.12}$$

In terms of (ω) we now have to solve

$$\frac{d(\omega)}{dt} = [A](\omega) \tag{B.1.13}$$

subject to

$$(\omega) = (\omega_0) = (x_0) + [A]^{-1}(b) \quad \text{at} \quad t = 0 \tag{B.1.14}$$

Equations B.1.16 and B.1.17 represent the system for which we already have the solution

$$(\omega) = [\exp[A]t](\omega_0) \tag{B.1.15}$$

Equation B.1.15 can be rewritten in terms of (x) as

$$(x) = [\exp[A]t](x_0) + [[\exp[A]t] - [I]][A]^{-1}(b) \tag{B.1.16}$$

B.2 THE METHOD OF SUCCESSIVE SUBSTITUTION

The method of successive substitution is useful for solving matrix differential equations in which the coefficients are functions of the independent variable t.

$$\frac{d(x)}{dt} = [A(t)](x) + (b(t)) \tag{B.2.1}$$

To illustrate the method of successive substitution, also known as Picard's method, we first consider the scalar differential equation

$$\frac{dx}{dt} = A(t)x \tag{B.2.2}$$

with the initial condition

$$x = x_0 \quad \text{at} \quad t = 0 \tag{B.2.3}$$

Separating variables and integrating Eq. B.2.2. gives

$$\int dx = \int A(t)x\,dt \tag{B.2.4}$$

Integrating from the initial condition at $t = 0$ gives

$$x = x_0 + \int_0^t A(t_1)x\,dt_1 \tag{B.2.5}$$

where we have introduce a dummy variable of integration t_1.

Equation B.2.5 gives x in terms of an integral of itself. Approximate expressions for x can be obtained by substituting any appropriate function of x under the integral on the right-hand side. For example, we could let $x = 0$ and the zeroth approximation to the solution to Eq. B.2.2 would be $x = x_0$. Obviously, this is not a very good approximation except when $t = 0$. A better one is obtained by substituting the entire right-hand side of Eq. B.2.5 for x under the integral

$$x = x_0 + \int_0^t A(t_1)\left\{ x_0 + \int_0^{t_1} A(t_2)x\,dt_2 \right\} dt_1$$

$$= x_0 + \int_0^t A(t_1)\,dt_1\,x_0 + \int_0^t A(t_1)\int_0^{t_1} A(t_2)x\,dt_2\,dt_1 \tag{B.2.6}$$

Now we have the first two terms of a series solution to Eq. B.2.4. The third term on the right-hand side of Eq. B.2.6 involves x; so, once again, we replace it using Eq. B.2.5 to give the next approximation

$$x = x_0 + \int_0^t A(t_1)\,dt_1\,x_0$$

$$+ \int_0^t A(t_1)\int_0^{t_1} A(t_2)\left\{ x_0 + \int_0^{t_2} A(t_3)x\,dt_3 \right\} dt_2\,dt_1$$

$$= x_0 + \int_0^t A(t_i)\,dt_1\,x_0 + \int_0^t A(t_1)\int_0^{t_1} A(t_2)\,dt_2\,dt_1\,x_0$$

$$+ \int_0^t A(t_1)\int_0^{t_1} A(t_2)\int_0^{t_2} A(t_3)x\,dt_3\,dt_2\,dt_1 \tag{B.2.7}$$

It should be obvious that we may continue indefinitely by substituting the right-hand side of Eq. B.2.5 for x in the last term of each series we generate. The final result is the solution of Eq. B.2.2

$$x = \Omega_0^t(A)x_0 \tag{B.2.8}$$

where

$$\Omega(A) = 1 + \int_0^t A(t_1)\,dt_1 + \int_0^t A(t_1)\int_0^{t_1} A(t_2)\,dt_2\,dt_1$$

$$+ \int_0^t A(t_1)\int_0^{t_1} A(t_2)\int_0^{t_2} A(t_3)\,dt_3\,dt_2\,dt_1$$

$$+ \cdots \tag{B.2.9}$$

The solution to the matrix generalization of Eq. B.2.2

$$\frac{d(x)}{dt} = [A(t)](x) \tag{B.2.10}$$

with

$$(x) = (x_0) \quad \text{at} \quad t = 0 \tag{B.2.11}$$

follows in exactly the same way. Separating variables in Eq. B.2.10 and integrating gives

$$(x) = (x_0) + \int_0^t [A(t_1)](x) \, dt_1 \tag{B.2.12}$$

The first approximation is obtained when we substitute for (x) under the integral using the entire right-hand side.

$$\begin{aligned}
(x) &= (x_0) + \int_0^t [A(t_1)] \left((x_0) + \int_0^{t_1} [A(t_2)](x) \, dt_2 \right) dt_1 \\
&= (x_0) + \int_0^t [A(t_1)] \, dt_1 (x_0) \\
&\quad + \int_0^t [A(t_1)] \int_0^{t_1} [A(t_2)](x) \, dt_2 \, dt_1
\end{aligned} \tag{B.2.13}$$

The second approximation is

$$\begin{aligned}
(x) &= (x_0) + \int_0^t [A(t_1)] \, dt_1 (x_0) \\
&\quad + \int_0^t [A(t_1)] \int_0^{t_1} [A(t_2)] \left\{ (x_0) + \int_0^{t_2} [A(t_3)](x) \, dt_3 \right\} dt_2 \, dt_1 \\
&= (x_0) + \int_0^t [A(t_1)] \, dt_1 (x_0) \\
&\quad + \int_0^t [A(t_1)] \int_0^{t_1} [A(t_2)] \, dt_2 \, dt_1 (x_0) \\
&\quad + \int_0^t [A(t_1)] \int_0^{t_1} [A(t_2)] \int_0^{t_2} [A(t_3)](x) \, dt_3 \, dt_2 \, dt_1
\end{aligned} \tag{B.2.14}$$

Continuing in this way we end up with

$$(x) = [\Omega_0^t(A)](x_0) \tag{B.2.15}$$

where the matrix $[\Omega_0^t(A)]$, termed the *matrizant* by Amundson (1966), is given by

$$\begin{aligned}
[\Omega_0^t(A)] &= [I] + \int_0^t [A(t_1)] \, dt_1 + \int_0^t [A(t_1)] \int_0^{t_1} [A(t_2)] \, dt_2 \, dt_1 \\
&\quad + \int_0^t [A(t_1)] \int_0^{t_1} [A(t_2)] \int_0^{t_2} [A(t_3)] \, dt_3 \, dt_2 \, dt_1 \\
&\quad + \cdots
\end{aligned} \tag{B.2.16}$$

The same approach can be used to solve the matrix differential equation (Eq. B.2.1). The solution is (Amundson, 1966)

$$(x) = [\Omega_0^t(A)]\left((x_0) + [\Omega_0^t(A)]\int_0^t [\Omega_0^\tau(A)]^{-1}(b(\tau))\, d\tau\right) \tag{B.2.17}$$

Let us demonstrate this method by reconsidering the matrix differential equation

$$\frac{d(x)}{dt} = [A](x) + (b) \tag{B.2.18}$$

with

$$(x) = (x_0) \quad \text{at} \quad t = 0 \tag{B.2.19}$$

that we solved in Section B.1. Even though $[A]$ and (b) are constant matrices we may use the method of successive substitution.

The solution can be written down immediately as

$$(x) = [\Omega_0^t(A)](x_0) + [\Omega_0^t(a)]\left(\int_0^t [\Omega_0^\tau(A)]^{-1}\, d\tau\right)(b) \tag{B.2.20}$$

where we have taken (b) out of the integral because it is constant.

The matrizant $[\Omega_0^t(A)]$ is found as follows:

$$[\Omega_0^t(A)] = [I] + \int_0^t [A]\, dt_1 + \int_0^t [A]\int_0^{t_1}[A]\, dt_2\, dt_1$$

$$+ \int_0^t [A]\int_0^{t_1}[A]\int_0^{t_2}[A]\, dt_3\, dt_2\, dt_1 + \cdots$$

$$= [I] + [A] + \tfrac{1}{2}[A]^2 t^2 + \tfrac{1}{6}[A]^3 t^3 + \cdots$$

$$= \exp[A]t \tag{B.2.21}$$

Thus, the matrizant of a constant matrix is the exponential matrix. Returning to Eq. B.2.20 we have

$$(x) = [\exp[A]t](x_0) + [\exp[A]t]\int_0^t [\exp[A]t_1]^{-1}(b)\, dt_1 \tag{B.2.22}$$

By using the definition of the exponential matrix (Eq. A.5.4), it is easy to show that

$$[\exp[A]]\exp[-[A]] = [I] \tag{B.2.23}$$

Hence,

$$[\exp[A]]^{-1} = \exp[-[A]] \tag{B.2.24}$$

which allows the integral in Eq. B.2.22 to be evaluated.

$$\int_0^t [\exp[A]t]^{-1}\, dt = \int_0^t \exp[-[A]t]\, dt$$

$$= \int_0^t \left[[I] - [A]t + \tfrac{1}{2}[t]^2 t^2 - \tfrac{1}{6}[A]^3 t^3 + \cdots\right] dt$$

$$= [[I] - \exp[-[A]t]][A]^{-1} \tag{B.2.25}$$

Substituting this result into Eq. B.2.20 we obtain

$$(x) = [\exp[A]t](x_0) + [\exp[A]t][[I] - \exp[-[A]t]][A]^{-1}(b)$$

$$= [\exp[A]t](x_0) + [[\exp[A]t] - [I]][A]^{-1}(b) \tag{B.2.26}$$

which, of course, is Eq. B.1.22. Although we did not really have to go through all of this calculus to obtain the above result, this example serves to illustrate the lengths to which one must go in order to evaluate the matrizant and the solution of a matrix differential equation by the method of successive substitution. This method is used to solve multicomponent mass transfer problems in Sections 8.3.5 (see, also, Sections 8.4 and 8.7), 9.3, and 10.4.

B.3   SOLUTION OF COUPLED DIFFERENTIAL EQUATIONS USING SIMILARITY TRANSFORMATIONS

An alternative to the methods described above can be used if the coefficient matrix is diagonalizable. Consider, once again, the matrix differential equation and its associated initial condition

$$\frac{d(x)}{dt} = [A](x) \tag{B.1.4}$$

with initial conditions

$$(x) = (x_0) \quad \text{at} \quad t = 0 \tag{B.1.5}$$

If $[A]$ is diagonalizable it is possible to find a nonsingular matrix $[E]$ such that

$$[E]^{-1}[A][E] = [\lambda] \tag{B.3.1}$$

where $[\lambda]$ is a diagonal matrix whose elements are the eigenvalues of $[A]$. Premultiplying Eq. B.1.4 by $[E]^{-1}$ and inserting the identity matrix $[E][E]^{-1}$ between $[A]$ and (x) gives

$$[E]^{-1}\frac{d(x)}{dt} = [E]^{-1}[A][E][E]^{-1}(x)$$

$$= [\lambda][E]^{-1}(x) \tag{B.3.2}$$

If $[A]$ is a constant matrix, that is, $[A] = [A(t)]$, then its eigenvalues λ_i and modal matrix $[E]$ will also be independent of t. Thus, we may bring $[E]^{-1}$ inside the derivative. Defining a column matrix of new variables $(\hat{x})$ by

$$(\hat{x}) = [E]^{-1}(x) \tag{B.3.3}$$

we have

$$\frac{d(\hat{x})}{dt} = [\lambda](\hat{x}) \tag{B.3.4}$$

Now, since $[\lambda]$ is diagonal, this result represents a set of m *uncoupled* differential equations

$$\frac{d\hat{x}_i}{dt} = \lambda_i \hat{x}_i \tag{B.3.5}$$

The solution to the ith equation is, of course,

$$\hat{x}_i = \exp(\lambda_i t)\hat{x}_{i0} \qquad i = 1, 2, \ldots, m \tag{B.3.6}$$

where $\hat{x}_{i0}$ is obtained from the transformation

$$(\hat{x}_0) = [E]^{-1}(x_0) \tag{B.3.7}$$

To obtain the solution in terms of the original set of variables (x) we arrange the set of Eqs. B.3.6 in matrix form

$$(\hat{x}) = [\exp[\lambda]t](\hat{x}_0) \tag{B.3.8}$$

and premultiply by $[E]$ to get

$$[E](\hat{x}) = [E][\exp[\lambda]t][E]^{-1}[E](\hat{x}_0) \tag{B.3.9}$$

or

$$(x) = [F](x_0) \tag{B.3.10}$$

where

$$[F] = [E][\exp[\lambda]t][E]^{-1} \tag{B.3.11}$$

This method is used in Chapter 5 to solve multicomponent diffusion problems.

APPENDIX C
Solution of Systems
of Algebraic Equations

C.1 SOLUTION OF SYSTEMS OF LINEAR EQUATIONS

Consider the system of simultaneous linear equations

$$[A](x) = (b) \tag{A.3.12}$$

If $[A]$ is nonsingular then the solution is given formally by

$$(x) = [A]^{-1}(b) \tag{A.3.13}$$

However, this method of solving a system of linear equations is not used in practice, largely because it is much too time consuming. A better method is based on the so-called LU decomposition in which the matrix $[A]$ is factorized as follows:

$$[A] = [L][U] \tag{C.1.1}$$

where $[L]$ is a *lower* triangular matrix

$$[L] = \begin{bmatrix} L_{11} & & & & \\ L_{21} & L_{22} & & & \\ \vdots & & & & \\ L_{n1} & L_{n2} & L_{n3} & \cdots & L_{nn} \end{bmatrix} \tag{C.1.2}$$

and $[U]$ is an *upper* triangular matrix

$$[U] = \begin{bmatrix} U_{11} & U_{12} & U_{13} & \cdots & U_{1n} \\ & U_{22} & U_{23} & \cdots & U_{2n} \\ & & & & \vdots \\ & & & & U_{nn} \end{bmatrix} \tag{C.1.3}$$

Following the computation of the factors $[L]$ and $[U]$ it is a simple matter to obtain the column matrix (x) as follows. The first step (forward substitution) is to compute a column matrix $(z) = [U](x)$ by solving the linear system $[L](z) = (b)$. The second step (backward substitution) involves the computation of (x) from $[U](x) = (z)$. We shall not go into the details of the decomposition process here. The interested reader can find them in almost any book on numerical methods. The entire method can be programmed in only a few lines of computer code (see, e.g., Press et al., 1992).

The LU decomposition method is particularly useful for solving linear systems of the form $[A][X] = [B]$, where $[X]$ and $[B]$ are rectangular (i.e., with more than one column).

For these systems, we simply perform the back substitution as many times as necessary (once for each column in $[X]$), but the factors of $[A]$, $[L]$ and $[U]$, need be computed only once.

C.2   SOLUTION OF SYSTEMS OF NONLINEAR EQUATIONS

A very large number of methods of solving systems of nonlinear algebraic equations has been devised (Ortega and Rheinbolt, 1970). However, just two methods are employed in the algorithms presented in this book: repeated substitution and Newton's method; we review these methods below.

C.2.1   Repeated Substitution

To use repeated substitution, the equations to be solved must be expressed in the form

$$(G((x))) = (x) \tag{C.2.1}$$

where (G) is a vector consisting of all the equations to be solved and (x) is, again, the vector of variables. The procedure for solving the equations is summarized in Algorithm C.1.

C.2.2   Newton's Method

To use Newton's method, the equations to be solved are written in the form

$$(F((x))) = (0) \tag{C.2.2}$$

where (F) is a vector consisting of all the equations to be solved and (x) is, again, the vector of variables. A Taylor series expansion of the function vector around the point (x_0) at which the functions are evaluated gives

$$(F((x))) = (F((x_0))) + [J]((x) - (x_0))) + O(|(x) - (x_0)|^2) \tag{C.2.3}$$

where $[J]$ is the *Jacobian* matrix of partial derivatives of (F) with respect to the independent variables (x)

$$J_{ij} = \partial F_i/\partial x_j \tag{C.2.4}$$

If (x) is the actual solution to the system of equations, then $(F((x))) = (0)$: Additionally, if the $O(|(x) - (x_0)|^2)$ term is neglected then Eq. C.2.3 can be rearranged to give

$$[J]((x) - (x_0)) = -(F((x_0))) \tag{C.2.5}$$

Here the only unknown is the value of the vector (x). Thus, by solving the linear system of Eqs. C.2.5 it is possible to obtain a value for this vector. If the new vector (x) obtained in this way does not actually satisfy the set of equations (F) then the procedure can be repeated using the calculated value as a new (x_0). The entire procedure is summarized in Algorithm C.2.

Algorithm C.1 Repeated Substitution

1. Set iteration counter k to zero. Estimate $(x)_0$.
2. Compute $(G((x)_k))$.

3. Set $(x)_{k+1} = (G((x)_k))$.
4. Check for convergence. If $\|(x)_{k+1} - (x)_k\|$ is less than some prescribed small number, stop. Otherwise, increment k and return to Step 2.

Algorithm C.2 Newton's Method

1. Set iteration counter k to zero. Estimate $(x)_0$.
2. Compute $(F(x)_k)$ and $[J]$.
3. Solve Eq. C.2.5 for $(x)_{k+1}$.
4. Check for convergence; if not obtained, increment k and return to Step 2.

APPENDIX D
Estimation of Thermodynamic Factors from Activity Coefficient Models

For concentrated nonideal liquid mixtures the thermodynamic factors can be calculated from models of the excess Gibbs energy G^{EX}. There are quite a number of models of G^{EX} available: the Margules, Van Laar, Wilson, NRTL, and UNIQUAC models are widely used. With the comprehensive data compilation edited by Gmehling and Onken (1977ff), an extremely large number of parameters of these solution models has become available and these can be used in the calculation of the thermodynamic factors. In the absence of experimental data, group contribution methods, such as ASOG (Analytical Solution of Groups) and UNIFAC (UNIQUAC Functional Activity Coefficients), can be used. There are many thermodynamics texts that discuss these models in great detail (see, e.g., Walas (1985) and Prausnitz et al., 1986). However, the necessary expressions for evaluating the thermodynamic factors are rarely needed in textbook treatments of thermodynamics. Taylor and Kooijman (1991) presented expressions for the thermodynamic factors for many of the above models. We summarize their results here.

D.1 ESTIMATION OF THE THERMODYNAMIC FACTOR FOR BINARY SYSTEMS

The thermodynamic factor Γ for a binary system is defined by

$$\Gamma = 1 + x_1 \frac{\partial \ln \gamma_1}{\partial x_1}\bigg|_{\Sigma} \tag{D.1.1}$$

The symbol Σ is used to indicate that the differentiation of $\ln \gamma_1$ with respect to mole fraction x_1 is to be carried out subject to the restriction that $x_1 + x_2 = 1$. For a regular solution, for example, the activity coefficient of component 1 is given by

$$\ln \gamma_1 = A x_2^2 \tag{D.1.2}$$

If we replace the mole fraction x_2 with $1 - x_1$ and differentiate with respect to x_1 we find

$$\frac{\partial \ln \gamma_1}{\partial x_1} = -2A(1 - x_1) = -2Ax_2 \tag{D.1.3}$$

and, therefore, Γ is given by

$$\Gamma = 1 - 2Ax_1 x_2 \tag{D.1.4}$$

Few real systems are adequately represented by the regular solution model. As a result, there are several more complicated models of G^{EX} that have been proposed. For these

534

other models we find that an alternative procedure, formally equivalent, is algebraically simpler to apply, especially for the multicomponent systems we consider later. In what follows we develop this procedure in detail and provide the results for a number of models of activity coefficients in solution.

Let Q be the dimensionless excess Gibbs energy

$$Q = Q(x_1, x_2) = G^{EX}(x_1, x_2)/RT$$
$$= x_1 \ln \gamma_1 + x_2 \ln \gamma_2 \tag{D.1.5}$$

The activity coefficients γ_1 and γ_2 may be expressed in terms of Q and its composition derivatives by (see Section D.2 for derivation)

$$\ln \gamma_1 = Q + Q_1 - x_1 Q_1 - x_2 Q_2 \tag{D.1.6}$$
$$\ln \gamma_2 = Q + Q_2 - x_1 Q_1 - x_2 Q_2 \tag{D.1.7}$$

where Q_1 and Q_2 are defined by

$$Q_1 \equiv \frac{\partial Q}{\partial x_1}\bigg|_{x_2} \qquad Q_2 \equiv \frac{\partial Q}{\partial x_2}\bigg|_{x_1} \tag{D.1.8}$$

That is, Q_1 is the partial derivative of Q with respect to mole fraction x_1, while holding mole fraction x_2 constant, and Q_2 is the partial derivative of Q with respect to mole fraction x_2, while holding mole fraction x_1 constant.

TABLE D.1   Regular Solution Model for Binary Systems

Dimensionless Gibbs excess energy

$$Q = A x_1 x_2$$

Parameter

$$A$$

Unconstrained first composition derivatives

$$Q_1 = A x_2 \qquad Q_2 = A x_1$$

Unconstrained second composition derivatives

$$Q_{11} = 0 \qquad Q_{12} = A$$
$$Q_{21} = A \qquad Q_{22} = 0$$

Activity coefficients

$$\ln \gamma_1 = A x_2^2$$
$$\ln \gamma_2 = A x_1^2$$

Thermodynamic factor

$$\Gamma = 1 - 2 A x_1 x_2$$

We may differentiate Eq. D.1.6 and D.1.7 to obtain the partial derivatives of $\ln \gamma_1$ and $\ln \gamma_2$ as

$$\left. \frac{\partial \ln \gamma_1}{\partial x_1} \right|_{x_2} = Q_{11} - x_1 Q_{11} - x_2 Q_{21} \tag{D.1.9a}$$

$$\left. \frac{\partial \ln \gamma_1}{\partial x_2} \right|_{x_1} = Q_{12} - x_1 Q_{12} - x_2 Q_{22} \tag{D.1.9b}$$

$$\left. \frac{\partial \ln \gamma_2}{\partial x_1} \right|_{x_2} = Q_{21} - x_2 Q_{21} - x_1 Q_{11} \tag{D.1.9c}$$

$$\left. \frac{\partial \ln \gamma_2}{\partial x_2} \right|_{x_1} = Q_{22} - x_2 Q_{22} - x_1 Q_{12} \tag{D.1.9d}$$

TABLE D.2 Margules Model for Binary Systems

Dimensionless Gibbs excess energy

$$Q = x_1 x_2 (A_{12} x_2 + A_{21} x_1)$$

Parameters

$$A_{12} \quad \text{and} \quad A_{21}$$

Unconstrained first composition derivatives

$$Q_1 = A_{12} x_2^2 + 2 A_{21} x_1 x_2$$
$$Q_2 = 2 A_{12} x_1 x_2 + A_{21} x_1^2$$

Unconstrained second composition derivatives

$$Q_{11} = 2 A_{21} x_2$$
$$Q_{12} = 2 A_{12} x_2 + 2 A_{21} x_1 = Q_{21}$$
$$Q_{22} = 2 A_{12} x_1$$

Other relations

$$Q = \tfrac{1}{3}(x_1 Q_1 + x_2 Q_2)$$

Activity coefficients

$$\ln \gamma_1 = Q_1 - 2Q$$
$$\ln \gamma_2 = Q_2 - 2Q$$

Thermodynamic factor

$$\Gamma = 1 + x_1 x_2 (Q_{11} + Q_{22} - 2Q_{12})$$
$$= 1 + 2 x_1 x_2 ((A_{21} - A_{12})(1 - 3x_1) - A_{12})$$

---

where the Q_{ij} are second partial derivatives defined by

$$Q_{11} \equiv \frac{\partial Q_1}{\partial x_1}\bigg|_{x_2} \qquad Q_{12} \equiv \frac{\partial Q_1}{\partial x_2}\bigg|_{x_1}$$

$$Q_{21} \equiv \frac{\partial Q_2}{\partial x_1}\bigg|_{x_2} \qquad Q_{22} \equiv \frac{\partial Q_2}{\partial x_2}\bigg|_{x_1}$$

(D.1.10)

TABLE D.3   Van Laar Model for Binary Systems

Dimensionless Gibbs excess energy

$$Q = A_{12}A_{21}x_1x_2/S$$
$$S = A_{12}x_1 + A_{21}x_2$$

Parameters

$$A_{12} \qquad \text{and} \qquad A_{21}$$

Unconstrained first composition derivatives

$$Q_1 = A_{12}A_{21}^2 x_2^2/S^2$$
$$Q_2 = A_{12}^2 A_{21} x_1^2/S^2$$

Unconstrained second composition derivatives

$$Q_{11} = -2A_{12}^2 A_{21}^2 x_2^2/S^3$$
$$Q_{12} = 2A_{12}^2 A_{21}^2 x_1 x_2/S^3 = Q_{21}$$
$$Q_{22} = -2A_{12}^2 A_{21}^2 x_1^2/S^3$$

Other relations

$$Q = x_1 Q_1 + x_2 Q_2$$
$$x_1 Q_{11} + x_2 Q_{21} = 0$$
$$x_1 Q_{12} + x_2 Q_{22} = 0$$

Activity coefficients

$$\ln \gamma_1 = Q_1 \qquad \ln \gamma_2 = Q_2$$

Unconstrainted composition derivatives of $\ln \gamma$

$$\frac{\partial \ln \gamma_1}{\partial x_1}\bigg|_{x_2} = Q_{11} \qquad \frac{\partial \ln \gamma_1}{\partial x_2}\bigg|_{x_1} = Q_{12}$$

$$\frac{\partial \ln \gamma_2}{\partial x_1}\bigg|_{x_2} = Q_{21} \qquad \frac{\partial \ln \gamma_2}{\partial x_2}\bigg|_{x_1} = Q_{22}$$

Thermodynamic factor

$$\Gamma = 1 + x_1(Q_{11} - Q_{12}) = 1 - 2x_1 x_2 A_{12}^2 A_{21}^2/S^3$$

TABLE D.4 Wilson Model for Binary Systems

Dimensionless Gibbs excess energy

$$Q = -x_1 \ln S_1 - x_2 \ln S_2$$
$$S_1 = x_1 + x_2 \Lambda_{12} \qquad S_2 = x_2 + x_1 \Lambda_{21}$$
$$\Lambda_{12} = (V_2/V_1)\exp(-(\mu_{12} - \mu_{11})/RT)$$
$$\Lambda_{21} = (V_1/V_2)\exp(-(\mu_{21} - \mu_{22})/RT)$$

Parameters

$$(\mu_{12} - \mu_{11}) \qquad (\mu_{21} - \mu_{22}) \qquad V_1, V_2$$

Unconstrained first composition derivatives

$$Q_1 = -\ln S_1 - x_1/S_1 - x_2\Lambda_{21}/S_2$$
$$Q_2 = -\ln S_2 - x_2/S_2 - x_1\Lambda_{12}/S_1$$

Unconstrained second composition derivatives

$$Q_{11} = -2/S_1 + x_1/S_1^2 + x_2\Lambda_{21}^2/S_2^2$$
$$Q_{12} = -\Lambda_{12}/S_1 - \Lambda_{21}/S_2 + x_1\Lambda_{12}/S_1^2 + x_2\Lambda_{21}/S_2^2$$
$$= Q_{21}$$
$$Q_{22} = -2/S_2 + x_2/S_2^2 + x_1\Lambda_{12}^2/S_1^2$$

Other relations

$$Q = x_1(1 + Q_1) + x_2(1 + Q_2)$$
$$x_1 Q_{11} + x_2 Q_{21} = -1$$
$$x_1 Q_{12} + x_2 Q_{22} = -1$$

Activity coefficients

$$\ln \gamma_1 = x_1 + x_2 + Q_1 \qquad \ln \gamma_2 = x_1 + x_2 + Q_2$$

Unconstrained composition derivatives of $\ln \gamma$

$$\left.\frac{\partial \ln \gamma_1}{\partial x_1}\right|_{x_2} = 1 + Q_{11} \qquad \left.\frac{\partial \ln \gamma_1}{\partial x_2}\right|_{x_1} = 1 + Q_{12}$$

$$\left.\frac{\partial \ln \gamma_2}{\partial x_1}\right|_{x_2} = 1 + Q_{21} \qquad \left.\frac{\partial \ln \gamma_2}{\partial x_2}\right|_{x_1} = 1 + Q_{22}$$

Thermodynamic factor

$$\Gamma = 1 + x_1(Q_{11} - Q_{12})$$

TABLE D.5   NRTL Model for Binary Systems

Dimensionless Gibbs excess energy

$$Q = x_1 x_2 (\tau_{21} G_{21}/S_1 + \tau_{12} G_{12}/S_2)$$
$$S_1 = x_1 + x_2 G_{21} \qquad S_2 = x_2 + x_1 G_{12}$$
$$G_{12} = \exp(-\alpha\tau_{12}) \qquad G_{21} = \exp(-\alpha\tau_{21})$$
$$\tau_{12} = (g_{12} - g_{11})/RT \qquad \tau_{21} = (g_{21} - g_{22})/RT$$

Parameters

$$(g_{12} - g_{11}), (g_{21} - g_{22}) \quad \text{and} \quad \alpha$$

Unconstrained first composition derivatives

$$Q_1 = -x_2 G_{21}\varepsilon_{11} + x_2 \varepsilon_{12}$$
$$Q_2 = -x_1 G_{12}\varepsilon_{22} + x_1 \varepsilon_{21}$$
$$\varepsilon_{11} = -C_1/S_1^2 \qquad \varepsilon_{12} = G_{12}(\tau_{12} - C_2/S_2)/S_2$$
$$\varepsilon_{21} = G_{21}(\tau_{21} - C_1/S_1)/S_1 \qquad \varepsilon_{22} = -C_2/S_2^2$$
$$C_1 = x_2 \tau_{21} G_{21} \qquad C_2 = x_1 \tau_{12} G_{12}$$

Unconstrained second composition derivatives

$$Q_{11} = 2x_2(G_{21}\varepsilon_{11}/S_1 - G_{12}\varepsilon_{12}/S_2)$$
$$Q_{12} = x_1(G_{12}\varepsilon_{12}/S_2 - G_{21}\varepsilon_{11}/S_1)$$
$$\qquad\qquad + x_2(G_{21}\varepsilon_{21}/S_1 - G_{12}\varepsilon_{22}/S_2) = Q_{21}$$
$$Q_{22} = 2x_1(G_{12}\varepsilon_{22}/S_2 - G_{21}\varepsilon_{21}/S_1)$$

Other relations

$$Q = x_1 Q_1 + x_2 Q_2$$
$$x_1 Q_{11} + x_2 Q_{21} = 0 \qquad x_1 Q_{12} + x_2 Q_{22} = 0$$

Activity coefficients

$$\ln \gamma_1 = Q_1 \qquad \ln \gamma_2 = Q_2$$

Unconstrained composition derivatives of $\ln \gamma$

$$\left.\frac{\partial \ln \gamma_1}{\partial x_1}\right|_{x_2} = Q_{11} \qquad \left.\frac{\partial \ln \gamma_1}{\partial x_2}\right|_{x_1} = Q_{12}$$

$$\left.\frac{\partial \ln \gamma_2}{\partial x_1}\right|_{x_2} = Q_{21} \qquad \left.\frac{\partial \ln \gamma_2}{\partial x_2}\right|_{x_1} = Q_{22}$$

Thermodynamic factor

$$\Gamma = 1 + x_1(Q_{11} - Q_{12})$$
$$= 1 - 2x_1 x_2 (\tau_{21} G_{21}^2/S_1^3 + \tau_{12} G_{12}^2/S_2^3)$$

TABLE D.6   UNIQUAC Model for Binary Systems

Dimensionless Gibbs excess energy

$$Q = Q^c + Q^r$$

$$Q^c = x_1 m_1 + x_2 m_2$$

$$m_1 = \left(1 - \frac{z}{2}q_1\right)\ln(r_1/r) + \frac{z}{2}q_1 \ln(q_1/q)$$

$$m_2 = \left(1 - \frac{z}{2}q_2\right)\ln(r_2/r) + \frac{z}{2}q_2 \ln(q_2/q)$$

$$\phi_1/x_1 = r_1/r \qquad \phi_2/x_2 = r_2/r$$

$$\theta_1/x_1 = q_1/q \qquad \theta_2/x_2 = q_2/q$$

$$r = x_1 r_1 + x_2 r_2$$

$$q = x_1 q_1 + x_2 q_2$$

$$z = 10$$

$$Q^r = -x_1 q_1 \ln S_1 - x_2 q_2 \ln S_2$$

$$S_1 = \theta_1 + \theta_2 \tau_{12} \qquad S_2 = \theta_2 + \theta_1 \tau_{21}$$

$$\tau_{12} = \exp(-(\lambda_{12} - \lambda_{11})/RT)$$

$$\tau_{21} = \exp(-(\lambda_{21} - \lambda_{22})/RT)$$

Parameters

$$(\lambda_{12} - \lambda_{11}), (\lambda_{21} - \lambda_{22}), r_1, q_1, r_2, q_2$$

Unconstrained first composition derivatives

$$Q_1 = Q_1^c + Q_1^r \qquad Q_2 = Q_2^c + Q_2^r$$

$$Q_1^c = m_1 - (r_1/r)(x_1 + x_2) + \frac{z}{2}q(r_1/r - q_1/q)$$

$$Q_2^c = m_2 - (r_2/r)(x_1 + x_2) + \frac{z}{2}q(r_2/r - q_2/q)$$

$$Q_1^r = q_1(1 - \ln S_1 - \theta_1/S_1 - \theta_2\tau_{12}/S_2)$$

$$Q_2^r = q_2(1 - \ln S_2 - \theta_1\tau_{21}/S_1 - \theta_2/S_2)$$

Unconstrained second composition derivatives

$$Q_{ij} = Q_{ij}^c + Q_{ij}^r$$

$$Q_{11}^c = -2r_1/r + (r_1/r)^2(x_1 + x_2) - \frac{z}{2}q(r_1/r - q_1/q)^2$$

$$Q_{12}^c = -r_1/r - r_2/r + (r_1 r_2/r^2)(x_1 + x_2)$$

$$\qquad - \frac{z}{2}q(r_2/r - q_2/q)(r_1/r - q_1/q)$$

$$\qquad = Q_{21}^c$$

$$Q_{22}^c = -2r_2/r + (r_2/r)^2(x_1 + x_2) - \frac{z}{2}q(r_2/r - q_2/q)^2$$

$$Q_{11}^r = q_1^2\left(1 - 2/S_1 + \theta_1/S_1^2 + \theta_2\tau_{12}^2/S_2^2\right)/q$$

$$Q_{12}^r = q_1 q_2\left(1 - \tau_{12}/S_2 - \tau_{21}/S_1 + \theta_1\tau_{21}/S_1^2 + \theta_2\tau_{12}/S_2^2\right)/q$$

$$\qquad = Q_{21}^r$$

$$Q_{22}^r = q_2^2\left(1 - 2/S_2 + \theta_1\tau_{21}^2/S_1^2 + \theta_2/S_2^2\right)/q$$

TABLE D.6 (*Continued*)

Other relations

$$Q = x_1(1 + Q_1) + x_2(1 + Q_2)$$
$$x_1 Q_{11} + x_2 Q_{21} = -1$$
$$x_1 Q_{12} + x_2 Q_{22} = -1$$

Activity coefficients

$$\ln \gamma_1 = x_1 + x_2 + Q_1 \qquad \ln \gamma_2 = x_1 + x_2 + Q_2$$

Unconstrained composition derivatives of $\ln \gamma$

$$\left. \frac{\partial \ln \gamma_1}{\partial x_1} \right|_{x_2} = 1 + Q_{11} \qquad \left. \frac{\partial \ln \gamma_1}{\partial x_2} \right|_{x_1} = 1 + Q_{12}$$

$$\left. \frac{\partial \ln \gamma_2}{\partial x_1} \right|_{x_2} = 1 + Q_{21} \qquad \left. \frac{\partial \ln \gamma_2}{\partial x_2} \right|_{x_1} = 1 + Q_{22}$$

Thermodynamic factor

$$\Gamma = 1 + x_1(Q_{11} - Q_{12})$$

Q_{12} and Q_{21} are equal

$$Q_{12} = Q_{21} \tag{D.1.11}$$

Equations D.1.9 provide the unconstrained (by $x_1 + x_2 = 1$) partial derivatives of the activity coefficients. It is essential to evaluate the second partial derivatives of Q before making any use of $x_1 + x_2 = 1$ to simplify the results of the first differentiation. Only after all differentiation has been carried out may we employ the summation relationship.

The partial derivative of $\ln \gamma_1$ needed in the evaluation of Γ is the constrained (by $x_1 + x_2 = 1$) derivative of $\ln \gamma_1$ and is related to the unconstrained derivatives by

$$\left. \frac{\partial \ln \gamma_1}{\partial x_1} \right|_\Sigma = \left. \frac{\partial \ln \gamma_1}{\partial x_1} \right|_{x_2} - \left. \frac{\partial \ln \gamma_1}{\partial x_2} \right|_{x_1} .$$
$$= (1 - x_1)Q_{11} - x_2 Q_{21} - (1 - x_1)Q_{12} + x_2 Q_{22}$$
$$= x_2(Q_{11} + Q_{22} - Q_{12} - Q_{21})$$
$$= x_2(Q_{11} + Q_{22} - 2Q_{12}) \tag{D.1.12}$$

The thermodynamic factor, defined by Eq. D.1.4, may now be expressed in terms of the Q_{ij} as

$$\Gamma = 1 + x_1 x_2(Q_{11} + Q_{22} - 2Q_{12}) \tag{D.1.13}$$

The results of applying this procedure to a number of activity coefficient models are given in Tables D.1–D.6. It should be noted that for some models there exist relations between Q and its derivatives that can simplify Eq. D.1.13. These special relations are also noted in the tables.

D.2   THERMODYNAMIC FACTORS FOR MULTICOMPONENT SYSTEMS

For multicomponent mixtures, the elements of the thermodynamic matrix are defined by

$$\Gamma_{ij} = \delta_{ij} + x_i \frac{\partial \ln \gamma_i}{\partial x_j}\bigg|_{T,P,\Sigma} \tag{D.2.1}$$

The symbol Σ is used to indicate that the differentiation of $\ln \gamma_i$ with respect to mole fraction x_j is to be carried out while keeping constant the mole fractions of all other species except the nth. The mole fraction of species n must be eliminated using the fact that the x_i sum to unity.

Of the models introduced in Section D.1, only the Wilson, NRTL, and UNIQUAC are extended to multicomponent systems. Below we show how this is done. At the same time we provide rigorous derivations of some of the relations presented in Eq. D.1.1.

For a multicomponent system G^{EX} is given by

$$Q = G^{EX}/RT = \sum_{i=1}^{n} x_i \ln \gamma_i \tag{D.2.2}$$

γ_i is defined by

$$\ln \gamma_i = \frac{\partial(n_t Q)}{\partial n_i}\bigg|_{T,P,n_k, k \neq i=1\cdots n} \tag{D.2.3}$$

where Q denotes the dimensionless excess Gibbs energy G^{EX}/RT, n_i is the number of moles of species i in solution and n_t is the total number of moles in the mixture

$$n_t = \sum_{i=1}^{n} n_i \tag{D.2.4}$$

The partial derivative in Eq. D.2.3 may be expanded to give

$$\ln \gamma_i = Q + n_t \frac{\partial Q}{\partial n_i}\bigg|_{T,P,n_k, k \neq i=1\cdots n} \tag{D.2.5}$$

We will find it useful to replace the partial derivative of Q with respect to the number of moles n_i in Eq. D.2.5 with the partial derivatives of Q with respect to the mole fractions x_i defined by

$$x_i = n_i/n_t \tag{D.2.6}$$

It is important to recognize that a change in the number of moles of species i, n_i, changes the mole fractions of all n species, not just the mole fraction of species i. Accordingly, we may express the partial derivatives in Eq. D.2.5 as

$$\frac{\partial Q}{\partial n_i} = \sum_{j=1}^{n} Q_j \frac{\partial x_j}{\partial n_i}\bigg|_{n_k} \tag{D.2.7}$$

where

$$Q_j \equiv \frac{\partial Q}{\partial x_j}\bigg|_{\not{\Sigma}} \tag{D.2.8}$$

is the partial derivative of Q with respect to the mole fraction x_j. The symbol $\not{\Sigma}$ is shorthand notation to emphasize that the mole fractions of all species except the jth are kept constant while performing the differentiation. The requirement that the mole fractions sum to unity is not used to eliminate any mole fraction before the differentiation has been carried out.

The mole fraction derivatives may be obtained by differentiating Eq. D.2.6 go give

$$\frac{\partial x_j}{\partial n_i}\bigg|_{n_k} = (\delta_{ij} - x_j)/n_t \tag{D.2.9}$$

where δ_{ij} is the Kronecker delta.

When we combine Eq. D.2.6 with Eqs. D.2.7 and D.2.8 we obtain the following expression for the activity coefficients γ_i

$$\ln \gamma_i = Q + Q_i - \sum_{k=1}^{n} x_k Q_k \tag{D.2.10}$$

Equation D.2.10 is the multicomponent generalization of Eqs. D.1.6 and D.1.7.

The derivatives of $\ln \gamma_i$ with respect to mole fraction x_j are obtained on differentiation of Eq. D.2.10 as

$$\frac{\partial \ln \gamma_i}{\partial x_j}\bigg|_{\not{\Sigma}} = Q_{ij} - \sum_{k=1}^{n} x_k Q_{kj} \tag{D.2.11}$$

where

$$Q_{ij} = \frac{\partial Q_i}{\partial x_j}\bigg|_{\not{\Sigma}} \tag{D.2.12}$$

are the partial derivatives of Q_i with respect to mole fraction x_j. The Q_{ij} are symmetric.

$$Q_{ij} = Q_{ji} \tag{D.2.13}$$

Equation D.2.11 provides the unconstrained mole fraction derivatives of $\ln \gamma_i$, and the differentiation of Q_i must be done prior to any use of the fact that mole fractions sum to unity in order to simplify the expression for Q_i. The constrained mole fraction derivatives (where only $n-1$ of the mole fractions are regarded as independent—i.e., the x_i sum to unity), needed in the evaluation of the Γ_{ij}, are given in terms of the unconstrained derivatives by

$$\frac{\partial \ln \gamma_i}{\partial x_j}\bigg|_{\Sigma} = \frac{\partial \ln \gamma_i}{\partial x_j}\bigg|_{\not{\Sigma}} - \frac{\partial \ln \gamma_i}{\partial x_n}\bigg|_{\not{\Sigma}}$$

$$= Q_{ij} - Q_{in} - \sum_{k=1}^{n} x_k(Q_{kj} - Q_{kn}) \tag{D.2.14}$$

TABLE D.7 Wilson Model for Multicomponent Systems

Dimensionless Gibbs excess energy

$$Q = -\sum_{i=1}^{n} x_i \ln(S_i)$$

$$S_i = \sum_{j=1}^{n} x_j \Lambda_{ij}$$

$$\Lambda_{ij} = (V_j/V_i)\exp\left(-(\mu_{ij} - \mu_{ii})/RT\right)$$

Parameters

$$(\mu_{ij} - \mu_{ii}), V_i$$

Unconstrained first composition derivatives

$$Q_i = -\ln(S_i) - \sum_{k=1}^{n} x_k \Lambda_{ki}/S_k$$

Unconstrained second composition derivatives

$$Q_{ij} = -\Lambda_{ij}/S_i - \Lambda_{ji}/S_j + \sum_{k=1}^{n} x_k \Lambda_{ki}\Lambda_{kj}/S_k^2$$

Other relations

$$Q = \chi + \sum_{i=1}^{n} x_i Q_i$$

$$\sum_{j=1}^{n} x_j Q_{ji} = -1 \qquad i = 1, 2, \ldots, n$$

$$\chi = \sum_{j=1}^{n} x_j$$

Activity coefficients

$$\ln \gamma_i = \chi + Q_i$$

Unconstrained composition derivatives of $\ln \gamma_i$

$$\left.\frac{\partial \ln \gamma_i}{\partial x_j}\right|_{\Sigma} = 1 + Q_{ij}$$

Thermodynamic factor

$$\Gamma_{ij} = \delta_{ij} + x_i(Q_{ij} - Q_{in})$$

TABLE D.8   NRTL Model for Multicomponent Systems

Dimensionless Gibbs excess energy

$$Q = \sum_{i=1}^{n} x_i C_i / S_i$$

$$C_i = \sum_{j=1}^{n} x_j G_{ji} \tau_{ji}$$

$$S_i = \sum_{j=1}^{n} x_j G_{ji}$$

$$G_{ij} = \exp(-\alpha_{ij} \tau_{ij})$$

$$\tau_{ij} = (g_{ij} - g_{ii})/RT$$

$$\tau_{ii} = 0 \quad \text{and} \quad G_{ii} = 1$$

Parameters

$$(g_{ij} - g_{ii}) \quad \text{and} \quad \alpha_{ij} \, (= \alpha_{ji})$$

Unconstrained first composition derivatives

$$Q_i = C_i / S_i + \sum_{k=1}^{n} x_k \varepsilon_{ik}$$

$$\varepsilon_{ij} = G_{ij}(\tau_{ij} - C_j/S_j)/S_j$$

Unconstrained second composition derivatives

$$Q_{ij} = \varepsilon_{ij} + \varepsilon_{ji} - \sum_{k=1}^{n} x_k (G_{ik}\varepsilon_{jk} + G_{jk}\varepsilon_{ik})/S_k$$

Other relations

$$Q = \sum_{i=1}^{n} x_i Q_i$$

$$\sum_{j=1}^{n} x_j Q_{ji} = 0 \quad i = 1, 2, \ldots, n$$

Activity coefficients

$$\ln \gamma_i = Q_i$$

Unconstrained composition derivatives of $\ln \gamma_i$

$$\left. \frac{\partial \ln \gamma_i}{\partial x_j} \right|_{\Sigma} = Q_{ij}$$

Thermodynamic factor

$$\Gamma_{ij} = \delta_{ij} + x_i(Q_{ij} - Q_{in})$$

TABLE D.9 UNIQUAC Model for Multicomponent Systems

Dimensionless Gibbs excess energy

$$Q = Q^c + Q^r$$

$$Q^c = \sum_{i=1}^{n} x_i m_i$$

$$m_i = \left(1 - \frac{z}{2} q_i\right) \ln(r_i/r) + \frac{z}{2} q_i \ln(q_i/q)$$

$$= \left(1 - \frac{z}{2} q_i\right) \ln(\phi_i/x_i) + \frac{z}{2} q_i \ln(\theta_i/x_i)$$

$$\phi_i/x_i = r_i/r$$

$$\theta_i/x_i = q_i/q$$

$$r = \sum_{i=1}^{n} x_i r_i$$

$$q = \sum_{i=1}^{n} x_i q_i$$

$$z = 10$$

$$Q^r = -\sum_{i=1}^{n} x_i q_i \ln(S_i)$$

$$S_i = \sum_{j=1}^{n} \theta_j \tau_{ji}$$

$$\tau_{ji} = \exp\left(-(\lambda_{ji} - \lambda_{jj})/RT\right)$$

$$\tau_{ii} = 1$$

Parameters

$$(\lambda_{ji} - \lambda_{jj}), r_i, q_i$$

Unconstrained first composition derivatives

$$Q_i = Q_i^c + Q_i^r$$

$$Q_i^c = m_i - (r_i/r)\chi + \frac{z}{2} q(r_i/r - q_i/q)$$

$$\chi = \sum_{j=1}^{n} x_j$$

$$Q_i^r = q_i \left(1 - \ln(S_i) - \sum_{k=1}^{n} \theta_k \varepsilon_{ik}\right)$$

$$\varepsilon_{ik} = \tau_{ik}/S_k$$

TABLE D.9 (*Continued*)

Unconstrained second composition derivatives

$$Q_{ij} = Q_{ij}^c + Q_{ij}^r$$

$$Q_{ij}^c = -r_i/r - r_j/r + \left(r_i r_j/r^2\right)\chi$$

$$\qquad - \frac{z}{2} q (r_j/r - q_j/q)(r_i/r - q_i/q)$$

$$Q_{ij}^r = q_i q_j \left(1 - \varepsilon_{ij} - \varepsilon_{ji} + \sum_{k=1}^n \theta_k \varepsilon_{ik} \varepsilon_{jk}\right) \Big/ q$$

Other relations

$$Q^c = \chi + \sum_{i=1}^n x_i Q_i$$

$$Q^r = \sum_{i=1}^n x_i Q_i$$

$$\sum_{j=1}^n x_j Q_{ji}^c = -1 \qquad i = 1, 2, \ldots, n$$

$$\sum_{j=1}^n x_j Q_{ji}^r = 0 \qquad i = 1, 2, \ldots, n$$

Activity coefficients

$$\ln \gamma_i^c = \chi + Q_i^c$$

$$\ln \gamma_i^r = Q_i^r$$

$$\ln \gamma_i = \ln \gamma_i^c + \ln \gamma_i^r = \chi + Q_i^c + Q_i^r$$

Unconstrained composition derivatives of $\ln \gamma_i$

$$\left.\frac{\partial \ln \gamma_i}{\partial x_j}\right|_{\Sigma} = 1 + Q_{ij}$$

Thermodynamic factor

$$\Gamma_{ij} = \delta_{ij} + x_i (Q_{ij} - Q_{in})$$

The elements of the thermodynamic factor follow from Eq. D.2.1 as

$$\Gamma_{ij} = \delta_{ij} + x_i \left\{ Q_{ij} - Q_{in} - \sum_{k=1}^n x_k (Q_{kj} - Q_{kn}) \right\} \tag{D.2.15}$$

The results of applying the above procedure to the multicomponent forms of the Wilson, NRTL, and UNIQUAC models are summarized in Tables D.7–D.9.

It is interesting to note that for the three models in Tables D.7–D.9 the summation in Eq. D.2.15 vanishes and the thermodynamic factor is given by the simpler equation

$$\Gamma_{ij} = \delta_{ij} + x_i \{ Q_{ij} - Q_{in} \} \qquad |\text{Special}| \tag{D.2.16}$$

D.3   THERMODYNAMIC FLUID STABILITY AND THE GIBBS FREE ENERGY

The total Gibbs free energy is given by (Walas, 1985; Prausnitz et al., 1986)

$$G = G^{ID} + G^{EX} \tag{D.3.1}$$

where G^{ID} is the ideal Gibbs free energy

$$G^{ID}/RT = \sum_{i=1}^{n} x_i \ln x_i \tag{D.3.2}$$

The Hessian matrix of the Gibbs free energy $[G]$ is useful in determining the stability of a given fluid phase. For a fluid phase to be stable the eigenvalues of $[G]$ must be positive. The elements of this matrix are defined by

$$G_{ij} \equiv \left. \frac{\partial^2 G}{\partial x_i \, \partial x_j} \right|_{\Sigma} \tag{D.3.3}$$

The G_{ij} may be expressed in terms of the Q_{ij} as

$$G_{ij} = \delta_{ij}/x_i + 1/x_n + Q_{ij} - Q_{in} - Q_{jn} + Q_{nn} \tag{D.3.4}$$

The first two terms are the contribution to the G_{ij} from the ideal Gibbs free energy, the last four terms represent the contribution from the excess Gibbs energy. For a binary system Eq. D.3.4 simplifies as follows:

$$\begin{aligned} G_{11}/RT &= 1/x_1 x_2 + Q_{11} + Q_{22} - 2Q_{12} \\ &= \Gamma/x_1 x_2 \end{aligned} \tag{D.3.5}$$

and we see that, for binary systems, G_{11} is closely related to the thermodynamic factor Γ.

APPENDIX E
About the Software

E.1 WHAT IS ON THIS DISK?

The accompanying diskette contains Mathcad implementations of the examples that are worked out in detail in the text. Some of these files require only minimum modification in order to be useful for solving some of the exercises. It is necessary to purchase the Mathcad program in order to load these examples.

The files on this disk may be used in conjunction with Mathcad for DOS version 2.5 or Mathcad for Windows versions 3.1 or 4.0. Mathcad is a product of MathSoft, Inc. The files on the disk were created with Mathcad for DOS version 2.5 and translated to the Mathcad for Windows 3.1 format.

To use these files you must run the installation program included on this diskette. Since all the example files are in a compressed format you will NOT be able to use them without running the installation program

The files on the diskette are given a name that corresponds to the numbering of the examples in the text. For example, the file named 04-2-5.MCD contains the solution to Example 4.2.5.

Please note that files created with the DOS version can be loaded into the Windows version but that the converse is not true. That is, files created by Mathcad for Windows cannot be loaded into Mathcad for DOS. To load these files into Macintosh Mathcad, consult your Mathcad documentation.

For more information, or to place an order for Mathcad, please contact MathSoft, Inc. at 1-800-688-8553 or send in the enclosed business reply postcard.

E.2 HARDWARE REQUIREMENTS

Mathcad requires the following hardware and software:

Mathcad for DOS

- IBM PC, PC/XT, PC/AT or compatible, including PS/2 series
- MS-DOS or PC-DOS version 2.x and above
- CGA, EGA, VGA, or Hercules Monochrome monitor and adapter
- 512K RAM required; 640K recommended
- Math coprocessor not required; but supported and recommended
- Printer (See Mathcad documentation for supported printers)

Mathcad for Windows:

- 80286/386/486 based IBM or compatible computer
- Math coprocessor not required
- MS-DOS or PC-DOS version 3.0 or later
- Microsoft Windows version 3.0 or later
- 2MB RAM minimum—all memory above 640K should be configured as extended memory
- Hard disk with at least 7MB free space
- Monitor and graphics card supported by Windows
- A mouse and printer supported by Windows

E.3   WHAT IS MATHCAD?

Mathcad is a program established to work with formulas, numbers, text, and graphs. Mathcad lets you enter equations expanded fully on your screen. The equation in Mathcad appears the way you might see it on a blackboard or in a reference book. You can even illustrate your work with graphics taken from another Windows application. Mathcad equations can be used to solve any math problem symbolically or numerically, and allows the user to present their work in two and three dimensional plots.

By combining text, graphics and equations in a single document Mathcad makes it easy to keep track of the most complex calculations. By printing the document exactly as it appears on the screen, Mathcad lets you make a permanent accurate record of your work.

E.4   MAKING A BACKUP COPY

Before you start to use the enclosed disk, we strongly recommend that you make a backup copy of the original. Remember, however, that a backup disk is for your own personal use only. Any other use of the backup disk violates copyright law. Please take the time now to make the backup, using the instructions below:

1. Assuming your floppy drive is "A", insert your DOS disk into drive A of your computer.
2. At the A: > , type DISKCOPY A: A: and press Return.

You will be prompted by DOS to place the disk to be copied into drive A.

3. Place the *Multicomponent Mass Transfer* disk into drive A.

Follow the directions on the screen to complete the copy. When you are through, remove the new copy of the disk and label it immediately. Remove the original disk and store it in a safe place.

```
Mass Transfer Installation Program
```

```
Choose each of the following menu selections to configure
the way in which Mass Transfer will be installed on your
system.
```

```
Edit destination paths   :      \TAYLOR
Select destination drive :      C:
Toggle overwrite mode    :      Overwrite All
Select groups to install
Start installation
```

```
                    ITEM DESCRIPTION
Allows you to edit each of the destination paths.
```

```
Press [ALT-X] to exit at any time
```

Figure E.1. *Multicomponent Mass Transfer* installation program startup screen.

E.5 INSTALLING THE DISK

The enclosed diskette contains 63 individual files in compressed format. In order to use the files, you must run the installation program from the diskette.

You can install the diskette onto your computer by following these simple steps:

1. Assuming your floppy drive is "A", insert the *Multicomponent Mass Transfer* disk in drive A of your computer.

2. At the A: > type INSTALL and press Return.

The installation program will be loaded. After the title screen appears, you will be given the options shown in Figure E.1.

To change any of the default settings, type the highlighted letter or move the menu bar to the desired option and press Enter.

3. To start the installation, type "S" or move the menu bar to the **Start installation** option, and press Enter.

After the installation is complete remove your original diskette and store it in a safe place.

```
File to load:   a:\mcad\02-2-1                                        0    0   auto
```

Figure E.2. Loading a file in Mathcad for DOS.

E.6   HOW TO USE THE FILES ON THIS DISK

E.6.1   Mathcad for DOS

To load a document from a disk:

1. Press [F5] or type [Esc] **load**. The top line menu will change to read "File to load:"
2. Type the name of the file you wish to open and press **Enter**. For example, to load Example 2.2.1, you can type A:\ MCAD \ 02-2-1 as shown in Figure E.2.

You will then be presented with the example file to edit, as shown in Figure E.3.

E.6.2   Mathcad for Windows

To load a document from a disk file, choose **Open Document** from the **File** menu. Mathcad prompts you for a name by displaying the dialog box shown in Figure E.4.

The current directory is shown in the "Directories" box. If the file you want to open is in this directory skip to the next step. Otherwise, move the pointer to the scrolling list of directories and double-click on the directory containing the file you want to open. This box

```
↓...mcad\02-2-1.MCD↓                                                 0    1   auto
```

"Example 2.2.1.   Diffusion of toluene in a binary mixture"

Consider diffusion in the system toluene (1) – n-tetradecane (2) –
n – hexane

The units for the diffusivities are 1e–09 m2/s

ORIGIN ≡ 1

$$
x2 := \begin{bmatrix} 1 \\ 0.803 \\ 0.672 \\ 0.501 \\ 0.336 \\ 0.215 \\ 0.113 \\ 0.0 \end{bmatrix}
\qquad
D1eff := \begin{bmatrix} 1.08 \\ 1.37 \\ 1.58 \\ 1.92 \\ 2.38 \\ 2.90 \\ 3.57 \\ 4.62 \end{bmatrix}
$$

i := 1 ..8

From the experimental data the infinite dilution diffusion coefficients can
be determined as follows

D120 := 1.08   D130 := 4.62

From the Maxwell–Stefan diffusion equations the effective diffusivity value
for toluene can be calculated from the equation

Figure E.3. Example 2.2.1 in Mathcad for DOS.

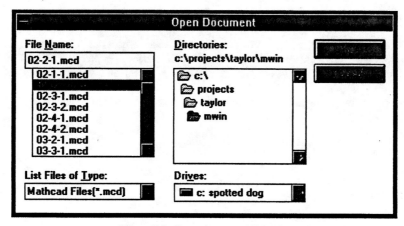

Figure E.4. Open document dialog box.

lists all subdirectories of the current directory. To switch to another drive, click on the arrow near the "Drive" box, then select the drive letter.

Next to the Directory is a scrolling list of all the files in the current directory with the extension "MCD". Double-click on one of these names to open it. You can also select a file name, then click "OK".

You will then be presented with the example file to edit. For instance, Example 2.2.1 is shown in Figure E.5.

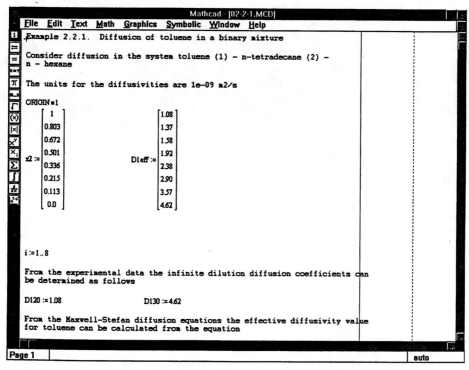

Figure E.5. Example 2.2.1 in Mathcad for Windows.

USER ASSISTANCE AND INFORMATION

John Wiley & Sons, Inc. is pleased to provide assistance to users of this software package. Should you have questions regarding the use of this package, please call our technical support number at (212) 850-6194 weekdays between 9 am and 4 pm Eastern Standard Time.

To place additional orders or to request information about other Wiley products, please call (800) 879-4539.

REFERENCES

Ackermann, G., "Wärmeübergang und Molekulare Stoffübertragung im gleichen Feld bei grossen Temperatur- und Partialdruckdifferenzen," *Forschungsheft V.D.I.*, **382**, 1–16 (1937).

AIChE, *Bubble Tray Design Manual: Prediction of Fractionation Efficiency*, AIChE, New York, 1958.

Aittamaa, J., "Estimating Multicomponent Plate Efficiencies in Distillation," *Kem.-Kemi*, **8**, 295–300, 317 (1981).

Alimadadian, A. and Colver, C. P., "A New Technique for the Measurement of Ternary Diffusion Coefficients in Liquid Systems," *Can. J. Chem. Eng.* **54**, 208–213 (1976).

Amundson, N. R., *Mathematical Methods in Chemical Engineering. I: Matrices and Their Applications*. Prentice-Hall, Englewood Cliffs, NJ, 1966.

Anderson, D. K., Hall, J. R., and Babb, A. L., "Mutual Diffusion in Nonideal Binary Liquid Mixtures," *J. Phys. Chem.*, **62**, 404–409 (1958).

Arnold, J. H., "Studies in Diffusion: III. Unsteady-State Vaporization and Absorption," *Trans. Am. Inst. Chem. Engrs.*, **40**, 361–378 (1944).

Arnold, K. R., *Unsteady-State Multicomponent Gaseous Diffusion*, Ph.D. Thesis in Chemical Engineering, Carnegie Institute of Technology, Pittsburgh, PA, 1965.

Arnold, K. R. and Toor, H. L., "Unsteady Diffusion in Ternary Gas Mixtures," *AIChE J*, **13**, 909–914 (1967).

Arwickar, K. J., *Mass Transfer in Packed Column Absorption and Multicomponent Distillation*, Ph.D. Thesis in Chemical Engineering, University of California, Santa Barbara, CA, 1981.

Austin, D. G. and Jeffreys, G. V., *The Manufacture of Methyl Ethyl Ketone from 2-Butanol*, The Institution of Chemical Engineers, Rugby, England, 1979.

Bandrowski, J. and Kubaczka, A., "On the Condensation of Multicomponent Vapors in the Presence of Inert Gases," *Int. J. Heat Mass Transfer*, **24**, 147–153 (1981).

Bandrowski, J. and Kubaczka, A., "On the Prediction of Diffusivities in Multicomponent Liquid Systems," *Chem. Eng. Sci.*, **37**, 1309–1313 (1982).

Bell, K. J. and Ghaly, M. A., "An Approximate Generalized Design Method for Multicomponent/Partial Condensers," *AIChE Symp. Ser.*, **69**, 72–79 (1972).

Bennett, D. L., Agrawal, R., and Cook, P. J., "New Pressure Drop Correlation for Sieve Tray Distillation Columns," *AIChE J*, **29**, 434–442 (1983).

Biddulph, M. W., and Kalbassi, M. A., "Distillation Efficiencies for Methanol/1-Propanol/Water," *Ind. Eng. Chem. Res.*, **27**, 2127–2135 (1988).

Biddulph, M. W. Kalbassi, M. A., and Dribicka, M. M., "Multicomponent Efficiencies in Two Types of Distillation Column," *AIChE J*, **34**, 618–625 (1988).

Billet, R., *Distillation Engineering*, Chemical Publishing Co., New York, 1979.

Bird, R. B., Stewart, W. E., and Lightfoot, E. N., *Transport Phenomena*, Wiley, New York, 1960.

Blom, J., *An Experimental Determination of the Turbulent Prandtl Number in a Developing Temperature Boundary Layer*, Ph.D. Thesis, Technical University of Eindhoven, The Netherlands, 1970.

Bolles, W. L., "Tray Hydraulics: Bubble-Cap Trays," Chap. 14, in *Design of Equilibrium Stage Processes*, Smith, B. D. (Ed.), McGraw-Hill, New York, 1963.

Bravo, J. L. and Fair, J. R., "Generalized Correlation for Mass Transfer in Packed Distillation Columns," *Ind. Eng. Chem. Process Des. Dev.*, **21**, 162–170 (1982).

Bravo, J. L., Rocha, J. A., and Fair, J. R., "Mass Transfer in Gauze Packings," *Hydrocarbon Processing*, 91–95 (January 1985).

Bravo, J. L., Rocha, J. A., and Fair, J. R., "Pressure Drop in Structured Packings," *Hydrocarbon Processing*, 45–48 (March 1986).

Bravo, J. L., Rocha, J. A., and Fair, J. R., "A Comprehensive Model for the Performance of Columns Containing Structured Packings," The Institution of Chemical Engineers Symposium Series No. 128, *Distillation and Absorption 1992*, A439–A457 (1992).

Bres, M. and Hatzfeld, C., "Three Gas Diffusion—Experimental and Theoretical Studies," *Pflügers Arch.*, **371**, 227–233 (1977).

Bröcker, S. and Schulze, W., "A New Method of Calculating Ternary Mass Transfer with a Non–Transferring Species Based on Gilliland's Parametric Solution of the Maxwell–Stefan Equations for the Film Model," *Chem. Eng. Commun.*, **107**, 163–172 (1991).

Buell, C. K. and Boatright, R. G., "Furfural Extractive Distillation for Separation and Purification of C4 Hydrocarbons," *Ind. Eng. Chem.*, **39**, 695–705 (1947).

Buffham, B. A. and Kropholler, H. W., "The Evaluation of the Exponential Matrix," *Proceedings of the Conference On Line Computer Methods Relevant to Chemical Engineering*, University of Nottingham, 64–70 (1971).

Bunke, C. M., Steel, E. R., and Sandall, O. C., "Mass Transfer in Multicomponent Distillation on Sieve Trays," The Institution of Chemical Engineers Symposium Series No. 104, *Distillation and Absorption 1987*, B423–B434 (1987).

Burchard, J. K. and Toor, H. L., "Diffusion in an Ideal Mixture of Three Completely Miscible Non-Electrolytic Liquids—Toluene, Chlorobenzene, Bromobenzene," *J. Phys. Chem.*, **66**, 2015–2022 (1962).

Burghardt, A., "On the Solutions of Maxwell–Stefan Equations for Multicomponent Film Model," *Chem. Eng. Sci.*, **39**, 447–453 (1984).

Burghardt, A., "Simulation of Binary Vapor Condensation in the Presence of an Inert Gas," *Int. Comm. Heat Mass Transfer*, **10**, 555–561 (1983).

Burghardt, A., "Transport Phenomena and Chemical Reactions in Porous Catalysts for Multicomponent and Multireaction Systems," *Chem. Eng. Process.*, **21**, 229–244 (1986).

Burghardt, A. and Krupiczka, R., "Wnikanie Masy W Ukladach Wielaskladnikowych-Teortyczna Analiza Zagadnienia i Okreslenie Wspolczynnikow Wnikania Masy," *Inz. Chem.*, **5**, 487–510 (1975).

Burghardt, A. and Warmuzinski, K., "Diffusional Methods of Calculation for Multicomponent Systems in Rectification Columns. I. Models of Equimolar and Nonequimolar Mass Transfer," *Int. Chem. Eng.*, **23**, 342–350 (1983).

Burghardt, A. and Warmuzinski, K., "Diffusional Methods of Calculation for Multicomponent Systems in Rectification Columns. II. A Model of Simultaneous Mass [and Heat] Transfer," *Int. Chem. Eng.*, **24**, 742–751 (1984).

Burghardt, A., Warmuzinski, K., Buzek, J., and Pytlik, A., "Diffusional Methods of Multicomponent Distillation and their Experimental Verification," *Chem. Eng. J.*, **26**, 71–84 (1983).

Caldwell, C. S. and Babb, A. L., "Diffusion in Ideal Binary Liquid Mixtures," *J. Phys. Chem.*, **60**, 51–56 (1956).

Carslaw, H. S. and Jaeger, J. C., *Conduction of Heat in Solids*, 2nd edition, Oxford University Press, Oxford, England, 1959.

Carty, R. and Schrodt, J. T., "Concentration Profiles in Ternary Gaseous Diffusion," *Ind. Eng. Chem. Fundam.*, **14**, 276–278 (1975).

Chan, H., *Tray Efficiencies for Multicomponent Distillation Columns*, Ph.D. Thesis in Chemical Engineering, The University of Texas at Austin, TX, 1983.

Chan, H. and Fair, J. R., "Prediction of Point Efficiencies on Sieve Trays. 1. Binary Systems," *Ind. Eng. Chem. Proc. Des. Dev.*, **23**, 814–819 (1984a).

Chan, H. and Fair, J. R., "Prediction of Point Efficiencies on Sieve Trays. 2. Multicomponent Systems," *Ind. Eng. Chem. Proc. Des. Dev.*, **23**, 820–827 (1984b).

Chapman, S. and Cowling, T. G., *The Mathematical Theory of Non-Uniform Gases*, 3rd ed. prepared in cooperation with D. Burnett, Cambridge University Press, Cambridge, England, 1970.

Christian, S. D., Neparko, E., and Affsprung, H. E., "A New Method for the Determination of Activity Coefficients of Components in Binary Liquid Mixtures," *J. Phys. Chem.*, **64**, 442–446 (1960).

Cichelli, M. T., Weatherford, W. D., and Bowman, J. R., "Sweep Diffusion Gas Separation Process, Part I," *Chem. Eng. Prog.*, **47**(2), 63–74 (1951).

Claesson, S. and Sundelöf, L. O., "Diffusion Libre au Voisinage de la Temperature Critique de Miscibilite," *J. Chim. Phys.*, **54**, 914–919 (1957).

Clarke, W. M. and Rowley, R. L., "The Mutual Diffusion Coefficient of Methanol–n-hexane Near the Consolute Point," *AIChE J*, **32**, 1125–1131 (1986).

Clift, R., Grace, J. R., and Weber, M. E., *Bubbles, Drops and Particles*, Academic Press, London, England, 1978.

Colburn, A. P. and Drew, T. B., "The Condensation of Mixed Vapors," *Trans. Am. Inst. Chem. Engrs.*, **33**, 197–215 (1937).

Coleman, B. D. and Truesdell, C., "The Reciprocal Relations of Onsager," *J. Chem. Phys.*, **33**, 28–31 (1960).

Cooper, A. R., "Vector Space Treatment of Multicomponent Diffusion," in *Geochemical Transport and Kinetics*, papers presented at a conference in Airlie House, Warrenton, Virginia, June 1973, Hofmann, A. W., Giletti, B. J., Yoder Jr., H. S., and Yund, R. A. (Eds.), Carnegie Institute of Washington, DC, 1974.

Cotrone, A. and De Giorgi, C., "A Rigorous Method for Evaluating Concentration Profiles and Mass Fluxes in Multicomponent Isothermal Gaseous Mixtures Diffusing Through Films of Given Thickness," *Ing. Chim. Ital.*, **7**(6), 84–97 (1971).

Crank, J., *"The Mathematics of Diffusion,"* 2nd ed., Clarendon Press, Oxford, England, 1975.

Cullinan, H. T., "Analysis of the Flux Equations of Multicomponent Diffusion," *Ind. Eng. Chem. Fundam.*, **4**, 133–139 (1965).

Cullinan, H. T. and coworkers "Prediction of Multicomponent Diffusion Coefficients," *Ind. Eng. Chem. Fundam.*, **5**, 281–283 (1966); **6**, 72–77 (1967); **6**, 616 (1967); **7**, 317–319 (1968); **7**, 331–332 (1968); **7**, 519–520 (1968); **9**, 84–93 (1970); **10**, 600–603 (1971); *Can. J. Chem. Eng.*, **45**, 377–381 (1967); **49**, 130–133 (1971); **49**, 632–636 (1971); **49**, 753–757 (1971); *AIChE J*, **13**, 1171–1174 (1967); **21**, 195–197 (1975).

Cullinan, H. T. and Lenczyk, J. P., "Determination of Chemical Potential Derivatives by Equilibrium Sedimentation," *Ind. Eng. Chem. Fundam.*, **8**, 819–822 (1969).

Cullinan, H. T. and Ram, S. K., "Mass Transfer in a Ternary Liquid-Liquid System," *Can. J. Chem. Eng.*, **54**, 156–159 (1976).

Cullinan, H. T. and Toor, H. L., "Diffusion in the Three-Component Liquid System Acetone–Benzene–Carbon Tetrachloride," *J. Phys. Chem.*, **69**, 3941–3949 (1965).

Cunningham, R. E. and Williams, R. J. J., *Diffusion in Gases and Porous Media*, Plenum Press, New York, 1980.

Cussler, E. L., *Multicomponent Diffusion*, Elsevier, Amsterdam, The Netherlands, 1976.

Cussler, E. L., *Diffusion: Mass Transfer in Fluid Systems*, Cambridge University Press, Cambridge, England, 1984.

Danckwerts, P. V., "Significance of Liquid Film Coefficients in Gas Absorption," *Ind. Eng. Chem.*, **43**, 1460–1467 (1951).

Danckwerts, P. V., *Insights into Chemical Engineering* (*Selected papers of PVD*), Pergamon Press, Oxford, England, 1981.

Daubert, T. E. and Danner, R. P., *Data Compilation Tables of Properties of Pure Compounds*, Design Institute for Physical Property Data, AIChE, New York, 1985.

Danner, R. P. and Daubert, T. E., *Manual for Predicting Chemical Process Design Data*, AIChE, New York, 1983.

De Groot, S. R. and Mazur, P., *Nonequilibrium Thermodynamics*, North-Holland, Amsterdam, The Netherlands, 1962.

DeLancey, G. B., "Analysis of Multicomponent Diaphragm Cell Data," *J. Phys. Chem.*, **73**, 1591–1593 (1969).

DeLancey, G. B., "The Effect of Thermal Diffusion in Multicomponent Gas Absorption," *Chem. Eng. Sci.*, **27**, 555–566 (1972).

DeLancey, G. B. and Chiang, S. H., "Dufour Effect in Liquid Systems," *AIChE J*, **14**, 664–665 (1968).

DeLancey, G. B. and Chiang, S. H., "Role of Coupling in Nonisothermal Diffusion," *Ind. Eng. Chem. Fundam.*, **19**, 138–144 (1970a).

DeLancey, G. B. and Chiang, S. H., "Analysis of Nonisothermal Multicomponent Diffusion with Chemical Reaction," *Ind. Eng. Chem. Fundam.*, **19**, 344–349 (1970b).

Deo, P. V., *Condensation of Mixed Vapours*, Ph.D. Thesis in Chemical Engineering, University of Manchester, Institute of Science and Technology, Manchester, England, 1979.

Diener, D. A. and Gerster, J. A., "Point Efficiencies in Distillation of Acetone–Methanol–Water," *Ind. Eng. Chem. Proc. Des. Dev.*, **7**, 339–345 (1968).

Dieter, K. and Hundertmark, F. G., "Calculation of the Efficiency of Rectifying Plates," *Chem. Ing. Tech.*, **35**, 620–627 (1963).

Dribicka, M. M. and Sandall, O. C., "Simultaneous Heat and Mass Transfer for Multicomponent Distillation in a Wetted Wall Column," *Chem. Eng. Sci.*, **34**, 733–739 (1979).

Dudley, G. J. and Tyrell, H. J. V., "Transport Processes in Binary and Ternary Mixtures Containing Water, Triethylamine and Urea," *J. Chem. Soc. Faraday Trans.*, **69**, 2188–2199, 2200–2208 (1973).

Dullien, F. A. L., "Statistical Test of Vignes' Correlation of Liquid Phase Diffusion Coefficients," *Ind. Eng. Chem. Fundam.*, **10**, 41–49 (1971).

Dullien, F. A. L. and Asfour, A-F. A., "Concentration Dependence of Mutual Diffusion Coefficients in Regular Binary Solutions: A New Predictive Equation," *Ind. Eng. Chem. Fundam.*, **24**, 1–7 (1985).

Duncan, J. B., *An Experimental Study of Three Component Gas Diffusion*, Ph.D. Thesis in Chemical Engineering, Carnegie Institute of Technology, Pittsburgh, PA, 1960.

Duncan, J. B. and Toor, H. L., "An Experimental Study of Three Component Gas Diffusion," *AIChE J*, **8**, 38–41 (1962).

Dunlop, P. J., Steel, B. J., and Lane, J. E., "Experimental Methods for Studying Diffusion in Liquids, Gases and Solids," in *Techniques of Chemistry, I, Physical Methods of Chemistry*, Weissberger, A. and Rossiter, B. W. (Eds.), Wiley, New York, 1972.

Elnashaie, S. S., Abashar, M. E., and Al-Ubaid, A. S., "Simulation and Optimization of an Industrial Ammonia Reactor," *Ind. Eng. Chem. Res.*, **27**, 2015–2022 (1988); **28**, 1267 (1989).

Erkey, C., Rodden, J. B., and Akgerman, A., "A Correlation for Predicting Diffusion Coefficients in alkanes," *Can. J. Chem. Eng.*, **68**, 661–665 (1990).

Ertl, H., Ghai, R. K., and Dullien, F. A. L., "Liquid Diffusion of Nonelectrolytes: Part II," *AIChE J*, **20**, 1–20 (1974).

Evans, D. F., Tominaga, T., and Davis, H. T., "Tracer Diffusion in Polyatomic Liquids," *J. Chem. Phys.*, **74**, 1298–1305 (1981).

Fair, J. R., "Liquid-Gas Systems," Section 18 in *Perry's Chemical Engineers' Handbook*, Perry, R. H. and Green, D. (Eds.), McGraw-Hill, New York, 1984.

Fair, J. R., "Distillation," in *Handbook of Separation Process Technology*, Rousseau, R. W. (Ed.), Wiley-Interscience, New York, 1987.

Fairbanks, D. F. and Wilke, C. R., "Diffusion Coefficients in Multicomponent Gas Mixtures," *Ind. Eng. Chem.*, **42**, 471–475 (1950).

Fick, A., "On Liquid Diffusion," *Phil. Mag.*, **10**, 30–39 (1855a).

Fick, A., "Über diffusion," *Poggendorff's Ann.*, **94**, 59–86 (1855b).

Fletcher, J. P., "A Method for the Rigorous Calculation of Distillation Columns Using a Generalized Efficiency Model," The Institution of Chemical Engineers Symposium Series No. 104, *Distillation and Absorption* 1987, A437–A447 (1987).

Fletcher, D. F., Maskell, S. J., and Patrick, M. A., "Theoretical Investigation of the Chilton–Colburn Analogy," *Trans. Inst. Chem. Engrs.*, **60**, 122–125, (1982).

Frey, D. D., "Prediction of Liquid Phase Mass Transfer Coefficients in Multicomponent Ion Exchange: Comparison of Matrix, Film–Model, and Effective Diffusivity Methods," *Chem. Eng. Commun.*, **47**, 273–293 (1986).

Froment, G. F. and Bischoff, K. B., *Chemical Reactor Analysis and Design*, Wiley, New York, 1979.

Fullarton, D. and Schlünder, E. U., "Diffusion Distillation: A New Separation Process for Azeotropic Mixtures," *Chem. Eng. Fundam.*, **2**, 53–66 (1983).

Fullarton, D. and Schlünder, E. U., "Diffusion Distillation—A New Separation Process for Azeotropic Mixtures. Part I: Selectivity and Transfer Efficiency," *Chem. Eng. Process*, **20**, 255–263 (1986a).

Fullarton, D. and Schlünder, E. U., "Diffusion Distillation—A New Separation Process for Azeotropic Mixtures. Part II: Dehydration of Isopropanol by Diffusion Distillation," *Chem. Eng. Process*, **20**, 265–270 (1986b).

Fuller, E. N., Ensley, K., and Giddings, J. C., "Diffusion of Halogenated Hydrocarbons in Helium. The Effect of Structure on Collision Cross Sections," *J. Phys. Chem.*, **73**, 3679–3685 (1969).

Fuller, E. N., Schettler, P. D., and Giddings, J. C., "A New Method for Prediction of Binary Gas-Phase Diffusion Coefficients," *Ind. Eng. Chem.*, **58**, 19–27 (1966).

Furno, J. S., Taylor, R., and Krishna, R., "Condensation of Vapor Mixtures. II—Simulation of Some Test Condensers," *Ind. Eng. Chem. Process Des. Dev.*, **25**, 98–101 (1986).

Geankoplis, C. J., *Mass Transport Phenomena*, Holt, Rinehart, and Winston, Inc., New York, 1972.

Gerster, J. A., Mizushina, T., Marks, T. N., and Catanach, A. W. "Plant Performance of a 13-ft.-diameter Extractive-Distillation Column," *AIChE J*, **1**, 536–543 (1955).

Gerster, J. A., Hill, A. B., Hochgraf, N. N., and Robinson, D. G., *Tray Efficiencies in Distillation Columns*, AIChE, New York, 1958.

Ghai, R. K., Ertl, H., and Dullien, F. A. L., "Liquid Diffusion of Nonelectrolytes: Part I," *AIChE J*, **19**, 881–900 (1973).

Gilliland, E. R., cited by Sherwood, T. K. and Pigford, R. L., *Absorption and Extraction*, 1st ed., McGraw-Hill, New York (1937).

Glasstone, S., *Sourcebook on Atomic Energy*, 2nd ed., Van Nostrand, Princeton, NJ, 1958.

Gmehling, J. and Onken, U. *Vapor-Liquid Equilibrium Data Collection*, DECHEMA Chemistry Data Series, Vol. 1ff, DECHEMA, 1977ff.

Goldstein, R. P. and Stanfield, R. B., "Flexible Method for the Solution of Distillation Design Problems using the Newton-Raphson Technique," *Ind. Eng. Chem. Process Des. Develop.*, **9**, 78–84 (1970).

Gorak, A., "Simulation Methods for Steady State Multicomponent Distillation in Packed Columns," The Institution of Chemical Engineers Symposium Series No. 104, *Distillation and Absorption 1987*, A413–A424 (1987).

Gorak, A., "Sensitivity of Film Mass Transfer Models in Multicomponent Distillation Calculations," *Inzynieria Chemiczna I Procesowa*, **1**, 93–109 (1988).

Gorak, A., *Berechnungsmethoden der Mehrstoffrektifikation—Theorie und Anwendunegn*, Habilitationschrift der RWTH Aachen, Germany, 1991.

Gorak, A., Kraslawski, A., and Vogelpohl, A., "Simulation and Optimization of Multicomponent Distillation," *Int. Chem. Eng.*, **30**, 1–15 (1990).

Gorak, A. and Vogelpohl, A., "Experimental Study of Ternary Distillation in a Packed Column," *Separ. Sci. Technol.*, **20**, 33–61 (1985).

Gorak, A., Wozny, G., and Jeromin, L., "Industrial Application of the Rate-Based Approach for Multicomponent Distillation Simulation," *Proceedings of the 4th World Congress in Chemical Engineering*, Karlsruhe, Germany, 1991.

Grew, K. E., "Thermal Diffusion," in *Transport Phenomena in Fluids*, Hanley, H. J. M. (Ed.), Marcel Dekker, New York, 1969.

Grew, K. E. and Ibbs, T. L., *Thermal Diffusion in Gases*, Cambridge University Press, Cambridge, England, 1962.

Grottoli, M. G., Biardi, G., and Pellegrini, L., "A New Simulation Model for a Real Trays Absorption Column," *Comput. Chem. Eng.*, **15**, 171–179 (1991).

Gupta, P. K. and Cooper, A. R., "The [D] Matrix for Multicomponent Diffusion," *Physica*, **54**, 39–59 (1971).

Haase, R., *The Thermodynamics of Irreversible Processes*, Addison-Wesley, London, England, 1969.

Haase, R. and Siry, M., "Diffusion im kritischen Entmischungsgebiet binärer flüssiger Systeme," *Z. Phys. Chem.*, Neue Folge, **57**, 56–73 (1968).

Happel, J., Cornell, P. W., Eastman, D. B., Fowle, M. J., Porter, C. A., and Schutte, A. H., "Extractive Distillation-Separation of C4 Hydrocarbons Using Furfural," *Trans. Am. Inst. Chem. Engrs.*, **42**, 189–215 (1946).

Hausen, H., "The Definition of the Degree of Exchange on Rectifying Plates for Binary and Ternary Mixtures," *Chem. Ingr. Tech.*, **25**, 595–597 (1953).

Hayduk, W., "Correlations for Molecular Diffusivities in Liquids," *Encyclopedia of Fluid Mechanics*, Cheremisinoff, N. P. (Ed.), Gulf Publishing Corp., Houston, TX, Vol. I, pp. 48–72, 1986.

Hayduk, W. and Laudie, H., "Prediction of Diffusion Coefficients for Nonelectrolytes in Dilute Aqueous Solutions," *AIChE J*, **20**, 611–615 (1974).

Hayduk, W. and Minhas, B. S., "Correlations for Prediction of Molecular Diffusivities in Liquids," *Can. J. Chem. Eng.*, **60**, 295–299 (1982); Correction, **61**, 132 (1983).

Higbie, R., "The Rate of Absorption of a Pure Gas into a Still Liquid During Short Periods of Exposure," *Trans. Am. Inst. Chem. Engrs.*, **31**, 365–383 (1935).

Helfferich, F., *Ion Exchange*, McGraw-Hill, New York, 1962.

Henley, E. J. and Seader, J. D., *Equilibrium Stage Separation Operations in Chemical Engineering*, Wiley, New York, 1981.

Hesse, D. and Hugo, P., "Untersuchung der Mehrkomponenten—Diffusion in Gemisch Wasserstoff, Athylen und Athan mit einem Stationarem Messrerfahren," *Chem. Ing. Tech.*, **44**, 1312–1318 (1972).

Hirschfelder, J. O., Curtiss, C. F., and Bird, R. B., *Molecular Theory of Gases and Liquids*, Second corrected printing, Wiley, New York, 1964.

Hofer, H., "Influence of Gas-phase Dispersion on Plate Column Efficiency," *Ger. Chem. Eng.*, **6**, 113–118 (1983).

Holland, C. D., *Fundamentals and Modelling of Separation Processes*, Prentice-Hall, Englewood Cliffs, NJ, 1975.

Holland, C. D., *Fundamentals of Multicomponent Distillation*, McGraw-Hill, New York, 1981.

Holland, C. D. and McMahon, K. S., "Comparison of Vaporization Efficiencies with Murphree–type Efficiencies in Distillation—I," *Chem. Eng. Sci.*, **25**, 431–436 (1970).

Holmes, J. T., Olander, D. R., and Wilke, C. R., "Diffusion in Mixed Solvents," *AIChE J*, **8**, 646–649 (1962).

Hougen, O. A. and Watson, K. M., *Chemical Process Principles*, Vol. III, 1st Ed., Wiley, New York, 1947.

Hsu, H. W. and Bird, R. B., "Multicomponent Diffusion Problems," *AIChE J*, **6**, 516–524 (1960).

Hung, J-S., *A Second Generation Nonequilibrium Model for Computer Simulation of Multicomponent Separation Processes*, Ph.D. Thesis in Chemical Engineering, Clarkson University, Potsdam, NY, 1991.

Jackson, R., *Transport in Porous Catalysts*, Elsevier, Amsterdam, The Netherlands, 1977.

Johns, L. E. and DeGance, A. E., "Diffusion in Ternary Ideal Gas Mixtures. I. On the Solution of the Stefan–Maxwell Equation for Steady Diffusion in Thin Films," *Ind. Eng. Chem. Fundam.*, **14**, 237–245 (1975).

Johnson, P. A. and Babb, A. L., "Liquid Diffusion of Non-Electrolytes," *Chem. Rev.*, **56**, 387–453 (1956).

Johnstone, H. F. and Pigford, R. L., "Distillation in a Wetted Wall Column," *Trans. Am. Inst. Chem. Engrs.*, **38**, 25–51 (1942).

Kaltenbacher, E., "On the Effect of the Bubble Size Distribution and the Gas Phase Diffusion on the Selectivity of Sieve Trays," *Chem. Eng. Fundam.*, **1**, 47–68 (1982).

Kato, S., Inazumi, H., and Suzuki, S., "Mass Transfer in a Ternary Gaseous Phase," *Int. Chem. Eng.*, **21**, 443–452 (1981).

Kern, D. Q., *Process Heat Transfer*, McGraw-Hill, New York, 1950.

Kett, T. R. and Anderson, D. K., "Ternary Isothermal Diffusion and the Validity of the Onsager Reciprocal Relations in Non-associating Systems," *J. Phys. Chem.*, **72**, 1268–1274 (1969).

Keyes, J. J. and Pigford, R. L. "Diffusion in a Ternary Gas System with Application to Gas Separation," *Chem. Eng. Sci.*, **6**, 215–226 (1957).

King, C. J., *Separation Processes*, 2nd ed., McGraw-Hill, New York, 1980.

King, C. J., Hsueh, L., and Mao, K-W., "Liquid Phase Diffusion of Nonelectrolytes at High Dilution," *J. Chem. Eng. Data*, **10**, 348–350 (1965).

Kirkaldy, J. S., "Isothermal Diffusion in Multicomponent Systems," *Advances in Materials Research*, **4**, 55–100 (1970).

Kister, H. Z., *Distillation Design*, McGraw-Hill, New York, 1992.

Kooijman, H. A. and Taylor, R., "Estimation of Diffusion Coefficients in Multicomponent Liquid Systems," *Ind. Eng. Chem. Res.*, **30**, 1217–1222 (1991).

Kooijman, H. A. and Taylor, R., "ChemSep—Another Software System for the Simulation of Separation Processes," *CACHE News*, **35**, 1–9 (1992).

Kosanovich, G. M., Ph.D. Dissertation in Chemical Engineering, State University of New York at Buffalo, Buffalo, NY, 1975.

Kosanovich, G. M. and Cullinan, H. T., "A Study of Molecular Transport in Liquid Mixtures Based on the Concept of Ultimate Volume," *Ind. Eng. Chem. Fundam.*, **15**, 41–45 (1976).

Krishna, R., "A Generalized Film Model for Mass Transfer in Non-Ideal Fluid Mixtures," *Chem. Eng. Sci.*, **32**, 659–667 (1977).

Krishna, R., "A Note on the Film and Penetration Models for Multicomponent Mass Transfer," *Chem. Eng. Sci.*, **33**, 765–767 (1978a).

Krishna, R., "Penetration Depths in Multicomponent Mass Transfer," *Chem. Eng. Sci.*, **33**, 1495–1497 (1978b).

Krishna, R., "A Simplified Film Model Description of Multicomponent Interphase Mass Transfer," *Chem. Eng. Commun.*, **3**, 29–39 (1979a).

Krishna, R., "Comparison of Models for Ternary Mass Transfer," *Letts. Heat Mass Transfer*, **6**, 73–76 (1979b).

Krishna, R., "Effect of Nature and Composition of Inert Gas on Binary Vapor Condensation," *Letts. Heat Mass Transfer*, **6**, 137–147 (1979c).

Krishna, R., "A Simplified Mass Transfer Analysis for Multicomponent Condensation," *Letts. Heat Mass Transfer*, **6**, 439–448 (1979d).

Krishna, R., "Ternary Mass Transfer in a Wetted-Wall Column—Significance of Diffusional Interactions. I–Stefan Diffusion," *Trans. Inst. Chem. Engrs.*, **59**, 35–43 (1981a).

Krishna, R., "An Alternative Linearized Theory of Multicomponent Mass Transfer," *Chem. Eng. Sci.*, **36**, 219–221 (1981b).

Krishna, R., "A Turbulent Film Model for Multicomponent Mass Transfer," *Chem. Eng. J.*, **24**, 163–172 (1982).

Krishna, R., "Model for Prediction of Point Efficiencies for Multicomponent Distillation," *Chem. Eng. Res. Des.*, **63**, 312–322 (1985).

Krishna, R., "A Unified Theory of Separation Processes Based on Irreversible Thermodynamics," *Chem. Eng. Commun.*, **59**, 33–64 (1987).

Krishna, R., "Problems and Pitfalls in the Use of the Fick Formulation for Intraparticle Diffusion," *Chem. Eng. Sci.*, **48**, 845–861 (1993a).

Krishna, R., "Analogies in Multiphase Reactor Hydrodynamics," in *Encyclopedia of Fluid Mechanics*, Supplement 2, Cheremisinoff, N. P. (Ed.), Gulf Publishing Corp., Houston, TX, in press (1993b).

Krishna, R., Low, C. Y., Newsham, D. M. T., Olivera-Fuentas, C. G., and Standart, G. L., "Ternary Mass Transfer in Liquid–Liquid Extraction," *Chem. Eng. Sci.*, **40**, 893–903 (1985).

Krishna, R., Martinez, H. F., Sreedhar, R., and Standart, G. L., "Murphree Point Efficiencies in Multicomponent Systems," *Trans. Inst. Chem. Engrs.*, **55**, 178–183 (1977).

Krishna, R. and Panchal, C. B., "Condensation of Binary Mixtures in the Presence of an Inert Gas," *Chem. Eng. Sci.*, **32**, 741–745 (1977).

Krishna, R., Panchal, C. B., Webb, D. R., and Coward, I., "An Ackermann–Colburn–Drew Type Analysis for Condensation of Multicomponent Mixtures," *Letts. Heat Mass Transfer*, **3**, 163–172 (1976).

Krishna, R., Salomo, R. M., and Rahman, M. A., "Ternary Mass Transfer in a Wetted Wall Column. Significance of Diffusional Interactions. Part II. Equimolar Diffusion," *Trans. Inst. Chem. Engrs.*, **59**, 44–53 (1981).

Krishna, R. and Standart, G. L., "A Multicomponent Film Model Incorporating an Exact Matrix Method of Solution to the Maxwell–Stefan Equations," *AIChE J*, **22**, 383–389 (1976a).

Krishna, R. and Standart, G. L., "Addition of Resistances for Non-Isothermal Multicomponent Mass Transfer," *Letts. Heat Mass Transfer*, **3**, 41–48 (1976b).

Krishna, R. and Standart, G.L., "Mass and Energy Transfer in Multicomponent Systems," *Chem. Eng. Commun.*, **3**, 201–275 (1979).

Krishna, R. and Taylor, R., "Multicomponent Mass Transfer—Theory and Applications," in *Handbook of Heat and Mass Transfer*, Cheremisinoff, N. P. (Ed.), Gulf Publishing Corp., Houston, TX, Chap. 7, Vol. II, pp. 259–432, 1986.

Krishnamurthy, R. and Andrecovich, M. J., "Application of a Nonequilibrium Stage Model to Cryogenic Process Simulation," *Proceedings of the International Symposium on Gas Separation Technology*, held in Antwerp, Belgium, Vansant, E. F. and Dewolfs, R. (Eds.), Elsevier, Amsterdam, The Netherlands, 1989.

Krishnamurthy, R. and Taylor R., "Calculation of Multicomponent Mass Transfer at High Transfer Rates," *Chem. Eng. J.*, **25**, 47–54 (1982).

Krishnamurthy, R. and Taylor, R., "A Nonequilibrium Stage Model of Multicomponent Separation Processes. I—Model Development and Method of Solution," *AIChE J*, **31**, 449–456 (1985a).

Krishnamurthy, R. and Taylor, R., "A Nonequilibrium Stage Model of Multicomponent Separation Processes. II—Comparison with Experiment," *AIChE J*, **31**, 456–465 (1985b).

Krishnamurthy, R. and Taylor, R., "A Nonequilibrium Stage Model of Multicomponent Separation Processes. III—The Influence of Unequal Component Efficiencies in Design," *AIChE J*, **31**, 1973–1985 (1985c).

Krishnamurthy, R. and Taylor, R., "Simulation of Packed Distillation and Absorption Columns," *Ind. Eng. Chem. Process Des. Dev.*, **24**, 513–524 (1985d).

Krishnamurthy, R. and Taylor, R., "Absorber Simulation and Design Using a Nonequilibrium Stage Model," *Can. J. Chem. Eng.*, **64**, 96–105 (1986).

Kronig, R. and Brink, J. C., "The Theory of Extraction From Falling Droplets," *Appl. Sci. Res.*, **A2**, 142–154 (1950).

Kubaczka, A. and Bandrowski, J., "An Explicit Approximate Solution of the Generalized Maxwell–Stefan Equations for the Multicomponent Film Model," *Chem. Eng. Commun.*, **95**, 89–97 (1990).

Kubaczka A. and Bandrowski, J., "Solutions of a System of Multicomponent Mass Transport Equation for Mixtures of Real Fluids," *Chem. Eng. Sci.*, **46**, 539–556 (1991).

Kubota, H., Yamanaka, Y., and Dalla Lana, I. G., "Effective Diffusivity of Multicomponent Gaseous Reaction System," *J. Chem. Eng. Jpn.*, **2**, 71–75 (1969).

Kumar, S., Wright, J. D., and Taylor, P. A., "Modeling and Dynamics of an Extractive Distillation Column," *Can. J. Chem. Eng.*, **62**, 780–789 (1984).

Laity, R. W., "Diffusion of Ions in an Electric Field," *J. Phys. Chem.*, **67**, 671–676 (1963).

Lancaster, P. and Tismenetsky, M., *The Theory of Matrices*, 2nd ed., Academic Press, New York, 1985.

Lao, M., Kingsley, J. P., Krishnamurthy, R., and Taylor, R., "A Nonequilibrium Stage Model of Multicomponent Separation Processes. VI: Simulation of Liquid–Liquid Extraction," *Chem. Eng. Commun.*, **86**, 73–89 (1989).

Launder, B. E. and Spalding, D. B., *Mathematical Models of Turbulence*, Academic Press, London, England, 1972.

Lee, H. L., Lightfoot, E. N., Reis, J. F. G., and Waissbluth, D. M., "The Systematic Description and Development of Separations Processes," in *Recent Developments in Separation Science*, Vol. III, Part A, Li, N. N. (Ed.), CRC Press, Cleveland, OH, 1977.

Leffler, J. and Cullinan, H. T., "Variation of Liquid Diffusion Coefficients with Composition in Binary Systems," *Ind. Eng. Chem.*, **9**, 84–88 (1970).

Lewis, W. K., "Rectification of Binary Mixtures. Plate Efficiency of Bubble Cap Columns," *Ind. Eng. Chem.*, **28**, 399–402 (1936).

Lightfoot, E. N., "Estimation of Heat and Mass Transfer Rates," *Lectures in Transport Phenomena*, *AIChE Continuing Education Series No. 4*, AIChE, New York, 1969.

Lightfoot, E. N., *Transport Phenomena and Living Systems*, McGraw-Hill, New York, 1974.

Lightfoot, E. N., Cussler, E. L., and Rettig, R. L., "Applicability of the Stefan–Maxwell Relations to Multicomponent Diffusion in Liquids," *AIChE J*, **8**, 708–710 (1962).

Lightfoot, E. N. and Scattergood, E. M., "Suitability of the Nernst–Planck Equations for Describing Electrokinetic Phenomena," *AIChE J*, **11**, 175–192 (1965).

Lockett, M. J., *Distillation Tray Fundamentals*, Cambridge University Press, Cambridge, England, 1986.

Lockett, M. J. and Ahmed, I. S., "Tray and Point Efficiencies from a 0.6 Metre Diameter Distillation Column," *Chem. Eng. Res. Des.*, **61**, 110–118 (1983).

Lockett, M. J. and Plaka, T., "Effect of Non-Uniform Bubbles in the Froth on the Correlation and Prediction of Point Efficiencies," *Chem. Eng. Res. Des.*, **61**, 119–124 (1983).

Löwe, A. and Bub, G., "Multiple Steady States for Isothermal Catalytic Gas–Solid Reactions with a Positive Reaction Order," *Chem. Eng. Sci.*, **31**, 175–178 (1976).

Ly, L-A. N., Carbonell, R. G., and McCoy, B. J., "Diffusion of Gases Through Surfactant Films: Interfacial Resistance to Mass Transfer," *AIChE J*, **25**, 1015–1024 (1979).

McCartney, R. F., *Plate Efficiencies in the Separation of C4 Hydrocarbons by Extractive Distillation*, M.S. Thesis, University of Delware, Delaware, 1948.

McDowell, J. K. and Davis, J. F., "A Characterization of Diffusion Distillation for Azeotropic Separation," *Ind. Eng. Chem. Res.*, **27**, 2139–2148 (1988).

McKay, W. N., "Experiments Concerning Diffusion of Multicomponent Systems at Reservoir Conditions," *J. Can. Petroleum Technol.*, April–June, 25–32 (1971).

McKeigue, K. and Gulari, E., "Affect of Molecular Association on Diffusion in Binary Liquid Mixtures," *AIChE J*, **35**, 300–310 (1989).

McNaught, J. M., "An Assessment of Design Methods for Condensation of Vapors from a Noncondensing Gas," in *Heat Exchangers—Theory and Practice*, Taborek, J., Hewitt, G. F., and Afgan, N. (Eds.), Hemisphere Publishing Corp., Washington, DC, 1983a.

McNaught, J. M., "An Assessment of Design Methods for Multicomponent Condensation Against Data from Experiments on a Horizontal Tube Bundle," The Institution of Chemical Engineers Symposium Series No. 75, *Condensers: Theory and Practice*, 447–458 (1983b).

McNulty, K. J. and Chatterjee, S. G., "Simulation of Atmospheric Crude Towers Including Packed Bed Pump-around Zones using a Rate-based Simulator," The Institution of Chemical Engineers Symposium Series No. 128, *Distillation and Absorption 1992*, A329–A344 (1992).

Marrero, T. R. and Mason, E. A., "Gaseous Diffusion Coefficients," *J. Phys. Chem. Ref. Data*, **1**, 3–118 (172).

Mason, E. A., Munn, R. J., and Smith, F. J., "Thermal Diffusion in Gases," *Advances in Atomic and Molecular Physics*, **2**, 33–87 (1966).

Mason, E. A. and Malinauskas, A. P., *Gas Transport in Porous Media: The Dusty-Gas Model*, Elsevier, Amsterdam, The Netherlands, 1983.

Mason, E. A. and Lonsdale, H. K., "Statistical Mechanical Theory of Membrane Transport," *J. Membrane Science*, **51**, 1–81 (1990).

Matthews, M. A. and Akgerman, A., "Infinite Dilution Diffusion Coefficients of Methanol and 2-Propanol in Water," *J. Chem. Eng. Data*, **33**, 122–123 (1988).

Maxwell, J. C., "On the Dynamical Theory of Gases," *Phil. Trans. Roy. Soc.*, **157**, 49–88 (1866); *Phil. Mag.*, **35**, 129–145, 185–217 (1868).

564   REFERENCES

Maxwell, J. C., *The Scientific Papers of James Clerk Maxwell*, Niven, W. D. (Ed.), Dover, New York, 1952.

Medina, A. G., Ashton, N., and McDermott, C., "Murphree and Vaporization Efficiencies in Multicomponent Distillation," *Chem. Eng. Sci.*, **33**, 331–339 (1978).

Medina, A. G., Ashton, N., and McDermott, C., "Hausen and Murphree Efficiencies in Binary and Multicomponent Distillation," *Chem. Eng. Sci.*, **34**, 1105–1112 (1979).

Merk, H. J., "The Macroscopic Equations for Simultaneous Heat and Mass Transfer in Isotropic, Continuous and Closed Systems," *Appl. Sci. Res.*, Sec. A, **8**, 73–99 (1960).

Modell, M. M. and Reid, R. C., *Thermodynamics and its Applications*, 2nd ed., Prentice-Hall, Englewood Cliffs, NJ, (1983).

Modine, A. D., *Ternary Mass Transfer*, Ph.D. Dissertation in Chemical Engineering, Carnegie Institute of Technology, Pittsburgh, PA, 1963.

Modine, A. D., Parrish, E. B., and Toor, H. L., "Simultaneous Heat and Mass Transfer in a Falling Laminar Film," *AIChE J*, **9**, 348–351 (1963).

Moler, C. and Van Loan, C., "Nineteen Dubious Ways to Compute the Exponential of a Matrix," *SIAM Rev.*, **20**, 801–836 (1978).

Mora, J-C. and Bugarel, R., "Application de la thermodynamique des processus irreversibles en genie chimique: Modelisation d'un plateau de rectification de grand diametre," *Entropie*, **69**, 4–11 (1976).

Muckenfuss, C., "Stefan–Maxwell Relations for Multicomponent Diffusion and the Chapman Enskog Solution of the Boltzmann Equations," *J. Chem. Phys.*, **59**, 1747–1752 (1973).

Murphree, E. V., "Rectifying Column Calculations," *Ind. Eng. Chem.*, **17**, 747–750, 960–964 (1925).

Myerson, A. S. and Senol, D., "Diffusion Coefficients Near the Spinodal Curve," *AIChE J*, **30**, 1004–1006 (1984).

Naphtali, L. M., "The Distillation Column as a Large System," Paper presented at AIChE National meeting, San Francisco, California, May 16, 1965.

Naphtali, L. M. and Sandholm, D. P., "Multicomponent Separation Calculations by Linearization" *AIChE J*, **17**, 148–153 (1971).

Neufeld, P. D., Janzen, A. R., and Aziz, R. A., "Empirical Relations to Calculate 16 of the Transport Collision Integrals $\Omega^{(l,s)*}$ for the Lennard–Jones (12-6) Potential," *J. Chem. Phys.*, **57**, 1100–1102 (1972).

Newman, J., *Electrochemical Systems*, 2nd ed., Prentice-Hall, Englewood Cliffs, NJ, 1991.

Ney, E. P. and Armistead, F. C., "The Self Diffusion Coefficient of Uranium Hexafluoride," *Phys. Rev.*, **71**, 14–19 (1947).

Nielsen, P. H. and Villadsen, J., "The Design and Operation of a Laboratory Ketene Generator," The Institution of Chemical Engineers Symposium Series No. 87, 287–294 (1984).

Ognisty, T. P. and Sakata, M., "Multicomponent Diffusion: Theory vs. Industrial Data," *Chem. Eng. Prog.*, **83**(3), 60–65 (1987).

Olivera–Fuentes, C. G. and Pasquel–Guerra, J., "The Exact Penetration Model of Diffusion in Multicomponent Ideal Gas Mixtures. Analytical and Numerical Solutions," *Chem. Eng. Commun.*, **51**, 71–88 (1987).

Onda, K., Takeuchi, H., and Okumoto, Y., "Mass Transfer Coefficients Between Gas and Liquid Phases in Packed Columns," *J. Chem. Eng. Jpn.*, **1**, 56–62 (1968).

Onsager, L., "Reciprocal Relations in Irreversible Processes," *Phys. Rev.*, **37**, 405–426, **38**, 2265–2279 (1931).

Onsager, L., "Theories and Problems of Liquid Diffusion," *Ann. New York Akad. Sci.*, **46**, 241–265 (1945).

Ortega, J. M. and Rheinbolt, W. C., *Iterative Solution of Nonlinear Equations in Several Variables*, Academic Press, New York, 1970.

Perkins, L. R. and Geankoplis, C. J., "Molecular Diffusion in a Ternary Liquid System with the Diffusing Component Dilute," *Chem. Eng. Sci.*, **24**, 1035–1042 (1969).

Plaka, T., Ehsani, M. R., and Korchinsky, W. J., "Determination of Individual Phase Transfer Units, N_G and N_L, for a 0.6 m Diameter Distillation Column Sieve Plate: Methylcyclohexane-Toluene System," *Chem. Eng. Res. Des.*, **67**, 316–328 (1989).

Ponter, A. B. and Au-Yeung, P. H., "Estimating Liquid Film Mass Transfer Coefficients in Randomly Packed Columns," in *Handbook of Heat and Mass Transfer*, Cheremisinoff, N. P. (Ed.), Gulf Publishing Corp., Houston, TX, Chap. 20, Vol. II, pp. 903–952, 1986.

Powers, M. F., Vickery, D. J., Arehole, A., and Taylor, R., "A Nonequilibrium Stage Model of Multicomponent Separation Processes—V. Computational Methods for Solving the Model Equations," *Comput. Chem. Eng.*, **12**, 1229–1241 (1988).

Prado, M. and Fair, J. R., "Fundamental Model for the Prediction of Sieve Tray Efficiency," *Ind. Eng. Chem. Res.*, **29**, 1031–1042 (1990).

Pratt, H. R. C., "The Design of Packed Absorption and Distillation Columns," *Ind. Chem.*, **26**, 470–475 (1950).

Prausnitz, J. M., Anderson, T. F., Grens, E. A., Eckert, C. A., Hsieh, R., and O'Connell, J. P., *Computer Calculations for Multicomponent Vapor–Liquid and Liquid–Liquid Equilibria*, Prentice–Hall, Englewood Cliffs, NJ, 1980.

Prausnitz, J. M., Lichtenthaler, R. N., and de Azevedo, E. G., *Molecular Thermodynamics of Fluid Phase Equilibria*, Prentice–Hall, Englewood Cliffs, NJ, 1986.

Present, R. D., *Kinetic Theory of Gases*, McGraw-Hill, New York, 1958.

Press, W. H., Teukolsky, S. A., Vetterling, W. T., and Flannery, B. P., *Numerical Recipes in Fortran: The Art of Scientific Computing*, 2nd ed., Cambridge University Press, Cambridge, England, 1992.

Price, B. C. and Bell, K. J., "Design of Binary Vapor Condensers Using the Colburn Equations," *AIChE Symp. Ser. 70*, 163–171, (1974).

Rathbun, R. E. and Babb, A. L., "Empirical Method for Prediction of The Concentration Dependence of Mutual Diffusivities in Binary Mixtures of Associated and Nonpolar Liquids," *Ind. Eng. Chem. Proc. Des. Dev.*, **5**, 273–275 (1966).

Reid, R. C., Prausnitz, J. M., and Poling, B., *The Properties of Gases and Liquids*, 4th ed., McGraw–Hill, New York, 1987.

Reid, R. C., Prausnitz, J. M., and Sherwood, T. K., *The Properties of Gases and Liquids*, 3rd ed., McGraw–Hill, New York, 1977.

Reijnhart, R., Piepers, J., and Toneman, H., "Vapour Cloud Dispersion and the Evaporation of Volatile Liquids in Atmospheric Wind Fields," *Atmosphere Environment*, **14**, 751–762 (1980).

Riede, Th. and Schlünder, E. U., "Diffusivities of the Ternary Liquid Mixture 2-Propanol–Water–Glycerol and Three Component Mass Transfer in Liquids," *Chem. Eng. Sci.*, **46**, 609–617 (1991).

Röhm, H. J., "The Simulation of Steady State Behavior of the Dephlegmation of Multicomponent Mixed Vapors," *Int. J. Heat Mass Transfer*, **23**, 141–146 (1980).

Röhm, H. J. and Vogelpohl, A., "Zur Berechnung der Stofftransportbeziehungen an der Phasengrenze von Zweiphasen-Mehrkomponenten-Systemen," *Wärme- und Stoffübertragung*, **13**, 231–239 (1980).

Rosner, D. E., "Thermal (Soret) Diffusion Effects on Interfacial Mass Transport Rates," *Physicochem. Thermodyn.*, **1**, 159–185 (1980).

Rosner, D. E., *Transport Processes in Chemically Reacting Flow Systems*, Butterworths, Boston, MA, 1986.

Rutten, Ph. W. M., *Diffusion in Liquids*, Ph.D. Thesis, Technical University Delft, Delft, The Netherlands, 1992.

Sage, F. E. and Estrin, J., "Film Condensation From a Ternary Mixture of Vapors upon a Vertical Surface," *Int. J. Heat Mass Transfer*, **19**, 323–333 (1976).

Sanni, S. A. and Hutchison, P., "Diffusivities and Densities for Binary Liquid Mixtures," *J. Chem. Eng. Data*, **18**, 317–322 (1973).

Sardesai, R. G., *Studies in Condensation*, Ph.D. Thesis in Chemical Engineering, University of Manchester Institute of Science and Technology, Manchester, England, 1979.

Sawistowski, H., "Mass Transfer Performance," Paper presented at the meeting of the European Federation of Chemical Engineers Working Party on Distillation, Absorption and Extraction, Linz, Austria, May, 1980.

Schlünder, E. U. (Ed.), *Heat Exchanger Design Handbook*, Hemisphere Publishing Corp., Washington, DC, 1983.

Schrodt, J. T., "Simultaneous Heat and Mass Transfer from Multicomponent Condensing Vapor—Gas Systems," *AIChE J*, **19**, 753–759 (1973).

Scriven, L. E., "Flow and Transfer at Fluid Interfaces," *Chem. Eng. Educ.*, 150–155 (Fall 1968); 26–29 (Winter 1969); 94–98 (Spring 1969).

Seader, J. D., "The B. C. (Before Computers) and A. D. of Equilibrium Stage Operations," *Chem. Eng. Educ.*, **19**(2), 88–103 (1985).

Seader, J. D., *Computer Modelling of Chemical Processes*, AIChE Monograph Series, No. 15, 81 (1986).

Seader, J. D., "The Rate-Based Approach for Modeling Staged Separations," *Chem. Eng. Progress*, 41–49 (October 1989).

Şentarli, I. and Hortaçsu, A., "Solution of the Linearized Equations of Multicomponent Mass Transfer with Chemical Reaction and Convection for a Film Model," *Ind. Eng. Chem. Res.*, **26**, 2409–2413 (1987).

Sethy, A. and Cullinan, H. T., "Transport of Mass in Ternary Liquid–Liquid Systems," *AIChE J*, **21**, 571–582 (1975).

Shah, A. K. and Webb, D. R., "Condensation of Single and Mixed Vapors from a Non-Condensing Gas in Flow Over a Horizontal Tube Bank," The Institution of Chemical Engineers Symposium Series No. 75, *Condensers: Theory and Practice*, 356–371 (1983).

Shain, S. A., "A Note on Multicomponent Diffusion," *AIChE J*, **7**, 17–19 (1961).

Sherwood, T. K., Pigford, R. L., and Wilke, C. R., *Mass Transfer*, McGraw–Hill, New York, 1975.

Sherwood, T. K. and Wei, J. C., "Ion Diffusion in Mass Transfer Between Phases," *AIChE J*, **1**, 522–527 (1955).

Siddiqi, M. A. and Lucas, K., "Correlations for Prediction of Diffusion in Liquids," *Can. J. Chem. Eng.*, **64**, 839–843 (1986).

Sideman, S. and Pinczewski, W. V., in *Topics in Transport Phenomena*, Gutfinger, C. (Ed.), Halstead Press, New York, 1975.

Sideman, S. and Shabtai, H., "Direct Contact Heat Transfer Between a Single Drop and an Immiscible Liquid Medium," *Can. J. Chem. Eng.*, **42**, 107–117 (1964).

Silver, L., "Gas Cooling with Aqueous Condensation," *Trans. Inst. Chem. Engrs.*, **25**, 30–42 (1947).

Slattery, J. C., *Momentum, Energy and Mass Transfer in Continua*, 2nd ed., Krieger Publishing Company, Huntington, NY, 1981.

Smith, B. D., *Design of Equilibrium Stage Processes*, McGraw–Hill, New York, 1964.

Smith, L. W. and Taylor, R., "Film Models for Multicomponent Mass Transfer—A Statistical Comparison," *Ind. Eng. Chem. Fundam.*, **22**, 97–104 (1983).

Spalding, D. B., *Heat Exchanger Design Handbook, I: Heat Exchanger Theory*, Schlünder, E. U. (Ed.), Hemisphere, Washington, DC, 1983.

Stainthorp, F. P. and Whitehouse, P. A., "General Computer Programs for Multi-Stage Counter-Current Separation Problems," *The Institution of Chemical Engineers Symposium Series No. 23*, 181–188, 189–192 (1967).

Standart, G. L., "The Mass, Momentum and Energy Equations for Heterogeneous Flow Systems," *Chem. Eng. Sci.*, **19**, 227–236 (1964).

Standart, G. L., "Studies on Distillation—V. Generalized Definition of a Theoretical Plate or Stage of Contacting Equipment," *Chem. Eng. Sci.*, **20**, 611–622 (1965).

Standart, G. L., "Comparison of Murphree Efficiencies with Vaporization Efficiencies," *Chem. Eng. Sci.*, **26**, 985–988 (1971).

Standart, G. L., Cullinan, H. T., Paybarah, A., and Louizos, N., "Ternary Mass Transfer in Liquid–Liquid Extraction," *AIChE J*, **21**, 554–559 (1975).

Standart, G. L., Taylor, R., and Krishna, R., "The Maxwell–Stefan Formulation of Irreversible Thermodynamics for Simultaneous Heat and Mass Transfer," *Chem. Eng. Commun.*, 3, 277–289 (1979).

Stefan, J., "Über das Gleichgewicht und die Bewegung, insbesondere die Diffusion von Gasmengen," *Sitzungsber. Akad. Wiss. Wien*, 63, 63–124 (1871).

Stewart, W. E., N. A. C. A. Tech. Note 3208 (1954).

Stewart, W. E., "Multicomponent Mass Transfer in Turbulent Flow," *AIChE J*, 19, 398–400 (1973).

Stewart, W. E. and Prober, R., "Matrix Calculation of Multicomponent Mass Transfer in Isothermal Systems," *Ind. Eng. Chem. Fundam.*, 3, 224–235 (1964).

Strigle, R. F., *Random Packings and Packed Towers: Design and Applications*, Gulf Publishing, Houston, TX, 1987.

Taitel, Y. and Tamir, A., "Film Condensation of Multicomponent Mixtures," *Int. J. Multiphase Flow*, 1, 697–714 (1974).

Tamir, A. and Merchuk, J. C., "Verification of a Theoretical Model for Multicomponent Condensation," *Chem. Eng. J.*, 17, 125–139 (1979).

Taylor, R., "On Exact Solutions of the Maxwell–Stefan Equations for the Multicomponent Film Model," *Chem. Eng. Commun.*, 10, 61–76 (1981a).

Taylor, R., "On Multicomponent Mass Transfer in Turbulent Flow," *Letts. Heat Mass Transfer*, 8, 397–404 (1981b).

Taylor, R., "More on Exact Solutions of the Maxwell–Stefan Equations for the Multicomponent Film Model," *Chem. Eng. Commun.*, 14, 361–362 (1982a).

Taylor, R., "Film Models for Multicomponent Mass Transfer: Computational Methods II—The Linearized Theory," *Comput. Chem. Eng.*, 6, 69–75 (1982b).

Taylor, R., "Solution of the Linearized Equations of Multicomponent Mass Transfer," *Ind. Eng. Chem. Fundam.*, 21, 407–413 (1982c).

Taylor, R., "Simulation of Binary Vapor Condensation in the Presence of an Inert Gas—A Sequel," *Int. Comm. Heat Mass Transfer*, 11, 429–437 (1984).

Taylor, R., "On an Explicit Approximate Solution of the Generalized Maxwell–Stefan Equations for Nonideal Fluids," *Chem. Eng. Commun.*, 103, 53–56 (1991).

Taylor, R. and Kooijman, H. A., "Composition Derivatives of Activity Coefficient Models (For the Estimation of Thermodynamic Factors in Diffusion)," *Chem. Eng. Commun.*, 102, 87–106 (1991).

Taylor, R., Kooijman, H. A., and Woodman, M. R., "Industrial Applications of a Nonequilibrium Model of Distillation and Absorption Operations," The Institution of Chemical Engineers Symposium Series No. 128, *Distillation and Absorption 1992*, A415–A427 (1992).

Taylor, R. and Krishnamurthy, R., "Film Models for Multicomponent Mass Transfer.—Diffusion in Physiological Gas Mixtures," *Bull. Math. Biol.*, 44, 361–376 (1982).

Taylor, R., Krishnamurthy, R., Furno, J. S., and Krishna, R., "Condensation of Vapor Mixtures. I—A Mathematical Model and a Design Method," *Ind. Eng. Chem. Proc. Des. Dev.*, 25, 83–97 (1985).

Taylor, R. and Noah, M. K., "Simulation of Binary Vapor Condensation in the Presence of an Inert Gas," *Letts. Heat Mass Transfer*, 9, 463–472 (1982).

Taylor, R., Powers, M. F., Lao, M., and Arehole, A., "The Development of A Nonequilibrium Model for Computer Simulation of Multicomponent Distillation and Absorpton Operations," The Institution of Chemical Engineers Symposium Series No. 104, *Distillation and Absorption 1987*, B321–B329 (1987).

Taylor, R. and Smith, L. W., "On Some Explicit Approximate Solutions of the Maxwell–Stefan Equations for the Multicomponent Film Model," *Chem. Eng. Commun.*, 14, 361–370 (1982).

Taylor, R. and Webb, D. R., "Stability of the Film Model for Multicomponent Mass Transfer," *Chem. Eng. Commun.*, 6, 175–189 (1980a).

Taylor, R. and Webb, D. R., "On the Relationship Between the Exact and Linearized Solutions of the Maxwell–Stefan Equations for the Multicomponent Film Model," *Chem. Eng. Commun.*, 7, 287–299 (1980b).

Taylor, R. and Webb, D. R., "Film Models for Multicomponent Mass Transfer: Computational Methods I—the Exact Solution of the Maxwell–Stefan Equations," *Comput. Chem. Eng.*, 5, 61–73 (1981).

Toor, H. L., "Diffusion in Three Component Gas Mixtures," *AIChE J*, 3, 198–207 (1957).

Toor, H. L., "Turbulent Diffusion and the Multicomponent Reynolds Analogy," *AIChE J*, 6, 525–527 (1960).

Toor, H. L., "Solution of the Linearized Equations of Multicomponent Mass Transfer," *AIChE J*, 10, 448–455, 460–465 (1964a).

Toor, H. L., "Prediction of Efficiencies and Mass Transfer on a Stage with Multicomponent Systems," *AIChE J*, 10, 545–547 (1964b).

Toor, H. L. and Burchard, J. K., "Plate Efficiencies in Multicomponent Distillation," *AIChE J*, 6, 202–206 (1960).

Toor, H. L. and Marchello, J. M., "Film-penetration Model for Mass and Heat Transfer," *AIChE J*, 4, 97–101 (1958).

Toor, H. L. and Sebulsky, R. T., "Multicomponent Mass Transfer. I—Theory," *AIChE J*, 7, 558–565 (1961a).

Toor, H. L. and Sebulsky, R. T., "Multicomponent Mass Transfer. II—Experiment," *AIChE J*, 7, 565–573 (1961b).

Toor, H. L. and Arnold, K. R., "Nature of the Uncoupled Multicomponent Diffusion Equations," *Ind. Eng. Chem. Fundam.*, 4, 363–364 (1965).

Toor, H. L., Seshadri, C. V., and Arnold, K. R., "Diffusion and Mass Transfer in Multicomponent Mixtures of Ideal Gases," *AIChE J*, 11, 746–747 (1965).

Truesdell, C., *Rational Thermodynamics*, McGraw-Hill, New York (1969).

Truesdell, C. and Toupin, R., "The Classical Field Theories," *Handbuch der Physik*, Flugge, S. (Ed.), Vol. III/1, Springer-Verlag, Berlin, Germany, 1960.

Tunison, M. E. and Chapman, T. W., "The Effect of a Diffusion Potential on the Rate of Liquid–Liquid Ion Exchange," *Ind. Eng. Chem. Fundam.*, 15, 196–201 (1976).

Turevskii, E. N., Aleksandrov, A. I., and Gorechenkov, V. G., "An Approximate Method for Calculating Nonequimolar Diffusion in Multicomponent Mixtures," *Int. Chem. Eng.*, 14, 112–115 (1974).

Tyn, M. T. and Calus, W. F., "Diffusion Coefficients in Dilute Binary Liquid Mixtures," *J. Chem. Eng. Data*, 20, 106–109 (1975a).

Tyn, M. T. and Calus, W. F., "Temperature and Concentration Dependence of Mutual Diffusion Coefficients of Some Binary Liquid Systems," *J. Chem. Eng. Data*, 20, 310–316 (1975b).

Tyrell, H. J. V. and Harris, K. R., *Diffusion in Liquids*, Butterworths, London, England, 1984.

Van Brocklin, L. P. and David, M. M., "Coupled Ionic Migration and Diffusion During Liquid Phase Controlled Ion Exchange," *Ind. Eng. Chem. Fundam.*, 11, 91–99 (1972).

Van Deemter, J. J., "Mixing and Contacting in Gas-Solid Fluidized Beds," *Chem. Eng. Sci.*, 13, 143–154 (1961).

Van Driest, E. R., "On Turbulent Flow Near a Wall," *J. Aeronaut. Sci.*, 23, 1007–1011 (1956).

Vermeer, D. and Krishna, R., "Hydrodynamics and Mass Transfer in Bubble Columns Operating in the Churn-Turbulent Regime," *Ind. Eng. Chem. Process Des. Dev.*, 20, 475–482 (1981).

Vickery, D. J., Taylor, R., and Gavalas, G. R., "A Novel Approach to the Computation of Multicomponent Mass Transfer Rates from the Linearized Equations," *Comput. Chem. Eng.*, 8, 179–184 (1984).

Vieth, W. R., Porter, J. H., and Sherwood, T. K., "Mass Transfer and Chemical Reaction in a Turbulent Boundary Layer," *Ind. Eng. Chem. Fundam.*, 2, 1–3 (1963).

Vignes, A., "Diffusion in Binary Solutions," *Ind. Eng. Chem. Fundam.*, 5, 189–199 (1966).

Vinograd, J. R. and McBain, J. W., "Diffusion of Electrolytes and Ions in their Mixtures," *J. Am. Chem. Soc.*, 63, 2008–2015 (1941).

Vitagliano, V., Sartorio, R., Chiavalle, E., and Ortona, O., "Diffusion and Viscosity in Water–Triethylamine Mixtures at 19 and 20°C," *J. Chem. Eng. Data*, 25, 121–124 (1980).

Vitagliano, V., Sartorio, R., Scala, S., and Spaduzzi, D., "Diffusion in a Ternary System and the Critical Mixing Point," *J. Sol. Chem.*, 7, 605–621 (1978).

Vogelpohl, A., "Murphree Efficiencies in Multicomponent Systems," The Institution of Chemical Engineers Symposium Series, No. 56, *Distillation 1979*, 2.1/25–2.1/31 (1979).

Voight, W., "Status and Plans of the DOE Uranium Enrichment Program," *AIChE Symp. Ser.*, 78, 1–9 (1982).

Von Behren, G. L., Jones, W. O., and Wasan, D. T., "Multicomponent Mass Transfer in Turbulent Flow," *AIChE J*, 18, 25–30 (1972).

Von Halle, E., "The Countercurrent Gas Centrifuge for the Enrichment of U-235," in Recent Advances in Separation Techniques—II, *AIChE Symp. Ser.*, 76, 82–87 (1980).

Walas, S. M., *Phase Equilibria in Chemical Engineering*, Butterworth, Stoneham, MA, 1985.

Walter, J. F. and Sherwood, T. K., "Gas Absorption in Bubble-Cap Columns," *Ind. Eng. Chem.*, 33, 493–501 (1941).

Webb, D. R., "Heat and Mass Transfer in Condensation of Multicomponent Vapors," *Proceedings of the Seventh International Heat Transfer Conference*, Munich, Germany, 5, 167–174 (1982).

Webb, D. R. and McNaught, J. M., "Condensers," in *Developments in Heat Exchanger Technology*, Chisholm, D. (Ed.), Applied Science Publishers, Barking, Essex, England, 1980.

Webb, D. R. and Panagoulias, D., "An Improved Approach to Condenser Design Using Film Models," *Int. J. Heat and Mass Transfer*, 30, 373–378 (1987).

Webb, D. R., Panchal, C. B., and I. Coward, "The Significance of Multicomponent Diffusional Interactions in the Process of Condensation in the Presence of a Non Condensable Gas," *Chem. Eng. Sci.*, 36, 87–95 (1981).

Webb, D. R. and Sardesai, R. G., "Verification of Multicomponent Mass Transfer Models for Condensation Inside a Vertical Tube," *Int. J. Multiphase Flow*, 7, 507–520 (1981).

Webb, D. R. and Taylor, R., "The Estimation of Rates of Multicomponent Condensation by a Film Model," *Chem. Eng. Sci.*, 37, 117–119 (1982).

Wesselingh, J. A., "Is Fick Fout," (in Dutch), *Procestechnologie* (2), 39–43 (1985).

Wesselingh, J. A., "How on Earth Can I Get Chemical Engineers to do their Multicomponent Mass Transfer Sums Properly?," *J. Membrane Sci.*, 73, 323–333 (1992).

Wesselingh, J. A. and Krishna, R., *Mass Transfer*, Ellis Horwood, Chichester, England, 1990.

Whitaker, S., "Role of Species Momentum Equation in the Analysis of the Stefan Diffusion Tube," *Ind. Eng. Chem. Res.*, 30, 978–983 (1991).

Whitehouse, P. A., *A General Computer Program Solution of Multicomponent Distillation Problems*, Ph.D. Thesis in Chemical Engineering, University of Manchester, Institute of Science and Technology, Manchester, England, 1964.

Wilke, C. R., "Diffusional Properties of Multicomponent Gases," *Chem. Eng. Prog.*, 46, 95–104 (1950).

Wilke, C. R. and Chang, P., "Correlation of Diffusion Coefficients in Dilute Solutions," *AIChE J*, 1, 264–270 (1955).

Wilke, C. R. and Lee, C. Y., "Estimation of Diffusion Coefficients for Gases and Vapors," *Ind. Eng. Chem.*, 47, 1253–1257 (1955).

Wilkinson, J. H., *The Algebraic Eigenvalue Problem*, Clarendon Press, Oxford, England, 1965.

Wong, C. F. and Hayduk, W., "Correlations for Prediction of Molecular Diffusivities in Liquids at Infinite Dilution," *Can. J. Chem. Eng.*, 68, 849–859 (1990).

Wozny, G., Neiderthund, M., and Gorak, A., "Ein neues Werkzeug zur rechnerunterstützten Simulation thermischer Trennverfahren in der fettchemischen Industrie," *Fat Sci. Technol.*, 93, 576–581 (1991).

Yao, Y. L., "Algebraical Analysis of Diffusion Coefficients in Ternary Systems," *J. Phys. Chem.*, 45, 110–115 (1966).

Young, T. C. and Stewart, W. E., "Comparison of Matrix Approximations for Multicomponent Transfer Calculations," *Ind. Eng. Chem. Fundam.*, 25, 476–482 (1986).

Young, T. C. and Stewart, W. E., "Collocation Analysis of a Boundary-Layer Model for Crossflow Fractionation Trays," *AIChE J*, 36, 655–664 (1990).

Zemaitis, Jr., J. F., Clark, D. M., Rafal, M., and Scrivner, N. C., *Handbook of Aqueous Electrolyte Thermodynamics*, AIChE, New York, 1986.

Zimmermann, A., Gourdon, C., Joulia, X., Gorak, A., and Casamatta, G., "Simulation of Multicomponent Extraction Process by a Nonequilibrium Stage Model Incorporating a Drop Population Model," *Comput. Chem. Eng.*, **16** (Suppl.), S403–S410 (1992).

Zogg, M., *Strömungs- und Stoffaustauschuntersuchungen an det Sulzer-Gewebepackung*, Ph.D. Thesis, ETH Zurich, Switzerland, 1972.

Zuiderweg, F. J., "Sieve Trays—A View of the State of the Art," *Chem. Eng. Sci.*, **37**, 1441–1464 (1982).

AUTHOR INDEX

SUBJECT INDEX